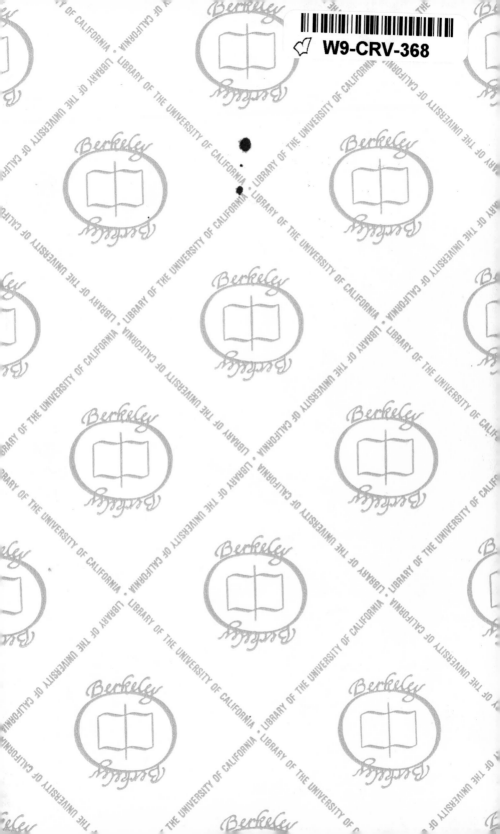

Thermodynamics and its Applications

PRENTICE-HALL INTERNATIONAL SERIES IN THE PHYSICAL AND CHEMICAL ENGINEERING SCIENCES

NEAL R. AMUNDSON, EDITOR, *University of Minnesota*

ADVISORY EDITORS

ANDREAS ACRIVOS, *Stanford University*
JOHN DAHLER, *University of Minnesota*
THOMAS J. HANRATTY, *University of Illinois*
JOHN M. PRAUSNITZ, *University of California*
L. E. SCRIVEN, *University of Minnesota*

AMUNDSON *Mathematical Methods in Chemical Engineering: Matrices and Their Application*

AMUNDSON AND ARIS *Mathematical Methods in Chemical Engineering: Vol. II, First Order Partial Differential Equations with Applications*

ARIS *Elementary Chemical Reactor Analysis*

ARIS *Introduction to the Analysis of Chemical Reactors*

ARIS *Vectors, Tensors, and the Basic Equations of Fluid Mechanics*

BALZHISER, SAMUELS, AND ELIASSEN *Chemical Engineering Thermodynamics*

BOUDART *Kinetics of Chemical Processes*

BRIAN *Staged Cascades in Chemical Processing*

CROWE et al. *Chemical Plant Simulation*

DOUGLAS *Process Dynamics and Control: Vol. I, Analysis of Dynamic Systems*

DOUGLAS *Process Dynamics and Control: Vol. II, Control System Synthesis*

FOGLER *The Elements of Chemical Kinetics and Reactor Calculations*

FREDRICKSON *Principles and Applications of Theology*

FRIEDLY *Dynamic Behavior of Processes*

HAPPEL AND BRENNER *Low Reynolds Number Hydrodynamics with Special Applications to Particulate Media*

HIMMELBLAU *Basic Principles and Calculations in Chemical Engineering*, 3rd edition

HOLLAND *Fundamentals and Modeling of Separation Processes*

HOLLAND *Multicomponent Distillation*

HOLLAND *Unsteady State Processes with Applications in Multicomponent Distillation*
KOPPEL *Introduction to Control Theory with Applications to Process Control*
LEVICH *Physicochemical Hydrodynamics*
MEISSNER *Processes and Systems in Industrial Chemistry*
MODELL and REID *Thermodynamics and its Applications*
NEWMAN *Electrochemical Systems*
OHARA AND REID *Modeling Crystal Growth Rates from Solution*
PERLMUTTER *Stability of Chemical Reactors*
PETERSEN *Chemical Reaction Analysis*
PRAUSNITZ *Molecular Thermodynamics of Fluid-Phase Equilibria*
PRAUSNITZ, ECKERT, ORYE, O'CONNELL *Computer Calculations for Multicomponent Vapor-Liquid Equilibria*
RUDD et al. *Process Synthesis*
SCHULTZ *Polymer Materials Science*
SEINFELD AND LAPIDUS *Mathematical Methods in Chemical Engineering: Vol. III, Process Modeling, Estimation, and Identification*
WHITAKER *Introduction to Fluid Mechanics*
WILDE *Optimum Seeking Methods*
WILLIAMS *Polymer Science and Engineering*

PRENTICE-HALL, INC.
PRENTICE-HALL INTERNATIONAL, INC., UNITED KINGDOM AND EIRE
PRENTICE-HALL OF CANADA, LTD., CANADA

Thermodynamics and its Applications

MICHAEL MODELL

ROBERT C. REID

Department of Chemical Engineering
Massachusetts Institute of Technology

PRENTICE-HALL, INC.

Englewood Cliffs, N. J.

Library of Congress Cataloging in Publication Data

Modell, Michael,
Thermodynamics and its applications.

(Prentice-Hall international series in the physical and chemical engineering sciences)
Includes bibliographical references.
1. Thermochemistry. I. Reid, Robert C., joint author. II. Title.
QD511.M68 541'.36 73-20291
ISBN 0-13-914861-2

10 9 8 7 6 5 4 3 2 1

Printed in the United States of America

PRENTICE-HALL INTERNATIONAL, INC., *London*
PRENTICE-HALL OF AUSTRALIA, PTY. LTD., *Sydney*
PRENTICE-HALL OF CANADA, LTD., *Toronto*
PRENTICE-HALL OF INDIA PRIVATE LIMITED, *New Delhi*
PRENTICE-HALL OF JAPAN, INC., *Tokyo*

Contents

Preface

As long as we can remember, our department has offered a one-semester, graduate level subject in classical thermodynamics. Traditionally, it has been applications-oriented; one of its primary objectives has been to develop competence and self-confidence in handling challenging applications in new and sometimes unusual situations. Half to two-thirds of the contact hours are usually devoted to problem-solving. Over the years, there accumulated many interesting, challenging problems—most of which originated from our consulting practice.

We have used a number of texts in conjunction with our graduate subject. None were completely satisfactory. We are convinced that a firm foundation in theory is essential for students who will be asked to fulfill the needs of tomorrow with an increasing demand for talents which are flexible and adaptable. On the other hand, the theory is useless unless the student can effectively bridge the gap between theory and application. Thus, we have attempted to develop a text with a rigorous theoretical and conceptual basis, interspersed with a relatively large number of examples and solutions. We have stressed to our students the desirability of working these examples before reviewing the solutions.

This text is intended to be a *learning text* rather than a teaching text. We have attempted to be thorough; but as a consequence of limited space and the short time a student spends in formal education, it is unreasonable to expect the student to appreciate all of the subtleties that will be apparent to the experienced reader. It is our hope that students will attain a basic level of understanding of theory and rationale of applications in their formal use of this text such that deeper insights can be gained in a self-instructional mode throughout their professional careers, as the need arises.

Following this philosophy, the text contains more material than one could hope to cover in one term—nor do we recommend a two-term sequence at the expense of the students' flexibility to shape their graduate curriculum to meet their individual needs. In three contact hours per week in a term, we have covered at a fairly rapid pace, all the chapters except for parts of Chapter 7 and Chapters 11 and 12; with four contact hours per week, we have covered thoroughly and at a more acceptable pace, ten chapters (excluding Chapters 12 and 7 or 11).

The theoretical basis of classical thermodynamics is developed in the first five chapters; that can be covered in one-third to one-half of a term. The flow of concepts is illustrated schematically in Figure P.1. The developments up to the introduction of the Fundamental Equation parallels the historical evolution of the classical body of knowledge (see Chapter 1).

The introduction of the formalism of the Fundamental Equation and Legendre transforms is a departure from traditional chemical engineering texts. (This route is becoming commonplace in physics and some other engineering fields, but these are often devoid of practical applications.) The Fundamental Equation is introduced because we believe it is of significant conceptual value in treating one of the central problems in engineering applications, namely, what are the minimum data required to reach a given objective and how does one manipulate available data to forms that are more appropriate to the problem at hand.

The Fundamental Equation in the energy representation, i.e., $U = f(S, V, N_1, \ldots, N_n)$, contains all thermodynamic information for a given single-phase, simple system. All other thermodynamic properties can be derived from it. Although we do not have available the Fundamental Equation for many materials, we can determine what other data sets have equivalent information content. Using Legendre transformations to preserve the information content, it is shown that, e.g., $H = f(S, P, N_1, \ldots, N_n)$ is also a Fundamental Equation and, thus, a Mollier diagram contains all thermodynamic information. Similarly, the Fundamental Equation of a pure material can be reconstructed from the equation of state and the heat capacity. Thus, any problem can be solved using P-V-T and C_p data; if these data are available, we need not search for any other data.

The last half of the text covers systems of increasing complexity. Following a discussion of single-phase systems of pure materials (Chapter 6), the criteria of equilibrium and stability are introduced (Chapter 7) so as to set the stage for treating mixtures and phase equilibrium.

Single-phase mixtures are synthesiezd from pure materials using the criteria of equilibrium to mix reversibly. Paralleling the pure material development, the type of data necessary to describe a mixture is discussed. The concept of ideal mixing is then appreciated as an idealization for which mixture properties can be synthesized from pure component data. Many common mixture properties (e.g., chemical potential, fugacity, activity coefficient) are explored as alternative methods of presenting similar information.

Phase equilibrium and chemical equilibrium are treated as progressively more complex applications of the building blocks covered previously. In these areas especially, it is stressed that thermodynamics is of little practical utility without sound engineering judgment. A phase diagram can only be constructed when there is prior knowledge of what phases do in fact exist and what properties (e.g., information equivalent to the Fundamental Equation) each phase exhibits. Similarly, the concept of chemical equilibria is of little use in the complex systems engineers generally face until there are data or insight into the kinetically feasible routes.

The last two chapters of the book deal with the thermodynamics of surfaces and nucleation (Chapter 11) and the thermodynamics of systems in electric, magnetic, stress, or other potential fields (Chapter 12). The approach used is parallel to that developed earlier, i.e., the applicable Fundamental Equation is found and Legendre transforms employed to relate the variables of interest in any real application.

It is impossible to acknowledge all who made this book a reality. We have been influenced by authors of previous articles and texts in thermodynamics and by our teachers. Professors J.M. Smith and H.P. Meissner excited our interest in this field and illustrated its power to attack and solve real and significant problems. Our students were critical and demanding and therefore a real delight. Sanjay Amin and Margaret Nemet provided significant comments as we approached completion. The typing was done in a superb manner by Ms. Judith Hawkins and Ms. Maria Tseng. To our wives and children, we are deeply indebted and grateful for their encouragement, understanding, and patience. Their confidence in this joint venture was a constant source of inspiration.

Cambridge, Mass.

M. Modell
R.C. Reid

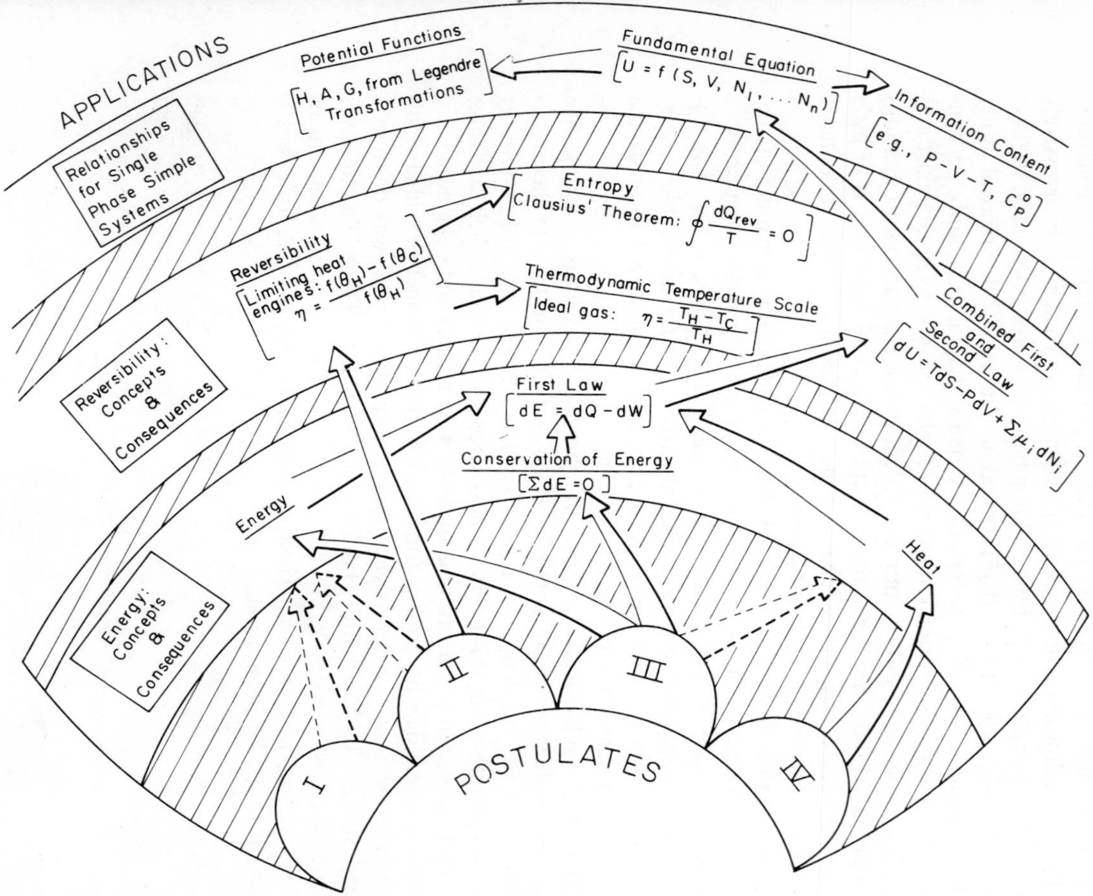
APPLICATIONS
Relationships for Single Phase Simple Systems
Potential Functions
[H, A, G, from Legendre Transformations]
Fundamental Equation
[U = f (S, V, N1, ... Nn)]
Information Content
[e.g., P-V-T, C°P]
Reversibility: Concepts & Consequences
Reversibility
[Limiting heat engines: η = (f(θH) - f(θC)) / f(θH)]
Entropy
[Clausius' Theorem: ∮ dQrev / T = 0]
Thermodynamic Temperature Scale
[Ideal gas: η = (TH - TC) / TH]
Combined First and Second Law
[dU = TdS - PdV + Σμi dNi]
Energy: Concepts & Consequences
Energy
First Law
[dE = dQ - dW]
Conservation of Energy
[ΣdE = 0]
Heat
POSTULATES
I
II
III
IV

Figure P-1

Thermodynamics

and its

Applications

Introduction 1

1.1 The Scope of Classical Thermodynamics

To the scientist, classical thermodynamics is one of a few mature fields epitomized by a rather well-defined, self-consistent body of knowledge. The essence of the theoretical structure of classical thermodynamics is a set of natural laws governing the behavior of macroscopic systems. The laws are derived from generalizations of observations and are largely independent of any hypothesis concerning the microscopic nature of matter. From these laws, a large number of corollaries and axioms are derivable by proofs based entirely on logic.

The scientist is sometimes at a loss to understand why the engineer has so much difficulty applying thermodynamics; after all, the theoretical development is rather straightforward. From the engineer's point of view, understanding the theory as developed by the chemist or physicist is not particularly difficult; however, the neat, self-contained presentation of the subject by the scientist is not necessarily amenable to practical application. Real-world processes are usually far from reversible, adiabatic, or well-mixed; very rarely are they isothermal or at equilibrium; few mixtures of industrial importance are ideal. Thus, the engineer must take a pragmatic approach to the application of thermodynamics to real systems. One of his major concerns is to redefine the real problem in terms of idealizations to which thermodynamics can be applied.

In the engineering context, almost all problems of thermodynamic importance can be classified into one of three types:

1. For a given process with prescribed (or idealized) internal constraints and boundary conditions, how do the properties of the system vary?
2. To effect given changes in system properties, what external interactions must be imposed? (The inverse of 1.)
3. Of the many alternative processes to effect a given change in a system, what are the efficiencies of each with respect to the resources at our disposal?

Problems of the first two classes require application of the First Law, which is developed in Chapter 3:

$$\Delta E = Q - W \tag{1-1}$$

where E is energy and Q and W are the heat and work interactions, respectively. The First Law may also be viewed as:

$$\text{internal changes} = \sum \text{interactions occurring at boundaries}$$

The change in energy can be related to variations of other internal properties of interest (e.g., T, P, V, etc.).

The third class of problems requires application of the Second Law, for which an idealization — the reversible process — is introduced as a standard for comparison.

There are basically only three steps required to develop a solution to any thermodynamic problem:

1. *Definition of the problem.* The real-world situation must be modelled by specifying the internal constraints and boundary conditions. For example, is a boundary permeable, semipermeable, or impermeable? Is heat transfer fast or slow relative to the time span of interest? Which chemical reactions are known to occur under the conditions of interest?
2. *Application of thermodynamic laws.* As described above, these either relate effects internal to the system with external interactions (the First Law) or they set limits on the extent of internal variations (the Second Law). The combined laws prescribe in part the relationships between property variations, but they do not uniquely specify the magnitude of the change in properties. For example, for a simple system undergoing a process in which the temperature and pressure are observed to change from T_1, P_1 to T_2, P_2, we might wish to calculate the energy change in order to specify the necessary heat and work interactions. We might employ the following analysis:
   a. From *thermodynamic reasoning*, ΔE is a unique function of T_1, P_1 and T_2, P_2 because E is a state function. Therefore, ΔE can be evaluated over any path between these end states.

b. From *mathematical reasoning*, over any path for which E is defined, dE may be expressed as an exact differential such as:

$$\Delta E = \int dE = \int_{T_1}^{T_2} \left(\frac{\partial E}{\partial T}\right)_{P_1} dT + \int_{P_1}^{P_2} \left(\frac{\partial E}{\partial P}\right)_{T_2} dP \qquad (1\text{-}2)$$

Note that $(\partial E/\partial T)_P$ and $(\partial E/\partial P)_T$ must be expressed as functions of T and P before Eq. (1-2) can be integrated.

c. Applying *thermodynamic reasoning*, E is defined as a function of T and P over a reversible path and, thus, $(\partial E/\partial T)_P$ and $(\partial E/\partial P)_T$ can be reduced to other variable sets that are more readily quantified:

$$\Delta E = \int_{T_1}^{T_2} \left[C_p - P\left(\frac{\partial V}{\partial T}\right)_P\right] dT - \int_{P_1}^{P_2} \left[T\left(\frac{\partial V}{\partial T}\right)_P + P\left(\frac{\partial V}{\partial P}\right)_T\right] dP \qquad (1\text{-}3)$$

where C_p is the constant-pressure heat capacity. Note that Eq. (1-3) is a *general result;* it must be satisfied by any material undergoing a change from T_1, P_1 to T_2, P_2. However, the value of ΔE is not unique; it differs from one material to the next, which leads us to the third and final step.

3. *Evaluation of property data*. There are property relationships that are unique characteristics of matter. For example, in Eq. (1-3), thermodynamics does not dictate the functions,

$$C_p = f_1(T, P); \qquad V = f_2(T, P) \qquad (1\text{-}4)$$

required for the integration. Evaluation of these property data lie outside of the scope of classical thermodynamics. However, they are essential to the solution of real problems and, hence, are within the scope of this text. The engineer must make recourse to a variety of methods (e.g., literature, experiments, correlations, or microscopic theories as developed with statistical mechanics) in order to determine or approximate these property relationships.

Before discussing the approach to classical thermodynamics (Section 1.3) used herein, it is instructive to review the historical evolution of this body of knowledge.

1.2 Preclassical Thermodynamics

The origin of classical thermodynamics can be traced back to the early 1600's. The laws, as we know them today, were not formalized until the late 1800's. The interim 250 to 300 years are called the *preclassical* period, during which many of our current concepts were developed.

The chronological development is a fascinating example of the application of scientific methodology. Experimentation (e.g., thermometry) led to the development of hypotheses and concepts (e.g., the adiabatic wall) which, in turn, suggested other experiments (e.g., calorimetry) followed by new concepts (the caloric theory), etc. The historical events also illustrate some of the potential pitfalls in scientific analysis, such as the overemphasis of intuitive images (e.g., the nature of heat) which go far beyond the existing body of observations and facts. Consequently, the preclassical period was marked with pedagogical controversy and much confusion.

The beginning of the preclassical period is usually associated with Galileo's attempts to quantify thermometry (ca. 1600). It is interesting to note that the seventeenth-century scientist was motivated primarily by a desire to understand phenomena that were perceived by his senses. In contrast, scientists today require very sensitive and elaborate instrumentation to detect phenomena that are far beyond the reaches of their senses. One of Galileo's principal objectives was to quantify the subjective experiences of hot and cold. The expansion of air upon heating was appreciated in the Hellenistic era, but it was never applied. Galileo used this phenomenon in his bulb-and-stem device with the stem submerged in water. The measurements would change with time, but in the early 1600's there was no reason to assume that they should not vary in such a manner. It was not until 1643, when a student of Galileo, Torricelli, developed the barometer, that it was appreciated that Galileo's device was more of a "barothermoscope" than a thermometer. As glass-blowing technology advanced, the availability of narrow capillaries led to the development of liquid thermometers in the 1630's. As might be expected, water was the first liquid used. Although difficulties that should be obvious to today's students were experienced, 10 years elapsed before the sealed alcohol thermometer gained acceptance. Gas thermometers did not reappear until the 1700's, when gas properties were better understood.

When any new experimental tool is developed, invariably there is the desire to quantify it so that results among different investigators can be compared. The quantification of thermometry required the introduction of at least two fiducial or fixed points. Stop for a moment and reflect on what fixed points you might have chosen had you been a scientist in 1640. The boiling point of water? It varies from day to day. The freezing point of water? It is difficult to find ice during most of the year. Furthermore, there was no reason to believe that materials like water had unique properties that would be reproducible. Thus, it is understandable that our ancestors turned to phenomenological references, such as the "warmest water the hand could stand" or the "most severe winter cold" or the "temperature of the human body." Later, selection of the melting point of butter or the freezing point of aniseed oil were steps toward objectivity, although these transition points were not very sharp. It was not until 1694 that the freezing and boiling points of water gained acceptance.

With the advance of quantitative thermometric experimentation it soon became apparent that different types of containers had different thermal properties. Hot liquids would cool less rapidly in mica or wood vessels than in metals. These observations led to the idealized concept of the adiabatic wall, which could be approached in practice; thus, the science of calorimetry was born.

If two portions of the same fluid were mixed in a calorimeter, the final temperature could be expressed as a weighted mean of the two initial temperatures:

$$t_f = \frac{a_1 t_1 + a_2 t_2}{a_1 + a_2} \tag{1-5}$$

The weighting factor could be mass or volume.

Although Eq. (1-5) is of the form of a conservation law, it is not at all clear *what* is conserved. A simplistic interpretation would have temperature as the conserved quantity. The conservation law was given more structure in the mid-1700's, when Eq. (1-5), with mass or volume as weighting factors, was shown to be invalid when different liquids were mixed. In the 1760's, Joseph Black suggested a modification that was consistent with mixing data for different fluids. The constants a_i in Eq. (1-5) were subdivided into a mass-related component and an intensity parameter, the specific heat, which was a unique property of the liquid. This was the first time that it was proposed that matter had distinctive properties in the thermodynamic sense.

Black's modification indicated that something other than temperature was conserved in the mixing process. This quantity was called heat, or caloric. The interpretation went far beyond the physical observations into the realm of metaphysics. The caloric theory attempted to define the microscopic nature of the conserved quantity.

Black's ingenious hypothesis led to a flurry of experimentation, during which specific heats were measured and reported in the then flourishing royal societies. The conservation law was repeatedly challenged, but by more exacting experimentation, the theory was enlarged to account for the variation of specific heat with temperature and, later, latent heats were introduced to account for phase transitions.

In the 1780's, no more than 20 years after Black's work, Count Rumford conducted his exhaustive experiments to show that mechanical work was an inexhaustible source of caloric. Hence, caloric could not be conserved and could not be of a material nature. Rumford suggested a revival of the mechanical concept of heat that had been abandoned 50 years earlier. Although we now know that Rumford's suggestion was closer to the truth, it is understandable why very few of his peers followed his lead. The statistical concepts necessary to relate micromechanical energy to the macroscopic energy of calorimetry were not to be introduced by Maxwell, Boltzmann, and Gibbs until a century later. Rumford's hypothesis failed to produce tangible phenomenological results.

Although the caloric theory remained in use for over 50 years after Rumford's work, he emphasized the dilemma between conservation and creation (or conversion) which was to perplex the best minds of the nineteenth century. The conversion crisis was firmly established by the work of Mayer and Joule in the 1840's.

In 1824, Carnot offered a partial reconciliation of the conversion and conservation phenomena with an argument based on the caloric theory (the results of which actually prolonged the life of the theory). Carnot introduced a step-change in the level of complexity and sophistication: he put forth a number of new concepts that were essential to the eventual clarification of preclassical ideas and that later led to the replacement of the caloric theory by the First and Second Laws. These firsts include the concepts of heat reservoirs and reversibility, and the requirement of a temperature difference to generate work from a heat interaction.

Carnot proposed that cyclic operation of an engine working between two heat reservoirs was analogous to water flowing over a dam. Some quantity, in being transferred from a high to a low potential, produced work but the quantity being transferred was conserved in the process. We know today that Carnot's hypothesis is incorrect because he assumed that the conserved quantity was caloric. He carried this reasoning further to prove that there must be a limiting efficiency of heat engines. If such a limitation did not exist, then two engines could be suitably operated in a cyclic process in order to bring the heat reservoirs back to their initial states, the net effect being the production of work. This, Carnot declared, was an impossibility.

We might ask on what basis Carnot ruled out the possibility of what we today call a "perpetual motion machine of the first kind." It was not until 1847 that Helmholtz advanced the hypothesis of the conservation of energy. With few notable exceptions, there appeared to be general agreement among scientists of the period that the basic powers of nature are uncreatable and indestructible. Thus, over a period of many years, we see the gradual acceptance of what we refer to today as a basic postulate.

Carnot's engine reconciled the conversion and conservation phenomena—at least in the reversible limit. But how is this consistent with conservation in the highly irreversible mixing process of calorimetry, or conversion in the equally irreversible process of generating heat through friction? Carnot was cognizant of these difficulties and called for further experimentation and also for reconsideration of the foundations of the theory.

The preclassical period drew to a close with the quantitative work of Joule who established the equivalence of mechanical, electrical, and chemical energy to heat. We can see that at that time there were a number of concepts yet to be clarified. Caloric had to be split into heat quantity, energy, and entropy. It had to be shown that heat and work were forms of energy transfer and that the interconvertibility was asymmetric. It is energy that is conserved in the calorimeter, in Carnot's cycle, and in frictional processes. Entropy is conserved only in the limit of the reversible process.

These developments occurred in relatively rapid succession beginning approximately in 1850 with the genius of Clausius, with important contributions from Kelvin, Maxwell, Planck, Duhem, and Poincaré, and terminating with the brilliance of Gibbs near the end of the century. Although we cannot cover these events here, we should not underestimate their historical importance. We will attempt to reconstruct these developments, applying the insight that we have gained over the last 70 years.

1.3 The Postulatory Approach

Almost all approaches to classical thermodynamics follow one of two extremes: the *historical* approach, which parallels closely the chronological development of concepts and misconceptions, and the *postulatory* approach, in which axioms that cannot be proved from first principles are stated. There are merits and drawbacks in each extreme.

Advocates of the historical approach contend that if we are to expect our students to evolve new concepts and theories we must expose them to the historical development of existing theories. Existing postulatory approaches make no reference to historical developments. The basis for the laws of thermodynamics is impersonally stated in a small number of postulates that cannot be proved and can only be disproved by showing that consequences derived from them are in conflict with experimental facts. The postulates tend to be mathematical and abstract, but the laws of thermodynamics are derivable from them. Many students are unimpressed because little insight is provided for the necessity to define new concepts or properties.

The approach we shall follow parallels the historical development in many respects. We shall begin by assuming the state of mind and body of knowledge available to the seventeenth-century scientist and by proceeding from that point to a logical development of a self-consistent set of rules that applies to the behavior of macroscopic bodies. In the process, we shall make use of many of the arguments put forth by our ingenious predecessors over the last 300 years, but we shall use hindsight to avoid the incorrect conclusions that prevailed at many points in the preclassical period and that resulted in much confusion (some of which is usually transferred to the student who studies thermodynamics by the historical approach).

At several junctures in our development we will face obstacles that cannot be obviated by invoking first principles. Our predecessors overcame these obstacles by trial-and-error experimentation until they amassed a large body of knowledge. In this manner, generalities were stated and rules established. We shall clearly identify those principles that our ancestors learned to accept without proof; these are stated as postulates, but in a form that could be understood by Black, Lavoisier, Kelvin, or Carnot. The ultimate verification of these postulates lies in the success of the formalism derived from them. In this vein, Schroedinger's equation is a basic postulate of quantum mechanics and Newton's laws of motion *were* basic postulates of classical mechanics. It is conceivable that, at some later date, new experimental information will be obtained that will necessitate revision or reformulation of the thermodynamic postulates, just as Newton's laws were found inapplicable to the motion of elementary particles of the atom.

Although the set of postulates presented in this text contains the same information content as those developed in other texts, the phrasing may appear significantly different. There are obviously many different sets of postulates[1] that are equally valid bases for the theoretical development. In developing a set of postulates, an attempt has been made to keep them as real as possible (as opposed to abstract) while still retaining a form that, hopefully, is readily acceptable to the chemical engineer.

SUGGESTED READING

ANDRADE, E. N. DA C. (1935). "Two Historical Notes—Humphrey Davy's Experiments on the Frictional Development of Heat; Newton's Early Notebook," *Nature*, **135**, 359.

BOMPASS, C. (1817). *An Essay on the Nature of Heat, Light, and Electricity*. London: T & C Underwood.

LARDNER, D. (1833). *A Treatise on Heat*. London: Longmans, Rees, Orme, Brown, Green, and Longmans.

LESLIE, J. (1804). *An Experimental Inquiry into the Nature and Propagation of Heat*. London: J. Mawman.

METCALFE, S. L. (1843). *Caloric* (2 volumes). London: William Pickering.

ROLLER, D. (1950). *The Early Development of the Concepts of Temperature and Heat*. Cambridge: Harvard University Press.

RUMFORD, B. T. (1798). "An Experimental Inquiry Concerning the Source of the Heat which is Supplied by Friction," *Phil. Trans.*, **88**, 80.

THOMSON, W. (1840). *An Outline of the Sciences of Heat and Electricity*. London: H. Bailliere.

TISZA, L., (1966). *Generalized Thermodynamics*, Paper 1. Cambridge: M. I. T. Press.

TYNDALL, J. (1880). *Heat, A Mode of Motion*. London: Longmans, Green, and Company.

URE, A. (1818). *New Experimental Researches on Some of the Leading Doctrines of Caloric*. London: William Bulmer and Company.

[1] See, for example, H. B. Callen, *Thermodynamics* (New York: Wiley, 1960), and G. W. Hatsopoulos and J. H. Keenan, *Principles of General Thermodynamics* (New York: Wiley, 1964).

Basic Concepts and Definitions

2

2.1 The System and Its Environment

If we are to develop a set of fundamental laws of nature without any preconceived notions, we must first develop the facility to perform some experiments. The subject of the experiment will be called the *system*, which will refer to a region that is clearly defined in terms of spatial coordinates. The surface surrounding this region will be referred to as the *boundary*. It may be an actual wall or it may be an imaginary surface whose position is defined during an experiment. The region of space external to the system and sharing a common boundary with the system will be referred to as the *environment* or *surroundings*.

As will be seen shortly, work and heat effects are defined in terms of events at system boundaries; thus, the choice of a boundary is usually dictated by the kind of information desired. In any given situation, there are often many different system boundaries one can choose, each having some advantages and disadvantages. Developing a facility for choosing those system boundaries that will result in the shortest path to the desired information is essential to the engineer. Some insight into the selection process can be gained by working a given problem using several different system boundaries.

2.2 Primitive Properties

To record events that occur within a system, we must devise experimental tools that are sensitive to changes in the system. *Primitive properties* will refer to characteristics of the system that can be determined or measured by performing a standardized experiment on the system. To insure that the measurement is a characteristic of only the system, we require that the experiment not disturb the system. The primitive property is of value because it is directly associated with the system at a particular time, and the observer need not know the history of the system to ascertain the value of the property.

A primitive property which is easily measured and which is particularly useful is the *thermometric* temperature (denoted by θ). The θ-temperature can be measured, for example, by noting the volume of a known mass of liquid in a sealed tube when the tube is brought into contact with the system. The mass of the thermometer should be small in relation to that of the system so that the measurement does not alter the system. (The effect of the thermometer on the system can be determined by inserting a second thermometer and noting any change in the reading of the first thermometer.)

It should be emphasized that the value of θ is completely arbitrary and depends on the type of fluid, the materials used in construction of the tube, and how the tube is notched and also on what labels are associated with the notches. Once the notches and labels have been made, however, the device can be used to observe *changes* occurring within a system. It could also be used to rank the θ-property of different systems. We shall return to a more detailed description of the thermometric temperature in Section 3.4.

Innumerable primitive properties could be defined by outlining suitable experiments. In this sense, volume, mass, pressure, index of refraction, color (given a standardized method of observation), etc., could be called primitive properties and scales for each property could be devised.

Primitive properties are useful in defining a system and in recording the occurrence of events in a system. In an *event*, at least one primitive property changes. An *interaction* is defined as events occurring simultaneously in the system and surroundings, at least one of which would not have occurred if the system were removed from the surroundings and placed in any other arbitrary environment.

2.3 Classification of Boundaries

The interactions between a system and its surroundings are governed by the nature of their common boundary. If the boundary is impermeable to mass flow, the system is called a *closed* system. An *open* system has a boundary that permits a mass flux of at least one component of the system through at least one point. In either case, the boundary may be *rigid* or *movable*.

There is one other set of conjugates needed to complete the classification of boundaries: the *adiabatic* and *diathermal* walls. These boundaries govern the extent of heat interaction between system and surroundings, but since we have not yet defined a heat interaction, this definition is inadmissible. The adiabatic boundary or wall is one of the key concepts in thermodynamics, and we will in fact use it to define work and heat interactions. In order to avoid a circular system of definitions, we will treat the adiabatic wall in detail in the next section.

In the absence of external force fields, specification of one of each of the three conjugate sets of boundaries (i.e., permeable and impermeable, rigid and movable, adiabatic and diathermal) is necessary in order to describe completely the external constraints placed on the system. Of the eight combinations[1] there is one that is of particular importance. A system enclosed by impermeable, rigid, and adiabatic walls is called an *isolated* system. It will become evident (see Section 3.4) that this system can have no interactions. Any events occurring within this system are independent of events in the environment.

2.4 The Adiabatic Wall

The concept of the adiabatic wall evolves from our experience; it can be illustrated by conducting a series of simple experiments. Consider a closed system in which a device to measure the thermometric temperature has been placed. Surround this system with a system having a higher thermometric temperature. If the initial system were constructed from aluminum, the variation of the thermometric temperature with time would look like curve 1 of Figure 2.1. Curves 2, 3, and 4 would result if the initial system were constructed from steel, glass, and asbestos, respectively. Finally, if the container were made of a double-walled Dewar, the variation of temperature over the time of the experiment would be quite small (curve 5). The adiabatic wall is an idealized concept representing the limiting case of curve *A* in Figure 2.1. In practice, adiabatic boundaries are approached in many situations, especially those in which events occur rapidly in relation to the time scale of the experiment.

The case diametric to that of adiabatic is the diathermal wall (curve *B*),

[1] It will become evident in discussing the criteria of equilibrium that only six of the eight combinations are meaningful. Of the four combinations involving permeable boundaries, there is no distinction between adiabatic and diathermal walls (see Section 7.3).

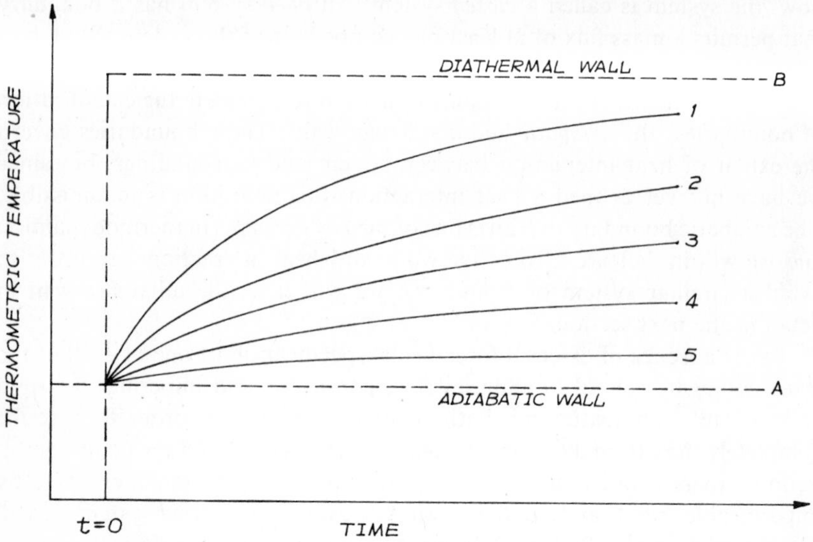

Figure 2.1

in which the change in thermometric temperature is rapid in relation to the time scale of the experiment (i.e., the thermometric temperature of the two systems is always identical).

2.5 Simple and Composite Systems

There is a special class of systems that plays a central role in the developments to follow. These systems, which will be referred to as *simple systems*, are devoid of any internal adiabatic, rigid, and impermeable boundaries and are not acted upon by external force fields or inertial forces.

A *phase* is defined as a region within a system throughout which all of the properties are uniform. A single-phase system is the simplest of simple systems. A system containing multiple phases is also a simple system provided that no phase acts as an adiabatic, rigid, or impermeable boundary to any other phase.

Composite systems are systems composed of two or more simple subsystems. There are no restrictions on the kinds of boundaries separating the subsystems of the composite.

Restraints are defined as barriers within a system that prevent some changes from occurring within the time span of interest. In simple systems, restraints of interest are barriers to chemical reaction or barriers to phase change. For example, the room–temperature reaction of hydrogen and oxygen to form water can be made to occur within milliseconds if a proper catalyst is incorporated in the system. In the absence of a catalyst, no noticeable reaction occurs within months or years. In the latter case, there is an internal restraint (the activation energy barrier) which, for all intents and purposes, prevents the occurrence of the reaction. For composite systems, internal boundaries that are adiabatic, rigid, or impermeable are also considered restraints.

Thermodynamics does not dictate the restraints that may be present in a given system. In any given situation, one must decide which restraints are present. Two factors that must be considered in making this decision are the laws of matter and the rates of the various conceivable processes: The former are (1) the law of continuity of matter (matter cannot move from one position to another without appearing at some time in the intervening space); (2) the law of conservation of electrical charge (net electrical charge must be conserved in all processes); and (3) the law of conservation of chemical elements (in the absence of nuclear transformations and relativistic effects, mass must be conserved). The application of some of these principles is illustrated in the following example.

Example 2.1

A closed vessel contains water, oil, and air at room temperature. If the system is synthesized by first adding the water and then layering the oil on the water, is this system a simple or composite system? If composite, define each simple subsystem and the internal restraints. If the vessel is shaken vigorously, is the system then a simple or composite system?

Solution

There are clearly no adiabatic or rigid walls within the vessel. Thus, it is only necessary to decide if there are impermeable walls. At room temperature, a system containing water and air should have appreciable water vapor present in the air. If the system is formed by layering the oil on the water without shaking, the law of continuity of matter requires that the water pass through the oil layer before it evaporates into the air. Thus, the water must diffuse through the oil layer. This process is slow and will not occur to any appreciable extent even after several hours. If our interest in this system did not extend beyond this time scale, the oil layer would have to be considered as an impermeable barrier to the water. In this case, the vessel would be considered a composite system of two simple subsystems: air + oil (assuming oil evaporates into the air within the time scale of the experiment), and water. If the system is shaken vigorously, water droplets will contact the air directly. Thus, the entire contents of the vessel would be considered a simple system.

2.6 States of a System

Now that we have the means for conducting experiments on a system, we would like to have a formal way of characterizing a system so that others could reproduce the experiments. For this purpose, we will identify the condition or *state* of the system by the values of those properties that are required to reproduce the system. Although this definition is functional, it is not very practical because we do not always know the number of properties that are required to specify the state of the system. We are fortunate, however, to have available a large body of experimental data, accumulated over several hundred years, which indicates that there are particular types of states that can be specified by delineating only a certain number of properties. These states, which are called *stable equilibrium states*, are defined in the next section. In general, nonequilibrium states can be specified (for the purpose of reproducing them) from a finite number of properties; the number of such properties, however, is not specified by the principles of classical thermodynamics.

2.7 Stable Equilibrium States

The aforementioned body of experimental data indicating the existence of these states is summarized in the following postulate.[2]

> ***Postulate I. For closed simple systems with given internal restraints, there exist stable equilibrium states which can be characterized completely by two independently variable properties in addition to the masses of the particular chemical species initially charged.***

By two independently variable properties,[3] it is meant that each of these properties could be varied (by at least a small amount) in at least one experiment during which the other property is held constant. For example, consider a closed vessel containing the simple system of a liquid and its vapor in a stable equilibrium state. Given the amount of material initially charged, the thermometric temperature, and the total volume of the vessel, the system could be reproduced at will because the volume and temperature are independently variable and, therefore, completely specify the system. If, however, the pressure

[2] This postulate is similar to a conclusion drawn by Duhem in 1899 and is sometimes referred to as Duhem's Theorem. See Prigogine and Defay as translated by Everett, *Chemical Thermodynamics* (London: Longmans, Ltd., 1954), p. 188.

[3] There is a class of thermodynamically trivial properties called *neutral* properties (e.g., the shape of a system of given volume) that cannot be used to determine the stable equilibrium state. See Hatsopoulos and Keenan, *Principles of General Thermodynamics* (New York: Wiley, 1965), pp. 30–32.

were specified instead of the total volume, the system could not be reproduced because, as we shall see later, pressure and temperature are not independently variable in this case.

We are now at an impasse. We do not know which two properties to choose to specify the state of the system because we do not yet know which properties are independently variable. For a given system with a given set of restraints, we find that the laws of thermodynamics (which we shall develop from our postulates) will result in relationships among certain variables. The relationships and the variables included therein will depend on the system and the restraints. Only from these relationships shall we be able to determine which sets of properties are not independently variable. By process of elimination, we shall then be able to determine which sets of properties are independently variable. For example, for the closed system of liquid and vapor in a stable equilibrium state, we can show that as a result of the requirement of phase equilibrium (Chapter 9), the vapor pressure is a unique function of temperature (the Clausius-Clapeyron equation). Thus, pressure and temperature cannot be independently variable in this case.

Knowing that stable equilibrium states exist is not nearly as informative as knowing when they exist. The second postulate is directed toward establishing this fact.

> ***Postulate II. In processes for which there is no net effect on the environment, all systems (simple and composite) with given internal restraints will change in such a way as to approach one and only one stable equilibrium state for each simple subsystem. In the limiting condition, the entire system is said to be at equilibrium.***

Postulate II is specific to systems which are in effect isolated; in other words, for processes which consist of a series of steps, the system may interact with the environment in two or more steps, but the net effect of these steps must leave the environment unaltered.

Since the stable equilibrium state is defined as a limiting condition toward which a simple system tends to change, it follows that none of the properties of this state varies with time. It follows from Postulate I that once a simple system has reached a stable equilibrium state, only two independently variable properties and the masses initially charged need to be specified to completely determine this state. Since all other properties are fixed in the stable equilibrium state, it follows that all other properties of the simple system are dependent variables that are determined by the two independently variable properties and the masses of the initial chemical species. Note that this conclusion is valid for each *simple* system at equilibrium, even if the simple systems are part of a composite system. The conclusion, however, does not apply to the composite system at equilibrium: that is, the state of a composite system at equilibrium cannot be specified by two independently variable properties plus the masses initially charged. The difficulty arises from the fact that all properties may vary from one subsystem to another within a composite system at equilibrium. For example, the thermometric temperature of a composite system has little significance if the subsystems of the composite are separated by adiabatic, impermeable walls. Primarily, Postulate I has been restricted to simple systems in order to avoid such difficulties.

In Postulate II, it is stated that for an isolated system, there exists one and only one set of stable equilibrium states (toward which the subsystems tend) *for a given set of internal restraints*. There will be different sets of stable equilibrium states for different sets of internal restraints. For example, with reference to Example 2.1, there will be a unique equilibrium state if we assume that there is an impermeable barrier preventing water from reaching the air space; there will also be a unique equilibrium state if we assume no such barrier exists. Although each of these states is unique, the properties of each will be different.[4] Thus, before the equilibrium state can be completely defined, the internal restraints must be specified. Specification of internal restraints is clearly an important part of specifying the system.

2.8 Thermodynamic Processes

A change of state[5] of a system is identified by a change in the value of at least one property. For systems initially in stable equilibrium states, changes of state will occur only when the system has an interaction with the environment *or* when internal restraints are altered. "Change-of-state" is usually applied to systems that are initially in one stable equilibrium state and are found after some event to be in another equilibrium state. The change of state is then fully described by the values of the properties in the two end states.

The *path* refers to the description all of the states that the system traverses during a change of state. Thus, the path is described in terms of the primitive properties that define the intermediate states. Paths for which all of the intermediate states are equilibrium states are termed *quasi-static* paths.[6] From Postulate I, quasi-static paths of closed simple systems can be completely described in terms of the values of only two independent properties.

[4] In some texts the former state would be referred to as metastable in the sense *that given sufficient time* the latter state (for which the term stable equilibrium state would be reserved) would be reached. The problem with this set of definitions is that almost all systems of interest to chemical engineers would then have to be classified as metastable even though the final stable equilibrium state may not be obtained in any time span of interest.

[5] In common usage, a change of state often is synonymous to a "change-of-phase." This is not the meaning we use in this book.

[6] Quasi-static paths are closely related to (and sometimes confused with) reversible processes. The distinction between the two will be considered in Section 4.6.

It also follows from Postulate II that if a system progressing along a quasi-static path is isolated at some point (for example, by temporarily altering a boundary condition), the values of all the properties will remain constant at the values observed just prior to isolation. It may, however, take more than two properties to describe a nonquasi-static path. If the system is isolated during such a path, some primitive properties will change after isolation as the system approaches a stable equilibrium state. For example, consider the system of a gas initially at two atmospheres which is contained in a cylinder fitted with a piston and stops. The stop holding the piston is removed, and the gas expands until the piston reaches a second stop. If the piston is lubricated, the expansion process will be rapid. At any instant during the process, there will be a pressure gradient within the gas phase. In order to describe such an intermediate state, it will be necessary to determine the pressure (in addition to other properties) at all points within the cylinder. The intermediate states are not stable equilibrium states; if such a state were isolated (by stopping the piston at an intermediate point), the pressure gradient would be damped out as the system approached a stable equilibrium state. Clearly, this frictionless process is not quasi-static. Alternatively, if there were external forces acting against the piston so that the expansion was very slow, no appreciable pressure gradient would be found. If the system were then isolated at an intermediate point, no properties would change because the system was, at all times, in some stable equilibrium state. Thus, the latter process was quasi-static.

The thermodynamic *process* involved in a change of state usually refers to a description of the end states, the phenomena occurring at the system boundaries (i.e., heat and work interactions, which are discussed in Chapter 3) during the process, and the path (which is usually described only for quasi-static processes). In many instances, however, the term "process" is loosely applied to describe the path without explicitly specifying the boundary conditions. Thus, an isothermal, isobaric, or isochoric process is one in which the temperature, pressure, or volume is constant. In such cases the boundary conditions are usually implied by the nature of the process or are immaterial to the problem in question.

2.9 Derived Properties

Primitive properties were defined in terms of an experiment or measurement made on the system at some point in time. Only experiments that do not disturb the system are allowed. By definition, primitive properties are not restricted to stable equilibrium states.

We have now established two basic postulates dealing with a particular class of states, the stable equilibrium states. These states have innumerable primitive properties associated with them. Each of these properties is measurable by definition. We must also ask ourselves if there are other *properties* of these stable equilibrium states that are not measurable by any method but could be used to characterize the system. No definite answer can be given at this point, but it will become obvious in Chapter 3 and those following that such properties do exist though no simple device has yet been made to measure them directly. We will find that we can define such properties in terms of changes in the system between initial and final stable equilibrium states. (Note the path need not be quasi-static.) In order to distinguish these properties from the *primitive* type, we will call them *derived* properties. As defined, these derived properties exist only for stable equilibrium states and, as such, may be used as variables to define a system as required in Postulate I.

Since derived properties are functions of state, and since any stable equilibrium state of simple systems can be characterized by the values of two independently variable properties plus the masses, any derived property can be expressed mathematically as a function of two other independently variable properties, derived or primitive.

It will be instructive at this point for the reader to review the mathematical relations of functions of state. A review of this topic can be found in Appendix B.

Energy: Concepts and Consequences 3

In this chapter we shall adopt the definition of work from the mechanics of rigid bodies and extend this concept to thermodynamic systems. We shall introduce energy as a unique measure of the work required to reach one stable state from another in certain processes. Since we cannot show *a priori* that this quantity is a function only of the end states, it will be necessary to postulate that energy is a derived property. The conservation law for energy will be a direct consequence of the new postulate. After defining work and postulating the existence of the energy–property, we shall present an operational definition of heat and discuss the directionality of heat interactions.

3.1 Work Interactions

The mechanical work associated with the movement of a rigid body is defined as

$$W = \int_{x_1}^{x_2} (\sum \mathbf{F}_s) \cdot d\mathbf{x} \tag{3-1}$$

or, in differential form,[1]

$$đW = (\sum \mathbf{F}_s) \cdot d\mathbf{x} \tag{3-2}$$

[1] As discussed in Appendix B, the symbol $đ$ is used throughout the text to denote differentials of functions that are not state variables.

where $\sum \mathbf{F}_s$ is the resultant force acting on the surface or boundary of the rigid body at a point where there is a differential displacement of the boundary, $d\mathbf{x}$. In keeping with the mechanical definition of work, *boundary forces* ($\mathbf{F}_s$) are distinguished from *body forces* or forces associated with external fields ($\mathbf{F}_b$) (i.e., centrifugal, gravitational, inertial, coulombic, etc.); in the absence of electric and magnetic fields, only boundary forces are used to calculate work.

For a rigid body acted upon by both boundary and body forces, Newton's second law of motion states that

$$\sum \mathbf{F}_s + \sum \mathbf{F}_b = 0 \tag{3-3}$$

In Eq. (3-3), the inertial force, $-M\mathbf{a}$ is considered a body force and is included in the second summation. For example, consider a weight suspended by a string, as in Figure 3.1. If the body is initially at rest, and if $\mathbf{F}_s$ and $\mathbf{g}$ are col-

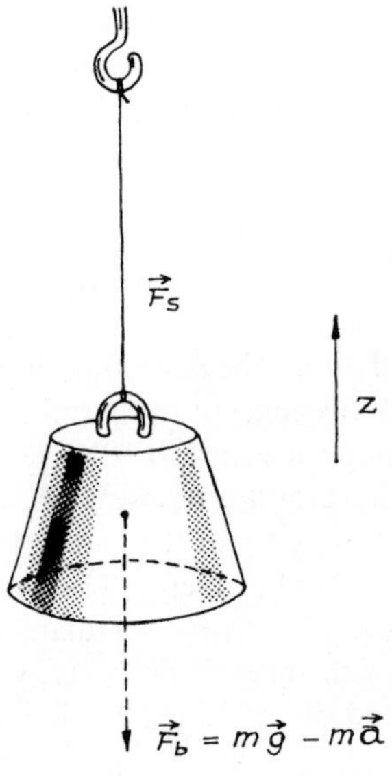

Figure 3.1

linear, then Eq. (3-3) becomes

$$F_s - Mg = 0 \tag{3-4}$$

If F_s is then increased so that the weight rises, at any instant during the motion, Eq. (3-3) is

$$F_s - Mg - M\frac{dv}{dt} = 0 \tag{3-5}$$

The differential work done on the weight at any instant of time is due only to the boundary force, F_s. Thus, using Eqs. (3-2) and (3-5),

$$đW = \mathbf{F}_s \cdot d\mathbf{x} = \left(Mg + M\frac{dv}{dt}\right) dz = Mg\,dz + Mv\,dv \tag{3-6}$$

The total work done on the weight between z_1 and z_2 is

$$W = Mg(z_2 - z_1) + \frac{M}{2}(v_2^2 - v_1^2) \tag{3-7}$$

On the right-hand side of Eq. (3-7), we call the first and second terms the difference in potential and kinetic energy, respectively. (Here we accept the terms of potential and kinetic energies as Mgz and $Mv^2/2$, but in no way have we defined the concept of energy.)

We are now in a position to define the work associated with systems of interest in thermodynamics; in particular, we are concerned with systems having nonrigid as well as rigid boundaries.

Consider the expansion of a gas contained in a cylinder fitted with a piston and surrounded by the atmosphere (Figure 3.2). If we choose the

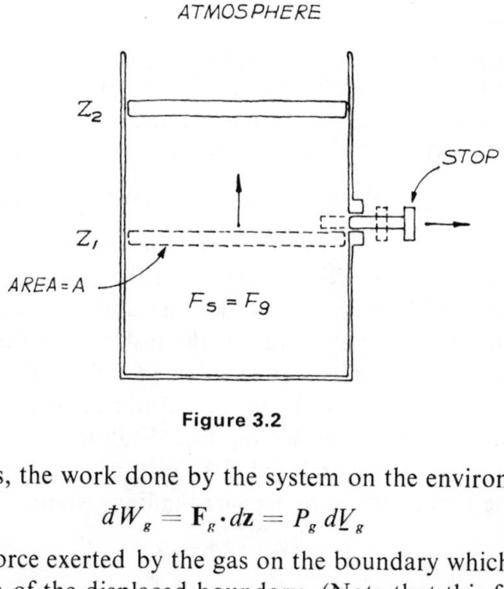

Figure 3.2

system as the gas, the work done by the system on the environment is

$$đW_g = \mathbf{F}_g \cdot d\mathbf{z} = P_g\,d\underline{V}_g \tag{3-8}$$

where $\mathbf{F}_g$ is the force exerted by the gas on the boundary which is displaced and A is the area of the displaced boundary. (Note that this force may not be the same as that exerted on the other boundaries of the gas system.) Thus, before the work can be calculated, $\mathbf{F}_g$ must be known as a function of $\mathbf{z}$. For the gas, the term F_g is, of course, equal to P_gA, where P_g is the pressure on the boundary that is displaced.

If neither F_g nor P_g is known, the work can still be determined by measuring the effects on the environment. For example, consider a differential element of time during the expansion. To illustrate, let us suppose that there is friction

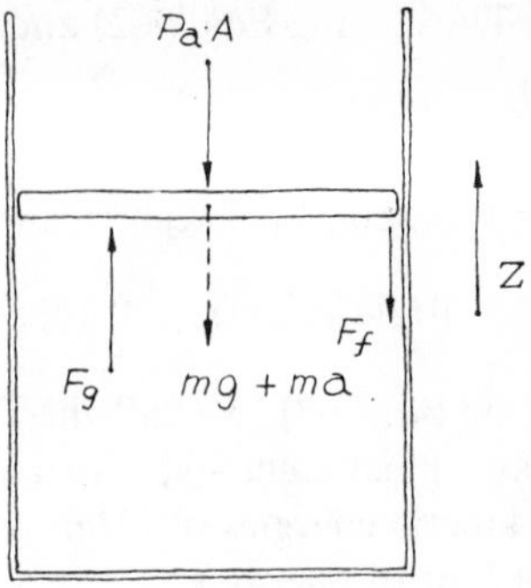

Figure 3.3

between the piston and the cylinder walls. As is shown in Figure 3.3, the unknown force F_g can be evaluated by making a force balance on the object which is acted upon by this force. In this case, the object is the piston. Thus,

$$F_g - P_a A - F_f - Mg - Mv\frac{dv}{dz} = 0 \tag{3-9}$$

where P_a is the pressure of the atmosphere, F_f is the frictional force, and M is the mass of the piston. Note that the inertial force acts in the direction opposite to that of the motion during acceleration.

Solving Eq. (3-9) for F_g and substituting into Eq. (3-8), we obtain

$$đW_g = P_a\, d\underline{V}_g + Mg\, dz + Mv\, dv + F_f\, dz \tag{3-10}$$

On the right-hand side of Eq. (3-10), the first term is work done by the gas in pushing back the atmosphere and increasing the volume of the gas by $d\underline{V}_g$. The second and third terms represent the work done by the system in increasing the potential and kinetic energy of the piston, and the last term is the work done on the cylinder wall to overcome friction. Note that if the frictional force were known, $đW_g$ could be calculated from Eq. (3-10), and P_g could be calculated directly by solving Eq. (3-9) for F_g.

A different point of view is found if the piston were chosen as the system. The work done by the piston on its surroundings would, then, be

$$đW_p = \mathbf{F}_p \cdot d\mathbf{z} \tag{3-11}$$

where $\mathbf{F}_p$ is the net boundary force exerted by the piston on the environment. According to Newton's third law of motion, the net force is *equal and opposite* to the net boundary force exerted by the environment on the piston. Therefore,

$$\mathbf{F}_p = \frac{(-P_g A + P_a A + F_f)\mathbf{z}}{|z|} \tag{3-12}$$

and

$$đW_p = -P_g\, d\underline{V}_g + P_a\, d\underline{V}_g + F_f\, dz \tag{3-13}$$

Finally, if the atmosphere were chosen as the system, the work done by the atmosphere on its surroundings would be

$$đW_a = \mathbf{F}_a \cdot d\mathbf{z} = P_a \, d\underline{V}_a = -P_a \, d\underline{V}_g \tag{3-14}$$

If we define $-F_f \, dz$ as the "work" done by the walls, $đW_w$, substitution of Eqs. (3-8) and (3-14) into Eq. (3-13) yields

$$đW_p = -đW_g - đW_a - đW_w \tag{3-15}$$

Thus, choosing any object in Figure 3.3 as the system, it is clear from Eq. (3-15) that the work done by the system on the environment is equal and opposite to the work done by the environment on the system. This result, which is valid for work interactions in general, is a consequence of the requirement that the sum of all body and boundary forces on a system is zero.

Returning to the problem of calculating the total work done by the gas in the expansion depicted in Figure 3.2, we see that it is clear from the discussion above that the work is equal to the total work done on the piston, atmosphere, and walls. If we neglect the frictional work for the time being (we shall return to this subject in Section 3.7), then

$$W_g = (P_a A + Mg)(z_2 - z_1) + \frac{M}{2}(v_2^2 - v_1^2) \tag{3-16}$$

There is one other method of measuring the work done by a system which, although not very practical, will be of help in visualizing a "thought" experiment which will be described shortly. We could remove the system of interest from its given environment, such as that shown in Figure 3.2, and place it in a cylinder covered with a weightless piston surrounded by a vacuum. The piston is balanced by weights placed on the top. If we then continually remove the weights so that the pressure–volume history of the gas during this process is identical to that which occurred in the original expansion, the work done by the gas during the hypothetical expansion is equal to the change in level (or potential energy) of the weights. Furthermore, this work is equal to the work done by the gas in the original expansion. Thus, the work done by a system can always be found by measuring the rise or fall of weights in the environment.

3.2 Adiabatic Work Interactions

Consider two closed systems that undergo an interaction through a common boundary. We shall call the interaction an *adiabatic work interaction* if the events occurring in each system could be repeated in such a way that the *sole* effect external to one system could be duplicated by the rise (or fall) of weights in a standard gravitational field and the *sole* effect external to the other system could be duplicated by an equivalent fall (or rise) of weights of equal magnitude. Some examples will help to illustrate the use of the definition.

Consider the situation illustrated in Figure 3.4(a). A vessel containing water initially at 0°C is bounded by adiabatic walls. A rough-surfaced disk is immersed in the water and attached to a drum by a shaft. The drum is rotated by allowing a weight to fall from the first to the second level. The first step in determining if this is an adiabatic work interaction is to designate the boundaries of the systems we wish to study. Let us choose the dashed line in Figure 3.4(a) as the boundary separating systems *A* and *B*. If we consider the events that occur external to system *A*, the sole effect is the fall of a weight. If the interaction is an adiabatic work interaction, we must somehow show that the events that occur external to *B* could be repeated solely by the raising of an identical weight by an identical amount. If, as in Figure 3.4(b), we

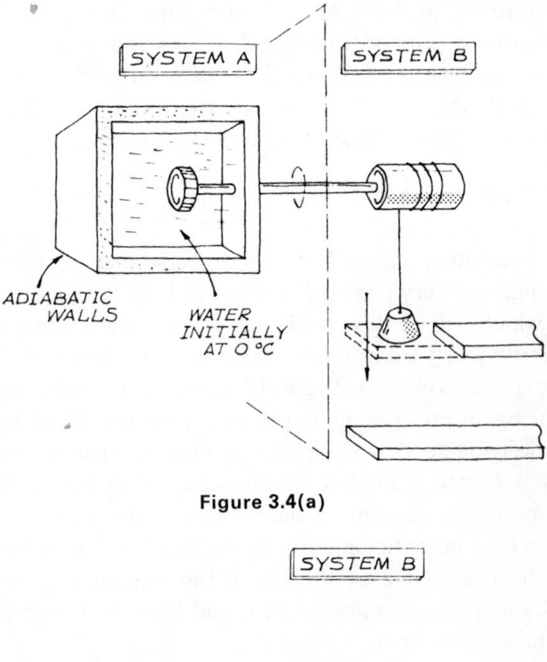

Figure 3.4(a)

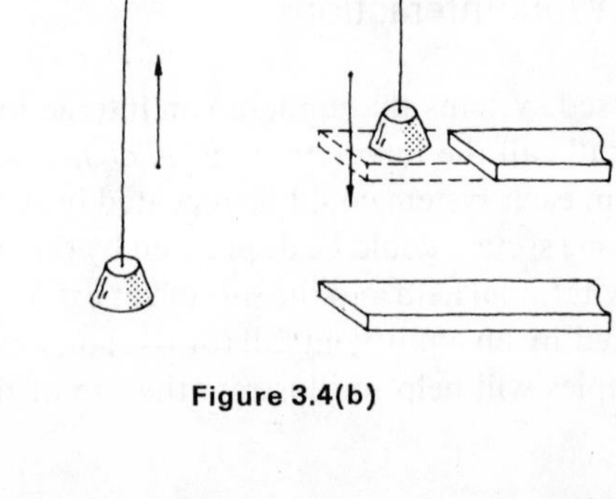

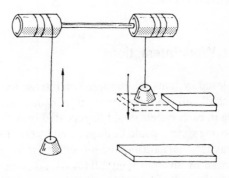

Figure 3.4(b)

replace system A with a drum that has a weight attached to it, and then repeat the event in system B, it is clear that the sole effect external to B is the required rise of a weight. Thus, the interaction is an adiabatic work interaction.

Now consider the slightly more complex situation illustrated in Figure 3.5(a). If we consider the interaction of system C with the composite system $A + B$, the situation is analogous to that of the previous example and we have an adiabatic work interaction. If, however, we consider the interaction of system A with the composite system $B + C$, the analysis is quite different. External to system A, the sole effect is the lowering of a weight since the final temperature of the water in system B will be 0°C as long as some ice remains in system A. (Note that we need only consider the initial and final states; whether or not the water remains at 0°C throughout the process is immaterial.) Although we can devise processes external to $B + C$ which

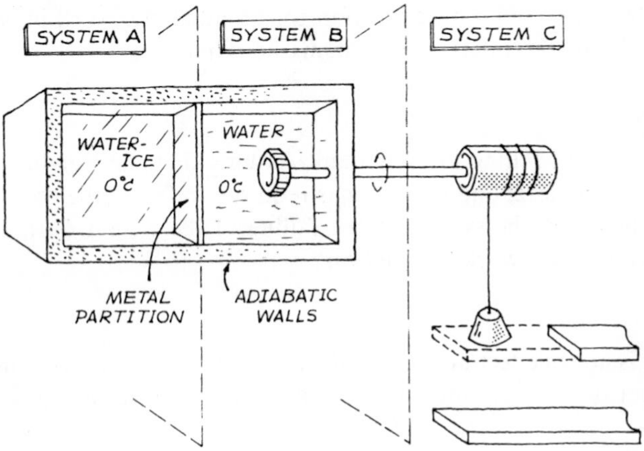

Figure 3.5(a)

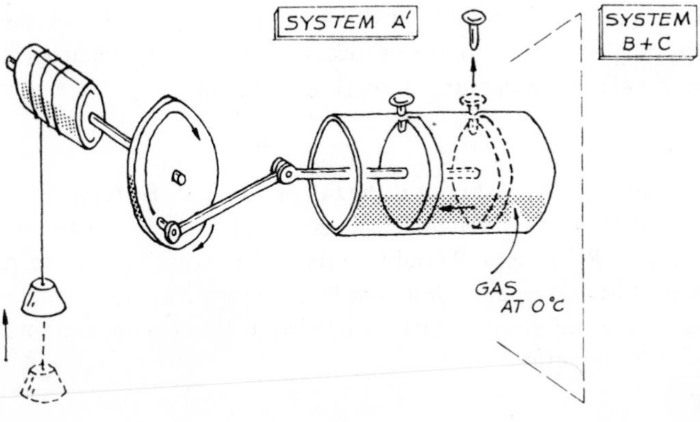

Figure 3.5(b)

would result in the rise of a weight, it does not appear possible to devise an experiment in which the *sole* effect external to $B + C$ is the rise of a weight. (At least no such experiment is known.) For example, we could replace system A with a cylinder filled with a gas at 0°C and fitted with a piston which in turn is attached to a flywheel (system A' in Figure 3.5(b)). As the weight in system C is lowered, the piston is freed to move to the left by removing a stop. As the piston moves out, the weight on the flywheel is raised. By judiciously choosing the components of the system in Figure 3.5(b), it would be possible to have the flywheel weight rise by the same amount as the weight in system C falls. However, the net effect external to system $B + C$ is the expansion of the gas in the cylinder in addition to the rise of the flywheel weight. Therefore, the interaction between A and $B + C$ is not an adiabatic work interaction.

It is clear that the classification of an interaction is only meaningful with respect to specified systems and boundaries. It should also be clear that a necessary (but not sufficient) condition for an adiabatic work interaction is that the walls of the interacting systems be adiabatic.

Example 3.1

Consider a weight falling in a vacuum. Assume that the weight is initially above the ground level and that 1 sec is required to fall to the ground. If the weight is chosen as the system, is the falling of the weight during the first 0.5 sec an adiabatic work interaction?

Solution

Since there is no net effect external to the falling weight (i.e., the falling weight has no effect on its surroundings), the interaction is not adiabatic work.

Example 3.2

Consider the situation illustrated in Figure 3.6, in which an electric generator is operated by a falling weight and in which the power generated is dissipated in a resistor. Neglect any dissipative processes such as I^2R line drop, friction in bearings, etc. Is this an adiabatic work interaction?

Solution

The sole effect external to system A is the fall of a weight. By replacing system A with a motor that has a weight attached to its shaft by a rope (i.e., the inverse of system B), system B could be made to execute the same process while the sole effect external to B would be an equivalent rise in the level of the weight. Hence, an electric current flowing between two systems is an adiabatic work interaction.

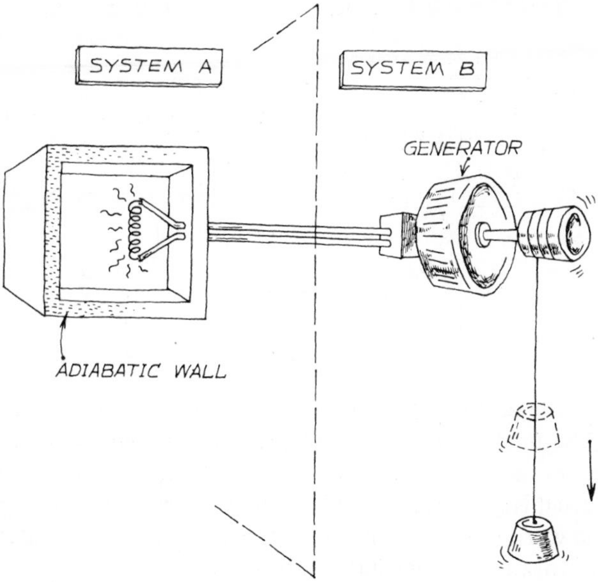

Figure 3.6

3.3 Energy

We have gone to the trouble to define an adiabatic work interaction because the magnitude of such interactions allows us to rank stable equilibrium states. The fact that adiabatic work interactions are always possible between stable equilibrium states cannot be developed from first principles, but the truth of such a statement has been borne out by a large body of experimental evidence. Thus, it is presented in the form of a postulate.

> ***Postulate III. For any states* (1) *and* (2), *in which a closed system is at equilibrium, the change of state represented by* (1) ⟶ (2) *and/or the reverse change* (2) ⟶ (1) *can occur by at least one adiabatic process and the adiabatic work interaction between this system and its surroundings is determined uniquely by specifying the end states* (1) *and* (2).**

As implied in the postulate, it is not always possible to go from state (1) to state (2) by an adiabatic process, but when this route is impossible, it must always be possible to find an adiabatic process from state (2) to state (1). For example, consider a container separated into two compartments with gas in one compartment and the other compartment initially evacuated

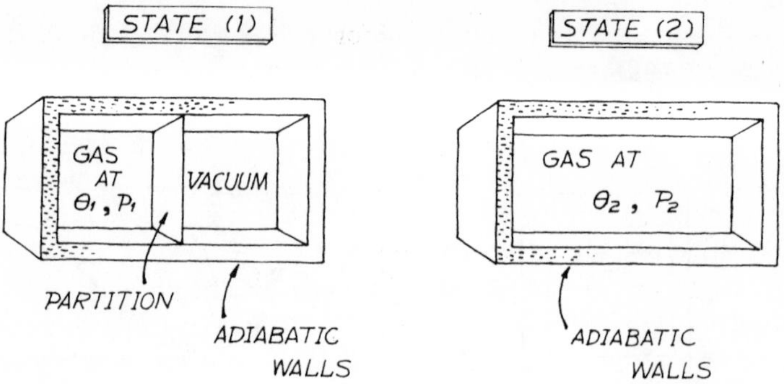

Figure 3.7

(Figure 3.7). The partition is removed and the gas expands to fill the entire container. The final temperature and pressure are measured to be θ_2 and P_2. If one takes the entire container as the system, the adiabatic work interaction between this system and its surroundings is clearly zero. If, as postulated, the adiabatic work interaction is only a function of the end states, it should also take no work to go from state (2) to state (1) by an adiabatic process if the reverse process were possible. However, it is clear that the change of state (2) $\longrightarrow$ (1) is not possible by an adiabatic process with zero work because we cannot use the environment without expending work and the gas will not compress itself spontaneously. (At least, as of this writing, no one has observed such a process. In Chapter 4 we will show that as a consequence of our postulates, certain processes—including self-compression of an isolated gas—are not possible.)

Since the postulate was stated for *any* states of a system at equilibrium, it follows that all stable states can be bridged by adiabatic processes originating from a given initial stable state. Thus, if state A is chosen as a reference state, any change to different states represented by $B_1, B_2, \ldots B_j, \ldots$ can be characterized by measuring experimentally the adiabatic work required for the change in state of A to B_j (or B_j to A, if the former change in state is not possible). Since the adiabatic work is only a function of the end states, the adiabatic work is a derived property of the system. We shall call this derived property the *energy*, $\underline{E}$, of the system and we shall follow the convention that the energy decreases when work is done *by* the system on the surroundings. That is,

$$\underline{E}_{B_j} - \underline{E}_A = -W^a_{A\to B_j} \tag{3-17}$$

where W^a is the adiabatic work and is always calculated as the work done by the system on the surroundings. Although by convention we could associate a value of $\underline{E}$ with each stable state, it is clear that only differences of energy have physical significance.

Eq. (3-17) appears as an abridged form of the First Law of thermodynamics. Eq. (3-17) leads directly to the conservation law for energy: Since the adiabatic work measured in the surroundings must be equal and opposite to the adiabatic work measured in the system (see Section 3.1), the energy change of the system must be equal and opposite to the energy change of the surroundings.

It should be noted that the definitions of adiabatic work and energy are not restricted to simple systems. They are valid for composite systems which may be in the fields of external forces. Throughout the text the symbol $\underline{E}$ is used to denote the energy of such composite systems. When a development is limited only to a system which is not acted upon by external force fields or inertial forces, the symbol $\underline{U}$ is used to denote the energy. The symbol $\underline{U}$ is used for composites and simple systems for which such forces are absent.

For a simple system, the energy can be defined with the aid of Postulate I. If each of the end states, A and B_j, of a simple system is uniquely specified by two independently variable properties plus the masses of the n components, then the energy associated with each stable state must also be a unique function of these $n + 2$ independent properties. For example, if θ and P are independently variable for a given system, then

$$\underline{U}_i = f(\theta_i, P_i, M_1, M_2, \ldots M_n) \tag{3-18}$$

It can also be shown that as a consequence of Postulate III, $\underline{U}$ must be first order in the total mass of the system. That is, the function of Eq. (3-18) is such that

$$a\underline{U}_i(\theta_i, P_i, M_1, M_2, \ldots M_n) = \underline{U}_i(\theta_i, P_i, aM_1, aM_2, \ldots aM_n) \tag{3-19}$$

where a is a constant. The proof can be developed by comparing the following two processes that have the same net effects: (1) a process in which two identical systems are acted upon simultaneously and (2) a process in which each system is acted upon separately.

Because energy is first order in mass, it can be shown that the energy of a composite of simple subsystems is equal to the sum of the energies of the subsystems of the composite. This conclusion will be used extensively in the following discussions.

The introduction of the concept of energy as given above illustrates the approach of classical thermodynamics. The treatment suffers because it does not give a physical picture of this property, energy. It has none of the connotations of the molecular or statistical thermodynamics concept of energy as a sum of molecular terms (i.e., translational, vibrational, rotational, electronic, and intermolecular energies). The classical approach coldly and impersonally limits one's view to the measurement of a change in a property by adiabatic work interactions.

3.4 Heat Interactions

Heat is an elusive entity that is only recognizable by its effect on material substances. For our discussion of work we were fortunately able to adopt definitions and procedures from mechanics. For a discussion of heat we have no precedent to follow since the onus of developing this concept lies within the realm of thermodynamics. Thus, we must endeavor to define heat by using the definitions and concepts already presented.

The key concept needed is that the energy difference between two states *can always* be determined by measuring the work in an adiabatic process connecting the two states (Postulate III). Now, with the same initial and final states, visualize *any* process (adiabatic or nonadiabatic), to go between these states. The energy difference is the same as that found for the adiabatic process because energy is a function of state only (i.e., it is independent of the path connecting the two states). If the process is not adiabatic, the work interaction will be different from that of the adiabatic process; however, the work can always be determined by one of the methods outlined in Section 3.1. *We then define heat as the sum of the energy change and the actual work performed.* That is,

$$Q = (\underline{E}_{\text{final}} - \underline{E}_{\text{initial}}) + W \tag{3-20}$$

where, by convention, W is the work done *by* the system on the surroundings and Q is the heat "added" to the system.

The above definition for heat, like that given previously for energy, is devoid of any microscopic significance. Nevertheless, it is of great practical utility. We will deduce shortly under what circumstances a heat interaction is to be expected and we will develop a method for ranking systems with respect to the direction of heat interactions.

In Eq. (3-20), heat is defined as the difference between the actual work in the process and the adiabatic work that would be required to effect the same change in state. Since any system completely enclosed by adiabatic walls can only undergo adiabatic work interactions, it follows that for such systems $Q = 0$. Alternatively, a system must have at least one diathermal wall if it is to undergo a heat interaction. The converse, however, is not necessarily true; namely, systems connected by a diathermal wall will not necessarily have a heat interaction.

In Section 2.3, an isolated system was defined as one having adiabatic, rigid,[2] and impermeable walls. Since the walls are adiabatic, the system cannot have any heat interactions. Since the walls are rigid, the system cannot have any work interactions. Hence, an isolated system can have no interactions with

[2] The term "rigid" is used in the generic sense throughout. That is, not only are the walls immovable, but no shafts or electrical conductors pass through them.

the environment and, therefore, the energy of an isolated system is invariant.

We will define a *pure heat interaction* as one for which the actual work is zero, and therefore $\Delta \underline{E} = Q$ where Δ is defined as final minus initial. For a system to undergo a pure heat interaction, the system must be surrounded by rigid walls, at least one of which is diathermal. As discussed below, the pure heat interaction is helpful in defining more specifically a thermometric temperature.

Consider two systems, *A* and *B*, which are closed, have rigid walls, and have adiabatic walls except for their common boundary, as indicated in Figure 3.8. When these two systems are first brought together, they may or may not

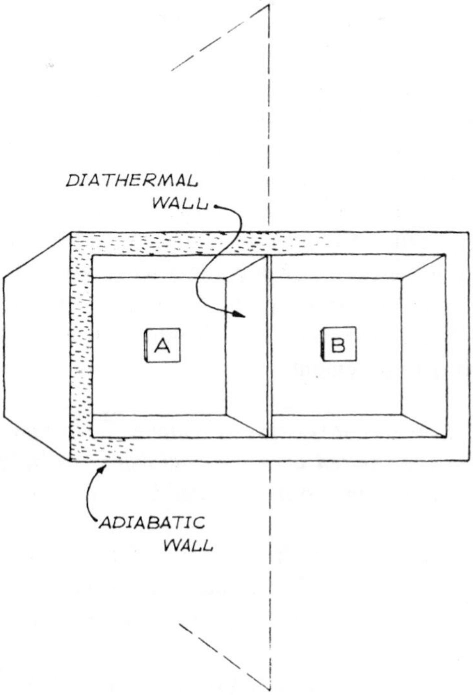

Figure 3.8

have a heat interaction. In this case, any heat interaction will be a pure heat interaction. We can tell if any interaction occurs by observing whether or not the primitive properties of *A* and *B* change after they are brought together.

If no heat interaction occurs when the systems are brought together, it follows from Postulate II that the composite is at equilibrium. Subsystems which are at equilibrium across a diathermal wall are said to be in *thermal equilibrium*. But if there is a heat interaction, it follows from Postulate II that the interaction must eventually cease since the subsystems of the isolated composite system will approach thermal equilibrium. If a heat interaction occurs, Q_A must be equal to $-Q_B$ because the total energy change of the isolated composite system, which is the sum of the energy changes of the subsystems, must be zero.

Let us now turn to the problem of determining the direction of a heat interaction. Let us redefine the procedure for measuring the θ-temperature. We construct a sealed liquid-in-glass thermometer in such a way that the walls are rigid and adiabatic except for one diathermal surface which is placed in contact with the system under study. Thus, the thermometer-system interaction is equivalent to the A–B interaction of Figure 3.8.[3] By heating or cooling the thermometer on a hot plate or in an ice bath, we can eventually find a liquid level for which no heat interaction occurs when the thermometer is placed in contact with system B and, hence, the measurement of the temperature of system B can be made without disturbing system B. (Alternatively, we can make the thermometer small in relation to the system and let the two come into thermal equilibrium with only a minor change in the properties of system B.) In this manner the liquid level can be used as a measure of the stable equilibrium state of the simple system, B. In order to give additional significance to the θ-property, an additional postulate is required. If we put the thermometer in contact with system B and then with system C, and if we find the same liquid level, there is still no way to prove that systems B and C are in thermal equilibrium with each other. Thus, we shall postulate such a requirement.

> ***Postulate IV. If the sets of systems* A, B *and* A, C *each have no heat interactions when connected across nonadiabatic walls, then there will be no heat interaction if systems* B *and* C *are also so connected.***

This postulate is sometimes referred to as the Zeroth Law of thermodynamics. It requires that systems with the same thermometric temperature be in thermal equilibrium and that no heat interaction occur between these systems.

It can be shown that the converse to Postulate IV is also true. If A, B are in thermal equilibrium, and A, C result in a heat interaction, then B, C must also result in a heat interaction. The proof involves assuming that B, C does not result in a heat interaction and then showing that this is contrary to the initial statement. Thus, composite systems attain thermal equilibrium only when the temperatures are uniform throughout the subsystems.

When systems A and B undergo a pure heat interaction such that $\Delta \underline{E}_A = -\Delta \underline{E}_B < 0$, we shall use the imprecise statement that "heat is transferred from

[3] Other primitive properties may be equally suitable for studying a heat interaction having no work or mass exchange. We chose the thermometric temperature as a convenient device that is easy to visualize.

A to B" instead of the longer, more precise statement that energy is transferred from A to B as a result of a pure heat interaction. In shorthand notation, we shall say $Q_{A \to B}$ is positive, which implies $Q_A = -Q_B < 0$.

We are now in a position to show that the thermometric temperature can be used to rank systems with respect to the direction of heat interactions. We shall prove that for the three systems, A, B, and C, if $Q_{A \to B}$ is positive (i.e., if heat is transferred from A to B) and if $Q_{B \to C}$ is positive, then θ_B must lie between θ_A and θ_C. Let us choose a thermometer scale so that $\theta_A > \theta_B$ and prove that θ_C cannot be equal to or greater than θ_B.

If $\theta_C = \theta_B$, then $Q_{B \to C} = 0$, which is in contradiction to the initial statement.

If $\theta_C > \theta_B$, then we could have chosen C so that $\theta_A = \theta_C$. If we then allow the three systems to interact in such a way that $Q_{A \to B} = Q_{B \to C}$ so that θ_B does not change in the process (i.e., $\Delta \underline{E}_B = 0$), the net result is a heat interaction from A to C. This, however, is not allowed because $\theta_A = \theta_C$ and, therefore, θ_C cannot be greater than θ_B. Thus, the θ-temperature can be used to rank different systems according to the directions in which heat interactions will occur.

It should be emphasized that the θ-temperature scale is still arbitrary. We could define another scale called the ξ-scale for which the freezing point of water is 100°ξ and the boiling point is 0°ξ. All that we require is that $\Delta\xi \longrightarrow 0$ when systems of different ξ undergo heat interactions. Of course, once we choose a temperature scale by some convention, we must be consistent throughout.

We shall adopt the *convention* that when $\theta_A > \theta_B$, the heat interaction is such that the energy of A decreases and the energy of B increases, or

$$\frac{d\underline{E}}{d\theta} > 0 \tag{3-21}$$

3.5 The Ideal Gas

Before exploring the applications of the First Law, it is convenient to point out here another temperature scale that will be very useful in later developments.

It has been found that for a gas, such as helium, a very simple experiment may be carried out to define a particular temperature scale. Suppose we confine a given quantity of gas in a container fitted with a piston and keep the pressure on this gas constant. We will immerse this gas system in boiling water at 1 atm pressure and measure the volume; we then repeat the experiment in an ice-water bath (the pressure above the bath should be that corresponding to water vapor, but for this discussion 1 atm pressure will not significantly affect the results). We denote the ideal gas temperature by T instead of θ.

We call the first temperature 373.16 and the second temperature 273.16 and then plot $\underline{V}$ against T, drawing a straight line between the two $\underline{V}$–T points and extrapolate to zero volume; it will be found that at $\underline{V} = 0$, our temperature scale also is zero. This $\underline{V}$–T relationship gives us an experimental method to measure temperature. Repeating the experiments at other (but low) pressures will give different lines but all will intersect the $\underline{V} = 0, T = 0$ singular point. Any of the various $\underline{V}$–T lines (each at a different pressure) can be used to define a temperature scale. This particular scale is called the "ideal gas" scale. If we were clever enough, we could bring all the $\underline{V}$–T lines (at different pressures) together by noting that if the same number of moles of gas were used in each experiment, then the product $P\underline{V}$ (rather than just $\underline{V}$) lines, which were drawn between the same two temperature points, fall on the same line and pass through the point $(P\underline{V}) = 0$, $T = 0$. This linearity is given analytically as $P\underline{V}/T =$ constant. Further, if we vary the moles of gas in the system, we can obtain by similar experiments the expression, $P\underline{V}/NT =$ constant. This latter constant is called the gas constant, R, and the relation when written as

$$P\underline{V} = NRT \tag{3-22}$$

is the ideal gas law.

Now review carefully the development of this particular temperature scale. We carried out two experiments to measure the volume (or $P\underline{V}$ or $P\underline{V}/N$ products) and then we defined temperatures at these two points as some numbers so that a straight line drawn between these two points intersects the origin.

If the ideal gas law is applied to any two experimental conditions, the expression $T_2 = T_1(P_2\underline{V}_2/P_1\underline{V}_1)$ results. In particular, if the two conditions are the freezing and normal boiling points of water, the ratio of $P_2\underline{V}_2/P_1\underline{V}_1$ is equal to 1.366. The ideal gas temperature scale is called the Kelvin scale (°K) if the number 273.16 is assigned to the freezing point of water; it is called the Rankine scale (°R) if the number 491.69 is assigned. If desired, other scales for which $T = 0$ at $\underline{V} = 0$ could be devised by assigning a different value to the freezing point of water and by assigning the normal boiling point of water a value of 1.366 times the freezing point.

An ideal gas will be defined as one that obeys the ideal gas law, Eq. (3-22) (where T is in °K or °R), and one for which the total energy is a function only of N and T, as given by

$$\underline{U} = N \int C_v \, dT + NU_0 \tag{3-23}$$

where C_v (the specific heat at constant volume) is only a function of temperature for an ideal gas and U_0 is some base value of specific energy (i.e., per mole). The simplification in working with the ideal gas is that $\underline{U}$ is a function of only N and T, whereas for other substances, $\underline{U}$ would have to be expressed as a function of N plus two other independently variable properties, in accordance with the general requirement stated in Postulate I.

3.6 The First Law for Closed Systems

Many authors refer to Eq. (3-24) as the First Law.

$$\Delta \underline{E} = Q - W \tag{3-24}$$

Since Q and W are defined only in terms of interactions at boundaries, Eq. (3-24) has significance only when applied to a specific system.

For a closed system interacting with the surroundings, the composite of system and surroundings can always be considered as a system of constant volume surrounded by adiabatic walls. Since such an isolated system can have no heat or work interactions with its environment, the energy of the composite system is invariant. Therefore,

$$\Delta \underline{E}_{\text{system}} = -\Delta \underline{E}_{\text{surroundings}} \tag{3-25}$$

where Δ is defined as final minus initial. Since we have already shown (Section 3.1) that

$$W_{\text{system}} = -W_{\text{surroundings}} \tag{3-26}$$

it follows that

$$Q_{\text{system}} = -Q_{\text{surroundings}} \tag{3-27}$$

For a differential change in state, the First Law can be written as

$$d\underline{E} = đQ - đW \tag{3-28}$$

where $đQ$ signifies that heat, like work, is a path function.

It should be noted that there may be many processes for going from one stable state to another; for each process, $\Delta \underline{E}$ is the same, but Q and W may be quite different.

Eqs. (3-24) and (3-28) apply to all processes of closed systems. The restriction of closed systems relates to the manner in which Postulate III was stated. Extension of the First Law to systems with permeable boundaries is presented in Section 3.8.

To the engineer the significance of the First Law may not be readily apparent from Eqs. (3-24) and (3-28). The fact that energy is conserved when two or more systems (forming an isolated composite) interact may have some aesthetic value, but since we do not usually measure energy directly, the statements of the First Law by themselves do not have real utility. We usually measure work effects and properties such as pressure, temperature, volume, and mass because these measurements are relatively simple. It is only when we relate energy (and other derived properties which we shall introduce later) to other measurable properties by using Postulate I that we obtain the full utility of the First Law. These relationships between properties (both derived and primitive) are not fixed by any principle or law of classical thermodynamics (although they place some restrictions on the form of the functionality, e.g., $\underline{U}$ must be first order in mass). In general, these relationships must be determined empirically, as discussed in Chapters 6 and 8. Eqs. (3-22) and (3-23) are examples of such relationships.

Almost all engineering applications of the First Law fall within two categories: (1) for given or measured interactions at the boundaries of a system, what are the corresponding changes in the properties of the system, and (2) for given changes in the properties, what interactions may occur at the boundaries?

3.7 Applications of the First Law for Closed Systems

The application of the First Law and the significance of many of the basic concepts are probably best illustrated by examples. In the following problems, we shall work exclusively with ideal gases in order to simplify the mathematics.

Example 3.3

Two well-insulated cylinders, each of 100 cm² area, are placed as shown in Figure 3.9. The pistons in both cylinders are of identical construction. The

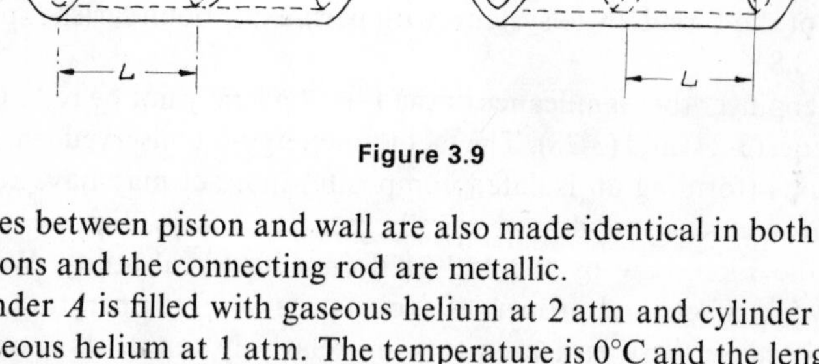

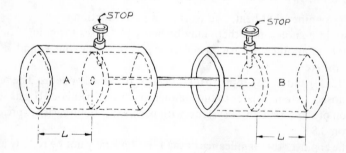

Figure 3.9

clearances between piston and wall are also made identical in both cylinders. The pistons and the connecting rod are metallic.

Cylinder A is filled with gaseous helium at 2 atm and cylinder B is filled with gaseous helium at 1 atm. The temperature is 0°C and the length L is 10 cm. Both pistons are only slightly lubricated.

The stops are removed. After all oscillations have ceased and the system is at rest, the pressures in both cylinders are, for all practical purposes, identical.

Assuming that the gases are ideal with a constant C_v and, for simplicity, assuming that the masses of cylinders and pistons are negligible (i.e., any energy changes of pistons and cylinders can be neglected), what are the final temperatures?

Solution

Let us choose the gases in compartments A and B as systems A and B, respectively. Since these are simple systems, we denote the energies by $\underline{U}$ instead of $\underline{E}$. Since the mass and initial temperature are known for the gas in each compartment, the final temperature can be calculated for each compartment, if the final energy—or energy change—of each compartment is first determined. Thus, from Eqs. (3-22) and (3-23) as applied for each system,

$$N_A = \frac{P_{A_i}\underline{V}_{A_i}}{RT_{A_i}} \tag{3-29}$$

$$N_B = \frac{P_{B_i}\underline{V}_{B_i}}{RT_{B_i}} \tag{3-30}$$

$$T_{A_f} = T_{A_i} + \frac{\underline{U}_{A_f} - \underline{U}_{A_i}}{N_A C_v} \tag{3-31}$$

and

$$T_{B_f} = T_{B_i} + \frac{\underline{U}_{B_f} - \underline{U}_{B_i}}{N_B C_v} \tag{3-32}$$

where $\underline{U}_{A_f}$ and $\underline{U}_{B_f}$ are the only unknowns ($\underline{U}_{A_i}$ and $\underline{U}_{B_i}$ can be chosen at will because $\underline{U}_0$ in Eq. (3-23) is an arbitrary constant).

Since the pistons and shaft have been assumed to be good conductors of heat, the final temperatures of compartments A and B will be equal:

$$T_{A_f} = T_{B_f} \tag{3-33}$$

To determine the energy change, let us apply the First Law, Eq. (3-24), to system A:

$$\underline{U}_{A_f} - \underline{U}_{A_i} = Q_A - W_A \tag{3-34}$$

The work done by this system is equal to the work done on the gas in B and the frictional work done on the walls of A and B, or

$$W_A = -W_B - W_{W_A} - W_{W_B} \tag{3-35}$$

where W_B is the work done *by* the gas in B and W_{W_A} is the work done *by* the wall of A, etc. Now consider the wall of A as the system and apply the First Law to this system:

$$\Delta \underline{U}_{W_A} = Q_{W_A} - W_{W_A} \tag{3-36}$$

If we neglect the change in energy of the wall, the frictional work done by the gas on the wall will be transmitted back to the gas in the form of a heat interaction; that is

$$Q_{W_A} = W_{W_A} \tag{3-37}$$

and furthermore,

$$Q_A = -Q_{W_A} - Q_{AB} \tag{3-38}$$

where Q_{AB} is the heat conducted from A to B through the pistons. Substituting Eqs. (3-35), (3-37), and (3-38) into Eq. (3-34),

$$\Delta \underline{U}_A = W_B + W_{W_B} - Q_{AB} \tag{3-39}$$

By analogy to Eqs. (3-37) and (3-38) as applied to the walls of B,

$$W_{W_B} = Q_{W_B} = -Q_B + Q_{AB} \tag{3-40}$$

and by analogy to Eq. (3-36),

$$W_B - Q_B = -\Delta \underline{U}_B \tag{3-41}$$

Eq. (3-39) becomes

$$\Delta \underline{U}_A = -\Delta \underline{U}_B \tag{3-42}$$

Of course, Eq. (3-42) could have been stated at the outset because the composite system $A + B$ is equivalent to an isolated system; this result, however, may not have been immediately obvious. Combining Eq. (3-42) with Eqs. (3-31) and (3-32), and making use of Eq. (3-33), we obtain

$$T_f = \frac{N_A T_{A_i} + N_B T_{B_i}}{N_A + N_B} \tag{3-43}$$

or, for this special case wherein $T_{A_i} = T_{B_i} = T_i$,

$$T_f = T_i = 0°\text{C}$$

Let us now use hindsight to reevaluate the problem. We had a qualitative feeling for the path in that we knew frictional work was involved, but we could not describe the path quantitatively because we did not know the coefficient of friction of either piston-cylinder. Nevertheless, we were able to determine the final conditions and, therefore, we did not have to describe the path to find the solution. Such a situation would be expected if the end state was independent of the path. This is obviously the case in the present example: Since the composite system $A + B$ is an isolated simple system, there is only one state to which it can go, and that is the one for which $\underline{U} = \underline{U}_{A_i} + \underline{U}_{B_i}$ and $T = T_{A_f} = T_{B_f}$. For an ideal gas, $\underline{U}$ is a unique function of T and N and, thus, the final temperature can be determined.[4]

Example 3.4

Consider the situation described in Example 3.3, but with well-insulated pistons and connecting rods of low thermal conductivity. What are the final

[4] If the gas were not ideal, $\underline{U} = f(T, \underline{V}, N)$. Thus, since the final energy, volume, and mass are known, the final temperature could still be determined.

temperatures after the oscillations have ceased and the pressures have equal-ized?

Solution

The composite system of $A + B$ is no longer a simple system because it contains an internal adiabatic wall. Therefore, the final composite cannot be described by a single equilibrium state; instead, the final conditions will depend on the path of the process.

   With the exception of Eq. (3-33), Eqs. (3-31) through (3-42) are still valid. Combining Eqs. (3-31) and (3-32) with Eq. (3-42) results in Eq. (3-44):

$$N_A C_v(T_{A_f} - T_{A_i}) + N_B C_v(T_{B_f} - T_{B_i}) = 0 \qquad (3\text{-}44)$$

which now gives us one equation in two unknowns, T_{A_f} and T_{B_f}. We could, of course, try to juggle the other equations to find another relationship be-tween T_{A_f} and T_{B_f}, but until we make some assumptions regarding the path, our efforts will be in vain.

   If we have no information on the coefficient of friction, we are forced to use our engineering judgment to simplify the situation while obtaining a close approximation to the actual conditions.

   Let us assume that there is friction only in compartment B. This will give us a lower bound for T_{A_f} and an upper bound for T_{B_f}. We can then treat the case of friction only in compartment A, which will give us an upper bound for T_{A_f}. In this manner, we can bracket the true solution.

   If there is no friction in compartment A and if we assume that the process is quasi-static (i.e., no pressure gradients within the compartment), we can write Eq. (3-28) for the simple system of the gas in A

$$d\underline{U}_A = -d W_A = -P_A d\underline{V}_A \qquad (3\text{-}45)$$

Since

$$d W_{W_A} = d Q_A = d Q_{AB} = 0$$

Substituting for $\underline{U}_A$ in Eq. (3-45) from Eq. (3-31), and for $\underline{V}_A$ from Eq. (3-22), Eq. (3-45) becomes

$$N_A C_v \, dT_A = -N_A R \left(dT_A - \frac{T_A}{P_A} dP_A \right) \qquad (3\text{-}46)$$

or

$$\left(\frac{C_v + R}{R} \right) \frac{dT_A}{T_A} = \frac{dP_A}{P_A} \qquad (3\text{-}47)$$

adiabatic compression.

Integrating between initial and final conditions, *how T-P behaves as internal energy changes due to work.*

$$\frac{T_{A_f}}{T_{A_i}} = \left(\frac{P_{A_f}}{P_{A_i}} \right)^{R/(C_v + R)} \qquad (3\text{-}48)$$

Eqs. (3-48) and (3-44) give us two equations in three unknowns, T_{A_f}, T_{B_f}, and $P_{A_f} = P_f$. The final pressure can be eliminated in the following manner.

   Since

$$\underline{V}_{A_f} + \underline{V}_{B_f} = \underline{V}_{A_i} + \underline{V}_{B_i} = \underline{V}_T \qquad (3\text{-}49)$$

Why?

by piston construction

and $\underline{V}_T$ is known, substitution of Eq. (3-22) into Eq. (3-49) gives

$$\underline{V}_T = (N_A T_{A_f} + N_B T_{B_f})\frac{R}{P_f} \tag{3-50}$$

Eq. (3-50) along with Eqs. (3-48) and (3-44) give us three equations in these unknowns. From Eqs. (3-50) and (3-44),

$$P_f = \frac{P_{A_i}\underline{V}_{A_i} + P_{B_i}\underline{V}_{B_i}}{\underline{V}_{A_i} + \underline{V}_{B_i}} = 1.5 \text{ atm} \tag{3-51}$$

From Eq. (3-48), with $C_v = 3$ cal/g-mol °K

$$T_{A_f} = \left(\frac{1.5}{2}\right)^{0.4} T_{A_i} = 243°\text{K}$$

From Eq. (3-44),

$$T_{B_f} = 334°\text{K}$$

If it is assumed that there is friction only in compartment A, then we would have found $T_{B_f} = 318°$K. Since in the actual case the friction is distributed between A and B, a better approximation might be $T_{B_f} = (318 + 334)/2 = 326°$K and from Eq. (3-44), $T_{A_f} = 246°$K.

The fact that the adiabatic wall in Example 3.4 prevents a direct solution to the problem in the absence of a complete description of the path is sometimes referred to as the "adiabatic dilemma." In fact, it is no dilemma at all, but results from the difference between heat and work interactions.

3.8 The First Law for Open Systems

A system was defined as being open if it has at least one region of its boundary which is permeable to mass. The First Law, as stated in Section 3.6, was restricted to closed systems. However, since an open system can always be considered as a closed system by redefining the boundaries, the extension of the First Law to open systems requires no additional postulates.

Consider an open system bounded by the σ-surface as illustrated in Figure 3.10.[5] Part of the σ-surface is diathermal and part is movable so that there may be heat and work interactions, Q_σ and W_σ, with the surroundings. The σ-surface also contains a region through which mass can enter. The boundary at that region may consist of a valve or a permeable membrane.

Consider a time, δt, during which a small quantity of mass, δm_e, enters the system bounded by the σ-surface. The properties of the entering mass are pressure, P_e, specific volume, v (volume per unit mass), and specific energy,

[5] Throughout the text, underlined capitals are used for *total* properties of the *system*, capitals without underline for *specific* properties of the *system*, and lower case letters for *specific* properties of entering and leaving streams of open systems.

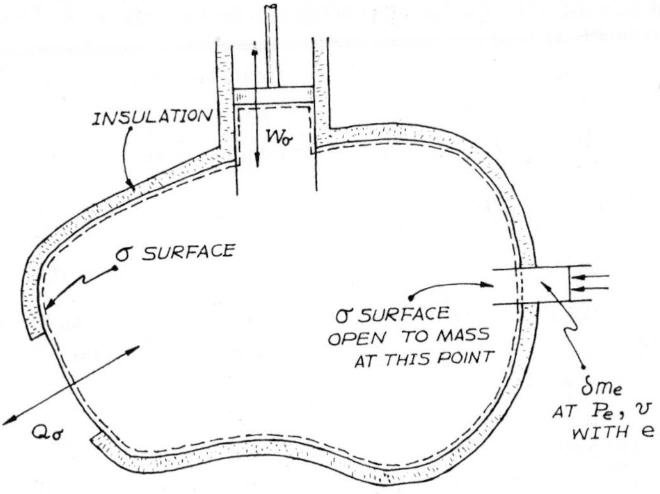

Figure 3.10

e (energy per unit mass). Although the region bounded by the σ-surface is an open system, the composite system of $\sigma + \delta m_e$ is a closed system. Defining $\underline{E}$ as the total energy of the system bounded by the σ-surface, let us apply the First Law to the closed composite system:

$$\underline{E}_2 - (\underline{E}_1 + e\,\delta m_e) = Q_\sigma - (W_\sigma - P_e v\,\delta m_e) \tag{3-52}$$

where subscripts 1 and 2 refer to initial and final conditions, respectively, and $P_e v\,\delta m_e$ is the P-$\underline{V}$ work required to push δm_e into the region bounded by the σ-surface.

Eq. (3-52) can be rearranged to a form similar to the First Law for a closed system:

$$\Delta\underline{E} = Q_\sigma - W_\sigma + (e + P_e v)\,\delta m_e \tag{3-53}$$

where $\Delta\underline{E}$, Q_σ, and W_σ apply only to the region bounded by the σ-surface (i.e., the open system), and the last term applies to the mass flux entering the system.

The differential form of Eq. (3-53) is

$$d\underline{E} = đQ_\sigma - đW_\sigma + (e + P_e v)\,dm_e \tag{3-54}$$

or, using specific energy and volume per mole of entering material,

$$d\underline{E} = đQ_\sigma - đW_\sigma + (e + P_e v)\,dn_e \tag{3-55}$$

where

$$d\underline{E} = d(ME) = M\,dE + E\,dM \tag{3-56}$$

or

$$d\underline{E} = d(NE) = N\,dE + E\,dN \tag{3-57}$$

depending on whether the specific energy of the open system is defined on a mass or mole basis.

For the general case wherein multiple streams enter and leave the system, the First Law for the open system is

$$d\underline{E} = đQ_\sigma - đW_\sigma + \sum_e (e_e + Pv_e)\, dm_e - \sum_\ell (e_\ell + Pv_\ell)\, dm_\ell \tag{3-58}$$

or, in the integrated form,

$$\Delta\underline{E} = Q_\sigma - W_\sigma + \sum_e \int (e_e + Pv_e)\, dm_e - \sum_\ell \int (e_\ell + Pv_\ell)\, dm_\ell \tag{3-59}$$

where the summations are taken over all entering and leaving streams, and where dm_e and dm_ℓ are both taken as positive quantities. That is,

$$dM = \sum_e dm_e - \sum_\ell dm_\ell \tag{3-60}$$

When the system defined by the σ-surface is a simple system, E may be replaced by U. Similarly, if the entering and leaving masses are simple systems, e may be replaced by u. (Note that the composite closed system is not a simple system if the boundaries at the points of mass flux are semipermeable.) Defining a term called the specific enthalpy (in units of energy per unit mass or mole) as

$$h = u + Pv \tag{3-61}$$

the general form of the First Law for such open systems is, in differential form,

$$d\underline{U} = đQ_\sigma - đW_\sigma + \sum_e h_e\, dm_e - \sum_\ell h_\ell\, dm_\ell \tag{3-62}$$

3.9 Application of the First Law for Open Systems

Example 3.5

A 100-ft³ storage tank containing 50 ft³ of liquid is to be pressurized with air from a large, high-pressure reservoir through a valve at the top of the tank to permit rapid ejection of the liquid (see Figure 3.11). The air in the reservoir is maintained at 100 atm and 100°F. The gas space above the liquid contains initially air at 1 atm and 50°F. The air flow into the tank is maintained at 0.05 moles/min. When the pressure in the tank reaches 5 atm, the liquid transfer valve is opened and the liquid is ejected at the rate of 5 ft³/min.

What is the air temperature when the pressure reaches 5 atm and when the liquid has been drained completely?

Neglect heat interactions at the gas-liquid and gas-tank boundaries. It may be assumed that the gas above the liquid is well mixed and that air is an ideal gas with $C_v = 5$ Btu/lb-mol°F.

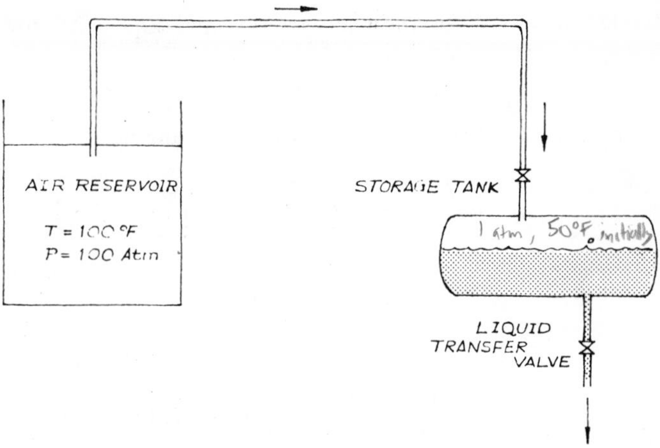

Figure 3.11

Solution

Let us treat the process in two steps: (a) the period during which the pressure rises from 1 to 5 atm and the volume of the gas in the tank is constant, and (b) the period during which liquid is drained.

Step (a): The most convenient system is the gas in the tank at any time, which is an open simple system with constant volume. The temperature of this system is related to the energy and moles by Eq. (3-23), and to the pressure, volume, and moles by Eq. (3-22). The energy may be related to the moles by using the First Law for an open system. These three relationships can be solved simultaneously for the temperature as a function of pressure.

Using the integrated form of Eq. (3-62), the First Law for an open system in the absence of body force fields,

$$\Delta \underline{U} = Q_\sigma - W_\sigma + \int h_e \, dn_e \tag{3-63}$$

where

$$Q_\sigma = 0$$
$$W_\sigma = 0$$
$$h_e = u + P_e v = \text{constant}$$

and

$$\int dn_e = N_2 - N_1$$

where it is assumed in the h_e relation that the volume of the reservoir is large relative to that of the tank. Thus,

$$\underline{U}_2 - \underline{U}_1 = (u + P_e v)(N_2 - N_1) \tag{3-64}$$

Substituting Eq. (3-23) for the energy terms and Eq. (3-22) for $P_e v$, and simplifying,

$$N_2 C_v T_2 - N_1 C_v T_1 = (N_2 - N_1)(C_v + R)T_e \tag{3-65}$$

Substituting Eq. (3-22) for N_2 and N_1, and rearranging,

$$T_2 = \frac{\kappa T_e}{1 + \frac{P_1}{P_2}\left[\kappa\left(\frac{T_e}{T_1}\right) - 1\right]} \tag{3-66}$$

where

$$\kappa = \frac{C_v + R}{C_v}$$

Thus,

$$T_2 = \frac{(1.4)(560)}{1 + (\frac{1}{5})[1.4(\frac{560}{510}) - 1]} = 709°\text{R or } 249°\text{F}$$

From Eq. (3-22),

$$N_2 = \frac{P_2 \underline{V}_2}{RT_2} = \frac{(5)(50)}{(0.730)(709)} = 0.48$$

It is interesting to note that, contrary to what might have been anticipated, the final temperature of the gas in higher than that of either the initial temperature or the temperature of the incoming gas. In the limit where $P_2 \gg P_1$, the temperature approaches κT_e independent of the initial conditions in the tank.

Step (*b*): If the system is again chosen as all of the gas in the tank at any time, the analysis is similar to that of Step (a) except that the volume of this system is changing continuously. As a result of this complication, $W_\sigma = \int P\, d\underline{V}$ cannot be evaluated until the pressure–volume history is determined. In this case, the differential form of the First Law for an open system, Eq. (3-62), is more convenient.

Thus,

$$d\underline{U} = đQ_\sigma - đW_\sigma + h_e\, dn_e \tag{3-67}$$

where

$$d\underline{U} = N\, dU + U\, dN$$
$$đQ_\sigma = 0$$
$$đW_\sigma = P\, d\underline{V}$$
$$dn_e = dN$$
$$h_e = u + P_e v = u + RT_e$$

so that

$$N\, dU + (U - u - RT_e)\, dN + P\, d\underline{V} = 0 \tag{3-68}$$

Substituting Eqs. (3-22) and (3-23) into Eq. (3-68), and simplifying,

$$\frac{dT}{T} + \left(1 - \frac{\kappa T_e}{T}\right)\frac{dN}{N} + \frac{R}{C_v}\frac{d\underline{V}}{\underline{V}} = 0 \tag{3-69}$$

from previous (a)

Since

N_2 *flow of air into tank*

$$dN = 0.05\, dt; \qquad N = 0.48 + 0.05\, t$$

and

flow of liquid,

$$d\underline{V} = 5\, dt; \qquad \underline{V} = 50 + 5\, t$$

↑ init head space

Eq. (3-69) becomes

$$\frac{dT}{T} + \left(1 - \frac{\kappa T_e}{T}\right)\left(\frac{dt}{9.6 + t}\right) + \left(\frac{R}{C_v}\right)\left(\frac{dt}{10 + t}\right) = 0 \qquad (3\text{-}70)$$

Integration of this equation from $t = 0$ to 10 min yields the final temperature; $T = 511°R$ or $51°F$.

PROBLEMS

3.1. A small, well-insulated cylinder and piston assembly (Figure P3.1) contains an ideal gas at 10 atm and 70°F. A mechanical lock prevents the piston from

Piston moves from A to B

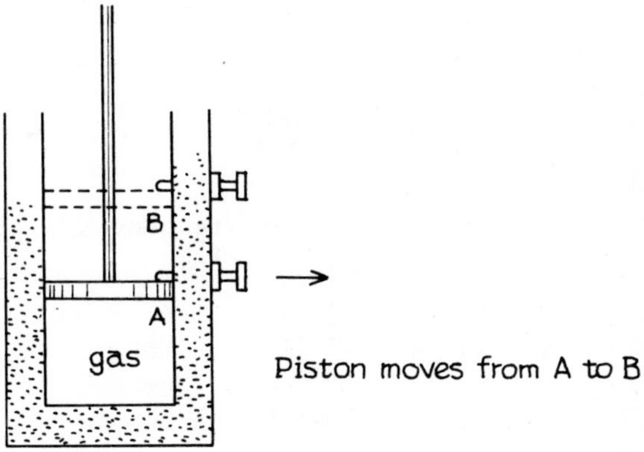

Figure P3.1

moving. The length of the cylinder containing the gas is 1 ft and the piston cross-sectional area 0.2 ft². The piston, which weighs 500 lb, is tightly fitted and when allowed to move, there are indications that considerable friction is present.

When the mechanical lock is released, the piston moves in the cylinder until it impacts and is engaged by another mechanical stop; at this point, the gas volume has just doubled.

As an engineer, can you estimate the temperature and pressure of the gas after such an expansion? Clearly state any assumptions.

Repeat the calculations if the cylinder were rotated both 90° and 180° before tripping the mechanical lock.

The heat capacity of the ideal gas is 5 Btu/lb-mol°R, independent of temperature and pressure. Consider the heat capacity of the piston and cylinder walls to be negligible.

3.2. An ideal gas is contained in a vertical cylinder fitted with a piston. The gas volume is 1 ft^3, the pressure 100 psig, and the temperature 25°C. The piston weighs 100 lb and has an area of 1 ft^2. The piston-cylinder assembly is immersed in a constant temperature bath at 25°C.

The gas is allowed to expand against the atmosphere even though the piston moves relatively slowly because of friction. In fact, because of the poor fit of the piston in the cylinder, no further movement of the piston occurs after the gas pressure has decreased to 25 psig.

Discuss the heat and work effects during this process.

3.3. (Refer to Figure P3.3 for notation.) A piston (*A*) and piston rod (*B*) are fitted inside a cylinder of length 20 in. and area 10 $in.^2$. Although the piston is quite

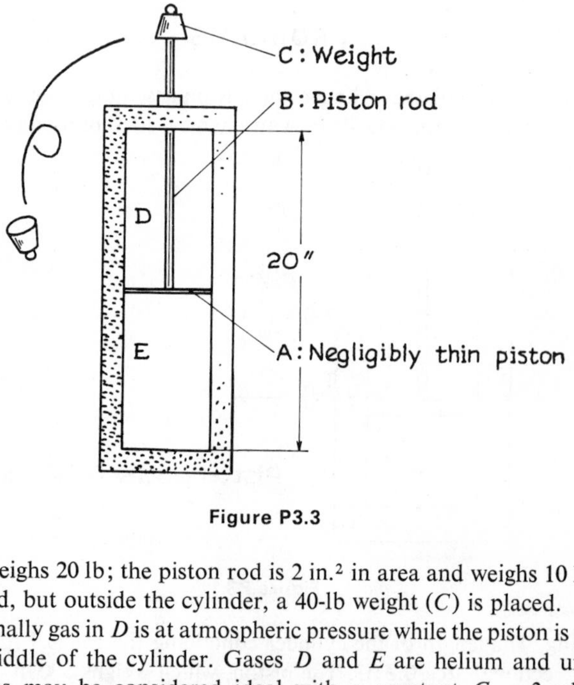

Figure P3.3

thin, it weighs 20 lb; the piston rod is 2 $in.^2$ in area and weighs 10 lb. On top of the rod, but outside the cylinder, a 40-lb weight (*C*) is placed.

Originally gas in *D* is at atmospheric pressure while the piston is positioned in the middle of the cylinder. Gases *D* and *E* are helium and under these conditions may be considered ideal with a constant $C_v = 3$ cal/g-mol°K. The initial temperature is everywhere 100°F.

Assuming the cylinder, piston, and piston rod to be nonconducting and having a negligible heat capacity, discuss any heat or work interactions if weight *C* should fall off. What is the final state of the system when the piston has stopped and there is a balance of forces across the piston? Do not neglect the fact that during motion there may be some friction between moving parts.

Consider cases in which the piston is (a) diathermal and (b) adiabatic.

3.4. A piston, latched into place in a cylinder (as shown in Figure P3.4), encloses helium at a pressure of 2 atm and a temperature of 60°F. A compressed spring is located in the right side of the piston, but this side is evacuated to a low

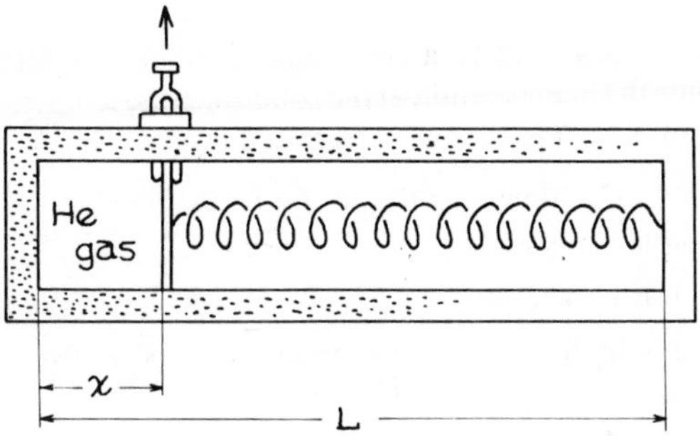

Figure P3.4

pressure. This spring exerts a force toward the left of a magnitude $F = 400\ x$ (pounds) where x is the distance between the piston and the far left end of the cylinder. Initially, as latched, $x = 1$ ft.

We release the latch and allow the piston to move until it finally ceases oscillating and stops. What is the value of x in this new position? What is the temperature and pressure of the helium at this time?

You may consider helium to be an ideal gas with a constant $C_v = 3.0$ cal/g-mol°K. Also assume that the temperature of the spring does not change during compression or expansion and that the piston, spring, and cylinder are nonconducting with negligible heat capacity. The area of the piston is 1 ft^2.

3.5. A horizontal cylinder 1.5 ft long is divided into two parts A and B (see Figure P3.5) by a latched piston. As illustrated, the volume of A is twice that of B and

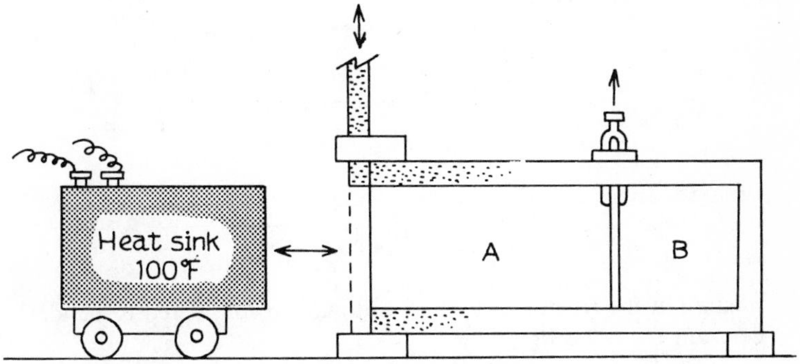

Figure P3.5

contains helium at 100°F and 10 atm. B contains hydrogen at 100°F and 1 atm. Both of these gases may be considered ideal with constant heat capacities as follows: C_v (helium) = 3 cal/g-mol°K, C_v (hydrogen) = 5 cal/g-mol°K.

Provision has also been made to connect volume A to a constant temperature reservoir at 100°F.

When the latch is removed, the piston is allowed to seek an equilibrium position so that there are equal pressures on each side.

Without neglecting friction but assuming no heat transfer to the cylinder walls, compute the final position of the piston and the temperature and pressure in both A and B for the following four cases:

No contact of volume A with the 100°F reservoir

piston is diathermal

piston is adiabatic

Volume A is in diathermal contact with the 100°F reservoir

piston is diathermal

piston is adiabatic

3.6. Our next research experiment is to be carried out in a vertical, cylindrical reactor 1 ft² in cross-sectional area and 2 ft long. (See Figure P3.6). A gas mixture in the reactor is at 60°F and may be considered to be ideal with a constant $C_p = 7$ cal/g-mol°K.

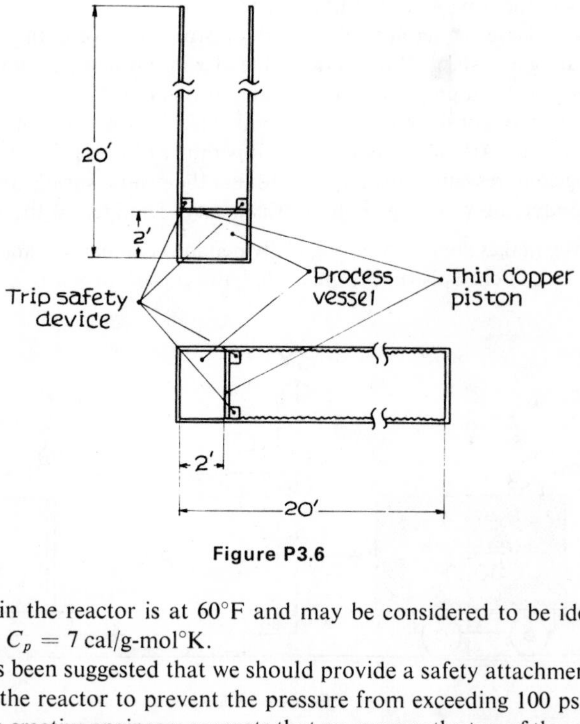

Figure P3.6

It has been suggested that we should provide a safety attachment of some kind on the reactor to prevent the pressure from exceeding 100 psia. One of our more creative engineers suggests that we remove the top of the reactor and weld on a pipe extension. This extension would be fitted with a heavy piston latched in place. We would also provide a pressure transducer and activation circuit in the reactor to unlatch the piston should the pressure exceed 100 psia.

The headroom in the laboratory is only 18 ft and, since we do not want any of our process gas escaping, we must select a piston of the correct mass so that it will not be blown out of the pipe extension. If the piston were made of copper (density = 0.32 lb/in.^{3}), what should its thickness be? Assume insignificant friction and neglect any heat transfer from the gas to the cylinder walls or piston.

Another engineer, however, has been advocating an alternate teachnique. He too wishes to remove the reactor top and weld to it a pipe extension with a latched piston. But he also wishes to put a cap on the top of this extension and rotate the cylinder on to its side. In this case, should the reactor pressure exceed 100 psia, the piston would move horizontally and compress the gas in the pipe extension. He also wishes to roughen the walls in the pipe so that there is friction between the moving piston and pipe walls but no gas leakage.

Again assuming negligible heat transfer, what is your best estimate of the final gas temperatures on both sides of the piston after pressures on both sides are equal? Assume initially that the gas in the closed pipe extension is air at 1 atm and 60°F.

Which of these two techniques would you select? Why?

3.7. Advertised is a small toy that will send up a signal flare and the operation "is so simple that it is amazing" (see Figure P3.7). Our examination of this

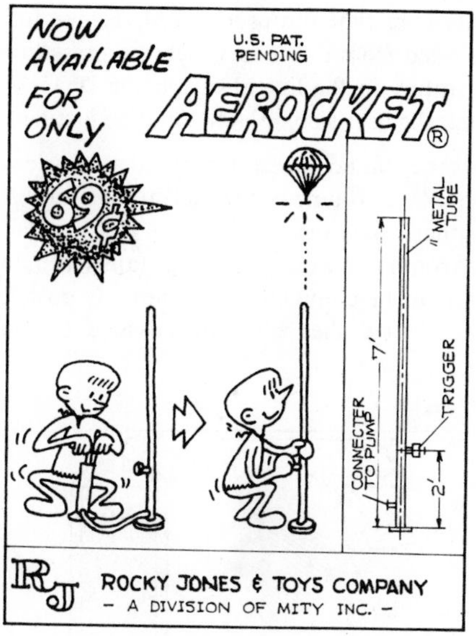

Figure P3.7

device indicates that it is a sheet metal tube 7 ft long and 1 in.^{2} in area. A plug shaped into the form of a piston fits into the tube and a mechanical trigger holds it in place 2 ft above the bottom. The piston weighs 3.46 lb and contains the necessary parachute and pyrotechnics to make the show exciting. To operate the device, the volume below the piston is pumped up to a pressure of about 4 atm (abs.) with a small hand pump, and then the trigger is depressed allowing the piston to fly out the top. The pyrotechnic and parachute devices are actuated by the acceleration force during ejection.

When we operated this toy last summer, the ambient temperature was 90°F.

(a) Assuming no friction in the piston and no heat transfer or other irreversibilities in the operation, how high would you expect the piston to go? What would be the time required from the start to attain this height?

(b) Since you are an engineer who is never satisfied with a commercial object, please suggest improvements to make the piston go even higher. What is the maximum height that could be obtained if it were limited to 4 atm pressure?

(c) Comment on the way you might analyze the expected performance if the restrictions in (a) were removed.

3.8. Hydrogen is being supplied to customers in cylinders containing 220 ft^3 of gas measured at 14.7 psia and 70°F. When the cylinders reach the users, the cylinders are guaranteed to be at least 2000 psia at 70°F. A customer complains that when the cylinders were received, they were not up to pressure. The supplier says that this is impossible because all cylinder valves were checked for leakage (it is agreed that leakage in valves or through cylinders does not exist), and to be absolutely sure that the user would receive full measure of gas, the cylinders were filled from a line in which the pressure never fell below 2100 psia, the temperature in the line always being 70°F $\pm$ 1°F.

Do you think the user's complaint could be justified or not? Why?

3.9 During an emergency launch operation, to fill a missile with RP-4 (a kerosene-based fuel), the ullage volume of the fuel storage tank is first pressurized with air from atmospheric pressure to a pressure of 150 psia (see Figure P3.9). The air is available from large external storage tanks at high pressure (1000 psia). This operation is to be completed as rapidly as possible. After the 150 psia pressure level is reached, the main transfer valve is opened and fuel flows at a

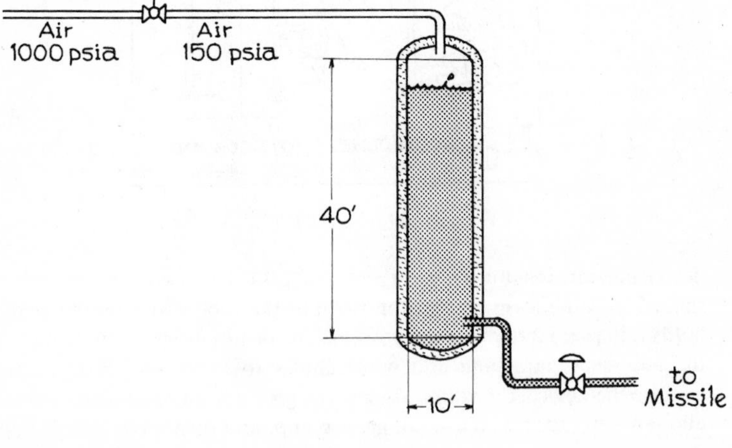

Figure P3.9

steady rate until the missile is loaded. It is necessary to maintain a constant gas pressure of 150 psia inside the fuel tank during transfer.

The fuel storage tank can be approximated as a right circular cylinder 40 ft tall and 10 ft in diameter and is originally filled to 90% of capacity. Transfer of fuel to a residual volume of 10% must be completed in 18 minutes.

(a) Comment on any safety hazards that might be encountered.
(b) What problems would you anticipate if the inlet gas control valve were to malfunction and the gas space above the fuel were to reach full storage tank pressure (1000 psia)? (The fuel tank has been hydrostatically tested to 4000 psia.)
(c) What is your estimate of the time-temperature history of the gas above the fuel during the entire operation? These data are needed to size the inlet air lines.

Assume ideal gases and that the operation is adiabatic and all hardware has negligible heat capacity. Initial temperatures may range from −25°F (arctic sites) to 140°F (equatorial sites) but for the purposes of a first estimate, use 70°F as an initial temperature.

3.10. In many installations in the chemical industry, occasions arise when compressed gas bottles are rapidly blown down. These bottles are constructed of carbon steel that becomes dangerously brittle at low temperatures. Certainly, for rapid blow-down situations the gas temperature could drop to such a low temperature that if rapid heat transfer with the cylinder were to occur, a hazardous operation would result. It is believed that the only place where very high heat transfer rates are possible is in the cylinder neck because velocities are highest in this region. In order to calculate neck wall temperatures, however, the time-variation of the bulk gas temperature of the bottle must be available.

Demonstrate your ability to estimate the bulk gas temperature of the bottle gas as a function of time for the first 2 minutes in the following case:

inside wall area = 100 ft^2
volume = 27.5 ft^3
thickness of wall = 0.33 in.
specific heat of wall = 0.1 Btu/lb°F
density of wall = 529 lb/ft^3
initial bottle pressure = 2000 psia
rate of pressure decay—bottle pressure is reduced by factor of 2 every 1.6 min
bottle gas—nitrogen (assume ideal gas behavior)
initial bottle gas and wall temperatures = 25°F

It may be assumed that heat transfer from the cylinder walls to the gas occurs by a natural convection process. For purposes of computation, assume the heat transfer coefficient is 8 Btu/hr-ft^2°F for all bottle pressures in excess of 1000 psia and 6 Btu/hr-ft^2°F for pressures less than 1000 psia.

3.11. A rigid laboratory gas cylinder of 2 ft^3 volume is charged with air at 2000 psia and 500°R. An experiment is to be carried out whereby the cylinder is rapidly vented by opening the valve on the cylinder top. The pressure in the cylinder is always so high that the flow is choked (i.e., sonic) in the valve throat. If one assumes that the gas is ideal and that there is negligible heat transfer between the gas and walls,

(a) Derive a relation to calculate the bottle pressure as a function of time.
(b) Calculate the time when the bottle pressure drops to 1000 psia.

For sonic flow through a round, sharp-edged orifice (assumed to apply to the valve throat),

$$\text{mass flow} = C_a AP\left[\frac{g\kappa m}{RT}\left(\frac{2}{\kappa+1}\right)^{(\kappa+1)/(\kappa-1)}\right]^{1/2}$$

where C_a = discharge coefficient (assume = 0.6)
A = orifice area = 0.01 ft^2
P = cylinder pressure
g = acceleration of gravity
R = gas constant
$\kappa = C_p/C_v = 1.4$ for air (assume constant)
m = molecular weight = 29

3.12. A well-insulated pipe of 1-in. inside diameter carries air at 2 atm pressure and 200°F. It is connected to a 1-ft^3 insulated "bulge," as shown in Figure P3.12.

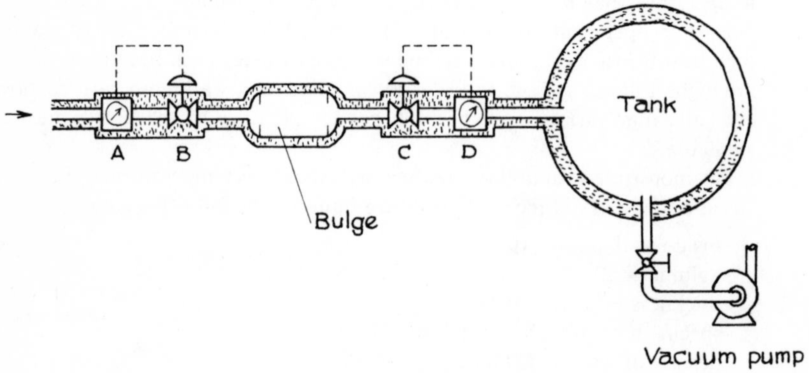

Figure P3.12

The air in the bulge is initially at 1 atm pressure and 100°F. *A* and *D* are flow meters which accurately measure the mass rate of air flow. Valves *B* and *C* control the air flow into and out of the bulge. Connected to the bulge is a 10-ft^3 rigid, adiabatic tank which is initially evacuated to a very low pressure.

At the start of the operation, valve *B* is opened to allow 0.01 lb of air/sec to flow into the bulge; simultaneously, valve *C* is operated in order to remove exactly 0.01 lb/sec from the bulge, into the tank. These flows are maintained constant as measured by the mass flow meters.

(a) What is the temperature and pressure of the gas in the bulge after 6 seconds?

(b) What is the temperature and pressure of the air in the large tank after 3 seconds?

Air may be assumed an ideal gas with a constant C_p of 7 Btu/lb-mol°R. Assume also that the gases, both in the bulge and large tank, are completely mixed so that there are no temperature or pressure gradients present.

3.13. A vessel containing a reactive compound is about 2 ft^3 in volume (gas space). There is an inert atmosphere of helium maintained at 1 atm at all times. If, however, the compound shows signs of decomposition, it is desired to increase very rapidly the helium over-pressure to 10 atm. This higher pressure will then be used to dump the reactive compound to a water-soak tank.

In order to accomplish this rapid pressurization, the vessel is connected by a short transfer line and valve to another vessel filled with high pressure helium. This vessel is 6 ft^3 and contains helium originally at 20 atm and 100°F.

What is the pressure of the helium supply vessel after a pressurization of the reactor? Assume ideal gases and adiabatic operation. How many reactors could the supply vessel serve simultaneously (each reactor is 2 ft^3 and is pressurized from 1 to 10 atm)?

3.14. In order to reduce gas storage costs, two companies, *A* and *B*, have built a common storage tank in the shape of a horizontal right circular cylinder 10 ft in diameter and 100 ft long (see Figure P3.14).

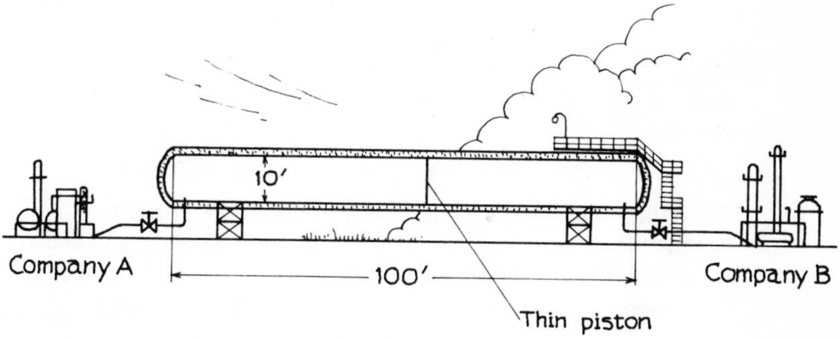

Figure P3.14

In order to decide how much gas each company uses between refills, a thin piston was placed in the tank. The piston moves freely, that is, there is essentially no friction present, and the pressure is the same on both sides. Thus, as company *A* uses gas, the piston moves left and as company *B* uses gas, the piston moves right.

When the gas company refills the tank, it must decide how much gas has been used by each company. It can easily measure the position of the piston and can, if necessary, install other instrumentation such as thermometers or pressure gauges in either or both ends of the tank. List the *minimum* instrumentation that you would recommend, and show from this list how the amount of gas consumed by both companies could be determined at the time of refilling.

Assume that: (1) the gas is ideal; (2) the piston is adiabatic; (3) the walls are well-insulated and have a low heat capacity; (4) at the start of each month after filling the tank, the gas company positions the piston in the center of the tank, equalizes the temperature in both ends, and carefully meters the total amount of gas added.

3.15. A thermodynamicist is attempting to model the process of balloon inflation by assuming that the elastic casing behaves like a spring opposing the expansion (see Figure P3.15).

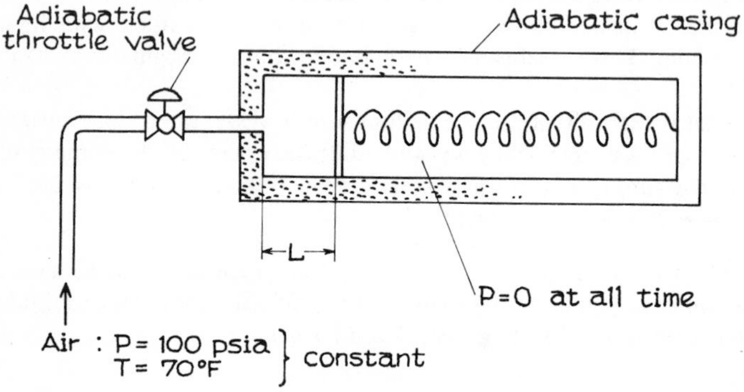

Figure P3.15

As air is admitted, the spring is compressed. The pressure in the gas space is given by:

$$P - P_i = k(L - L_i), \qquad k = \text{constant} = 20 \text{ psia/ft}$$

The initial conditions in the gas space are: $P_i = 14.7$ psia, $T_i = 70°\text{F}$, and $L_i = 0.5$ ft. The piston area is 0.2 ft². The air is an ideal gas and $C_v = 5$ cal/g-mol°K, independent of temperature.

What is the air temperature in the gas space when $L = 2$ ft?

3.16. A large, externally insulated hydrogen liquid storage tank made of stainless steel has just been drained and the walls and residual gas are at 36°R (the atmospheric boiling point of hydrogen). The pressure is 1 atm.

The tank is connected to a supply of high pressure hydrogen gas at 70°F and very rapidly pressurized to 115 psia. The tank is then held at this pressure by allowing additional gas to enter until the gas and walls are at the same temperature. What is this temperature?

Data:

Tank: $\underline{V} = 10$ ft³, wall mass = 285 lb

Assume that hydrogen is an ideal gas with $C_p = 2.6$ Btu/lb°R, independent of temperature.

The heat capacity of stainless steel varies with temperature as shown below:

$\int_{36}^{T(°R)} C_p \, dT$ (Btu/lb)	T, °R
0	36
0.02	50
0.24	80
0.60	100
1.74	137
2.27	150
3.1	170

3.17. An all quartz Dewar* flask is filled initially with liquid hydrogen at 1 atm (see Figure P3.17a). The inner walls cool to the normal boiling point of hydrogen, 20.6°K. The liquid is then quickly poured out and the flask evacuated to a very low pressure. Assume that at the end of the evacuation the walls are still at 20.6°K.

The Dewar is then connected to a large tank of helium gas at 2 atm and 300°K and pressurized very rapidly to 2 atm. After pressurization, the connecting line is left open to allow additional flow to occur in order to maintain a pressure of 2 atm. There is heat transfer between the helium gas and inner Dewar walls, but assume no heat transfer by radiation, convection, or conduction across the walls of the Dewar.

How many g-mols of helium are there in the Dewar after all flow has ceased?

Data:
C_p (helium) = 5 cal/g-mol°K
C_v (helium) = 3 cal/g-mol°K
helium is an ideal gas
Dewar flask volume = 8.206 l
inner walls of Dewar = 1000 g
temperature of the environment = 300°K

*A Dewar flask is a double-walled vessel with the annulus evacuated so that the rate of heat transfer from the inside to the environment is very small.

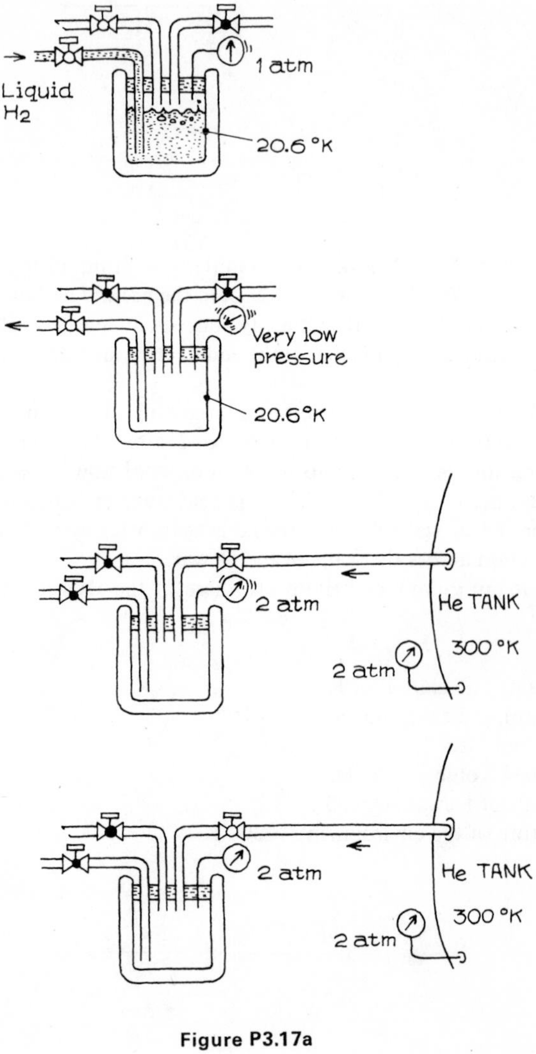

Figure P3.17a

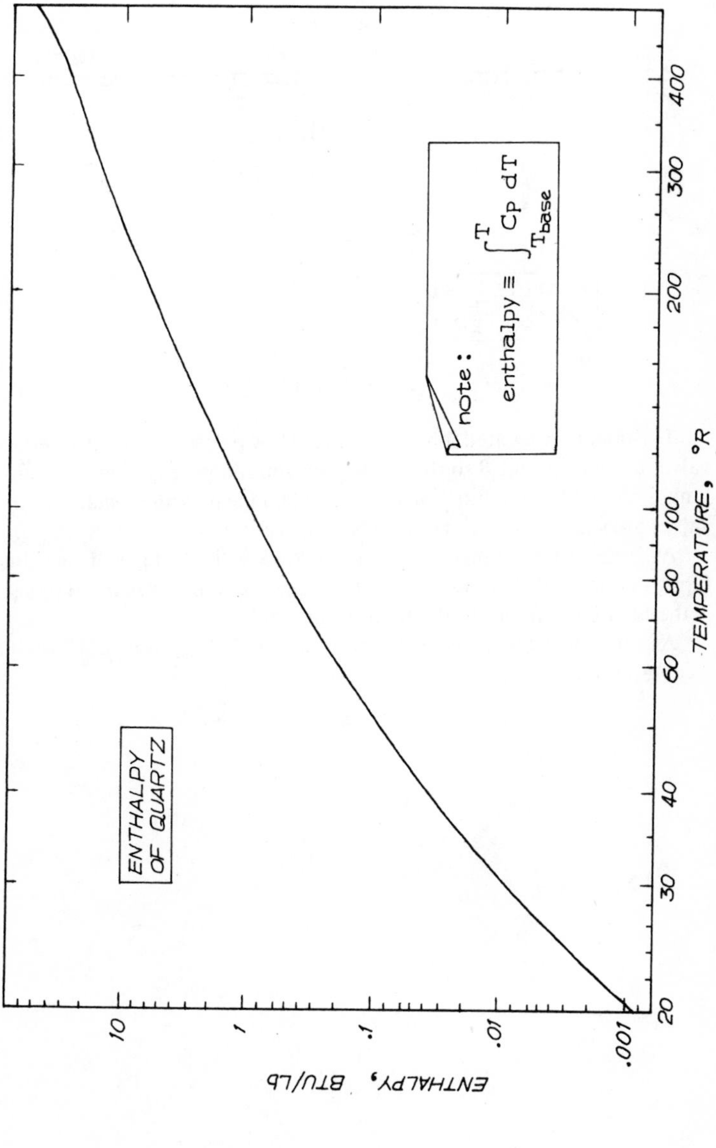

Figure P3.17b

3.18. A rigid, adiabatic tank with a volume of 10 ft³ has connected to it two supply lines, as shown in Figure P3.18. Line *A* leads from a large helium tank at 3 atm and 25°C. Line *B* leads from a large nitrogen tank also at 3 atm, 25°C.

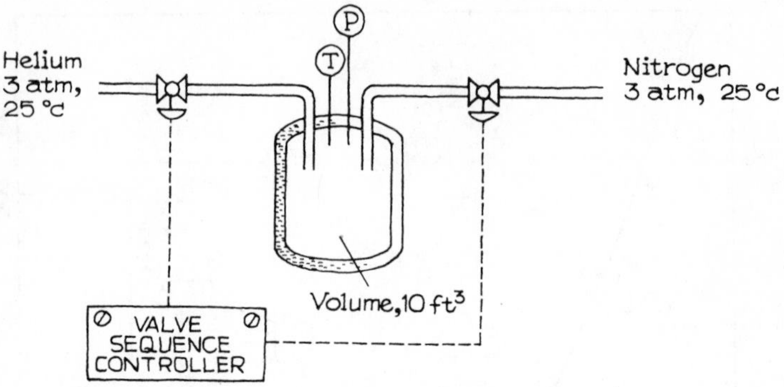

Figure P3.18

The tank is evacuated initially to a very low pressure. A controller operates valves on lines *A* and *B* so that they open simultaneously. The pressure in the tank is monitored as a function of time. When the pressure reaches 2 atm, both valves are shut. At this time, the gas temperature in the tank is 195°C.

Assume no heat transfer between the tank walls and gas during this short filling period, and also assume that the gases are well-mixed at all times. What is the final mole fraction of helium in the tank?

Assume ideal gases and let C_p (nitrogen) = 7 cal/g-mol°K, C_p (helium) = 5 cal/g-mol°K.

Reversibility: Concepts and Consequences 4

In Postulate III, we alluded that only certain processes between stable equilibrium states may be possible. We have in fact already stated all the postulates necessary for determining which processes are and are not possible. In this chapter, we shall explore in greater depth the consequences of these postulates.

One of the underlying concepts in the developments to follow is the reversible process, which is discussed in Section 4.2. This process has not been found to occur in reality; it is in fact a limiting condition that cannot be attained, but it can be closely approached. Consequently, this concept cannot be developed fully by examining real systems in the laboratory; instead, we must utilize thought experiments. In these thought experiments, we shall use a device called a heat engine, which is described in the next section. Although many kinds of heat engines (also referred to as power cycles) have been in use for centuries, we will employ almost exclusively the fictitious reversible heat engine.

The approach we shall follow is in many ways similar to the historical development of thermodynamics in that it was the introduction of a reversible heat engine by Carnot in 1824 that formed the basis of the introduction of the concept of entropy by Clausius in 1850.

4.1 Heat Engines

A *heat engine* is a closed device that undergoes heat interactions with one or more systems and work interactions with a *work reservoir*. The work reservoir is a device that operates adiabatically and quasi-statically and is used for storing energy. For example, a system of weights at different levels in a gravitational field can be used as a work reservoir. The heat engine always undergoes a *cyclic* process such that any net effects appear only in the external

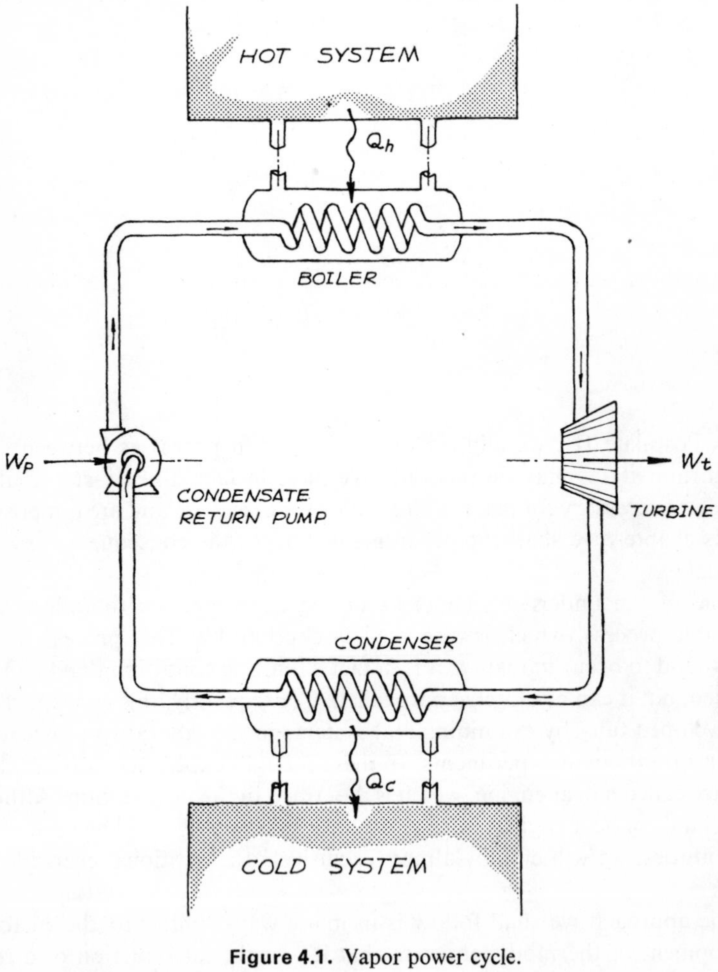

Figure 4.1. Vapor power cycle.

systems and in the work reservoir. The above definition excludes various open system-power cycles such as the internal combustion engine.

There is great diversity in kinds of heat engines. Although we need not specify any particular one for the developments to follow, it may be instructive to illustrate the operation with the Rankine cycle, shown in Figure 4.1. The internal working fluid is vaporized at a high temperature and high pressure in a boiler. Useful work is obtained by expanding the vapor to a low pressure in a turbine. The low-pressure vapor is liquefied in a condenser, and the liquid is pressurized and returned to the boiler.

The heat engine is a convenient device for evaluating processes (real and imaginary). We shall use heat engines to determine whether or not a process is consistent with the postulates and concepts which have been developed previously. A process will be considered allowable if we know of a real case in which the process occurs. (If such a process violates one of our postulates, we would have to revise the postulates.) If we have no prior experience for a given process, the process will be considered *impossible* if we can prove that the process leads to a violation of one or more of our postulates. If, however, we cannot prove that the process violates a postulate, the process may be possible or impossible since we may not have been clever enough to show that a violation exists.

In the processes shown in Figure 4.2, we will choose systems A and B that are in stable equilibrium states prior to their participation in the heat engine interaction. We will also choose them so that $\theta_A > \theta_B$ (recall that we have

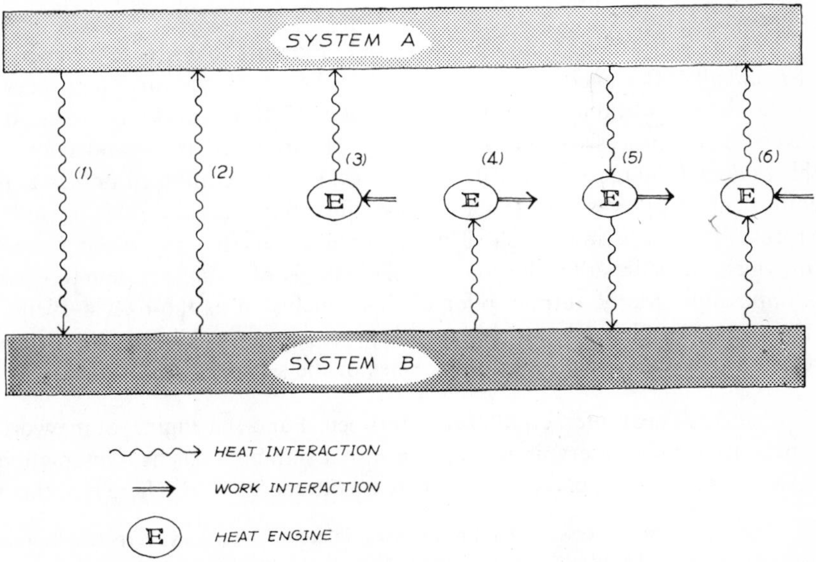

Figure 4.2

chosen by convention that a rise in θ corresponds to an increase in energy). Now let us consider a number of conceivable processes.

Case 1. A heat interaction occurs from A to B (i.e. , the energy of A decreases and the energy of B increases) without any work being performed. Since we know of real cases in which such interactions exist, case (1) is clearly allowable.

Case 2. A heat interaction occurs from B to A without any work being performed. Since we have notched our thermometer so that $dE/d\theta > 0$, and since this process leads to an increase in energy of A and a decrease in energy of B, the net effect of this process is to increase $\Delta\theta = \theta_A - \theta_B$. This process is in violation of Postulate II since the composite system $A + B$ does not tend to a state of equilibrium (i.e., $\Delta\theta$ does not tend to zero). Since we have no prior knowledge of real cases in which such processes occur, this process is impossible. Thus, it is concluded that any process in which the net effect is the transfer of heat from a cooler to a hotter system is impossible. A similar conclusion was drawn by Clausius over a hundred years ago, and some authors refer to this result as Clausius' statement of the Second Law.

Case 3. A work interaction occurs whereby work from the reservoir passes to the engine and results in a heat interaction with A. This process is allowable since we know of real cases in which this process occurs (for example, in Figure 3.5(a) consider B as the engine and C as the work reservoir). Note that the heat interaction could also have been to system B instead of A.

Case 4. A heat interaction involving only B occurs to decrease the energy of B and all of this energy appears as work in the work reservoir. This process can be shown to lead to a violation of Postulate II. If process (4) could occur, we could use process (3) to extract the work produced in (4) and convert this to a heat interaction with A. The net result of the combined processes is equivalent to process (2), which is impossible. Since process (3) is possible, process (4) must be impossible. Thus, it is concluded that any cyclic process for which the net result is the conversion of energy of a single system to work is impossible. Many authors refer to this conclusion as the Kelvin–Planck statement of the Second Law. Impossible processes of this kind are sometimes referred to as perpetual-motion machines of the second kind (PMM2).[1]

Case 5. A heat interaction occurs between A and the engine. Some work is produced in the reservoir and there is also a simultaneous heat interaction between the engine and system B to increase the energy of B. There is nothing

[1] Note that these processes do not necessarily violate the First Law. Perpetual-motion machines of the first kind (PMM1) refer to processes that lead to a net change in the energy of the universe.

in the postulates to prevent such an occurrence, and processes of this kind are well-known; the vapor power cycle (Figure 4.1) is an example. Note that although the direction of the work interaction can be reversed, altering either of the heat interaction vectors leads to an impossible process.

Case 6. All of the arrows in (5) are reversed, the net effects being extraction of work from the reservoir, decrease in energy of B and increase in energy of A. Again, there is no violation of the postulates and real cases of process (6) are known as refrigerators.

Of all the cases discussed, (5) and (6) are of the most immediate interest. We note that (5) and (6) are opposites, and any enterprising person could immediately conjure up some interesting combined processes. As yet we have not placed any quantitative value on the heat and work interactions except to insure that we have not violated the First Law which necessitates the conservation of energy. Why couldn't we use case (5) and take 100 units of energy from A, put 90 of them in the work reservoir, and reject 10 to B? Then, following this, we could use case (6) to take 50 units from B, 50 units from the work reservoir, and reject 100 units to A. In this sequence, system A has undergone a cycle—but the work reservoir has gained a net 40 units and system B has lost 40 units. This particular combination of cases (5) and (6), carried out as prescribed, is in reality case (4), which we found to be impossible. Thus, some combinations of cases (5) and (6) lead to impossible processes, whereas other combinations are allowable. In order to avoid combinations that lead to a violation of our postulates, we must specify some limitations to the way in which the heat and work effects are split. We will find out later that the split depends on the temperature of systems A and B, but as of now we simply recognize that there is a limitation.

A convenient way to delineate the split is to specify an efficiency, η, for case (5) as

$$\eta_5 \equiv \frac{\text{work done by engine}}{\text{heat transferred from hot system}} \tag{4-1}$$

or,

$$\eta_5 = -\frac{W_E}{Q_A} \qquad \text{(4-2)[2]}$$

That is, the more efficient we are, the more work we can get out of a given heat interaction between system A and the engine.

We found from case (4) that the efficiency cannot equal unity; what, then, does limit the efficiency? In a practical sense, we realize that factors such as

[2] Following the conventions described in Chapter 3, W_E is the work for the system comprising the engine and is positive when work is done *by* the engine. Q_A is the heat interaction for system A and is positive when the interaction *increases* the energy of A. In all allowable processes, only one of the two terms is negative and, therefore, η is positive.

friction and other resistances will decrease engine efficiency, but we have made no mention of such factors in deciding that there exists a limiting value of the efficiency. In fact, we shall assume in the thought experiments to follow that we can construct an engine that is not plagued by friction and other resistances; such engines will be referred to as *reversible* heat engines.

The efficiency of a heat engine must be less than one

$$\eta_5 < 1 \tag{4-3}$$

If we define the efficiency for case (6) by Eq. (4-1),[3] then to avoid a combination of processes (5) and (6) which violates our postulates (as discussed above), we require that

$$\eta_5 \leqslant \eta_6 < 1 \tag{4-4}$$

Note that Eq.(4-4) must be satisfied regardless of the kind of engine.

4.2 Reversible Processes

It can be seen by inspection of Eq. (4-4) that the most efficient engines that could be conceived would correspond to $\eta_5 = \eta_6$. For these engines, we could operate process (5) to obtain the maximum work and then use process (6) to restore systems *A* and *B and* the work reservoir to their original states. The combination is an example of a *reversible cycle*, and the engines involved are referred to as reversible engines.

In the general case, *a process will be called reversible if a second process could be performed in* at least one way *so that the system and all elements of its environment can be restored to their respective initial states, except for differential changes of second order.*[4]

It can readily be shown that each step in the path of a reversible process must be reversible. It can also be shown that in a reversible process, all systems must be in states of equilibrium at all times (i.e., all subsystems must traverse quasi-static paths.) The proof follows from Postulate II: if a system in a nonequilibrium state is isolated, it will tend toward a state of equilibrium. Since there is no way to transform a system from a state of equilibrium to the nonequilibrium state without removing it from isolation, any process

[3] The efficiency of refrigeration cycles such as case (6) are more commonly measured by the coefficient of performance, ω, where

$$\omega \equiv -\frac{\text{heat transferred from cold system}}{\text{work done by engine}}$$

[4] It can be proved that the maximum work obtainable from, for example, an expansion process corresponds to the hypothetical case in which the boundary is moved at an infinitesimal rate for an infinite time. Thus, the difference in the maximum work obtained in an expansion and the minimum work required for the reverse compression is of the order of $(dP)(d\underline{V})$.

involving an intermediate nonequilibrium state is irreversible. Many useful corollaries follow directly from this last conclusion. For example, it can be shown that simple systems involved in reversible processes can have no internal pressure or temperature gradients.

Finally, it can be shown that friction and similar resistances must not be present if a process is to be reversible. The proof follows from the fact that the work required in such processes exceeds the minimum (or the work obtained is less than the maximum) because a finite unbalance of boundary forces is required to effect the changes involved.

4.3 Thermodynamic Temperature

One of the major results of the previous section can be summarized as follows: The efficiency of all cycles involving reversible heat engines that operate between two given systems with different thermometric temperatures is a constant. It is a simple matter to extend this reasoning to show that the efficiency of any reversible engine is dependent on the thermometric temperatures of both systems with which it interacts. For example, we can prove that the efficiency must depend on the temperature of the cold system in the following manner. Let us assume that η is only a function of the temperature of the hot system and then show that this assumption cannot be valid. If we were to operate a reversible engine as in case (5) between two systems at θ_A and θ_B, and also operate a reversible refrigeration cycle as in case (6) between two systems at θ_A and θ_C, the efficiencies of the two processes will be the same (if the initial assumption is correct). If we had chosen systems such that $\theta_A > \theta_B > \theta_C$, the net effect of the combined process would have been a transfer of heat from system C to system B. But since θ_B was chosen to be greater than θ_C, the net effect, which is equivalent to case (2), is in violation of our postulates. Therefore, our assumption was incorrect; the efficiency of the reversible engines must involve the temperature of the cold system. In this manner, it can be shown that the efficiency is a function of the temperatures of both systems. Thus,

$$\eta = -\frac{W_E}{Q_A} = f_1(\theta_A, \theta_B) \tag{4-5}$$

Since

$$W_E = -(Q_A + Q_B) \tag{4-6}$$

where Q_B is positive for heat transferred *to* system B, Eq. (4-5) could be expressed in the equivalent form

$$\frac{Q_B}{Q_A} = f_2(\theta_A, \theta_B) = \eta - 1 \tag{4-7}$$

As a result of the foregoing deductions, the primitive property, thermometric temperature, has assumed a role of prime significance. Of course, we could have used any number of clever techniques to define properties that have the same information content as the thermometric temperature (e.g., electrical resistance, thermocouples, thermal electron emission detectors, infrared emission analyzers, etc.),[5] and we would have arrived at the same result expressed by Eqs. (4-5) and (4-7). In fact, until we specify the form of the function f_1 or f_2, any primitive temperature measurement would be equally acceptable.

Let us now look at the problem of determining the form of the function f_1 or f_2. We imagine our systems A and B (with θ_A greater than θ_B) operating as in case (5) to produce work with a reversible heat engine, E_1 [see Figure 4.3(a)]. Imagine that these systems are so large that they do not change in state by a significant amount during the process. Now connect to system A *another* heat engine, E_2, which in this case rejects heat to a new system C. System C has a temperature intermediate between A and B. Also, any heat transferred to C is immediately transferred to a third reversible engine, E_3, which rejects heat to system B [see Figure 4.3(b)].

If each of the processes shown in Figures 4.3(a) and 4.3(b) reduces the energy of system A by an equal amount, and if there is no accumulation of energy in system C, then it is obvious that the work obtained from engine E_1 is equal to the sum of the work from engines E_2 and E_3. If this were not so, then the more efficient of the two procedures could be reversed, resulting in a violation of Postulate II.

Using the nomenclature shown in Figure 4.3,

$$Q_{A_1} = Q_{A_2} \tag{4-8}$$

$$Q_{B_1} = Q_{B_3} \tag{4-9}$$

$$Q_{C_2} = -Q_{C_3} \tag{4-10}$$

$$W_1 = W_2 + W_3 \tag{4-11}$$

Using Eq. (4-7) for each engine,

$$\frac{Q_{B_1}}{Q_{A_1}} = f_2(\theta_A, \theta_B) \tag{4-12}$$

$$\frac{Q_{C_2}}{Q_{A_2}} = f_2(\theta_A, \theta_C) \tag{4-13}$$

$$\frac{Q_{B_3}}{Q_{C_3}} = f_2(\theta_C, \theta_B) \tag{4-14}$$

Multiplying Eq. (4-13) by Eq. (4-14) and equating the result to Eq. (4-12),

$$[f_2(\theta_A, \theta_C)][f_2(\theta_C, \theta_B)] = -f_2(\theta_A, \theta_B) \tag{4-15}$$

[5] Recall the definition of a "temperature" measuring device. It is a system, closed and rigid, which shows a variation in at least one primitive property when allowed to undergo a finite heat interaction with another system whose temperature is being measured.

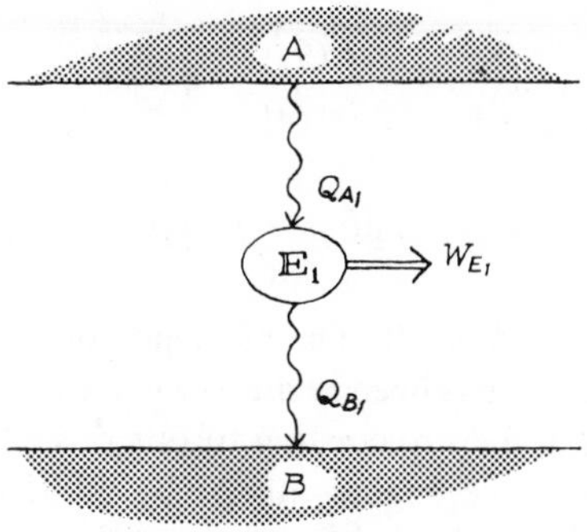

Figure 4.3(a)

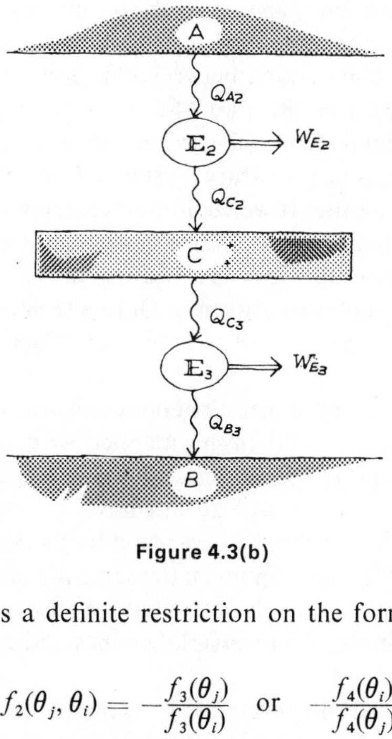

Figure 4.3(b)

This result places a definite restriction on the form of the function f_2; that is,

$$f_2(\theta_j, \theta_i) = -\frac{f_3(\theta_j)}{f_3(\theta_i)} \quad \text{or} \quad -\frac{f_4(\theta_i)}{f_4(\theta_j)} \tag{4-16}$$

If we adopt the former form, Eq. (4-12) reduces to

$$\frac{Q_{B_1}}{Q_{A_1}} = -\frac{f_3(\theta_A)}{f_3(\theta_B)} = \eta - 1 \tag{4-17}$$

and

$$\eta = \frac{f_3(\theta_B) - f_3(\theta_A)}{f_3(\theta_B)} \tag{4-18}$$

If we adopt the latter form,

$$\frac{Q_{B_1}}{Q_{A_1}} = -\frac{f_4(\theta_B)}{f_4(\theta_A)} = \eta - 1 \tag{4-19}$$

and

$$\eta = \frac{f_4(\theta_A) - f_4(\theta_B)}{f_4(\theta_A)} \tag{4-20}$$

This is a very profound result. The efficiency of a reversible, but otherwise arbitrary, heat engine has been related to some function of the temperatures of the systems A and B. According to our preceding arguments, since the efficiency of a reversible engine cannot be multivalued, it is apparent that f_3 and f_4 must be single-valued functions of θ.

Let us recapitulate the progress we have made thus far. We have concluded that any heat engine or refrigerator operating reversibly must have a single-valued efficiency that is a function only of the primitive property, temperature (thermometric or any other empirical temperature scale which has the same information content). The relation between efficiency and temperature must be of the form given by Eq. (4-18) or Eq. (4-20). It is clear that for given systems, A and B, the experimental technique used to define temperature is arbitrary; the efficiency, however, is not arbitrary. Systems A and B having been chosen, the efficiency is set by nature. If we could build a reversible engine, we could measure experimentally the value of the efficiency. Since the efficiency is not arbitrary and the temperature scale is arbitrary, the form of the functions in Eqs. (4-18) and (4-20) cannot be arbitrary. Once a temperature scale has been decided upon, there will be one and only one valid form of each of the functions f_3 and f_4.

We have said that we could measure empirically the efficiency if we could construct a reversible engine and, in this manner, we could empirically determine the functions f_3 and f_4. Nevertheless, we still do not have access to any real reversible engines and we still do not have any reason to believe that one can be constructed. We can, however, conduct a thought experiment for a hypothetical reversible engine in much the same manner as did Carnot over a century ago. It is only through such analysis that we are able to define unequivocally the efficiency of reversible engines and a consistent temperature scale.

Thus, let us construct our reversible heat engine cycle using an ideal gas as a working fluid. Such a cycle, involving heat input from system A at T_A and rejection of heat to system B at T_B, is called a Carnot cycle. There are four steps to this cycle:

1. Heat flows to an ideal gas contained in a piston-cylinder device at temperature T_A.
2. The ideal gas system is then isolated from system A and allowed to expand adiabatically and reversibly to a lower pressure so that the temperature is T_B.

3. At this lower pressure, the ideal gas system is connected to system B and the gas compressed, heat being rejected to system B at a constant temperature, T_B.
4. At a particular point, the ideal gas system is isolated from system B and compressed adiabatically and reversibly to the original pressure. The point of initiation of this step is determined so that the final temperature after compression is T_A.

The heat engine (i.e., the ideal gas system) has undergone a cycle: work has been produced in the work reservoir, the energy of system A has decreased, and the energy of system B has increased. By calculating the actual work and heat flows, and assuming that the ideal gas has the properties such that $P\underline{V} = NRT$ and the energy of the gas is not a function of pressure but only of temperature, it is then possible to show that

$$\eta = \frac{T_A - T_B}{T_A} \tag{4-21}$$

The proof is left as an exercise.

From Eq. (4-21) it is clear that the functions f_3 and f_4 are of a very simple form if the temperature is measured by the ideal gas scale.[6] The functions would, of course, have been much more complicated if we had used a conventional sealed mercury thermometer. Fortunately, the ideal gas temperature scale has been adopted as *the* thermodynamic temperature scale. As discussed in Section 3.5, there are many ideal gas temperature scales, the more common ones being the Kelvin and Rankine scales.

We can measure approximately an ideal gas temperature since empirically we have found that simple gases at low pressures will obey the two relations specified above to a high degree of approximation. From this point on we will use the T-scale instead of the θ-scale and assume that we can accurately measure T.

4.4 The Theorem of Clausius

We are now in a position to extend our analysis of heat engines to the development of an additional derived property, the entropy. There are, perhaps, as many ways to proceed logically to infer this property as there are textbooks in thermodynamics. In whichever way we choose, however, we must limit ourselves to using the stated postulates or the conclusions obtained from them. We will first derive a quantitative definition of entropy and then proceed to show that it has a very significant bearing on the concept of equilibrium.

Let us first rephrase the results of the last section. Combining Eqs. (4-7)

[6] We developed Eq. (4-21) without specifying *a priori* any functional form for f_3 or f_4. We shall have no further need for these functions since we now have a practical temperature scale and a consistent expression for the efficiency.

and (4-21) for a reversible heat engine cycle, we have

$$\frac{Q_B}{Q_A} = -\frac{T_B}{T_A} \tag{4-22}$$

or

$$\frac{Q_A}{T_A} + \frac{Q_B}{T_B} = 0 \tag{4-23}$$

The temperatures of A and B are assumed not to change in the heat engine cycle; if they were to change, one could still write the equations but in a differential form, i.e.,

$$\frac{đQ_A}{T_A} + \frac{đQ_B}{T_B} = 0 \tag{4-24}$$

Let us now shift our attention to the interior of the closed, reversible engine. We imagine that it is charged with some material which we shall call the working substance. The only restriction we have placed thus far on the internals of the engine is that it operate reversibly. The engine may be a composite or simple system.

We shall assume, for simplicity, that there are no external body force fields and that only P-$\underline{V}$ work need be considered; the results, however, are valid for systems in external body force fields. Since we are considering *reversible* heat engine cycles, and since all steps within a reversible process must be reversible, it follows that we cannot have any pressure gradients across moving boundaries or any temperature gradients across diathermal boundaries. Therefore, in any $P\,d\underline{V}$ work terms, P is both the external pressure and the internal pressure at the region of the moving boundary. Similarly, T_A is the temperature of external system A and also the temperature of that portion of the working substance at the diathermal boundary with system A. A similar conclusion holds for T_B. We may have many different events occurring in our system (i.e., adiabatic compressions and expansions, isothermal compressions and expansions, etc.). In Figure 4.4(a) we show a path representing an arbitrary reversible change in state of our system from i to f (i.e., a change that might occur during some part of the heat engine cycle). In Figure 4.4(b) the same path is shown and we have drawn through points i and f curves that represent those paths which would occur if adiabatic expansions or compressions were to take place starting from either i or f. Next, in Figure 4.4(c) we repeat (a) and (b) but add another path curve to represent the behavior of our system if it were at point g and were expanded isothermally (and reversibly) to h. This path g–h, of course, represents a particular temperature level. The exact position of point g is established as shown below.

The work proceeding along the real path i–f is the integral under the path curve. We choose point g so that the area under the path curve $ighf$ is equal to the actual work.

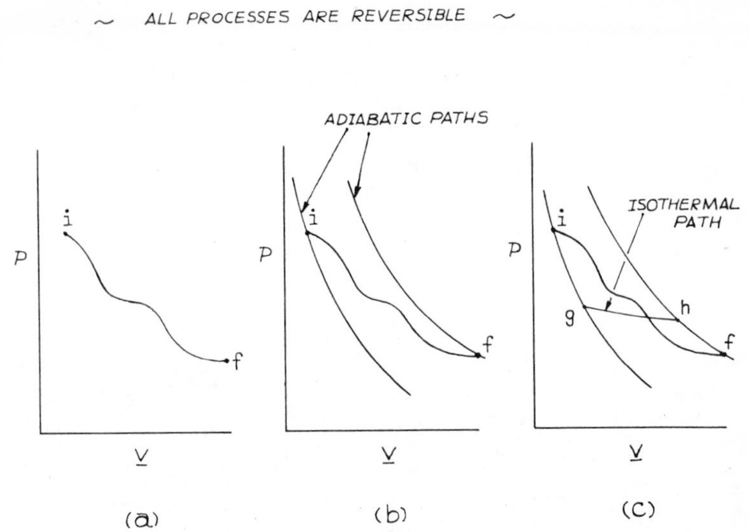

Figure 4.4

For these two alternative paths, $\Delta \underline{E}_{if} = \Delta \underline{E}_{ighf}$ and $W_{if} = W_{ighf}$. Therefore, $Q_{if} = Q_{ighf}$. Since it was specified that $Q_{ig} = Q_{hf} = 0$, then $Q_{if} = Q_{gh}$. This simple little scheme is known as the Theorem of Clausius: *Given any reversible process in which the temperature changes in any prescribed manner, it is always possible to find a reversible zigzag process consisting of adiabatic-isothermal-adiabatic steps such that the heat interaction in the isothermal step is equal to the heat interaction in the original process.*

4.5 Entropy

The Theorem of Clausius is useful in analyzing the entire cycle carried out by a reversible heat engine or, for that matter, any system undergoing a reversible, cyclic process. This is shown in Figure 4.5 by the curve *jifkdcj*. The unusual curve is drawn to emphasize that an actual P–$\underline{V}$ path is not necessarily simple. The heat interactions in various portions of the cycle may result from contacts with heat reservoirs at different temperatures. Choose for particular examination a portion of the cycle represented by the terminal points *i* and *f*. Draw path lines through these points representing adiabatic processes starting from *i* and *f*. These adiabatic path lines cut the cycle at points *c* and *d* respectively. From Clausius' theorem we know that the path curves *i*–*f* and *c*–*d* can be broken down into a series of adiabatic and isothermal paths with the same work and heat interactions as occur in the original path. Let the heat interaction in path *i*–*f* be from a reservoir at T_A and in

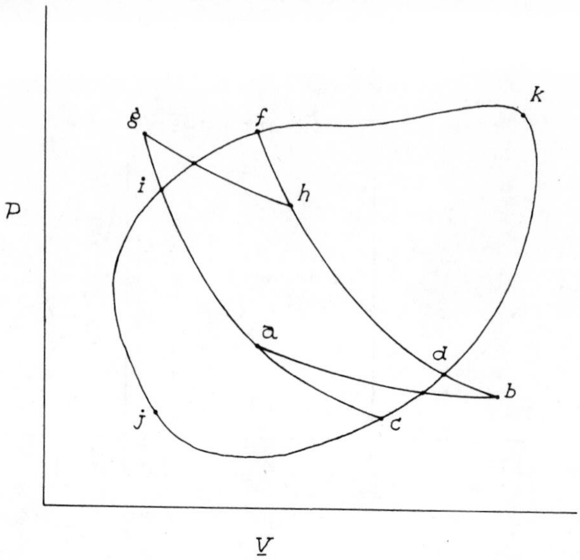

Figure 4.5

path c–d from a reservoir at T_B. We note that this net inner cycle *ighfdbac* is, in reality, a Carnot cycle. Thus,

$$\frac{Q_{if}}{T_{if}} + \frac{Q_{cd}}{T_{cd}} = 0 \tag{4-25}$$

If the paths i–f and c–d are made infinitesimal,

$$\frac{đQ_{if}}{T_{if}} + \frac{đQ_{cd}}{T_{cd}} = 0 \tag{4-26}$$

where T_{if} and T_{cd} become equal to the temperature of that part of the system in contact with the diathermal boundary at the respective points in the *actual* path. (The heat interactions in the differential cycle must still be equal to the heat interactions in the actual path.)

An infinite number of infinitesimal cycles, the first beginning at j and the last ending at k (see Figure 4.5), cover the entire cycle; summing,

$$\oint \left(\frac{đQ}{T}\right)_{\text{rev}} = 0 \tag{4-27}$$

where T and $đQ$ are the actual values associated with the original path, and the subscript *rev* reminds us that this result is only valid if the original path is reversible.

Eq. (4-27) could be written in the equivalent form,

$$\int_j^k \left(\frac{đQ}{T}\right)_{\text{rev}} = -\int_k^j \left(\frac{đQ}{T}\right)_{\text{rev}} \tag{4-28}$$

This result does not appear very profound. If, however, we carry our reasoning one step further, we can formalize our results in a manner that will be most useful.

If we proceed from j to k via the top path shown in Figure 4.5 and then return from k to j via any other conceivable reversible path, it is readily shown that for this new cycle Eqs. (4-27) and (4-28) are still valid. Thus, we can conclude that there is a quantity, $\int (đQ/T)_{\text{rev}}$, which is conserved in *any* reversible cyclic process. This quantity has all of the characteristics associated with a derived property of the system; we shall call this property the *entropy*, and we shall represent it by $\underset{\sim}{S}$. Thus,

$$\Delta \underset{\sim}{S} \equiv \int \left(\frac{đQ}{T}\right)_{\text{rev}} \tag{4-29}$$

or

$$d\underset{\sim}{S} \equiv \left(\frac{đQ}{T}\right)_{\text{rev}} \tag{4-30}$$

Eqs. (4-29) and (4-30) were derived for systems in which only force-displacement work occurs. The derivation can be extended to include systems that are acted on by body and inertial forces as well. The result is not changed by including other forces. Clausius' theorem is still valid because we can always find a reversible adiabatic-isothermal-adiabatic path having the same work interaction as the actual path. Furthermore, Eqs. (4-29) and (4-30) follow directly from Clausius' theorem. Thus, the defining equations for entropy, Eqs. (4-29) and (4-30), are valid for any system (simple or composite) undergoing a reversible process. In the case of a composite system consisting of two or more subsystems at different temperatures (which are separated from one another by internal adiabatic boundaries), the change in entropy of the composite is

$$\sum_i \frac{đQ_i}{T_i}$$

where $đQ_i$ is the heat transferred to subsystem i at T_i.

Note that in the derivation given above, only differences in entropy take on any significance, and these differences apply only between states at equilibrium.

If we follow the same line of reasoning used when energy was introduced as a property (see Section 3.3), it can be shown that entropy must also be first order in mass. Unlike the energy, there is nothing in our previous developments or in our postulates to require that the entropy of the universe should always be conserved. On the contrary, we shall soon prove that the entropy of the universe must increase in any irreversible process and that entropy is conserved only in reversible processes. In fact, the entropy change is of great use as a means for determining how closely real processes approach reversible processes.

Finally, we present an operational method for determining the entropy difference between two stable equilibrium states of a closed, simple system. There are no entropy meters available, and there are no meters to measure directly the quantity of heat transferred. Thus, in order to determine an entropy change between two stable equilibrium states, we must visualize any convenient *reversible* path and calculate the reversible work that would be done. Then, using Eqs. (3-28) and (4-30), we have

$$T\,d\underline{S} = d\underline{E} + đW_{\text{rev}} \tag{4-31}$$

which, for a closed simple system, reduces to

$$T\,d\underline{S} = d\underline{U} + P\,d\underline{V} \tag{4-32}$$

or, over the entire reversible path,

$$\Delta\underline{S} = \int \left(\frac{1}{T}\right) d\underline{U} + \int \left(\frac{P}{T}\right) d\underline{V} \tag{4-33}$$

The units of either entropy or temperature are completely arbitrary; all that we require is that the product of the entropy and temperature units be equivalent to the units of energy. Hence, we are free to choose one or the other at will, but, once we choose one, the units of the other are fixed. The accepted convention is to specify temperature in Kelvin or Rankine units and entropy in units of energy per unit temperature.

Example 4.1

A 10-lb block of copper is at 850°F and 1556 ft above a tank of water (40 lb at 40°F originally). It falls into the water. Assuming no splashing, what is the change in entropy of the universe? The heat capacity of copper may be taken as 0.1 Btu/lb°R.

Solution

We must first define the conditions of the initial and final stable equilibrium states of the systems for which the entropy change is to be calculated. The only systems within the universe that are affected by the events are the copper block and the water. The initial conditions are stated in the problem. The conditions of the final equilibrium that is attained after the block has fallen and the temperatures have equilibrated can be determined by an analysis of the composite system of water plus copper block. Thus,

$$\Delta\underline{E}_w + \Delta\underline{E}_c = Q - W \tag{4-34}$$

where subscripts w and c refer to water and block, respectively. In the actual process, the composite system has no heat or work interactions with the surroundings. Thus,

$$M_w C_w (T_{w_f} - T_{w_i}) + M_c C_c (T_{c_f} - T_{c_i}) + M_c \left(\frac{g}{g_c}\right) \Delta h_c = 0 \qquad (4\text{-}35)$$

or

$$(40)(1)(T_{w_f} - 40) + (10)(0.1)(T_{c_f} - 850) - \frac{(10)(1556)}{(778)} = 0$$

Since

$$T_{w_f} = T_{c_f} = T_f$$

$$T_f = 60°\text{F} = 520°\text{R}$$

Now that the conditions of the final equilibrium state have been defined, we can calculate the entropy change of each subsystem of the composite. To do so, we must disregard completely the actual process (because it is clearly irreversible) and devise a reversible process for effecting the change from initial to final states.

The net process for the water is simply a temperature rise from 40 to 60°F. We could accomplish this reversibly by successively bringing the water into contact with an infinite number of heat reservoirs varying in temperature from 40° to 60°F and by transferring an infinitesimal quantity of heat from each of the reservoirs. For this reversible process, the entropy change in the water is given by Eq. (4-29):

$$\Delta \underline{S}_w = \int_{T_{w_i}}^{T_{w_f}} \left(\frac{đQ}{T}\right)_{\text{rev}} \qquad (4\text{-}36)$$

where $đQ$ is the heat transferred from the reservoir at T. Since

$$đQ = M_w C_w \, dT \qquad (4\text{-}37)$$

$$\Delta \underline{S}_w = M_w C_w \int_{500°\text{R}}^{520°\text{R}} \frac{dT}{T} \qquad (4\text{-}38)$$

or

$$\Delta \underline{S}_w = 1.56 \text{ Btu/°R}$$

In order to effect a reversible change of the copper block, we could lower the block at an infinitesimal rate using a frictionless pulley with a 10-lb counterweight. In lowering the block, we would raise the counterweight an equivalent amount. This step is clearly reversible and adiabatic so that there is no entropy change. Then, we could cool the block from 850°F to 60°F in an analogous manner to that used for heating the water. Therefore,

$$\Delta \underline{S}_c = M_c C_c \int_{1310°\text{R}}^{520°\text{R}} \frac{dT}{T} = -0.92 \text{ Btu/°R}$$

Thus,

$$\Delta \underline{S}_{\text{universe}} = 1.56 - 0.92 = 0.64 \text{ Btu/°R}$$

Note that if the entire process were carried out reversibly, $\Delta \underline{S}_{\text{universe}}$ would be zero. Thus, for the reversible processes described above, the total entropy change of the heat reservoirs is -0.64 Btu/°R.

Example 4.2

Prove that for an *adiabatic* process between defined states I and II of a closed system, the entropy change of the system must be equal to or greater than zero.

Solution

If the process were reversible, Eq. (4-29) would show that $\Delta \underline{S}_{\text{system}}$ must be zero. If the process were irreversible, we could follow this process by a second process to bring the system back to state I. Let us make this second process reversible and let us make it such that all heat transfer between system and surroundings occurs during an isothermal step. Such a reversible process can always be found (see Clausius' theorem, Section 4.4). In the combination of the two processes, the system has undergone a complete cycle so that the overall energy and entropy changes must be zero. In the reversible portion of the cycle, there may have been work and heat interactions; in the combined processes, clearly, the net work must have been equal to the net heat interaction. The net work done by the system and the net heat added to the system could not have been positive since this would be a PMM2. Thus, the net work and net heat must have been negative. (The net work and net heat could not have been zero since this would correspond to a reversible adiabatic process.) Since the heat interaction (which occurs during the return reversible step only) is negative, the entropy change of the system must also be negative in the return step. Therefore, the entropy change of the closed system in the irreversible adiabatic step must have been positive.

It should be clear from Example 4.2 that the entropy change of any closed system in any adiabatic process (reversible or irreversible) cannot be negative. Since an isolated system is a special case of a closed, adiabatic system, it follows that *the entropy change of an isolated system in any process must be equal to or greater than zero*. Furthermore, since the universe can be considered an isolated system, we are led to the same conclusion first stated by Clausius over a century ago: "*Die Entropie der Welt strebt einen Maximum zu.*"

4.6 Internal Reversibility

In Section 4.2, it was shown that in a reversible process all interacting systems must be in states of equilibrium at all times. Hence, a process involving a temperature gradient across a diathermal boundary separating two systems is an irreversible process because the composite system is not in thermal equilibrium. Thus, case (1) illustrated in Figure 4.2 is an irreversible process. Let us focus, however, our attention on system A in Figure 4.2. By altering the environment external to system A from that of case (1) to that of case (5), we could make the process reversible if we were to use a reversible heat engine. Since we could have effected the same changes *in system A* through either of these processes, the irreversibility in case (1) must be external to system A. If, however, during the heat interaction a temperature gradient had been established within system A, the process would have been irreversible and the irreversibility would have occurred within system A.

In many engineering applications it may be necessary to minimize irreversibilities in a process in order to, for example, reduce power costs. Here it would be helpful to know where the irreversibilities occur in a process. In order to identify subsystems that proceed along reversible paths within a process that is irreversible *in toto, we will define a process of a system as being internally reversible if the process can be performed in at least one way with another environment selected such that the system and all elements of this environment can be restored to their respective initial conditions, except for differential changes of second order in external work reservoirs.*

As discussed in Section 4.2, it then follows that a system undergoing an internally reversible process must traverse a quasi-static path. Nevertheless, an entire process is not necessarily reversible even if all subsystems traverse quasi-static paths because there may be irreversibilities occurring at the *boundaries* between the subsystems.

It should be obvious that for internally reversible systems there is no need to devise artificial processes to calculate entropy changes for the system in question; the actual interactions will suffice for this purpose.

Example 4.3

Consider a heat engine operating between two systems, 1 and 2, at T_1 and T_2, respectively, with $T_1 > T_2$. If the two systems are internally reversible, and if the overall process is irreversible, show that the entropy change of the universe must be greater than zero.

Solution

If the overall process were reversible, then for a complete cycle of the engine,

$$\Delta \underline{S}_{\text{universe}} = \Delta \underline{S}_1 + \Delta \underline{S}_2 = \frac{\underline{Q}_1}{T_1} + \frac{\underline{Q}_2}{T_2} = 0 \tag{4-39}$$

If we denote the work produced by a reversible and an irreversible engine as W_R and W_I, it can be shown that W_I must be less than W_R in order to avoid the possibility of a PMM2. If we run the reversible and irreversible engines in a manner such that the heat transferred from system 1 was the same in both cases, it is clear that Q_2 must be greater for the irreversible case than for the reversible case. Therefore,

$$\Delta \underline{S}_{2_I} > \Delta \underline{S}_{2_R}$$

while

$$\Delta \underline{S}_{1_I} = \Delta \underline{S}_{1_R}$$

so that for the irreversible case,

$$\Delta \underline{S}_{\text{universe}} = \Delta \underline{S}_{1_I} + \Delta \underline{S}_{2_I} > 0 \tag{4-40}$$

4.7 Criteria of Equilibrium and Stability

It was shown in Example 4.2 that for any permissible process occurring within a closed, adiabatic (as well as isolated) system, the change in total entropy must be equal to or greater than zero. The total entropy is, of course, equal to the sum of the entropies of all subsystems of the isolated composite.

In Postulate II it was stated that any isolated system will tend to change in such a way that it will approach one and only one stable equilibrium state for each simple subsystem of the composite. When this limiting set of conditions is established, it follows that at no later time can other processes occur. Therefore, *for an isolated system, the total entropy of the system must be a maximum when the system is at a state of equilibrium.*

In other words, if we have an isolated system, which may consist of many subsystems with given internal restraints, there are various conceivable sets of conditions for each subsystem; each set of conditions may be consistent with the requirements of isolation (total energy, volume, and mass are constant), but only one set of conditions will correspond to a state of equilibrium of the composite, and that particular set of conditions is the one for which the total entropy is a maximum.

These results can be summarized in the following manner:

1. For any allowable processes to occur,

$$\Delta \underline{S}_{\underline{E},\underline{V},M} \geqslant 0 \tag{4-41}$$

2. For all virtual[7] processes that originate from a state of equilibrium,

$$\Delta \underline{S}_{\underline{E},\underline{V},M} < 0 \tag{4-42}$$

where the subscripts in Eqs. (4-41) and (4-42) denote that total energy, total volume, and total mass are invariant in any of the processes considered.

[7] Any process that can be conceived—whether or not the process is allowable—is called a *virtual* process. This process, of course, must be consistent with any internal and external restraints placed on the system.

Alternatively, these results can be expressed as

$$d\underline{S}_{\underline{E},\underline{V},M} = 0 \tag{4-43}$$

and

$$d^2\underline{S}_{\underline{E},\underline{V},M} < 0 \tag{4-44}$$

for any virtual processes originating from a state of equilibrium[8]. Eqs. (4-43) and (4-44) are each necessary conditions of equilibrium, although both must be satisfied for sufficiency. Eq. (4-43) is sometimes referred to as the *criterion of equilibrium* and Eq. (4-44) as the *criterion of stability*.

It is appropriate at this point in our development to elaborate on the interdependence of properties alluded to in Postulate I. In Section 2.7 it was stated that from the laws of thermodynamics we could derive relationships between certain properties and could thence decide which properties were not independent for a given situation. At this point, we have already developed the laws, but we have yet to show which interdependencies of properties follow from them. Although this topic is treated in detail in subsequent chapters, it is instructive to outline the steps involved. For example, consider the simple case of a closed system of a pure liquid and vapor. At equilibrium, we shall find that if the entropy is to be a maximum, the temperature, pressure, and chemical potential (a property to be introduced later), must be equal in both phases. As shown in Chapter 9, a relationship between the pressure and temperature is obtained from these requirements and, therefore, the temperature and pressure are not independently variable for this system.

As we shall see in the developments to follow, there are no such relationships regarding the properties of single-phase, simple systems. That is, *any two thermodynamic*[9] *properties (in addition to the masses initially charged) of a single-phase, simple system are independently variable.*

Example 4.4

For an isolated simple system containing a single phase in a stable equilibrium state, prove that for any infinitesimal internal variation that can be

[8] Eqs. (4-43) and (4-44) are more restrictive than Eq. (4-42); although they both imply that $\underline{S}$ is a maximum at equilibrium, the former equations also imply that $\underline{S}$ is a continuous and differentiable function and that $d^2\underline{S}$ is nonvanishing. We shall see shortly (Sections 4.8 and 4.9) that the partial derivatives of $\underline{S}$ with respect to $\underline{E}$, $\underline{V}$, and M do exist and, hence, $\underline{S}$ is a continuous and differentiable function. The general case, which allows the possibility of $d^2\underline{S} = 0$, is treated in Chapter 7.

[9] The restriction to *thermodynamic* properties is necessary in order to exclude the possibility of choosing two directly related properties. For example, it would be possible to choose the electrical potential observed at a thermocouple junction placed in the system and the level of liquid in a bulb thermometer in contact with the system as the two properties; however, both of these properties are dependent on one thermodynamic property, the temperature T.

conceived (e.g., a change in composition caused by a chemical reaction):

$$\sum_{i=1}^{n} \left(\frac{\partial \underline{U}}{\partial N_i}\right)_{\underline{S},\underline{V},N_1,\ldots,N_n \text{ (except } N_i)} dN_i = 0 \tag{4-45}$$

if $(\partial \underline{U}/\partial \underline{S})_{\underline{V},N_1,\ldots,N_n}$ is finite.

Solution

For the isolated system at equilibrium, Eq. (4-43) is a necessary condition. For a single-phase simple system of n components, any property can be expressed as a function of any two other properties in addition to the masses or mole numbers. Choosing $\underline{U}, \underline{V}, N_1, \ldots N_n$ as the $n + 2$ independent variables, $\underline{S}$ can be expressed as a function of these variables. Therefore,

$$d\underline{S} = \left(\frac{\partial \underline{S}}{\partial \underline{U}}\right)_{\underline{V},N} d\underline{U} + \left(\frac{\partial \underline{S}}{\partial \underline{V}}\right)_{\underline{U},N} d\underline{V} + \sum_{i=1}^{n} \left(\frac{\partial \underline{S}}{\partial N_i}\right)_{\underline{U},\underline{V},N_j[i]} dN_i \tag{4-46}$$

where subscript N indicates that all N_i are constant in the partial differentiation and subscript $N_j[i]$ indicates that all N_j are constant except for $j = i$. Since the system is isolated, $\underline{U}$ and $\underline{V}$ must be constant for the infinitesimal process described by Eq. (4-46). Therefore,

$$d\underline{U} = 0 \tag{4-47}$$

and

$$d\underline{V} = 0 \tag{4-48}$$

However, the mole numbers are not necessarily invariant; for an isolated system, the total mass is constant, but the mole numbers of the components may change if we visualize a process involving one or more chemical reactions. For such virtual processes, Eq. (4-43) must still be satisfied. (In the derivation of Eq. (4-43), such processes were in no way excluded.) Thus, substituting Eqs. (4-47) and (4-48) into Eq. (4-46), and using Eq. (4-43),

$$d\underline{S} = \sum_{i=1}^{n} \left(\frac{\partial \underline{S}}{\partial N_i}\right)_{\underline{U},\underline{V},N_j[i]} dN_i = 0 \tag{4-49}$$

Since for any set of variables, x, y, and z, where $f(x, y, z) = 0$

$$\left(\frac{\partial x}{\partial y}\right)_z \left(\frac{\partial y}{\partial z}\right)_x \left(\frac{\partial z}{\partial x}\right)_y = -1 \tag{4-50}$$

it follows that

$$\left(\frac{\partial \underline{S}}{\partial N_i}\right)_{\underline{U},\underline{V},N_j[i]} = -\frac{\left(\frac{\partial \underline{U}}{\partial N_i}\right)_{\underline{S},\underline{V},N_j[i]}}{\left(\frac{\partial \underline{U}}{\partial \underline{S}}\right)_{\underline{V},N}} \tag{4-51}$$

iff f_x, f_y, f_z are nowhere zero (or, at least, nonzero at the point of evaluation of the identity

Substituting Eq. (4-51) into Eq. (4-49), and simplifying (note that $(\partial \underline{U}/\partial \underline{S})_{\underline{V},N}$ is not a variable in the summation over i),

$$\frac{1}{\left(\frac{\partial \underline{U}}{\partial \underline{S}}\right)_{\underline{V},N}} \sum_{i=1}^{n} \left(\frac{\partial \underline{U}}{\partial N_i}\right)_{\underline{S},\underline{V},N_j[i]} dN_i = 0 \tag{4-52}$$

If $(\partial \underline{U}/\partial \underline{S})_{\underline{V},N}$ is finite, Eq. (4-52) implies Eq. (4-45).

4.8 The Combined Law for Single-Phase, Closed Systems

For any closed, simple system undergoing a reversible process, it was shown in Section 4.5 [Eq. (4-32)] that

$$T\,d\underline{S} = d\underline{U} + P\,d\underline{V}$$

It should be apparent from the discussion in Section 4.6 that Eq. (4-32) is equally valid for all internally reversible or quasi-static processes of closed, simple systems.

Rearranging Eq. (4-32),

$$d\underline{U} = T\,d\underline{S} - P\,d\underline{V} \tag{4-53}$$

Eq. (4-53) is commonly referred to as the *combined First and Second Law of thermodynamics* for a *closed, simple system.* It represents the special case of the First Law, Eq. (3-28), applied to an infinitesimal quasi-static path of a closed, simple system.

We shall now show that Eq. (4-53) is an exact differential for a single-phase, closed, simple system and, consequently, the temperature and pressure are related to partial differential quantities.

For a single-phase simple system in a stable equilibrium state, we know that any property could be expressed as a function of any two other properties in addition to the masses or mole numbers. Thus, we could express the energy as

$$\underline{U} = f(\underline{S}, \underline{V}, N_1, \ldots, N_n) \tag{4-54}$$

Over an infinitesimal quasi-static path for this sytem, we could express Eq. (4-54) in differential form:

$$d\underline{U} = \left(\frac{\partial \underline{U}}{\partial \underline{S}}\right)_{\underline{V},N} d\underline{S} + \left(\frac{\partial \underline{U}}{\partial \underline{V}}\right)_{\underline{S},N} d\underline{V} + \sum_{i=1}^{n} \left(\frac{\partial \underline{U}}{\partial N_i}\right)_{\underline{S},\underline{V},N_j[i]} dN_i \tag{4-55}$$

The summation in Eq. (4-55) must be zero for a stable equilibrium state of a closed system (see Example 4.4). It must also be zero for a single-phase system undergoing a quasi-static process, which is simply a traverse of stable equilibrium states. Hence, Eq. (4-55) reduces to:

$$d\underline{U} = \left(\frac{\partial \underline{U}}{\partial \underline{S}}\right)_{\underline{V},N} d\underline{S} + \left(\frac{\partial \underline{U}}{\partial \underline{V}}\right)_{\underline{S},N} d\underline{V} \tag{4-56}$$

It then follows that since Eqs. (4-53) and (4-56) are valid for the infinitesimal process under consideration (i.e., a quasi-static process for a single-phase, closed, simple system), the coefficients of the differentials must be equal. Therefore,

$$T = \left(\frac{\partial \underline{U}}{\partial \underline{S}}\right)_{\underline{V},N} \tag{4-57}$$

and

$$-P = \left(\frac{\partial \underline{U}}{\partial \underline{V}}\right)_{\underline{S},N} \tag{4-58}$$

4.9 The Combined Law for Single-Phase, Open Systems

The combined law can be extended to open systems in a manner analogous to that used to develop the First Law for open systems (see Section 3.8). Consider a one-component, single-phase system bounded by the σ-surface in Figure 4.6. There are, in general, work and heat interactions that occur at the bound-

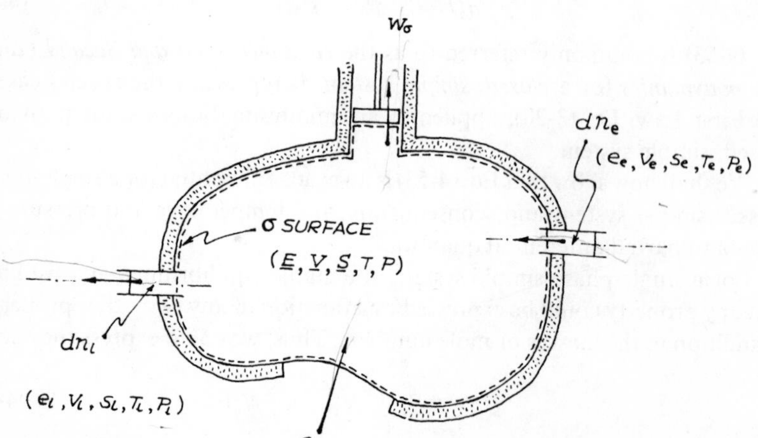

Figure 4.6

ary of this system. Now consider an infinitesimal time, dt, during which an infinitesimal quantity of material enters (dn_e), and leaves (dn_l), the system bounded by the σ-surface.

The First Law for the open system bounded by the σ-surface is

$$d\underline{E} = đQ - đW + (e_e + Pv_e)\, dn_e - (e_l + Pv_l)\, dn_l \tag{4-59}$$

where $\underline{E}$ is the energy of the open system.

If we are to apply the Second Law to this process, we must restrict our attention to processes that are reversible. For the one-component system under consideration, if mass is to be added and withdrawn reversibly, the properties of the entering and leaving mass must be identical to those of the open system. It is only under these conditions that we could reverse an infinitesimal process and obtain the original conditions. For example, with reference to Figure 4.6, if $P_e \neq P$ or $T_e \neq T$, the pressure or temperature of the open system would be altered by the addition of mass. If we then reversed the flows and extracted the same quantity of mass, the pressure and temperature of this mass would be something other than P_e and T_e. For the reversible case, we then have

$$T_e = T_\ell = T \tag{4-60}$$

$$P_e = P_\ell = P \tag{4-61}$$

$$e_e = e_\ell = E \tag{4-62}$$

$$s_e = s_\ell = S \tag{4-63}$$

$$v_e = v_\ell = V \tag{4-64}$$

and

$$dn_e - dn_\ell = dN \tag{4-65}$$

Let us now consider how the entropy of the open system changes during the process. The total entropy changes as a result of a change in mass of the open system and as a result of any heat interactions. Thus,

$$d\underline{S} = đQ_{\text{rev}}/T + s_e\, dn_e - s_\ell\, dn_\ell \tag{4-66}$$

which reduces to

$$d\underline{S} = đQ_{\text{rev}}/T + S\, dN \tag{4-67}$$

Solving Eq. (4-67) for $đQ_{\text{rev}}$ and substituting into Eq. (4-59), and using Eqs. (4-60) through (4-65), we have for the reversible process,

$$d\underline{E} = T\, d\underline{S} - đW_{\text{rev}} + (E + PV - TS)\, dN \tag{4-68}$$

Note that all of the quantities appearing in Eq. (4-68) refer to the open system bounded by the σ-surface.

If the single-phase system under consideration is not acted upon by external force fields or inertial forces, it is then a simple system and

$$đW_{\text{rev}} = P\, d\underline{V} \tag{4-69}$$

where $\underline{V}$ is the volume of the system bounded by the σ-surface. Eq. (4-68) can then be written as

$$d\underline{U} = T\, d\underline{S} - P\, d\underline{V} + (H - TS)\, dN \tag{4-70}$$

where H, the specific enthalpy, is $U + PV$. The coefficient of dN is called the *chemical potential*, μ. That is,

$$\mu \equiv H - TS = U + PV - TS \tag{4-71}$$

so that Eq. (4-70) can be written as

$$d\underline{U} = T\,d\underline{S} - P\,d\underline{V} + \mu\,dN \tag{4-72}$$

Eq. (4-72) is the form of the combined laws for a single-phase, single-component open system. It is applicable to all internally reversible (i.e., quasi-static) paths of such systems. For this case, the differential equation describing the variation of properties, Eq. (4-55), reduces to

$$d\underline{U} = \left(\frac{\partial \underline{U}}{\partial \underline{S}}\right)_{\underline{V},N} d\underline{S} + \left(\frac{\partial \underline{U}}{\partial \underline{V}}\right)_{\underline{S},N} d\underline{V} + \left(\frac{\partial \underline{U}}{\partial N}\right)_{\underline{S},\underline{V}} dN \tag{4-73}$$

Comparing Eq. (4-72) with Eq.(4-73), we have

$$T = \left(\frac{\partial \underline{U}}{\partial \underline{S}}\right)_{\underline{V},N} \tag{4-74}$$

$$P = -\left(\frac{\partial \underline{U}}{\partial \underline{V}}\right)_{\underline{S},N} \tag{4-75}$$

$$\mu = \left(\frac{\partial \underline{U}}{\partial N}\right)_{\underline{S},\underline{V}} \tag{4-76}$$

Example 4.6

For an isolated composite system consisting of a single-phase simple subsystem of pure A and a single-phase simple subsystem of a mixture of A and B separated by a rigid, diathermal membrane permeable only to A (see Figure

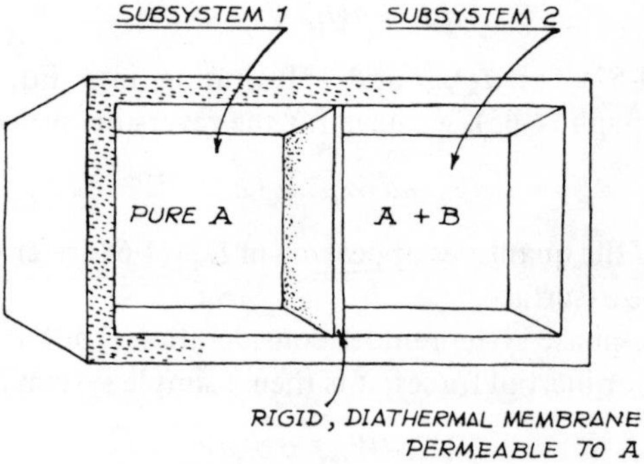

Figure 4.7

4.7), show that at equilibrium $(\partial \underline{U}/\partial \underline{S})_{\underline{V},N}$ and $(\partial \underline{U}/\partial N_A)_{\underline{S},\underline{V},N_B}$ must be the same for each subsystem. Assume that A and B do not react.

Solution

The criterion of equilibrium for the composite system is

$$d\underline{S} = d\underline{S}_1 + d\underline{S}_2 = 0 \tag{4-77}$$

The external restraints are

$$\left.\begin{aligned} d\underline{U} &= d\underline{U}_1 + d\underline{U}_2 = 0 \\ d\underline{V} &= d\underline{V}_1 + d\underline{V}_2 = 0 \\ dN_A &= dN_{A_1} + dN_{A_2} = 0 \\ dN_B &= 0 \end{aligned}\right\} \tag{4-78}$$

The internal restraints are

$$\left.\begin{aligned} d\underline{V}_1 &= d\underline{V}_2 = 0 \\ dN_B &= dN_{B_1} = 0 \end{aligned}\right\} \tag{4-79}$$

Expressing the entropy of each subsystem in the form of Eq. (4-46), substituting into Eq. (4-77), and making use of Eq. set (4-78),

$$\begin{aligned} d\underline{S} = &\left[\left(\frac{\partial \underline{S}_1}{\partial \underline{U}_1}\right)_{\underline{V}_1, N_{A1}} - \left(\frac{\partial \underline{S}_2}{\partial \underline{U}_2}\right)_{\underline{V}_2, N_{A2}, N_{B2}}\right] d\underline{U}_1 \\ &+ \left[\left(\frac{\partial \underline{S}_1}{\partial \underline{V}_1}\right)_{\underline{U}_1, N_{A1}} - \left(\frac{\partial \underline{S}_2}{\partial \underline{V}_2}\right)_{\underline{U}_2, N_{A2}, N_{B2}}\right] d\underline{V}_1 \\ &+ \left[\left(\frac{\partial \underline{S}_1}{\partial N_{A_1}}\right)_{\underline{U}_1, \underline{V}_1} - \left(\frac{\partial \underline{S}_2}{\partial N_{A_2}}\right)_{\underline{U}_2, \underline{V}_2, N_{B2}}\right] dN_{A_1} \\ &+ \left(\frac{\partial \underline{S}_2}{\partial N_{B_2}}\right)_{\underline{U}_2, \underline{V}_2, N_{A2}} dN_{B_2} = 0 \end{aligned} \tag{4-80}$$

The second and fourth terms must be zero by virtue of the internal restraints, Eq. set (4-79). For the first and third terms we can conceive of virtual processes in which $d\underline{U}_1$ and dN_{A_1} both change in either positive or negative directions. Therefore, Eq. (4-80) can be satisfied only if each coefficient of these differentials vanishes. That is,

$$\left(\frac{\partial \underline{S}_1}{\partial \underline{U}_1}\right)_{\underline{V}_1, N_{A1}} = \left(\frac{\partial \underline{S}_2}{\partial \underline{U}_2}\right)_{\underline{V}_2, N_{A2}, N_{B2}} \tag{4-81}$$

and

$$\left(\frac{\partial \underline{S}_1}{\partial N_{A_1}}\right)_{\underline{U}_1, \underline{V}_1} = \left(\frac{\partial \underline{S}_2}{\partial N_{A_2}}\right)_{\underline{U}_2, \underline{V}_2, N_{B2}} \tag{4-82}$$

Applying Eq. (4-51) to both sides of Eq. (4-82), and making use of Eq. (4-81),

$$\left(\frac{\partial \underline{U}_1}{\partial N_{A_1}}\right)_{\underline{S}_1, \underline{V}_1} = \left(\frac{\partial \underline{U}_2}{\partial N_{A_2}}\right)_{\underline{S}_2, \underline{V}_2, N_{B2}} \tag{4-83}$$

which is what we set out to prove.

For multicomponent, single-phase simple systems, Eq. (4-55) is applicable over any quasi-static path:

$$d\underline{U} = \left(\frac{\partial \underline{U}}{\partial \underline{S}}\right)_{\underline{V},N} d\underline{S} + \left(\frac{\partial \underline{U}}{\partial \underline{V}}\right)_{\underline{S},N} d\underline{V} + \sum_{i=1}^{n} \left(\frac{\partial \underline{U}}{\partial N_i}\right)_{\underline{S},\underline{V},N_j[i]} dN_i$$

From consideration of equilibria between multicomponent and one-component systems, such as illustrated in Example 4.6, it can be shown that for multicomponent systems as well as for one-component systems, Eqs. (4-57) and (4-58) are still applicable.

For multicomponent systems, the chemical potential of component i in the mixture is defined as

$$\mu_i \equiv \left(\frac{\partial \underline{U}}{\partial N_i}\right)_{\underline{S},\underline{V},N_j[i]} \tag{4-84}$$

so that the combined law for any single-phase simple system is commonly written as

$$d\underline{U} = T\,d\underline{S} - P\,d\underline{V} + \sum_{i=1}^{n} \mu_i\,dN_i \tag{4-85}$$

Since $\underline{U}$ is a function of $\underline{S}$, $\underline{V}$, $N_1, \ldots N_n$, any partial derivative of $\underline{U}$ can also be expressed as a function of these same variables. Thus, μ_i can be expressed as a function of $\underline{S}$, $\underline{V}$, $N_1, \ldots N_n$. In fact, μ_i can be expressed as a function of any $n + 2$ variables that form an independent set for the energy.

Example 4.7

Consider an isolated composite system consisting of two subsystems, each being a single phase with components A and B. If the boundary separating the two phases is movable, diathermal, and permeable to both A and B, show that at equilibrium the temperature, pressure, and chemical potential of each component must be the same in each phase.

Solution

Employing the criterion of equilibrium in the form of Eq. (4-77), expressing the entropy of each phase by Eq. (4-85), and applying the constraints of isolation (i.e., $d\underline{U} = d\underline{V} = dN_A = dN_B = 0$),

$$\begin{aligned} d\underline{S} = {} & \left(\frac{1}{T_1} - \frac{1}{T_2}\right) d\underline{U}_1 + \left(\frac{P_1}{T_1} - \frac{P_2}{T_2}\right) d\underline{V}_1 \\ & - \left(\frac{\mu_{A_1}}{T_1} - \frac{\mu_{A_2}}{T_2}\right) dN_{A_1} - \left(\frac{\mu_{B_1}}{T_1} - \frac{\mu_{B_2}}{T_2}\right) dN_{B_1} = 0 \end{aligned} \tag{4-86}$$

If the boundary between phases is movable, diathermal, and permeable, there are no internal restraints. Therefore, each of the coefficients in Eq. (4-86) must be zero because $d\underline{U}_1$, $d\underline{V}_1$, dN_{A_1}, and dN_{B_1} can each be varied independently in some virtual processes. Thus, $T_1 = T_2$, $P_1 = P_2$, $\mu_{A_1} = \mu_{A_2}$, and $\mu_{B_1} = \mu_{B_2}$.

4.10 The General Form of the Combined Law for Simple Systems

We have seen that Eq. (4-85) is applicable to all single-phase simple systems traversing quasi-static paths. We can extend this development to any simple system in the following manner. Consider a multiphase simple system divided into π subsystems each of which is a multicomponent single phase. Eq. (4-85) is applicable to each of the subsystems. That is,

$$d\underline{U}^k = T^k\, d\underline{S}^k - P^k\, d\underline{V}^k + \sum_{i=1}^{n} \mu_i^k\, dN_i^k \tag{4-87}$$

for $k = 1, 2, \ldots, \pi$.

If the composite system is in a state of equilibrium, it can be shown that T^k, P^k, and μ_i^k (where $k = 1, 2, \ldots, \pi$) must be the same for all subsystems (see Example 4.7). Therefore,

$$d\underline{U} = \sum_{k=1}^{\pi} d\underline{U}^k = T \sum_{k=1}^{\pi} d\underline{S}^k - P \sum_{k=1}^{\pi} d\underline{V}^k + \sum_{i=1}^{n} \mu_i \sum_{k=1}^{\pi} dN_i^k \tag{4-88}$$

Since $\underline{S}$, $\underline{V}$, and N_i are first order in mass, the entropy, volume, and mole numbers of the composite are the sum of these properties over all subsystems. Thus, for the composite simple system,

$$d\underline{U} = T\, d\underline{S} - P\, d\underline{V} + \sum_{i=1}^{n} \mu_i\, dN_i \tag{4-89}$$

Eq. (4-89) represents an extremely important result which we have deduced from the criterion of equilibrium. By inspection of this equation, we see that *for a composite simple system which may include many phases as subsystems, the properties $\underline{S}$, $\underline{V}$, $N_1, \ldots, N_n$ must be independently variable.* Prior to this development, we had no basis for deciding which variables could be chosen as an independent set for a *multiphase* system. We knew only that any two thermodynamic properties plus the masses of the species could be used for a *single-phase* system, but we had no generalities applicable to multiphase systems. We now have established that $\underline{S}$, $\underline{V}$, $N_1, \ldots, N_n$ represent an independent set for all multiphase simple systems. Of course, there may be other sets of independently variable properties, and we shall define criteria for determining these other sets in Chapter 5.

4.11 Reversible Work of Expansion and Compression in Flow Systems

The minimum work required to pressurize a flowing fluid or the maximum work obtainable in expanding a flowing fluid is often required in process design calculations. In such an analysis, it should be obvious that to prevent a PMM2 a reversible engine must be employed. Since all such engines have the same efficiency, an analysis need be carried out only for one particular type. We choose here, for convenience, the reciprocating engine, although the results are applicable to all other reversible engines, pumps, turbines, etc.

Consider the cyclic operation of a compressor which processes δn moles of fluid in one cycle. The upstream and downstream conditions are P_1, T_1, V_1, and P_2, T_2, V_2, respectively. The compression cycle can be divided into the following four steps, which are illustrated in Figure 4.8.

Step 1. With the cylinder containing initially N_1 moles of fluid at P_1 and T_1 (condition *A*), δn moles are introduced at constant pressure and temperature during the upstroke of the compressor.

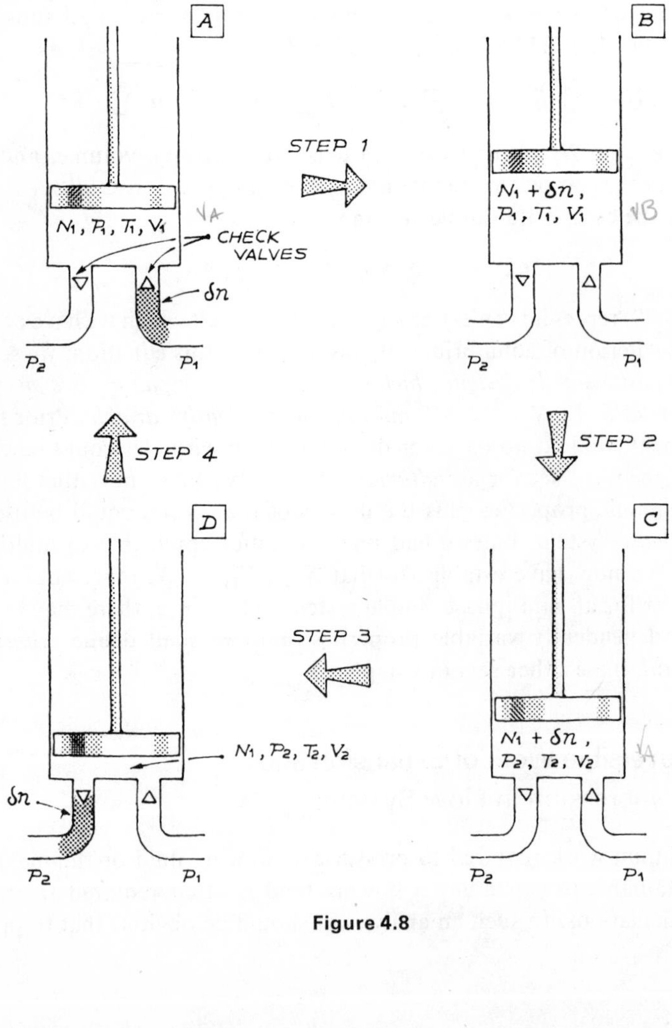

Figure 4.8

Step 2. The contents of the cylinder $(N_1 + \delta n)$ are compressed from P_1, T_1 (condition B) to P_2, T_2 (condition C).

Step 3. The outlet check valve opens and δn moles are expelled at constant pressure and temperature as the piston completes the downstroke.

Step 4. The contents of the cylinder are expanded from P_2, T_2 (condition D) to P_1, T_1 (condition A).

Let us now analyze the work done *by the fluid on the compressor* in each step. Let us choose the open system of the fluid in the cylinder at any time, and let us assume that the entire process is reversible. We will signify the work done by the fluid by δW, where δ represents a finite quantity corresponding to operation on the finite mass, δn.

specific volume!

Step 1: $\delta W_1 = \int_A^B P \, d\underline{V} = P_1(\underline{V}_B - \underline{V}_A) = P_1 V_1 \, \delta n$

Step 2: $\delta W_2 = \int_B^C P \, d\underline{V} = (N + \delta n) \int_B^C P \, dV$ *specific volume*

Step 3: $\delta W_3 = \int_C^D P \, d\underline{V} = P_2(\underline{V}_D - \underline{V}_C) = -P_2 V_2 \, \delta n$

Step 4: $\delta W_4 = \int_D^A P \, d\underline{V} = N \int_D^A P \, dV$

The total work done by the fluid is

$$\delta W = \sum \delta W_i = \delta n \left(P_1 V_1 - P_2 V_2 + \int_B^C P \, dV \right) + N\left(\int_B^C P \, dV + \int_D^A P \, dV \right)$$

$\int_B^C V \, dP = VP \big|_B^C - \int_B^C P \, dV$

(4-90)

Since

int. by parts.

$$P_1 V_1 - P_2 V_2 + \int_B^C P \, dV = -\int_B^C V \, dP \tag{4-91}$$

$$\delta W = -\delta n \int_B^C V \, dP + N\left(\int_B^C P \, dV + \int_D^A P \, dV \right) \tag{4-92}$$

In many cases, the functional dependence of P on V during compression (B to C) and expansion (D to A) is identical. For example, it was found in Example 3.4 that for quasi-static, frictionless (i.e., reversible) adiabatic compression of an ideal gas, the temperature-pressure relationship is given by Eq. (3-48), or

$$\frac{T_2}{T_1} = \left(\frac{P_2}{P_1} \right)^{R/(C_v+R)} = \left(\frac{P_2}{P_1} \right)^{(1-1/\kappa)} \tag{4-93}$$

Since $T_2/T_1 = P_2 V_2 / P_1 V_1$, Eq. (4-93) becomes

$P_1 \underline{V}_1 = n_1 RT_1$

$P_1 V_1 = RT_1$

$$\frac{V_2}{V_1} = \left(\frac{P_2}{P_1} \right)^{-(1/\kappa)} \tag{4-94}$$

or

$P_2 V_2 = RT_2$

$$PV^\kappa = \text{constant} \tag{4-95}$$

In cases such as this,

$$\int_B^C P\,dV = \int_A^D P\,dV \tag{4-96}$$

because $V_C = V_D = V_2$ and $V_B = V_A = V_1$. Therefore, Eq. (4-92) simplifies to

$$\frac{\delta W}{\delta n} = -\int_B^C V\,dP = -\int_{P_1}^{P_2} V\,dP \tag{4-97}$$

The work done by the compressor on the fluid is, of course, the negative of Eq. (4-97). This quantity, however, is not equal to the *total* work done on the fluid in passing from the upstream to the downstream lines. The total work is

$$-\delta n \int_{P_1}^{P_2} P\,dV$$

(where P is the same function of V as is found in step 2), as can be verified by choosing the closed systems of δn moles and calculating the work for this system through each step. The difference between the total work done on the fluid and the work done by the compressor is $P_1V_1 - P_2V_2$, which is the work done on the δn moles by the fluid behind it in the upstream line and the fluid ahead of it in the downstream line.

It should be noted that the limits of integration in Eq. (4-97) will vary from one cycle to the next unless the process under consideration is operated under steady-state conditions.

Example 4.8

A 1-ft³ tank that initially contains air at 1 atm and 70°F is to be evacuated by pumping out the contents, as illustrated in Figure 4.9. The tank contents are maintained at 70°F throughout the operation by heat transfer through

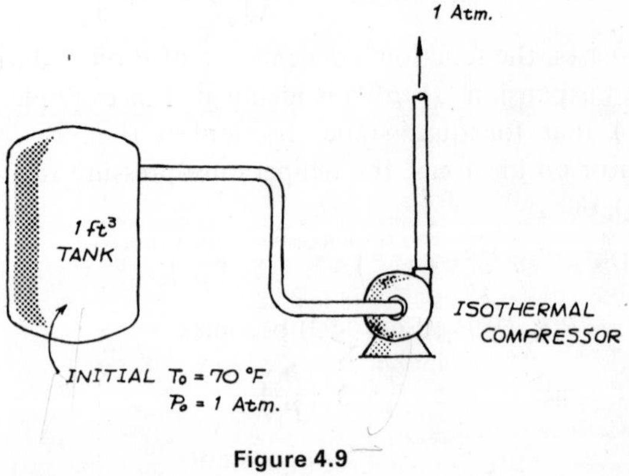

Figure 4.9

the walls. The compressor discharges the air at 1 atm and is operated isothermally at 70°F. What is the total work done by the compressor?

Assume that the compressor operates reversibly and that air is an ideal gas.

Solution

If it is assumed that the amount of air processed during one cycle of the compressor is small in relation to the total quantity of air expelled, then the work of any one cycle can be treated as a differential and the properties within the tank can be assumed to vary smoothly with the amount of air expelled. Thus, for a differential amount processed, the work done by the gas on the compressor is given by Eq. (4-97)

$$đW = -dn \int_{P_t}^{P_a} V\, dP \tag{4-98}$$

where P_a is the discharge pressure (1 atm) and P_t is the tank pressure during the cycle under consideration.

Introducing the ideal gas law and integrating,

$$đW = RT \ln\left(\frac{P_t}{P_a}\right) dn \tag{4-99}$$

Since the amount of gas processed, dn, is equal to the decrease in gas in the tank, $-dN$, and since

$$dN = \frac{V_t}{RT}\, dP_t \tag{4-100}$$

Eq. (4-99) becomes

$$đW = -V_t \ln\left(\frac{P_t}{P_a}\right) dP_t \tag{4-101}$$

Integrating between $P_t = 1$ atm and $P_t = 0$,

$$W = -V_t[P_t \ln P_t - P_t]_1^0 = -V_t \cdot (1 \text{ atm}) \tag{4-102}$$

Thus, the work done by the compressor is 2117 ft-lb.

PROBLEMS

4.1. Prove that the following processes are irreversible.

(a) A closed, thermally insulated cylinder is separated into two parts by a piston fixed with a locking mechanism. Air is contained in one half and the other half is evacuated. Unlock the piston.

(b) A viscous fluid in a pot is rotated slowly with a stirrer. Remove the stirrer and allow the fluid to come to rest.

(c) Two closed, rigid systems A and B are at different temperatures. Place these two systems in contact through a diathermal wall.

4.2. Most of us have seen, in novelty stores, small glass birds that appear to enjoy taking an endless series of drinks of water from a glass as illustrated in Figure P4.2. If we looked closely, we could see that these toys are simply two hollow glass bulbs separated by a tube and mounted on a swivel joint. The lower bulb is partially filled with a volatile liquid such as ethyl ether.

Figure P4.2

Ether boils at room temperature and some of the vapor condenses in the upper bulb which is kept cold by evaporation of water on a wick placed over the bulb. In the drinking step the wick is moistened.

The ether condensed in the upper bulb is prevented from returning to the lower bulb by the upward flow of vapor. Also, some liquid is pumped to the upper bulb by a "coffee percolator action." When the upper bulb is nearly full, the bird's center of gravity shifts, the bird swings to a horizontal position (and "drinks"), thereby allowing the ether to flow back down the tube to the lower bulb. The bird then becomes upright and the cycle is repeated.

We would like to connect this bird to some mechanism and allow it to pump water from a lower level. If we assume that *all* the water pumped is eventually evaporated from the bird's head, what is the greatest height from which it can pump water for steady-state operation? Assume normal ambient conditions and neglect any heat losses or other irreversibilities.

4.3. A Hilsh vortex tube for sale commercially is fed with air at 27°C and 5 atm into a tangential slot near the center (point A in Figure P4.3). Stream B leaves from the left end at 1 atm and −23°C; stream C leaves at the right end at 1 atm and 37°C.

These two streams then act as a sink and source for a Carnot engine and both streams leave the engine at 1 atm and T_D.

(a) What is T_D?

(b) If stream A is 1 g-mol/sec, what are the flow rates of streams B and C?

(c) What is the Carnot work output per mole of stream A?

(d) What is the entropy change of the overall process per mole of A?

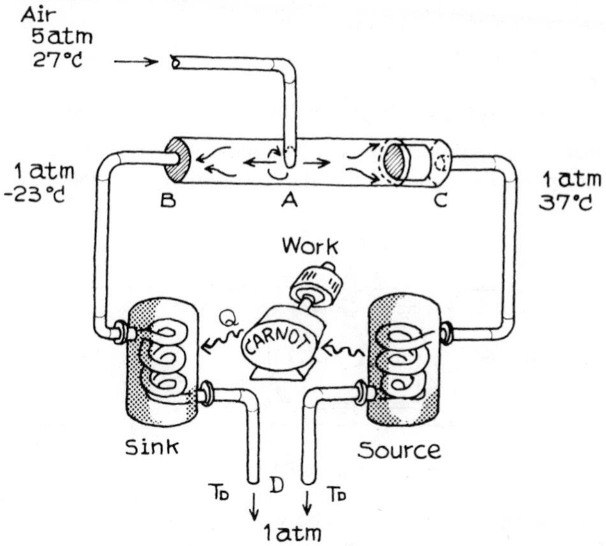

Figure P4.3

(e) What is the entropy change in the Hilsh tube per mole of A?

(f) What is the maximum work that one could obtain *by any process* per mole of A if all heat were rejected or absorbed from an isothermal reservoir at T_D?

Assume ideal gases which have a constant heat capacity, $C_p = 7$ Btu/lb-mol °R.

4.4. A bar of aluminum is placed in a large bath of ice and water (Figure P4.4). Current is passed through the bar until at steady state there is a power dissipation of 1000 watts. A thermocouple on the surface of the aluminum reads 640°F. Film boiling is assumed to occur at the interface with subsequent

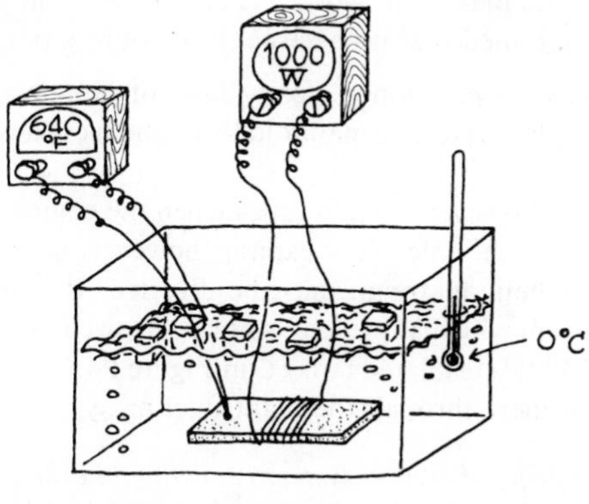

Figure P4.4

collapse of the steam bubbles. What is the entropy change in the bar, water, and universe during two minutes of operation at steady state for this highly *irreversible* operation? Assume that there is ice remaining at the end of the two minutes.

Also show that the electrical conductivity is always a positive number.

4.5. A well-insulated container having a volume of 6 ft^3 is divided internally into two equal parts by a rigid, adiabatic partition (Figure P4.5). Through this

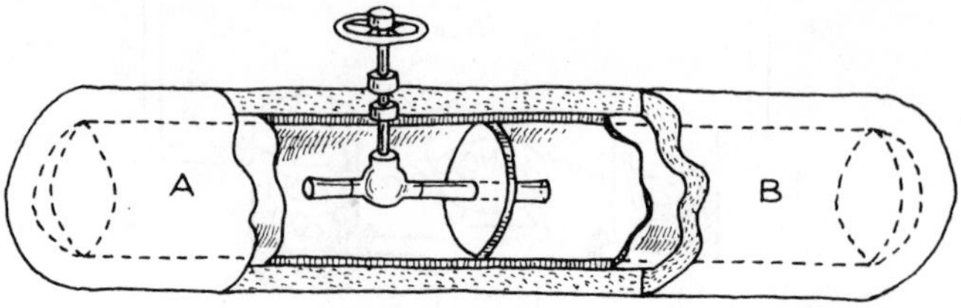

Figure P4.5

partition is a small connecting tube with a valve accessible to the outside. Side A contains nitrogen gas at 100°F and 20 psia, and side B also contains nitrogen but at 100 psia and 200°F. Both gases are ideal with constant heat capacities of $C_p = 7$ Btu/lb-mol°R.

(a) The valve noted above is quickly opened and pressures are rapidly equalized on both sides. No conduction of heat occurs. What are the temperatures and pressures on both sides? What is the entropy change of the universe?

(b) The valve is left open and eventually the temperatures become equal on both sides. What is this final temperature and pressure? In this case, what is the entropy change of the universe?

(c) Repeat (a) and (b) if oxygen were substituted for nitrogen in side B. In part (b) when making any calculations, assume that concentrations also have equalized.

(d) Repeat (a) and (b) if side A originally contained N^{14} and side B, N^{15}.

(e) What is the maximum work one could obtain in parts (a) through (d) if it were assumed that no external heat sources or sinks were available?

4.6. Under ordinary operation a steady flow of helium gas equal to 0.1 lb/sec passes from a large storage manifold through an expansion engine as shown in Figure P4.6.

There are, however, certain times when the engine must be shut down for short intervals; the inlet flow cannot, however, be decreased at such times and thus the helium stream must be diverted. It is proposed to employ an adjacent system which at present is not being used. This latter system consists of a large (150 ft^3) insulated tank (C in Figure P4.6) with a safety valve venting to the atmosphere through the plant ducting system.

The emergency diverting system is operated in the following manner. If the expansion engine must be shut down (or operates improperly), valve B

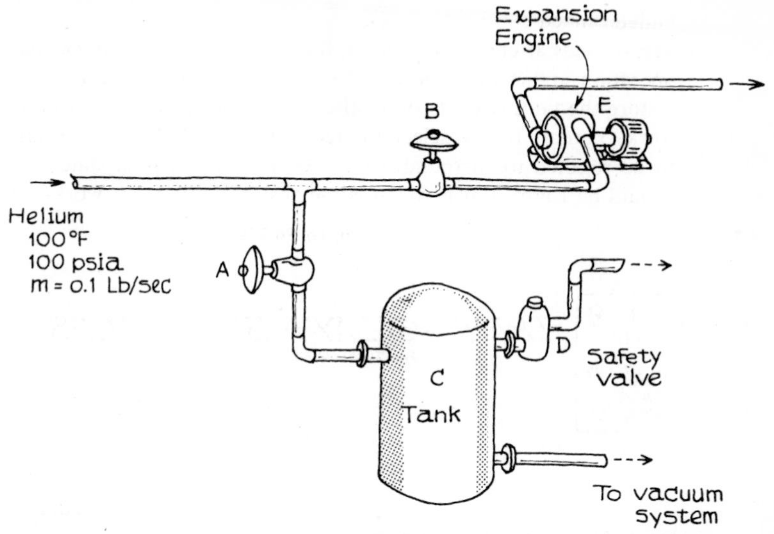

Figure P4.6

is shut and A is opened, letting helium into tank C. When the pressure in C rises to 40 psia, safety valve D operates, venting gas.

Initially tank C is evacuated to a very low pressure. When gas enters this tank, we will assume that it is well-mixed but has negligible heat transfer with the tank walls. Also assume that helium is an ideal gas with a constant $C_v =$ 3 cal/g-mol°K.

(a) How long will flow enter tank C before the pressure increases to 40 psia and the safety valve opens?

(b) If the safety valve on tank C should operate, we would like to maintain a constant mass of gas in this tank equal to the mass at the time the safety tripped. The flow into the tank is, as noted above, 0.1 lb/sec; the flow out of the safety valve may be expressed as:

$$\text{flow, lb/sec} = \frac{KAP}{T}$$

where K is a constant, A the valve throat area, and P, T the pressure and temperature in tank C. To keep the tank mass constant, the valve flow area, A, will be varied. What are the pressure and temperature in tank C ten sec after the safety valve opens? What is the variation of A with time necessary to keep the mass of gas in tank C constant?

(c) What is the total entropy change of the gas, the surroundings, and the universe during the time between the opening of valve A and just prior to the opening of relief valve D?

(d) What is the total entropy change of the gas, the surroundings, and the universe during the time between the opening of relief valve D and a time 10 sec later? The vented gas mixes with an infinite amount of air exterior to the tank and cools to 100°F. Leave the entropy change of mixing as an undetermined constant.

(e) Ten seconds after venting begins, the original expander comes back into operation so that valves A and D are shut and B is opened. It is desired to restore the tank C to its original evacuated state with the least possible work. You are free to select any technique you deem feasible; all heat is to be rejected to surroundings at 100°F, and the final state of the gas should be 1 atm, 100°F. What is the absolute minimum work required?

4.7. **Figure P4.7(a)**

TRANS-GALAXY-SPACELINES

— A DIVISION OF MITY INC.—

June 13, 1984

Welcome New Summer Employee:

As your first job with our famous old spaceline, we want you to answer a small technical problem that has arisen in our new line of boosters known to you as the Super Dodos.

We have a number of small jets on this series that will be used in attitude control in space. These jets will be powered by low-pressure nitrogen gas heated by an arc at the jet nozzle. Your problem deals with the nitrogen storage system.

The nitrogen is stored in a large well-insulated, 14.5 ft^3 sphere at 1 atm pressure. At take-off the temperature is 40°F. The mass rate of flow of N_2 will be constant and be equal to 0.01 lb/sec. Since the pressure inside the sphere must always be kept at 1 atm, a heater will be used inside the sphere.

(a) Under these conditions, what will be the temperature of the N_2 in the sphere, the instantaneous rate of heat flow to the heater, and the total heat required after 10 sec of operation?

(b) The energy requirement of the heater may be difficult to meet. Rocky Jones, our boy genius, has made a suggestion that we would like you to evaluate. Rocky wishes to take the nitrogen from the sphere that leaves at a pressure of 1 atm and sphere-gas temperature and put it through a "black-box" to extract the *maximum* work possible before sending it to the arc-jets. The interior metal of the booster may act as a heat source or sink at 40°F, and the final nitrogen pressure must be no less than 0.02 atm as fed to the jets. The work from the black-box will be degraded to heat and fed back to supply the sphere-heat energy.

What is the instantaneous additional heat required (Btu/sec) at the start of operation? Will the black-box supply *all* of the sphere-heat required—and, if so, for how long?

Data:

Nitrogen is an ideal gas with a $C_p = 7$ Btu/lb-mol°R, independent of temperature. Assume that there are no thermal lags in the system in obtaining the work in part (b).

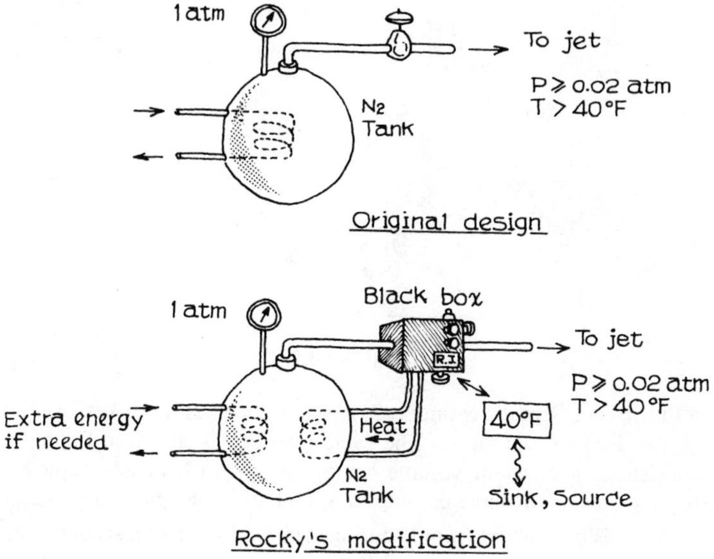

Figure P4.7

4.8. A missile environmental tank is to be used for space experiments at low pressures. Initially, the tank contains one lb-mol of air at 70°F and 1 atm. Evacuation is to be carried out using a small vacuum pump that has a volumetric *intake* capacity of 80 ft^3/min,* irrespective of intake pressure.

(a) How long will it take before the tank pressure has dropped to 0.1 atm if the pump operates reversibly and isothermally and if the air in the tank remains at 70°F by heat transfer with the environment which is also at 70°F? What is the total work required?

(b) Can this pump be used to evacuate completely the air in the tank? If so, state the time and energy required. If not, state clearly which law of thermodynamics is violated if a perfect vacuum is to be attained.

(c) What are the entropy changes in the gas, surroundings, and in the universe for (a) and (b)?

4.9. The reversible engine E shown in Figure P4.9 receives heat from a reversible Carnot engine *C-I* at temperature T_0. The engine E consists of a piston enclosing 0.01 lb-mol of an ideal gas of constant heat capacity, $C_p =$

* Strokes per minute times displacement per stroke = 80 ft^3/min.

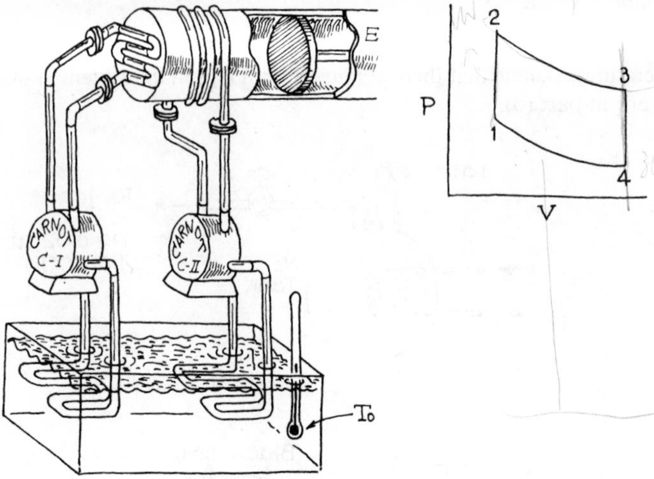

Figure P4.9

5 Btu/lb-mol°R in the cylinder. Since the engine has no valves, no flow of gas occurs into or out of E. The cycle operated is shown in Figure P4.9 and consists of a constant volume heating from 1 to 2, an isentropic expansion to 3, a constant volume cooling to 4, and an isentropic compression to 1.

Heat is discharged from E during step 3-4 by use of a second differential-size reversible Carnot engine *C-II*, to the same heat reservoir at T_0.

(a) How much work is done by the engine E per cycle?

(b) How much work is done by the Carnot engines as E completes one cycle?

Data:

$T_0 = 0°C$, $T_2 = 800°C$, $T_4 = 30°C$, $P_2 = 5$ atm, $\underline{V}_4 = 4\underline{V}_1$

4.10. A square metal bar has an axial temperature distribution given by $T(°R) = 500(1 + x^2)$. The variable x is the distance, in feet, from one end. The temperature is uniform at any cross-section. The sides of the bar are 1 in. and the length 1 ft, and the density and heat capacity are 576 lb/ft^3 and $0.0002T$ Btu/lb°R.

(a) What would be the change in internal energy of the bar if it were cooled to 0°C?

(b) What is the maximum amount of work that could have been obtained in (a), given a large heat sink at 0°F?

(c) What is the thermal efficiency of step (b)?

4.11. In several parts of the world, there exist ocean currents of differing temperatures that come into contact. An example of this takes place off the coast of Southern Africa, where the warm Agulhas and the cold Benguela currents meet at Cape Point, near Cape Town (see Figure P4.11). It has been proposed that work may be obtained from these ocean currents if a heat engine could operate between the warm current as a source and the cold current as a sink.

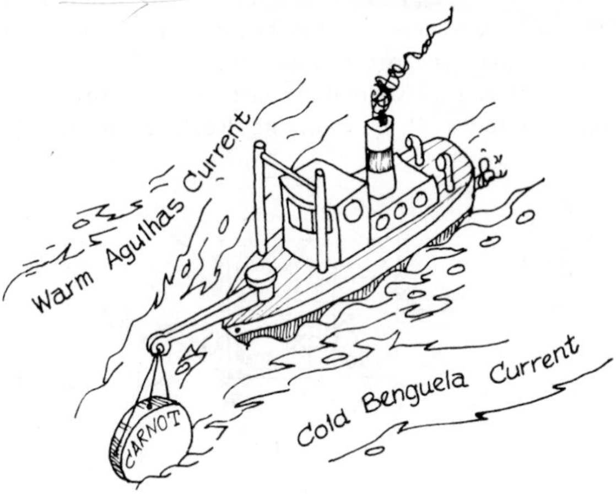

Figure P4.11

Assume that the system may be simplified into two channels of water in contact and flowing cocurrently. Furthermore, assume that no mixing occurs between the streams and that the effect of heat conduction between the streams is negligible.

(a) Derive a general expression for the maximum amount of power that could be obtained from the system. Express your result in terms of temperatures, flow rates, and physical properties of the streams.

(b) Repeat (a) if the two streams flow countercurrent rather than cocurrent. What is the pinch temperature and where does it occur?

(c) It has been estimated that the Benguela current is 16 million meters^3/sec, and its initial temperature is 40°F. The Agulhas current is 20 million meters^3/sec and its initial temperature is 80°F. Calculate and compare the power obtainable from these two currents when they flow cocurrently and also countercurrently.

4.12. Three identical 10-lb blocks of metal are available. Each has a constant heat capacity of 0.5 Btu/lb°R. Initially, they are at temperatures of 200, 300, and 400°K.

(a) If no net heat or work interactions are allowed between the blocks and the environment, what is the maximum temperature that one can reach in any one of the blocks?

(b) Under similar restraints as in (a), what is the minimum temperature that one could attain?

(c) Derive an equation to allow one to predict the maximum and minimum temperature attainable for one block, starting with n blocks each at T_k with masses m_k and with heat capacities C_k.

4.13. Our highly efficient plant has a nitrogen stream at 2.5 atm and 90°F which is presently vented to the atmosphere. The management would like to use this stream to satisfy some of the heating or cooling requirements of other processes. Our old friend, Rocky Jones, boy genius, has devised a black-box that will produce equal amounts of a hot stream at 440°F and a cold stream at −260°F and, thus, satisfy simultaneously some heating *and* cooling requirements (see Figure P4.13). Furthermore, Rocky claims that his device will be

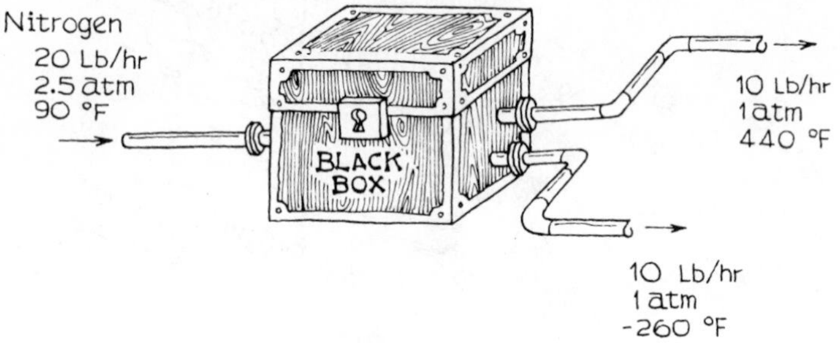

Figure P4.13

self-sustaining because no additional heat or work need be supplied to the device.

One of our new engineers, Barry Goldfinder, claims that he has a black-box device that will produce equal amounts of a hot stream of 500°F and a cold stream of −320°F. His device will also be self-sustaining.

(a) Are either (or both) of these devices possible? Explain.

(b) Present a process that will satisfy the requirements of Rocky's black-box. Calculate all heat and work interactions for each device used in the process and indicate how the devices should be arranged so that no additional heat or work need be supplied by the environment.

4.14. We wish to fill some nitrogen cylinders from a supply manifold that is only at 70°F and 1 atm (Figure P4.14). Nitrogen will be fed to the suction side of a

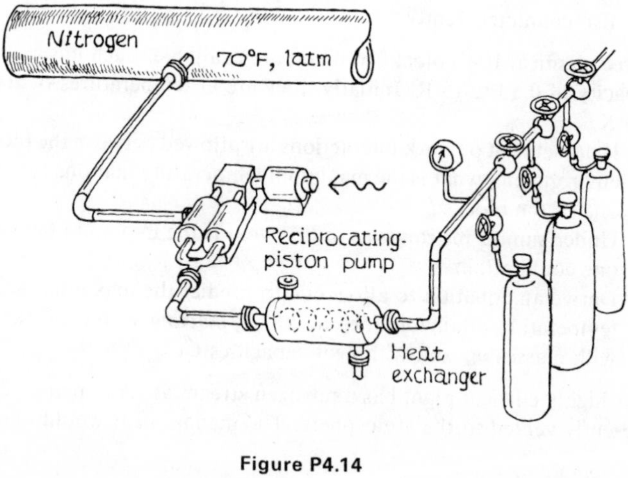

Figure P4.14

compressor and discharged through a small heat exchanger to the cylinders.

Initially, each cylinder contains nitrogen at 29.4 psia. The final pressure after filling will be 1500 psia. Assume that the nitrogen entering a cylinder is at 70°F and, because of the high heat capacity of the cylinder, the gas and the cylinder remain essentially at 70°F during the filling. Each cylinder has a volume of 1 ft³.

(a) With no intercooling in the compressor, what is the minimum horsepower motor required to fill one cylinder in 3 minutes? What is the work? Assume that the compressor operates isothermally and reversibly.

(b) Repeat (a) if the compressor operates adiabatically and reversibly.

Nitrogen behaves as an ideal gas with a constant $C_p = 7$ Btu/lb-mol°R and $C_v = 5$ Btu/lb-mol°R.

4.15. A reciprocating engine with a frictionless piston is equipped with the usual intake and exhaust valves (Figure P4.15). Clearance volume (i.e., the volume

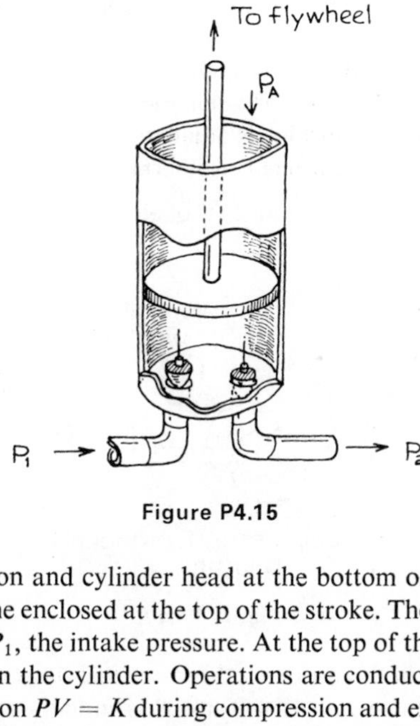

Figure P4.15

between the piston and cylinder head at the bottom of the stroke) is exactly half of the volume enclosed at the top of the stroke. The exhaust pressure, P_2, is two thirds of P_1, the intake pressure. At the top of the stroke 0.02 lb of gas is enclosed within the cylinder. Operations are conducted so that the process follows the relation $PV = K$ during compression and expansion, where P and V are pressure and specific volume, respectively, and K is a constant.

(a) If the engine is operating perfectly, how much net work is done per lb of gas passing through the engine? Assume a constant pressure, P_a, on the back side of the piston.

(b) Repeat (a), assuming zero pressure on the back side of the piston.

(c) Repeat (a), assuming that the engine is rebuilt so that the clearance is now zero. This was achieved by lengthening the stroke so that 0.02 lb of gas, at P_2, was again present at the top of the stroke.

(d) The valve mechanism fails and the engine is rebuilt as in part (c). It operates as follows: the intake valve remains open to the end of the stroke, at which time this valve closes and the exhaust valve opens. The pressure falls to the exhaust pressure, whereupon the piston returns with the exhaust valve open to the end of this stroke. The valves now reverse and the cycle is repeated. The relation $PV = K$ is still obeyed throughout the operation. How much net work is done per lb of gas passing through the engine?

4.16. It is proposed that a device be constructed to operate as follows: An evacuated tank of 30 ft^3 is attached to the exhaust of an air turbine-driven grinding wheel. Air at atmospheric pressure and 70°F will be allowed to enter the turbine inlet, the pressure drop between the atmosphere and the tank serving to operate the turbine-driven grinder. When the pressure in the originally evacuated tank has risen to atmospheric, the turbine, of course, stops.

Estimate (as closely as possible), clearly stating all assumptions, how many horsepower-hours of work may be obtained up to the time the pressure in the tank has risen to atmospheric. The ideal gas law may be assumed. Determine also the air temperature in the tank at the end of the filling process.

Consider four cases:

(a) Tank is adiabatic, turbine is isothermal.
(b) Tank is adiabatic, turbine is adiabatic.
(c) Tank is isothermal, turbine is adiabatic.
(d) Tank is isothermal, turbine is isothermal.

4.17. A tank A (see Figure P4.17) holds 1 ft^3 of air at 100 psia and 500°R. It is desired

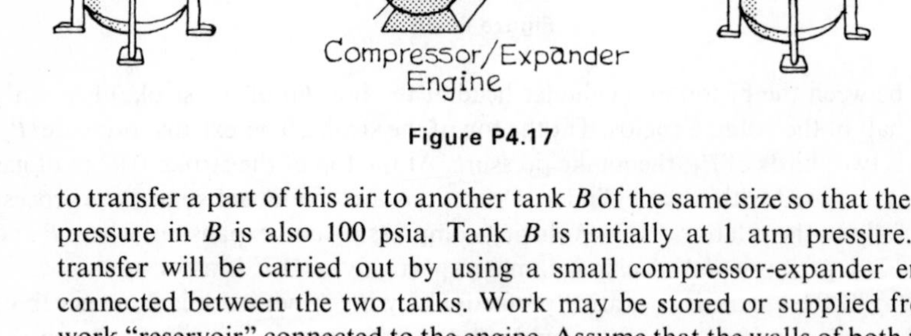

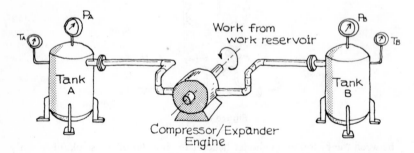

Figure P4.17

to transfer a part of this air to another tank B of the same size so that the final pressure in B is also 100 psia. Tank B is initially at 1 atm pressure. The transfer will be carried out by using a small compressor-expander engine connected between the two tanks. Work may be stored or supplied from a work "reservoir" connected to the engine. Assume that the walls of both tanks and connecting lines are rigid and adiabatic, and that the operation of the engine is reversible and adiabatic.

(a) Set up a differential equation relating the rate of energy transfer to or from the reservoir (energy/mol of gas flowing) to the initial conditions and to the pressures in the two tanks at any time t; what is the rate just as the engine is started?

(b) Set up a differential equation relating the temperature and pressure in tank B at any time t.

(c) Clearly indicate how these equations (and any others needed) may be solved in order to yield the total energy gain or depletion from the reservoir.

Air may be assumed to be an ideal gas with a constant heat capacity of $C_p = 7$ Btu/lb-mol°R,

4.18. Two very large gas storage spheres each contain air at 30 psia (Figure P4.18).

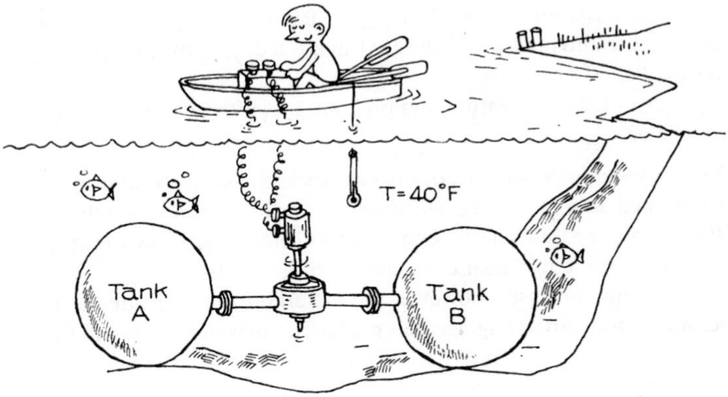

Figure P4.18

They are connected across a small reversible compressor with stroke volume negligible in relation to the volume of the gas spheres. The tanks, connecting lines, and compressor are immersed in a constant temperature bath at 40°F.

The compressor will take suction from one sphere, compress the gas, and discharge to the other sphere. The gas is at 40°F at all times.

What is the work requirement to compress the gas in one sphere to 50 psia? What is the heat interaction to the constant temperature bath?

Assume that air is an ideal gas with $C_p = 7$ Btu/lb-mol°R.

4.19. A patent on a new adiabatic device known as the "vacuum energizer" (Figure P4.19) may be filed soon.

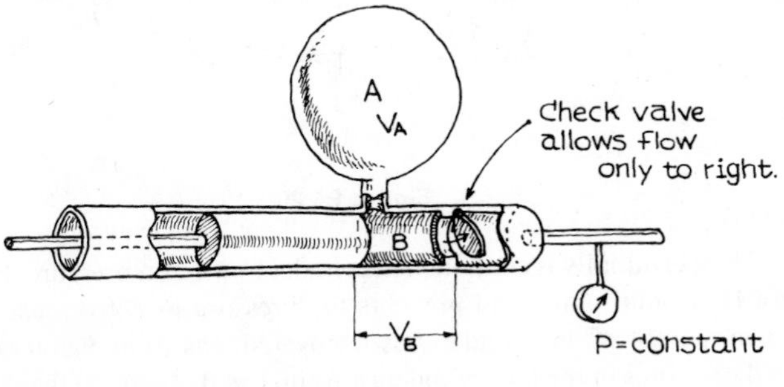

Figure P4.19

An ideal gas is stored in bulb A at some initial pressure P_{A_0}. A long solid piston starts at the right end of the cylinder B and is pulled horizontally to the left. It uncovers the port so that gas from A can pass into cylinder B until pressures in A and B are equalized. The piston then pushes the gas out through the check valve at constant pressure. The piston is then drawn back, creating a vacuum in B until the port is again uncovered, and the cycle is repeated. The gas is air and is ideal; no heat transfer occurs from the gas to the walls or between compartments A and B.

If $P_{A_0} = 10$ atm, $T_{A_0} = 500°$R, $\underline{V}_A = 0.8$ ft^3, and $\underline{V}_B = 0.2$ ft^3, what is the temperature and pressure of the gas leaving B during the first cycle and during the nth cycle?

What is the temperature and pressure in A after the first cycle and after the nth cycle?

4.20. In some experiments with a newly synthesized gaseous compound, it has been determined that it is very sensitive and will explode violently at pressures slightly above atmospheric. Other experiments have shown that the material may be unstable in a heat exchanger with hot walls.

For some purposes, however, the material must be heated to a high temperature. Our project engineer has proposed the system shown in Figure P4.20.

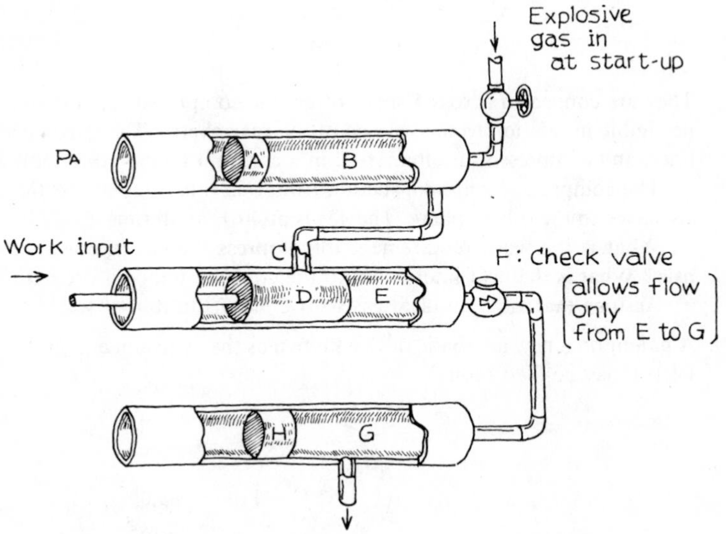

Figure P4.20

The gas initially is contained in cylinder B at 1 atm pressure. Piston A floats in a horizontal plane and prevents the pressure in B from exceeding 1 atm. When the port C in cylinder E is uncovered, gas from B enters the cylinder until pressures in the two cylinders are equalized. Piston D then moves, covers port C, and pushes the gas out slowly through check valve F into cylinder G

which in turn is fitted with a floating piston H to prevent the pressure from exceeding 1 atm. The piston D then is withdrawn to uncover port C and the cycle is repeated. Valve F only allows flow from E to G, but not the reverse. The pistons have essentially no friction.

It is claimed that the gas in G is hotter than in B. Any desired temperature may be attained by cascading such devices in series (i.e., feeding gas out from G through J into another tube similar to the outlet of B).

What is the temperature of the gas after *two* cascaded steps?

Data:

Cylinders are 3 in. inside diameter.
Assume no clearance in cylinder E.
Initial mass of gas in B is 1 lb-mol and has a temperature of 100°F.
Assume ideal gases with a constant $C_p = 10$ Btu/lb-mol°R.

4.21. Prove that in an adiabatic, reversible multistage compressor operation, the total work requirement is minimized when the compression ratio for each stage is the same, provided also that between each stage the gas is cooled to the initial feed temperature.

4.22. In a laboratory experiment you must evacuate as rapidly as possible some process equipment. Initially, the equipment contains nitrogen gas at 40°F and 14.69 psia pressure. The volume is 5 ft³.

You have at your disposal a typical rotary vacuum pump. It is rated to pump 1 ft³/min of actual volume.* There is one complication. The equipment is adiabatic except for a small heater that must be left on at all times and liberates 2 Btu/min at a constant rate.

(a) What is the pressure inside the equipment after 2 minutes of pumping?
(b) What is the temperature of the nitrogen remaining in the process equipment after 2 minutes of pumping?
(c) One of your friends kindly offers to lend you his vacuum pump (which is exactly like yours) and he says that you can reach the same pressure in half the time if you connect it up in parallel with yours. Do you agree?
(d) What is the lowest pressure you can pump on your system using one pump? Explain.
(e) If you wish only to pump down to 4 psia, what minimum horsepower motor do you recommend? The vacuum pump is neither adiabatic nor isothermal, but some tests indicate that the relationship

$$PV^{1.3} = \text{constant}$$

is applicable in the compressor. Base your *minimum* estimate on the assumption that the pump is reversible.

For N_2, $C_p = 7$ Btu/lb-mol°R, independent of temperature.

4.23. We have been requested to design a small system that will allow a manufacturer to test helium vacuum pumps both under transient conditions and also for long steady-state periods. The system we now propose is shown in Figure P4.23.

* Note: This pump may be considered to be a reciprocating one, with no clearance, and with an intake stroke of $\delta \underline{V}$ ft³ where $(\delta \underline{V})$(RPM) = 1 ft³/min.

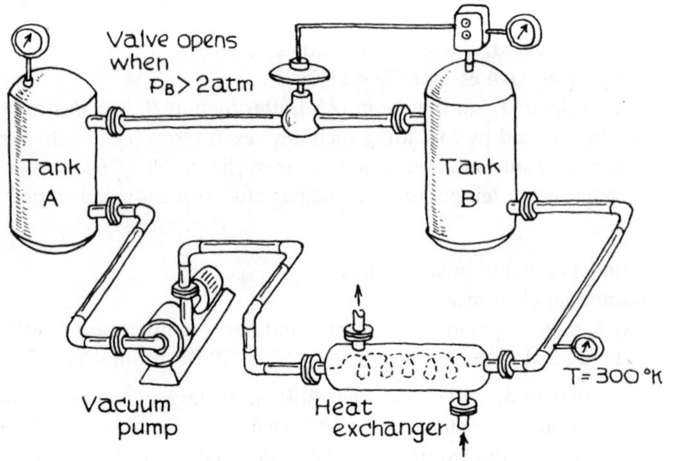

Figure P4.23

Tanks A and B are thin-walled, well-insulated tanks, each with a volume of 100 liters. The vacuum pump will take suction from A and discharge into B. Between the pump and B there is a heat exchanger to cool the gases to a constant temperature of 300°K.

Initially, both tanks A and B are charged with helium at 1 atm and 300°K. The vacuum pump is started and gas flows through the heat exchanger into B until the pressure in B reaches 2 atm. At this time valve C automatically opens and vents gas from B to A, always keeping the pressure in B at 2 atm.

(a) Just at the time the pressure in B reaches 2 atm, what is the temperature in both A and B and what is the pressure in A?
(b) After the system has been running for a long time, with P_B = constant = 2 atm, what is the temperature in B and what is the pressure in A?
(c) What is the change in entropy *of all the helium gas* in the system from the start until the pressure in B just reaches 2 atm?
(d) Can you suggest any improvements in the system to simplify it and still test the vacuum pumps over transient and longtime periods?
(e) What is the work supplied to the vacuum pump from the start of the test until the pressure in B just reaches 2 atm?
(f) What is the heat load in the heat exchanger during the same period as described in (e)?

Data:

Helium is an ideal gas with a constant C_p = 5 cal/g-mol°K.

The volume of the connecting lines and pump is negligible compared to the volume of spheres A and B.

Since the vacuum pumps to be tested are adiabatic and reversible, we can consider that each element of helium gas flowing through the pump obeys the relationship PV^{κ} = constant.

4.24. Our old friend, Rocky Jones, has been tinkering in the laboratory and has produced an interesting device that he calls an "integral pulsed shock tube." As yet, we do not see a large commercial market. In fact, we are not sure what it really does! Can you analyze this device and answer the brief questions given later?

A long, insulated tube (Figure P4.24) is divided into chambers of equal

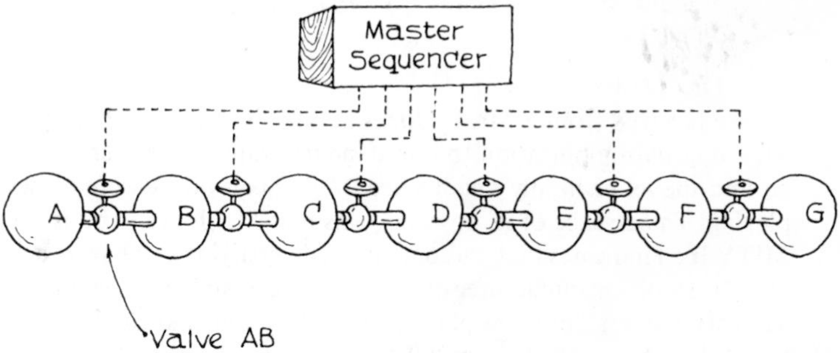

Figure P4.24

volume by rigid, adiabatic partitions. Each partition has a fast-operating valve to allow flow of gas between compartments when the valve is open. The operation of this device is a follows: Compartment A is initially filled with helium gas at 40°F and 64 atm. The remaining compartments are evacuated to zero pressure. All valves are closed. At time zero, valve AB is quickly opened and gas flows from A to B. Just when the *pressures* are equalized in A and B, valve AB is shut and valve BC opened. As before, just at the time the pressures in B and C are equal, valve BC is shut and valve CD opened. This sequence is continued until gas enters the end compartment G.

There is no axial heat conduction and no heat conduction across the valves or partitions. Helium is an ideal gas with a $C_p = 5$ cal/g-mol°K and $C_v =$ 3 cal/g-mol°K.

(a) When valve FG has just closed, what is the temperature and the pressure in compartment G?

(b) When the sequencing is completed, all valves are opened and the pressure allowed to equalize in all compartments. Slow axial conduction also equalizes the temperature in all compartments. What is the equilibrium temperature and pressure?

(c) Rocky is uncertain how much work it requires to prepare his integral pulsed shock tube for firing. Estimate the minimum work per mole of helium initially in compartment A.

In this calculation, assume that all compartments of the shock tube initially contain helium at 1 atm, 40°F. Any helium that is removed must be pumped into a large pipeline containing helium at a constant pressure of 2 atm, 40°F. Also, any helium used to charge compartment A to 64 atm is to be taken from the same pipeline.

If, in your calculations, any heat transfer occurs, assume that you have a large heat sink or source at 40°F.

(d) Can you suggest any use for Rocky's new device?

4.25. **Figure P4.25(a)**

The Establishment

— Manufacturer of fine, old processes —

Dear Dr. Reader:

We received from a Mr. R. Jones an interesting letter that poses a somewhat unusual modification to our standard water condenser design. I won't go into the letter in any detail since, to me, parts of it were somewhat perplexing. The essence of it, however, was that Mr. Jones (who is president of MITY Inc.) indicates that his company is diversifying and has recently bought out Carnoco, a manufacturer of diminutive reversible heat engines. Mr. Jones suggests that in all our new plants, wherein we cool and condense stream S-13 (see Figure A, attached) from 300°F to 80°F, we eliminate the heat exchangers entirely and, instead, substitute a set of Carnoco engines. If I understand Mr. Jones correctly, his process diagram would appear as in Figure B, attached. Mr. Jones claims that we could reduce the cooling water requirement (and still maintain the outlet cooling water temperature at 60°F), obtain a great deal of "free" work, and eliminate completely the cost of our heat exchanger.

Using the following data, would you please evaluate Mr. Jones' proposal and indicate the maximum cost you think that we could afford to pay for the purchase and installation of the Carnoco reversible engines.

cooling water cost:	3.0¢/1000 gal
value of any work output:	1.0¢/KWHr
present cost of heat exchanger, installed:	$30,000
operational time per year	7200 hr
investment charges	15%

Your attention to this request should be given priority.

Sincerely yours,

Godfrey Cross

Godfrey Cross

Boss

Figure A: Present Design

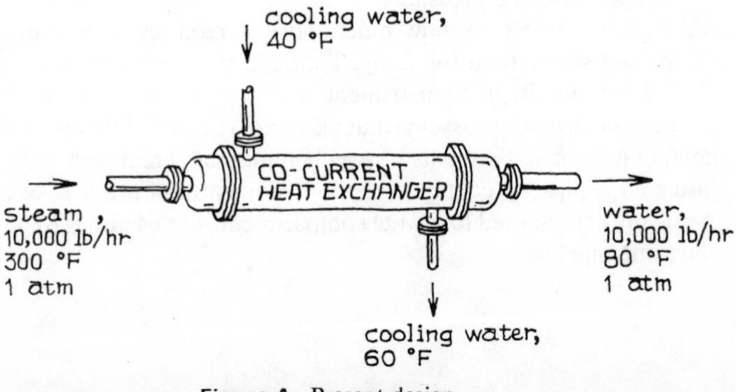

Figure A. Present design

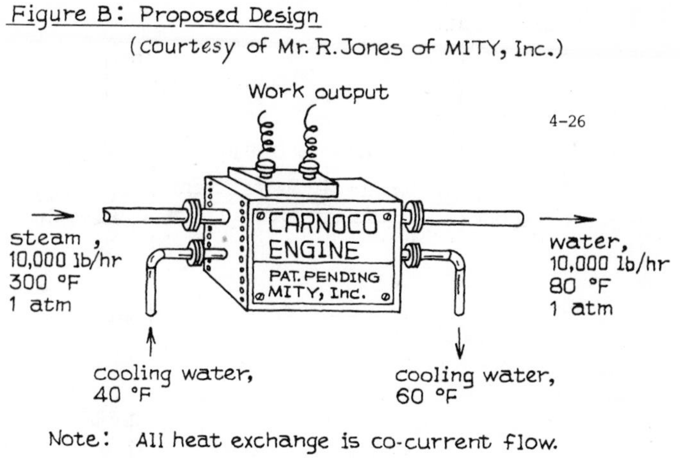

Figure B. Proposed design

4.26. A bar of pure copper is bonded to two large isothermal reservoirs at 212°F and 32°F, respectively (Figure P4.26). The bar is square in cross section, 0.25 in. on a side, and well-insulated, except at the ends. It is 1 ft long. The thermal conductivity of pure copper is 220 Btu/hr-ft°F.

This irreversible process must result in an entropy increase of the universe. After steady state is attained:

(a) Determine the *rate* at which this entropy increases.

(b) If the center six inches of the bar were cut out and replaced with an alloy with a conductivity of 120 Btu/hr-ft°F, how would the entropy production rate be affected?

(c) The 212°F bath is removed and, in its place, an electric resistance heater is inserted. We measure a current of 1 ampere and a resistance of 5.05 ohms in the heater. What is the steady-state entropy production rate of the universe now?

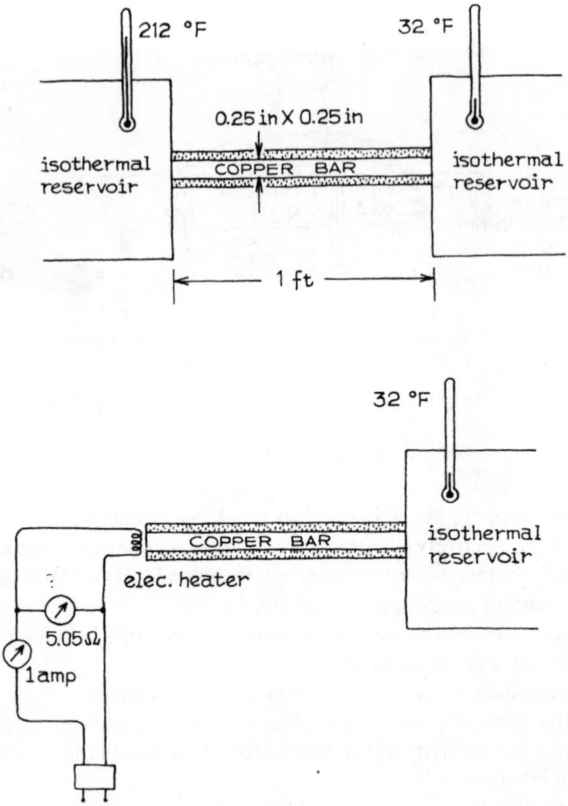
212 °F
32 °F
0.25 in X 0.25 in
isothermal reservoir
COPPER BAR
isothermal reservoir
1 ft
32 °F
COPPER BAR
isothermal reservoir
elec. heater
?
5.05 Ω
1 amp

Figure P4.26

Thermodynamic Relations for Simple Systems 5

In this chapter we shall treat simple systems and develop general property-relationships that are consequences of the postulates and laws discussed in the previous chapters. These relationships, which must be obeyed by all simple systems, should be distinguished from the *physical* property relationships. Physical properties refer to relationships that are specific for each substance and are not dictated by the laws of classical thermodynamics: they must be obtained by experimentation, extrapolation of existing data for similar materials, or development from molecular models using statistical mechanics. Determination of physical properties is discussed in subsequent chapters and Appendices.

In this chapter we shall restrict our attention to simple systems. For such systems we have seen in Section 4.10 that the properties $\underset{\sim}{S}, \underset{\sim}{V}, N_1, \ldots, N_n$ form an independently variable set for the energy, $\underset{\sim}{U}$. We shall explore the nature of this relationship and the many equivalent forms that this relationship assumes.

5.1 The Fundamental Equation

As will be seen presently, the relationship

$$\underset{\sim}{U} = f_U(\underset{\sim}{S}, \underset{\sim}{V}, N_1, \ldots, N_n) \tag{5-1}$$

completely describes all of the stable equilibrium states of a simple system. By solving Eq. (5-1) explicitly for $\underset{\sim}{S}$, an alternative form is obtained:

$$\underset{\sim}{S} = f_S(\underset{\sim}{U}, \underset{\sim}{V}, N_1, \ldots, N_n) \tag{5-2}$$

Either relationship is called the *Fundamental Equation:* Eq. (5-1) is termed the *energy representation* and Eq. (5-2) the *entropy representation.*

The Fundamental Equation represents a surface in $n + 3$ dimensional space. The points on this surface represent stable equilibrium states of the simple system. Quasi-static processes can be represented by a curve on this surface. Processes that are not quasi-static are not identified with points on this surface. (Recall that derived properties such as $\underset{\sim}{U}$ and $\underset{\sim}{S}$ are not defined for nonequilibrium states.)

The $n + 2$ first-order partial derivatives of the Fundamental Equation are the traces of the tangent planes taken parallel to one of the coordinate axes. The significance of these tangent plane traces can be seen by expressing the Fundamental Equation in differential form. For the energy representation, Eq. (5-1),

$$d\underset{\sim}{U} = \left(\frac{\partial f_U}{\partial \underset{\sim}{S}}\right)_{\underset{\sim}{V},N} d\underset{\sim}{S} + \left(\frac{\partial f_U}{\partial \underset{\sim}{V}}\right)_{\underset{\sim}{S},N} d\underset{\sim}{V} + \sum_{i=1}^{n}\left(\frac{\partial f_U}{\partial N_i}\right)_{\underset{\sim}{S},\underset{\sim}{V},N_j[i]} dN_i \tag{5-3}$$

where subscript $N_j[i]$ indicates that all N_j except $j = i$ are held constant in the differentiation. If we compare Eq. (5-3) to the combined law for a simple system (see Section 4.10), namely

$$d\underset{\sim}{U} = T\,d\underset{\sim}{S} - P\,d\underset{\sim}{V} + \sum_{i=1}^{n} \mu_i\,dN_i \tag{5-4}$$

it is clear that

$$\left(\frac{\partial f_U}{\partial \underset{\sim}{S}}\right)_{\underset{\sim}{V},N} = T = g_T(\underset{\sim}{S}, \underset{\sim}{V}, N_1, \ldots, N_n) \tag{5-5}$$

$$-\left(\frac{\partial f_U}{\partial \underset{\sim}{V}}\right)_{\underset{\sim}{S},N} = P = g_P(\underset{\sim}{S}, \underset{\sim}{V}, N_1, \ldots, N_n) \tag{5-6}$$

and

$$\left(\frac{\partial f_U}{\partial N_i}\right)_{\underset{\sim}{S},\underset{\sim}{V},N_j[i]} = \mu_i = g_i(\underset{\sim}{S}, \underset{\sim}{V}, N_1, \ldots, N_n) \tag{5-7}$$

where the functions g_T, g_P, and g_i could be obtained directly from the Fundamental Equation if it were available. Eqs. (5-5) through (5-7) are called the *equations of state.* As shown in Sections 5.2 and 5.3, only two of the three equations of state of a pure substance are independent.

The second-order partial derivatives of the Fundamental Equation are also related to quantities that can be measured experimentally. For example, for a *pure material*, there are four second-order partial derivatives at constant mass or moles:

$$\frac{\partial^2 \underline{U}}{\partial \underline{S}^2} = \frac{\partial}{\partial \underline{S}}\left[\left(\frac{\partial \underline{U}}{\partial \underline{S}}\right)_{\underline{V},N}\right]_{\underline{V},N} = \left(\frac{\partial T}{\partial \underline{S}}\right)_{\underline{V},N} \tag{5-8}$$

$$\frac{\partial^2 \underline{U}}{\partial \underline{V}^2} = \frac{\partial}{\partial \underline{V}}\left[\left(\frac{\partial \underline{U}}{\partial \underline{V}}\right)_{\underline{S},N}\right]_{\underline{S},N} = -\left(\frac{\partial P}{\partial \underline{V}}\right)_{\underline{S},N} \tag{5-9}$$

$$\frac{\partial^2 \underline{U}}{\partial \underline{S}\,\partial \underline{V}} = \frac{\partial}{\partial \underline{S}}\left[\left(\frac{\partial \underline{U}}{\partial \underline{V}}\right)_{\underline{S},N}\right]_{\underline{V},N} = -\left(\frac{\partial P}{\partial \underline{S}}\right)_{\underline{V},N} \tag{5-10}$$

$$\frac{\partial^2 \underline{U}}{\partial \underline{V}\,\partial \underline{S}} = \frac{\partial}{\partial \underline{V}}\left[\left(\frac{\partial \underline{U}}{\partial \underline{S}}\right)_{\underline{V},N}\right]_{\underline{S},N} = \left(\frac{\partial T}{\partial \underline{V}}\right)_{\underline{S},N} \tag{5-11}$$

Of these four derivatives, only three are independent because the last two are related by the reciprocity theorem of Maxwell. That is, for any smoothly varying function for which $X = f(Y, Z)$, the order of differentiation is immaterial and, therefore,

$$\frac{\partial}{\partial Z}\left[\left(\frac{\partial X}{\partial Y}\right)_Z\right]_Y = \frac{\partial}{\partial Y}\left[\left(\frac{\partial X}{\partial Z}\right)_Y\right]_Z \tag{5-12}$$

Thus,

$$-\left(\frac{\partial P}{\partial \underline{S}}\right)_{\underline{V},N} = \left(\frac{\partial T}{\partial \underline{V}}\right)_{\underline{S},N} \tag{5-13}$$

The Fundamental Equation for an n-component mixture is:

$$\underline{U} = f_U(\underline{S}, \underline{V}, N_1, \ldots, N_n) \tag{5-14}$$

We shall show that for an *n-component mixture* there are $n + 1$ first-order and $(n + 2)(n + 1)/2$ second-order *independent* partial derivatives of the Fundamental Equation. These derivatives are particularly important because they form a basis for all other partial derivatives involving thermodynamic properties. That is, any partial derivative can be expressed in terms of an independent set of first-order and second-order derivatives of any form of the Fundamental Equation. Proof of this statement for a pure material is given in Section 6.3.

It should be clear that Fundamental Equations would be of great use if they were available. Unfortunately, Fundamental Equations of very few materials have been developed to date. The problem is that the complete form of the Fundamental Equation is not specified by classical thermodynamics; each substance has its own peculiarities that are reflected in different functionalities of the Fundamental Equation. Thus, there is no single Fundamental Equation governing the properties of all materials.

The postulates of classical thermodynamics place some restrictions on the form of the Fundamental Equation. Let us examine the differential form of the Fundamental Equation in the energy representation, Eq. (5-3). Since $\underline{U}$ must be first order in mass or mole number, we can apply Euler's theorem (see Appendix C) to obtain the integrated form of Eq. (5-3):

$$\underline{U} = \left(\frac{\partial f_U}{\partial \underline{S}}\right)_{\underline{V},N} \underline{S} + \left(\frac{\partial f_U}{\partial \underline{V}}\right)_{\underline{S},N} \underline{V} + \sum_{i=1}^{n} \left(\frac{\partial f_U}{\partial N_i}\right)_{\underline{S},\underline{V},N_j[i]} N_i \qquad (5\text{-}15)$$

Eq. (5-15) is a linear partial differential equation of the first order. Therefore, the solution must be of the form

$$\underline{U} = x\left[g\left(\frac{y}{x}, \frac{z}{x}, \ldots\right)\right] \qquad (5\text{-}16)$$

where $x, y, z, \ldots$ can be $\underline{S}, \underline{V}, N_1, \ldots, N_n$ or any permutation of these variables. For a one-component system, it is most convenient to choose $x = N$, $y = \underline{S}$, and $z = \underline{V}$; we then obtain

$$\underline{U} = N\left[g\left(\frac{\underline{S}}{N}, \frac{\underline{V}}{N}\right)\right] \qquad (5\text{-}17)$$

or, since $\underline{U} = NU$,

$$U = g(S, V) \qquad (5\text{-}18)$$

The only other requirements our prior developments place on the form of the Fundamental Equation are that $\underline{U}$ should be a single-valued function of $\underline{S}$, $\underline{V}$, and N (see Postulate III), and that $(\partial f_U/\partial \underline{S})_{\underline{V},N} = T$ should be non-negative.

5.2 Intensive and Extensive Properties

At this point, let us digress to a special case of Postulate I. In Eq. (5-18), we see that for a one-component system, only two properties, S and V, are required to obtain the specific energy of a system. This is in no way a violation of Postulate I; by delineating the *specific* properties (expressed in terms of unit mass or mole number), we have determined the "intensity" of the system but not the "extent" of the system. To completely specify the system (e.g., so that it can be reproduced by others), we must specify the mass of the system in addition to S and V.

The variables that express intensity of the system are zero order in mass and are called *intensive variables*. Variables that relate to extent of the system are first order in mass and are called *extensive variables*.

We shall now prove that for a single-phase simple system of n components, any intensive property can be defined by the values of $n + 1$ other intensive properties.[1] Let us call $b, c_1, c_2, \ldots, c_{n+1}$ *intensive* properties of a single-

[1] The proof is restricted to a single-phase system in order to allow us to choose any $n + 2$ variables as an independent set. The proof can be extended to specific cases of composite simple systems provided that the $n + 2$ variables chosen form an independently variable set.

phase simple system containing n components. In general, we can express b as a function of $n + 2$ other properties according to Postulate I. Let us choose these $n + 2$ as $c_1, c_2, \ldots, c_{n+1}$, and the total moles (or mass) N. Thus,

$$db = \left(\frac{\partial b}{\partial c_1}\right)_{c_j[1],N} dc_1 + \ldots \left(\frac{\partial b}{\partial c_{n+1}}\right)_{c_j[n+1],N} dc_{n+1} + \left(\frac{\partial b}{\partial N}\right)_{c_1,\ldots,c_{n+1}} dN \qquad (5\text{-}19)$$

Integrating Eq. (5-19) by using Euler's theorem (see Appendix C), we have

$$\left(\frac{\partial b}{\partial N}\right)_{c_1,\ldots,c_{n+1}} N = 0 \qquad (5\text{-}20)$$

Since N can be nonzero, $(\partial b/\partial N)_{c_1,\ldots,c_{n+1}}$ must be zero. Therefore, Eq. (5-19) reduces to a function of $n + 1$ intensive variables. Of course, these $n + 1$ intensive variables must be independent so that we clearly cannot use all of the n mole fractions $x_1, \ldots, x_n$. We could, however, use $n - 1$ mole fractions in addition to two other intensive variables to obtain the required $n + 1$.

Note that this result is valid because we limited the original set of $n + 2$ variables to include only one extensive variable; if we had included two extensive variables in the original $n + 2$ set, no partial derivative would have to be zero. Thus, we could state as a corollary to Postulate I: *For a single-phase simple system, the change of any intensive variable can be expressed as a function of any* n + 1 *other independent intensive variables.* We shall use this corollary frequently in Chapter 8 in dealing with the properties of mixtures.

At this point, a word of caution is in order when dealing with intensive and extensive variables in partial derivatives. For example, if for a pure material we express U as a function of S and V, then using Eqs. (5-19) and (5-20), we find $(\partial U/\partial N)_{S,V} = 0$. But $(\partial \underline{U}/\partial N)_{\underline{S},\underline{V}}$ is not zero; from Eq. (5-7) applied to a pure material,

$$\left(\frac{\partial \underline{U}}{\partial N}\right)_{\underline{S},\underline{V}} = \mu \qquad (5\text{-}21)$$

Since $\underline{U} = NU$, we also have

$$\left(\frac{\partial \underline{U}}{\partial N}\right)_{\underline{S},\underline{V}} = U + N\left(\frac{\partial U}{\partial N}\right)_{\underline{S},\underline{V}} = \mu \qquad (5\text{-}22)$$

Since U is not equal to μ, $(\partial U/\partial N)_{\underline{S},\underline{V}}$ is not equal to zero. Thus, each of the three derivatives, $(\partial U/\partial N)_{S,V}$, $(\partial \underline{U}/\partial N)_{\underline{S},\underline{V}}$, and $(\partial U/\partial N)_{\underline{S},\underline{V}}$ have different connotations. The first represents the change in the specific energy as we add more material while maintaining constant specific entropy and specific volume. Since we are holding two intensive variables constant during the process, all other intensive variables (e.g., T, P, etc.) for the pure material must remain unchanged. The only way to conduct the process is to enlarge the system in direct proportion to the added mass. The second and third cases, however, represent changes in the total and specific energy during a process in which we maintain constant total entropy and total volume. Since we are adding mass to the system, the only way to keep total entropy and total volume constant is to change the specific entropy and specific volume (e.g., by varying T and P during the addition of mass). Thus, the specific energy changes as the state of the system is varied. The total energy changes because both the specific energy and mass vary.

Example 5.1

In the entropy representation, the Fundamental Equation for a monatomic ideal gas is:

$$\underline{S} = N\left[S^0 + R\ln\left\{\left(\frac{U}{U^0}\right)^{3/2}\left(\frac{V}{V^0}\right)\right\}\right] \tag{5-22a}$$

where S^0, U^0, and V^0 are constants representing values in a reference or base state. From the energy representation in the form of Eq. (5-1), determine the three equations of state in the form of Eqs. (5-5) through (5-7).

Solution

Solving Eq. (5-22a) explicitly for U yields

$$U = U^0\left(\frac{V^0}{V}\right)^{2/3} e^{(2/3)(S-S^0)/R} \tag{5-22b}$$

and, thus,

$$\underline{U} = NU^0\left(\frac{V^0}{V}\right)^{2/3} e^{(2/3)(S-S^0)/R} \tag{5-22c}$$

The equations of state can be found directly by partial differentiation of Eq. (5-22b) or (5-22c)

$$T = \left(\frac{\partial \underline{U}}{\partial \underline{S}}\right)_{\underline{V},N} = \left(\frac{\partial U}{\partial S}\right)_V = \frac{2}{3}\left(\frac{U^0}{R}\right)\left(\frac{V^0}{V}\right)^{2/3} e^{(2/3)(S-S^0)/R} \tag{5-22d}$$

$$-P = \left(\frac{\partial \underline{U}}{\partial \underline{V}}\right)_{\underline{S},N} = \left(\frac{\partial U}{\partial V}\right)_S = -\frac{2}{3}\,\frac{U^0(V^0)^{2/3}}{V^{5/3}}\,e^{(2/3)(S-S^0)/R} \tag{5-22e}$$

$$\begin{aligned}\mu = \left(\frac{\partial \underline{U}}{\partial N}\right)_{\underline{S},\underline{V}} &= \frac{\partial}{\partial N}\left[NU^0\left(\frac{V^0}{V}\right)^{2/3} e^{(2/3)(S-S^0)/R}\right]_{\underline{S},\underline{V}} \\ &= U^0\left(\frac{V^0}{V}\right)^{2/3} e^{(2/3)(S-S^0)/R}\left(\frac{5}{3} - \frac{2}{3}\frac{S}{R}\right)\end{aligned} \tag{5-22f}$$

5.3 Reconstruction of the Fundamental Equation

We shall now consider the minimum information necessary for reconstruction of a Fundamental Equation for a given material. Although we are not necessarily interested in knowing the Fundamental Equation (in general, they are much too cumbersome to use), we are interested in knowing what information content is equivalent to the Fundamental Equation. Once we have this, we need not look for any more because all other information of thermodynamic interest can be obtained therefrom.

As shown in Section 5.1, if the Fundamental Equation were known, the properties T, P, μ_i, could be determined by partial differentiation as expressed in the equations of state in Eqs. (5-5) through (5-7). Alternatively, the Fundamental Equation can be recovered, if all of the equations of state were known, by substituting these equations into Eq. (5-15) or the more common form:

$$\underline{U} = T\underline{S} - P\underline{V} + \sum_{i=1}^{n} \mu_i N_i \tag{5-23}$$

As shown in Section 5.2, the $n + 2$ intensive variables, T, P, μ_i, which are expressed explicitly by the equations of state, are not all independently variable. Any one of these variables can be expressed in differential form in terms of the other $n + 1$ variables and, upon integration, an expression between the $n + 2$ variables can be determined to within an arbitrary constant. It thus follows that only $n + 1$ equations of state are necessary to determine the Fundamental Equation to within an arbitrary constant.[2]

As an example, let us consider a pure material. The chemical potential, μ, can be expressed as a function of T and P:

$$d\mu = \left(\frac{\partial \mu}{\partial T}\right)_P dT + \left(\frac{\partial \mu}{\partial P}\right)_T dP \tag{5-24}$$

In Section 5.4, it is shown that $(\partial \mu/\partial T)_P = -S$ and $(\partial \mu/\partial P)_T = V$. If Eqs. (5-5) and (5-6) were known, we could solve these simultaneously to obtain

$$S = g\,(T, P) \tag{5-25}$$

and

$$V = g'(T, P) \tag{5-26}$$

These equations are usually available, although not necessarily in analytical form, but in many cases the entropy is known only to within an arbitrary constant. Let us assume that we have available Eqs. (5-26) and (5-27):

$$S = S^0 + g''(T, P) \tag{5-27}$$

where S^0 is an arbitrary constant. Substitution into Eq. (5-24) and integration from an arbitrary reference state for which $T = T^0$, $P = P^0$, $V = V^0$, and $\mu = \mu^0$ leads to

$$\mu = \mu^0 - S^0(T - T^0) - \int_{T^0}^{T} g''(T, P)\, dT + \int_{P^0}^{P} g'(T, P)\, dP \tag{5-28}$$

which is the desired relationship, μ as a function of T and P. Although this

[2] Since only differences in $\underline{U}$ have physical significance, we need not specify the energy representation more definitively than to within an arbitrary constant.

equation contains two arbitrary constants (S^0 and μ^0), when Eqs. (5-5), (5-6), and (5-28) are substituted into Eq. (5-23) in order to obtain the Fundamental Equation, these two arbitrary constants appear as a sum; in particular, we would find

$$U = f_U(S, V) + (T^0S^0 - P^0V^0 + \mu^0) \tag{5-29}$$

or

$$U^0 = T^0S^0 - P^0V^0 + \mu^0 \tag{5-30}$$

Of the three arbitrary constants in Eq. (5-30) (i.e., U^0, S^0, and μ^0), it is clear that only two can be chosen independently. Thus, we can set base values of U and S at the reference state for which $T = T^0$, $P = P^0$, and $V = V^0$, but having done so, the base value for μ is uniquely specified by Eq. (5-30).

Example 5.2

Express the chemical potential of a monatomic ideal gas as a function of T and P starting with the Fundamental Equation given in Example 5.1.

Solution

To use Eq. (5-24) in the form

$$d\mu = -S\,dT + V\,dP \tag{5-30a}$$

we must first find the functions $g(T, P)$ or $g''(T, P)$ and $g'(T, P)$ in Eqs. (5-25) through (5-27). Using the two equations of state found in Example 5.1, namely, Eqs. (5-22d) and (5-22e), we can solve simultaneously for these functions. First eliminating S by dividing Eq. (5-22d) by Eq. (5-22e) we obtain

$$\frac{T}{P} = \frac{V}{R} \tag{5-30b}$$

or

$$V = g' = \frac{RT}{P} \tag{5-30c}$$

Then, solving Eq. (5-22d) for S and using Eq. (5-30c) to eliminate V yields:

$$S = g = S^0 + \frac{3}{2}R\ln\left[\frac{3(RT)^{5/3}}{2U^0(PV^0)^{2/3}}\right] \tag{5-30d}$$

or

$$g'' = \frac{3}{2}R\ln\left[\frac{3(RT)^{5/3}}{2U^0(PV^0)^{2/3}}\right] \tag{5-30e}$$

Substituting Eqs. (5-30c) and (5-30e) into Eq. (5-28) and integrating, we obtain the desired result:

$$\mu = \mu^0 - S^0(T - T^0) - \frac{3}{2}R\int_{T_0}^{T}\left\{\ln\left[\frac{3(RT)^{5/3}}{2U^0(PV^0)^{2/3}}\right]\right\}_{P=P^0} dT + R\int_{P_0}^{P}\left(\frac{T}{P}\right)_{T=T} dP \tag{5-30f}$$

where we have chosen to maintain pressure constant at P^0 in the first integration and temperature constant at T in the second.

Integrating and simplifying, we find

$$\mu - \mu^0 = -S^0(T - T^0) - \frac{3}{2}RT \ln\left[\frac{3(RT)^{5/3}}{2U^0(P^0V^0)^{2/3}}\right] + \frac{3}{2}RT^0 \ln\left[\frac{3(RT^0)^{5/3}}{2U^0(P^0V^0)^{2/3}}\right] + \frac{5}{2}R(T - T^0) + RT \ln\frac{P}{P^0} \quad (5\text{-}30g)$$

The last term can be combined with the second term on the right-hand side since

$$RT \ln\frac{P}{P^0} = -\frac{3}{2}RT \ln\left(\frac{P^0}{P}\right)^{2/3} \quad (5\text{-}30h)$$

The third term on the right-hand side can be shown to be zero in the following manner. From Eq. (5-22d) evaluated at the base state for which $T = T^0$, $U = U^0$, $S = S^0$, and $V = V^0$,

$$T^0 = \frac{2}{3}\frac{U^0}{R} \quad (5\text{-}30i)$$

or

$$U^0 = \frac{3}{2}RT^0 \quad (5\text{-}30j)$$

Since $RT^0 = P^0V^0$, we have

$$\ln\left[\frac{3(RT^0)^{5/3}}{2U^0(P^0V^0)^{2/3}}\right] = \ln\left[\frac{3(RT^0)^{5/3}}{2(\frac{3}{2}RT^0)(RT^0)^{2/3}}\right] = \ln 1 = 0$$

With these simplifications, Eq. (5-30g) reduces to

$$\mu - \mu^0 = -S^0(T - T^0) + \frac{5}{2}R(T - T^0) - \frac{3}{2}RT \ln\left[\frac{3(RT)^{5/3}}{2U^0(PV^0)^{2/3}}\right] \quad (5\text{-}30k)$$

5.4 Legendre Transformations

In the energy representation of the Fundamental Equation, Eq. (5-1), the properties $\underline{S}$, $\underline{V}$, $N_1, \ldots, N_n$ are treated as independent variables. This is not always a convenient set of independent parameters. For example, since temperature can be measured much more conveniently than entropy, we might like to use T, $\underline{V}$, $N_1, \ldots, N_n$ as the independent variables. For a single-phase simple system, we can always express a property such as $\underline{U}$ in terms of $n + 2$ other properties such as T, $\underline{V}$, $N_1, \ldots, N_n$.
Thus, we know that a function f exists such that

$$\underline{U} = f(T, \underline{V}, N_1, \ldots, N_n) \quad (5\text{-}31)$$

and

$$d\underline{U} = \left(\frac{\partial f}{\partial T}\right)_{\underline{V},N} dT + \left(\frac{\partial f}{\partial \underline{V}}\right)_{T,N} d\underline{V} + \sum_{i=1}^{n} \left(\frac{\partial f}{\partial N_i}\right)_{T,\underline{V},N_j[i]} dN_i \qquad (5\text{-}32)$$

Given the Fundamental Equation, the function of Eq. (5-31) can be found by differentiating Eq. (5-1) to obtain Eq. (5-5),

$$T = g_T(\underline{S}, \underline{V}, N_1, \ldots, N_n)$$

and then solving Eqs. (5-1) and (5-5) simultaneously in order to eliminate $\underline{S}$. The result is an equation of the form

$$\underline{U} = f(T, \underline{V}, N_1, \ldots, N_n) = f\left[\left(\frac{\partial f_U}{\partial \underline{S}}\right)_{\underline{V},N}, \underline{V}, N_1, \ldots, N_n\right] \qquad (5\text{-}33)$$

Although Eq. (5-33) is of the form desired [i.e., Eq. (5-31)], the information content of Eq. (5-33) is less than that of the Fundamental Equation. Eq. (5-33) is a partial differential equation that can be integrated to yield the Fundamental Equation only to within an arbitrary function of integration.

We must now ask whether or not there are other functions with the same information content as that of the Fundamental Equation but with independent variables other than $\underline{S}, \underline{V}, N_1, \ldots, N_n$. The answer is that there are such functions if we are willing to restrict ourselves to a set of independent variables in which we choose only one from each of the following pairs: $\underline{S}, T$; $\underline{V}, P$; N_i, μ_i. (These pairs of variables are usually referred to as *conjugate coordinates.*) That is, we can replace any variable, z_i, of the set $\underline{S}, \underline{V}, N_1, \ldots,$ N_n by the slope of the tangent in the $\underline{U} - z_i$ plane, $(\partial \underline{U}/\partial z_i)$.

The proof of this statement is a rather lengthy exercise in line geometry known as the Legendre transformation. We will omit the proof here and only outline the logic and the mechanics.

Let us first demonstrate the logic by treating the two-dimensional case. Given a curve described by $y = g(z)$, we would like to describe this curve in terms of the variable $\xi = dy/dz$ instead of the variable z. It can be shown that if y is a unique function of z, the curve can be completely described by specifying the family of line tangents to this curve at each value of z (see Figure 5.1). A single line tangent is of the form

$$y = \xi z + \psi \qquad (5\text{-}34)$$

where ψ is the intercept of the tangent on the y-axis. The complete family of line tangents can be expressed as $\psi = f(\xi)$. Thus, this relation has the same information content as the original relation, $y = g(z)$.

The relation $\psi = f(\xi)$ is called the Legendre transform, or simply, the transform of y with respect to z. The mechanics of obtaining $\psi = f(\xi)$ from $y = g(z)$ are extremely simple. Starting with

$$y = g(z) \qquad (5\text{-}35)$$

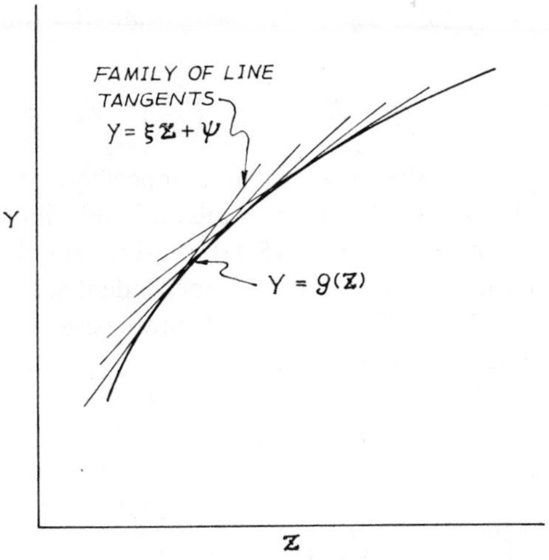

Figure 5.1

and differentiating with respect to z, we obtain

$$\xi = g'(z) \tag{5-36}$$

By definition,

$$\psi = y - \xi z \tag{5-37}$$

Eqs. (5-35) to (5-37) can be solved simultaneously in order to eliminate y and z, thereby obtaining the transform, $\psi = f(\xi)$.

Now let us consider the m-dimensional case. Given the function

$$y = g(z_1, \ldots, z_k, z_{k+1}, \ldots, z_m) \tag{5-38}$$

let us seek the transform having the same information content but with the independent variables $\xi_1, \ldots, \xi_k, z_{k+1}, \ldots, z_m$ where $\xi_i = (\partial y/\partial z_i)_{z_j[i]}$. That is, we seek the partial transform of y with respect to $z_1, \ldots, z_k$, which is

$$\psi = f(\xi_1, \ldots, \xi_k, z_{k+1}, \ldots, z_m) \tag{5-39}$$

In this case ψ is defined as

$$\psi \equiv y - \sum_{i=1}^{k} \xi_i z_i \tag{5-40}$$

Note that the summation extends over only those variables that are being transformed into their conjugate coordinates.

Differentiation of Eq. (5-40) yields

$$d\psi = dy - \sum_{i=1}^{k} \xi_i\, dz_i - \sum_{i=1}^{k} z_i\, d\xi_i \tag{5-41}$$

Expressing Eq. (5-38) in differential form,

$$dy = \sum_{i=1}^{m} \left(\frac{\partial y}{\partial z_i}\right)_{z_j[i]} dz_i = \sum_{i=1}^{k} \xi_i\, dz_i + \sum_{i=k+1}^{m} \left(\frac{\partial y}{\partial z_i}\right)_{z_j[i]} dz_i \tag{5-42}$$

Substituting Eq. (5-42) into Eq. (5-41), we obtain the differential of the transform:

$$d\psi = -\sum_{i=1}^{k} z_i\, d\xi_i + \sum_{i=k+1}^{m} \left(\frac{\partial y}{\partial z_i}\right)_{z_j[i]} dz_i \tag{5-43}$$

Eqs. (5-40) and (5-43) are the ones of major importance; they describe the procedure for obtaining the transform in integral and differential forms.

It is important to note that in Eq. (5-43), the differentials are $\xi_1, \ldots, \xi_k, z_{k+1}, \ldots, z_m$. Since these variables are the independent set for the transform in question [i.e., ψ in Eq. (5-39)], Eq. (5-43) must be an exact differential. Thus, we can conclude immediately that

$$-z_i = \left(\frac{\partial \psi}{d\xi_i}\right)_{\xi_1,\ldots,(\text{no } \xi_i)\ldots\xi_k, z_{k+1},\ldots,z_m} \tag{5-44}$$

and

$$\left(\frac{\partial y}{\partial z_i}\right)_{z_j[i]} = \left(\frac{\partial \psi}{\partial z_i}\right)_{\xi_1,\ldots,\xi_k, z_{k+1},\ldots,(\text{no } z_i)\ldots z_m} \tag{5-45}$$

Furthermore, we can apply the Maxwell reciprocity theorem to Eq. (5-43) to show that, for example,

$$\left(\frac{\partial z_i}{\partial \xi_j}\right)_{\xi_1,\ldots,(\text{no } \xi_j)\ldots\xi_k, z_{k+1},\ldots,z_m} = \left(\frac{\partial z_j}{\partial \xi_i}\right)_{\xi_1,\ldots,(\text{no } \xi_i)\ldots\xi_k, z_{k+1},\ldots,z_m} \tag{5-46}$$

where i and j are from the set $1, \ldots, k$ and $i \neq j$.

Example 5.3

Determine the integral and differential forms of the partial transform of $\underline{U}$ with respect to $\underline{S}$, i.e., find ψ and $d\psi$ where $\psi = \psi(T_1\, \underline{V}, N_1, \ldots, N_n)$.

Solution

With respect to Eq. (5-38), we have $y = \underline{U}$, $z_1 = \underline{S}$, $z_2 = \underline{V}$, $z_3 = N_1$, etc., and we want to transform $\underline{S}$ into $\xi_1 = T$. From Eq. (5-40),

$$\psi = y - \xi_1 z_1 = \underline{U} - T\underline{S} \tag{5-47}$$

From Eq. (5-43),

$$d\psi = -z_1\, d\xi_1 + \sum_{i=2}^{n+2} \left(\frac{\partial y}{\partial z_i}\right)_{z_j[i]} dz_i = -\underline{S}\, dT + \left(-P\, d\underline{V} + \sum_{i=1}^{n} \mu_i\, dN_i\right) \tag{5-48}$$

There are three Legendre transformations of the Fundamental Equation that are of particular importance, only one of which was illustrated in Example 5.3. These are the enthalpy, $\underline{H}$ (which was introduced in another context in Section 3.8), the Helmholtz free energy, $\underline{A}$, and the Gibbs free energy, $\underline{G}$. These properties, together with the energy, $\underline{U}$, are commonly referred to as the thermodynamic *potential functions*. The integrated and differential forms of each of these functions are given below.

$$\underline{U} = \psi(\underline{S}, \underline{V}, N_1, \ldots, N_n) \tag{5-49}$$

$$\underline{H} = \psi(\underline{S}, P, N_1, \ldots, N_n) = \underline{U} + P\underline{V} \tag{5-50}$$

$$\underline{A} = \psi(T, \underline{V}, N_1, \ldots, N_n) = \underline{U} - T\underline{S} \tag{5-51}$$

$$\underline{G} = \psi(T, P, N_1, \ldots, N_n) = \underline{U} - T\underline{S} + P\underline{V} \tag{5-52}$$

$$d\underline{U} = T\,d\underline{S} - P\,d\underline{V} + \sum_{i=1}^{n} \mu_i\,dN_i \tag{5-53}$$

$$d\underline{H} = T\,d\underline{S} + \underline{V}\,dP + \sum_{i=1}^{n} \mu_i\,dN_i \tag{5-54}$$

$$d\underline{A} = -\underline{S}\,dT - P\,d\underline{V} + \sum_{i=1}^{n} \mu_i\,dN_i \tag{5-55}$$

$$d\underline{G} = -\underline{S}\,dT + \underline{V}\,dP + \sum_{i=1}^{n} \mu_i\,dN_i \tag{5-56}$$

Since Eqs. (5-53) through (5-56) are exact differentials, it follows that

$$T = \left(\frac{\partial \underline{U}}{\partial \underline{S}}\right)_{\underline{V},N} = \left(\frac{\partial \underline{H}}{\partial \underline{S}}\right)_{P,N} \tag{5-57}$$

$$-P = \left(\frac{\partial \underline{U}}{\partial \underline{V}}\right)_{\underline{S},N} = \left(\frac{\partial \underline{A}}{\partial \underline{V}}\right)_{T,N} \tag{5-58}$$

$$-\underline{S} = \left(\frac{\partial \underline{A}}{\partial T}\right)_{\underline{V},N} = \left(\frac{\partial \underline{G}}{\partial T}\right)_{P,N} \tag{5-59}$$

$$\underline{V} = \left(\frac{\partial \underline{H}}{\partial P}\right)_{\underline{S},N} = \left(\frac{\partial \underline{G}}{\partial P}\right)_{T,N} \tag{5-60}$$

and

$$\mu_i = \left(\frac{\partial \underline{U}}{\partial N_i}\right)_{\underline{S},\underline{V},N_j[i]} = \left(\frac{\partial \underline{H}}{\partial N_i}\right)_{\underline{S},P,N_j[i]} = \left(\frac{\partial \underline{A}}{\partial N_i}\right)_{T,\underline{V},N_j[i]}$$

$$= \left(\frac{\partial \underline{G}}{\partial N_i}\right)_{T,P,N_j[i]} \tag{5-61}$$

If we apply Euler's theorem to Eq. (5-56) for a one-component system, we obtain

$$\underline{G} = \mu N \tag{5-62}$$

or

$$G = \mu \tag{5-63}$$

For a multicomponent system,

$$\underline{G} = \sum_{i=1}^{n} \mu_i N_i \tag{5-64}$$

or

$$G = \sum_{i=1}^{n} \mu_i x_i \tag{5-65}$$

where x_i is the mole fraction of component i.

There is one additional Legendre transform of the energy that we shall use often, the total transform of $\underline{U}$ with respect to $\underline{S}, \underline{V}, N_1, \ldots, N_n$ or $\psi(T, P, \mu_1, \ldots, \mu_n)$. Applying Eq. (5-40),

$$\psi = \underline{U} - T\underline{S} + P\underline{V} - \sum_{i=1}^{n} \mu_i N_i \tag{5-66}$$

Substitution of Eq. (5-23) into Eq. (5-66) yields

$$\psi(T, P, \mu_1, \ldots, \mu_n) = 0 \tag{5-67}$$

From this result, together with Eq. (5-43), we find

$$d\psi = 0 = -\underline{S}\, dT + \underline{V}\, dP - \sum_{i=1}^{n} N_i\, d\mu_i \tag{5-68}$$

or

$$\sum_{i=1}^{n} x_i\, d\mu_i = -S\, dT + V\, dP \tag{5-69}$$

Eq. (5-69) is called the Gibbs-Duhem equation. It represents the relationship between the $n + 2$ intensive variables alluded to in Section 5.3. For a one-component system, Eq. (5-69) simplifies to

$$d\mu = -S\, dT + V\, dP \tag{5-70}$$

Since this equation results from a transform of the Fundamental Equation, it must be an exact differential. Therefore, $-S = (\partial\mu/\partial T)_P$ and $V = (\partial\mu/\partial P)_T$. ?? how about $dP = \frac{d\mu}{V} + \left(\frac{S}{V}\right) dT$, which also "results" from a transformation etc

---

Example 5.4

Apply Legendre transformations to the Fundamental Equation of a monatomic ideal gas (see Example 5.1) to determine the three potential functions, $H = f(S, P)$, $A = f(T, V)$, and $G = f(T, P)$. Show that G is consistent with the expression for μ derived in Example 5.2.

Solution

(a) $H = f(S, P)$. Starting with the energy representation, Eq. (5-22b), the three equations analogous to Eqs. (5-35) through (5-37) are:

$$\underline{U} = NU^0\left(\frac{V^0}{V}\right)^{2/3} e^{(2/3)(S-S^0)/R} \tag{5-70a}$$

$$P = \frac{2}{3}\,\frac{U^0(V^0)^{2/3}}{V^{5/3}}\, e^{(2/3)(S-S^0)/R} \tag{5-70b}$$

$$\underline{H} = \underline{U} + P\underline{V} \tag{5-70c}$$

We must solve these simultaneously in order to eliminate $\underline{U}$ and $\underline{V}$. Using Eqs. (5-70a) and (5-70b) to eliminate first V and then U, we obtain:

$$\underline{U} = \frac{3}{2} NP\left[\frac{2}{3}\,\frac{U^0(V^0)^{2/3}}{P}\, e^{(2/3)(S-S^0)/R}\right]^{3/5} \tag{5-70d}$$

and

$$\underline{V} = N\left[\frac{2}{3}\frac{U^0(V^0)^{2/3}}{P}e^{(2/3)(S-S^0)/R}\right]^{3/5} \qquad \text{(5-70e)}$$

Thus, Eq. (5-70c) becomes

$$\underline{H} = \frac{5}{2}NP\left[\frac{2}{3}\frac{U^0(V^0)^{2/3}}{P}e^{(2/3)(S-S^0)/R}\right]^{3/5} \qquad \text{(5-70f)}$$

or

$$H = \frac{5}{2}P\left[\frac{2}{3}\frac{U^0(V^0)^{2/3}}{P}e^{(2/3)(S-S^0)/R}\right]^{3/5}$$

which is the desired function of S and P.

(b) $A = f(T, V)$. Again using Eq. (5-70a), but replacing Eqs. (5-70b) and (5-70c) by:

$$T = \frac{2U^0}{3R}\left(\frac{V^0}{V}\right)^{2/3}e^{(2/3)(S-S^0)/R} \qquad \text{(5-70g)}$$

$$\underline{A} = \underline{U} - T\underline{S} \qquad \text{(5-70h)}$$

we obtain a set of three equations that must be solved simultaneously in order to eliminate $\underline{U}$ and $\underline{S}$. Solving for $\underline{U}$ and $\underline{S}$ in terms of T and V, we obtain:

$$\underline{U} = \frac{3}{2}NRT \qquad \text{(5-70i)}$$

$$\underline{S} = N\left[S^0 + \frac{3}{2}R\ln\left\{\frac{3RT}{2U^0}\left(\frac{V}{V^0}\right)^{2/3}\right\}\right] \qquad \text{(5-70j)}$$

Substitution in Eq. (5-70h) yields

$$A = \frac{3}{2}RT\left[1 - \ln\left\{\frac{3RT}{2U^0}\left(\frac{V}{V^0}\right)^{2/3}\right\}\right] - TS^0 \qquad \text{(5-70k)}$$

(c) $G = f(T, P)$. Since G requires transformation of two variables, we must solve four simultaneous equations, namely Eqs. (5-70a), (5-70b), (5-70g), and the defining equation for $\underline{G}$:

$$\underline{G} = \underline{U} - T\underline{S} + P\underline{V} \qquad \text{(5-70l)}$$

These yield the three desired equations:

$$\underline{U} = \frac{3}{2}NRT, \qquad \underline{V} = \frac{NRT}{P} \qquad \text{(5-70m)}$$

and

$$\underline{S} = N\left[S^0 + \frac{3}{2}R\ln\left\{\frac{3(RT)^{5/3}}{2U^0(PV^0)^{2/3}}\right\}\right] \qquad \text{(5-70n)}$$

which, upon substitution into Eq. (5-70l), gives us the desired result:

$$G = RT\left[\frac{5}{2} - \frac{3}{2}\ln\left\{\frac{3(RT)^{5/3}}{2U^0(PV^0)^{2/3}}\right\}\right] - TS^0 \qquad \text{(5-70o)}$$

To transform Eq. (5-70o) into a form analogous to Eq. (5-30k), let us evaluate G^0 at T^0 and P^0:

$$G^0 = RT^0\left[\frac{5}{2} - \frac{3}{2}\ln\left\{\frac{3(RT^0)^{5/3}}{2U^0(P^0V^0)^{2/3}}\right\}\right] - T^0S^0 \tag{5-70p}$$

As shown in Example 5.2, the ln term vanishes. Thus,

$$G^0 = \frac{5}{2}RT^0 - T^0S^0 \tag{5-70q}$$

Subtracting Eq. (5-70q) from Eq. (5-70o) yields Eq. (5-70k):

$$G - G^0 = -S^0(T - T^0) + \frac{5}{2}R(T - T^0) - \frac{3}{2}RT\ln\left\{\frac{3(RT)^{5/3}}{2U^0(PV^0)^{2/3}}\right\}$$

5.5 Applications of Legendre Transformations

In many cases, we are concerned with calculating an entropy difference between two stable equilibrium states. For example, suppose we have a process between states 1 and 2 for a single-phase, one-component, closed simple system. Let us assume that the end states have been characterized in terms of the properties T and V, and we would like to know the entropy change of the system. Let us first express the intensive property S in terms of the $n + 1$ intensive properties T and V. That is,

$$dS = \left(\frac{\partial S}{\partial T}\right)_V dT + \left(\frac{\partial S}{\partial V}\right)_T dV \tag{5-71}$$

Since Eq. (5-71) is an exact differential, it can be integrated over any path provided that the variables are defined over that path. Since the entropy is only defined for stable equilibrium states, the path chosen for the integration must be reversible or at least internally reversible. Although the actual process may have been irreversible, we can still calculate the entropy change by integrating Eq. (5-71)—*provided the integration is conducted over a reversible path*. Specifically, we must be certain to use values of $(\partial S/\partial T)_V$ and $(\partial S/\partial V)_T$ that would have been observed if the process were conducted quasi-statically.

We can always reduce partial derivatives of the form $(\partial S/\partial y)_T$, where y is P or V (or $\underline{V}$) to partial derivatives involving only P, V, and T by applying Maxwell's reciprocity relation to one of the transforms of $\underline{U}$. For example, if we desire $(\partial S/\partial V)_{T,N}$, we need only apply the reciprocity relation to the differential form of $\psi(T, \underline{V}, N_1, \ldots, N_n)$. Thus, we could immediately write

$$d\psi = -\underline{S}\,dT - P\,d\underline{V} + \sum_{i=1}^{n} \mu_i\,dN_i$$

From which it follows that

$$\left(\frac{\partial \underline{S}}{\partial \underline{V}}\right)_{T,N} = \left(\frac{\partial S}{\partial V}\right)_{T,N} = \left(\frac{\partial P}{\partial T}\right)_{V,N} \tag{5-72}$$

In this manner, we could evaluate Eq. (5-71) if we knew the equation of state and the specific heat at constant volume, $C_v \equiv T(\partial S/\partial T)_V$, for then Eq. (5-71) reduces to

$$dS = \frac{C_v}{T}\, dT + \left(\frac{\partial P}{\partial T}\right)_V dV \tag{5-73}$$

Since Eq. (5-73) is an exact differential, it can be integrated to yield (see appendix B)

$$S_2 - S_1 = \int_{T_1}^{T_2} \left(\frac{C_v}{T}\right)_{V_1} dT + \int_{V_1}^{V_2} \left[\left(\frac{\partial P}{\partial T}\right)_V\right]_{T_2} dV \tag{5-74}$$

The method of Legendre transforms is also useful in clarifying our thinking on the information content of various pieces of data. For example, any one of the three relationships

$$\underline{H} = f(\underline{S}, P, N_1, \ldots, N_n) \tag{5-75}$$

$$\underline{A} = f(T, \underline{V}, N_1, \ldots, N_n) \tag{5-76}$$

$$\underline{G} = f(T, P, N_1, \ldots, N_n) \tag{5-77}$$

has the same information content as that of the Fundamental Equation. Thus, if we have any one of these relationships available for a given problem, we need not search for any additional data for the system in question. This point is illustrated in the following example.

Example 5.5

Given a Mollier diagram, as shown in Figure 5.2, for a one-component, single-phase simple system, show how lines of constant P can be constructed on a U-T diagram.

Solution

Let us first show that we could evaluate the equation of state (at least in tabular form) and C_p $[\equiv T(\partial S/\partial T)_P]$ from the Mollier diagram, and then show how we would determine $U = f(T, P)$ from these data.

From the slopes of the isobars in the Mollier diagram, we obtain the temperature at each set of S and P since

$$T = \left(\frac{\partial H}{\partial S}\right)_P = f_1(S, P) \tag{5-78}$$

By replotting the diagram as lines of constant S using H and P as axes, we obtain from the slope

$$V = \left(\frac{\partial H}{\partial P}\right)_S = f_2(S, P) \tag{5-79}$$

From Eqs. (5-78) and (5-79), we can construct a table of P at each T and V,

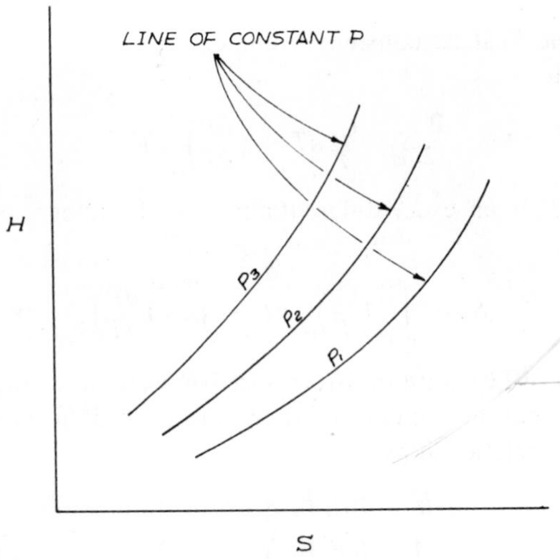

Figure 5.2. Typical Mollier diagram for a one-component gas

which is equivalent to the equation of state, $P = f_3(T, V)$. If, using Eq. (5-78), we plot lines of constant P where S and T are axes, we can evaluate C_P from the slope.

Using C_p and $P = f_3(T, V)$, we can evaluate U as a function of T and P as follows: First expressing U as a differential function of T and P, we have

$$dU = \left(\frac{\partial U}{\partial T}\right)_P dT + \left(\frac{\partial U}{\partial P}\right)_T dP \tag{5-80}$$

For a one-component system,

$$dU = T\,dS - P\,dV \tag{5-81}$$

so that

$$\left(\frac{\partial U}{\partial T}\right)_P = T\left(\frac{\partial S}{\partial T}\right)_P - P\left(\frac{\partial V}{\partial T}\right)_P \tag{5-82}$$

and

$$\left(\frac{\partial U}{\partial P}\right)_T = T\left(\frac{\partial S}{\partial P}\right)_T - P\left(\frac{\partial V}{\partial P}\right)_T \tag{5-83}$$

Substituting Eqs. (5-82) and (5-83) into Eq. (5-80), we have

$$dU = \left[C_p - P\left(\frac{\partial V}{\partial T}\right)_P\right] dT + \left[T\left(\frac{\partial S}{\partial P}\right)_T - P\left(\frac{\partial V}{\partial P}\right)_T\right] dP \tag{5-84}$$

With the exception of $(\partial S/\partial P)_T$, all the partial derivatives in Eq. (5-84) can be evaluated directly from the data. The other derivative, $(\partial S/\partial P)_T$, can be evaluated from $P = f_3(T, V)$ by using the reciprocity relation applied to

$\psi(P, T, N_1, \ldots, N_n)$. Thus,

$$d\psi = -\underline{S}\, dT + \underline{V}\, dP + \sum_{i=1}^{n} \mu_i\, dN_i \tag{5-85}$$

and

$$\left(\frac{\partial \underline{S}}{\partial P}\right)_{T,N} = -\left(\frac{\partial \underline{V}}{\partial T}\right)_{P,N} \tag{5-86}$$

or

$$\left(\frac{\partial S}{\partial P}\right)_T = -\left(\frac{\partial V}{\partial T}\right)_P \tag{5-87}$$

Substituting Eq. (5-87) into Eq. (5-84) and integrating,

$$U_2 - U_1 = \int_{T_1}^{T_2} \left[C_p - P\left(\frac{\partial V}{\partial T}\right)_P\right]_{P_1} dT - \int_{P_1}^{P_2} \left[T\left(\frac{\partial V}{\partial T}\right)_P + P\left(\frac{\partial V}{\partial P}\right)_T\right]_{T_2} dP \tag{5-88}$$

Choosing a set of values for T_1 and P_1 as a reference point and setting $U_1 = 0$, we could calculate U_2 at any other set of conditions.

Although we might be tempted to reverse the procedure to construct a Mollier diagram from data consisting of $U = f(T, P)$, it should be clear that such an attempt is doomed to failure. The function $U = f(T, P)$ is not a Fundamental Equation and, consequently, a Mollier diagram could not be constructed without additional data.

5.6 Legendre Transformations of Partial Derivatives

Legendre transformations have been used to demonstrate the equivalence and symmetry of alternate forms of the Fundamental Equation. In theory, we can approach a given problem using various sets of independent properties, depending on which properties are known, available, or of interest. For example, the $\underline{A}$-representation is convenient if $T, \underline{V}, N_1, \ldots, N_n$ are at our disposal. In practice, there are many situations in which we must evaluate a property associated with one representation given the information content equivalent to another representation.

To accomplish such tasks, we shall now describe a general procedure for transforming properties associated with one representation to functions of properties of any other representation.

In the energy representation, the Fundamental Equation of a single-phase mixture is given by Eq. (5-1). In general, there are $n + 2$ first-order partial derivatives, which are the equations of state given by Eqs. (5-5) through (5-7). These are the coefficients of the differential form of $\underline{U}$, Eq. (5-4), which can be written as:

$$d\underline{U} = U_S\, d\underline{S} + U_V\, d\underline{V} + \sum_{i=1}^{n} U_{N_i}\, dN_i \tag{5-89}$$

where we employ the short-hand notation: $U_z = \partial \underline{U}/\partial z$. Of the $n + 2$ first-order partial derivatives, only $n + 1$ are independently variable; the Gibbs-Duhem equation, Eq. (5-68) relates any one to the other $n + 1$ derivatives. In short-hand notation, Eq. (5-68) is

$$\underline{S}\, dU_S + \underline{V}\, dU_V + \sum_{i=1}^{n} N_i\, dU_{N_i} = 0 \tag{5-90}$$

There are $(n + 2)$ second-order partial derivatives of the form $\partial^2 \underline{U}/\partial z_i^2$ and $(n + 2)(n + 1)/2$ of the form $\partial^2 \underline{U}/\partial z_i\, \partial z_j$ (the number of combinations of $(n + 2)$ variables taken two at a time). Thus, there are a total of $(n + 3)(n + 2)/2$ second-order partial derivatives of the Fundamental Equation. For example, for a binary mixture of components A and B, there are ten:

$$U_{SS} = \left(\frac{\partial^2 \underline{U}}{\partial \underline{S}^2}\right)_{\underline{V},N} = \left(\frac{\partial T}{\partial \underline{S}}\right)_{\underline{V},N} \tag{5-91}$$

$$U_{VV} = \left(\frac{\partial^2 \underline{U}}{\partial \underline{V}^2}\right)_{\underline{S},N} = -\left(\frac{\partial P}{\partial \underline{V}}\right)_{\underline{S},N} \tag{5-92}$$

$$\begin{aligned} U_{SV} &= \frac{\partial}{\partial \underline{V}}\left[\left(\frac{\partial \underline{U}}{\partial \underline{S}}\right)_{\underline{V},N}\right]_{\underline{S},N} = \frac{\partial}{\partial \underline{S}}\left[\left(\frac{\partial \underline{U}}{\partial \underline{V}}\right)_{\underline{S},N}\right]_{\underline{V},N} \\ &= \left(\frac{\partial T}{\partial \underline{V}}\right)_{\underline{S},N} = -\left(\frac{\partial P}{\partial \underline{S}}\right)_{\underline{V},N} \end{aligned} \tag{5-93}$$

$$U_{N_AN_A} = \left(\frac{\partial^2 \underline{U}}{\partial N_A^2}\right)_{\underline{S},\underline{V},N_B} = \left(\frac{\partial \mu_A}{\partial N_A}\right)_{\underline{S},\underline{V},N_B} \tag{5-94}$$

$$U_{N_BN_B} = \left(\frac{\partial^2 \underline{U}}{\partial N_B^2}\right)_{\underline{S},\underline{V},N_A} = \left(\frac{\partial \mu_B}{\partial N_B}\right)_{\underline{S},\underline{V},N_A} \tag{5-95}$$

$$\begin{aligned} U_{SN_A} &= \frac{\partial}{\partial N_A}\left[\left(\frac{\partial \underline{U}}{\partial \underline{S}}\right)_{\underline{V},N}\right]_{\underline{S},\underline{V},N_B} = \frac{\partial}{\partial \underline{S}}\left[\left(\frac{\partial \underline{U}}{\partial N_A}\right)_{\underline{S},\underline{V},N_B}\right]_{\underline{V},N} \\ &= \left(\frac{\partial T}{\partial N_A}\right)_{\underline{S},\underline{V},N_B} = \left(\frac{\partial \mu_A}{\partial \underline{S}}\right)_{\underline{V},N} \end{aligned} \tag{5-96}$$

$$U_{SN_B} = \left(\frac{\partial T}{\partial N_B}\right)_{\underline{S},\underline{V},N_A} = \left(\frac{\partial \mu_B}{\partial \underline{S}}\right)_{\underline{V},N} \tag{5-97}$$

$$\begin{aligned} U_{VN_A} &= \frac{\partial}{\partial N_A}\left[\left(\frac{\partial \underline{U}}{\partial \underline{V}}\right)_{\underline{S},N}\right]_{\underline{S},\underline{V},N_B} = \frac{\partial}{\partial \underline{V}}\left[\left(\frac{\partial \underline{U}}{\partial N_A}\right)_{\underline{S},\underline{V},N_B}\right]_{\underline{S},N} \\ &= -\left(\frac{\partial P}{\partial N_A}\right)_{\underline{S},\underline{V},N_B} = \left(\frac{\partial \mu_A}{\partial \underline{V}}\right)_{\underline{S},N} \end{aligned} \tag{5-98}$$

$$U_{VN_B} = -\left(\frac{\partial P}{\partial N_B}\right)_{\underline{S},\underline{V},N_A} = \left(\frac{\partial \mu_B}{\partial \underline{V}}\right)_{\underline{S},N} \tag{5-99}$$

$$\begin{aligned} U_{N_AN_B} &= \frac{\partial}{\partial N_B}\left[\left(\frac{\partial \underline{U}}{\partial N_A}\right)_{\underline{S},\underline{V},N_B}\right]_{\underline{S},\underline{V},N_A} = \frac{\partial}{\partial N_A}\left[\left(\frac{\partial \underline{U}}{\partial N_B}\right)_{\underline{S},\underline{V},N_A}\right]_{\underline{S},\underline{V},N_B} \\ &= \left(\frac{\partial \mu_A}{\partial N_B}\right)_{\underline{S},\underline{V},N_A} = \left(\frac{\partial \mu_B}{\partial N_A}\right)_{\underline{S},\underline{V},N_B} \end{aligned} \tag{5-100}$$

These derivatives are the coefficients in the second-order differential of $\underset{\sim}{U}$:

$$d^2\underset{\sim}{U} = U_{SS}(d\underset{\sim}{S})^2 + 2U_{SV}(d\underset{\sim}{S})(d\underset{\sim}{V}) + U_{VV}(d\underset{\sim}{V})^2 + 2\sum_{i=1}^{n}[U_{SN_i}(d\underset{\sim}{S})(dN_i) + U_{VN_i}(d\underset{\sim}{V})(dN_i)] + \sum_{i=1}^{n}\sum_{j=1}^{n} U_{N_iN_j}(dN_i)(dN_j) \tag{5-101}$$

Among these $(n + 3)(n + 2)/2$ second-order derivatives there are $(n + 2)$ relationships that must be satisfied. These result from division of Eq. (5-90) by the differential of each of the $(n + 2)$ independent variables, $\underset{\sim}{S}, \underset{\sim}{V}, N_1, \ldots, N_n$. For example,

$$\underset{\sim}{S}\left(\frac{\partial U_S}{\partial \underset{\sim}{S}}\right)_{\underset{\sim}{V},N} + \underset{\sim}{V}\left(\frac{\partial U_V}{\partial \underset{\sim}{S}}\right)_{\underset{\sim}{V},N} + \sum_{i=1}^{n} N_i\left(\frac{\partial U_{N_i}}{\partial \underset{\sim}{S}}\right)_{\underset{\sim}{V},N} = 0 \tag{5-102}$$

or

$$\underset{\sim}{S}U_{SS} + \underset{\sim}{V}U_{SV} + \sum_{i=1}^{n} N_iU_{SN_i} = 0 \tag{5-103}$$

Similarly,

$$\underset{\sim}{S}U_{SV} + \underset{\sim}{V}U_{VV} + \sum_{i=1}^{n} N_iU_{VN_i} = 0 \tag{5-104}$$

and for each N_j,

$$\underset{\sim}{S}U_{SN_j} + \underset{\sim}{V}U_{VN_j} + \sum_{i=1}^{n} N_iU_{N_iN_j} = 0 \tag{5-105}$$

Thus, the total number of independent second-order partial derivatives is $[(n + 3)(n + 2)/2] - (n + 2)$ or $(n + 2)(n + 1)/2$.

Example 5.6

Of the ten second-order partial derivatives of a binary system, as given in Eqs. (5-91) through (5-100), find a set of six that are independent and express the other four in terms of the independent set.

Solution

There are $(n + 2)$ or four equations, formed from the Gibbs-Duhem equation, which are equivalent to Eqs. (5-103) through (5-105):

$$\underset{\sim}{S}U_{SS} + \underset{\sim}{V}U_{SV} + N_AU_{SN_A} + N_BU_{SN_B} = 0 \tag{5-106}$$

$$\underset{\sim}{S}U_{SV} + \underset{\sim}{V}U_{VV} + N_AU_{VN_A} + N_BU_{VN_B} = 0 \tag{5-107}$$

$$\underset{\sim}{S}U_{SN_A} + \underset{\sim}{V}U_{VN_A} + N_AU_{N_AN_A} + N_BU_{N_AN_B} = 0 \tag{5-108}$$

$$\underset{\sim}{S}U_{SN_B} + \underset{\sim}{V}U_{VN_B} + N_AU_{N_AN_B} + N_BU_{N_BN_B} = 0 \tag{5-109}$$

To insure that these equations are always satisfied, no more than three second-order derivatives from each equation can be included in the independent set. Thus, we could eliminate any column or row of derivatives in the set of

Eqs. (5-106) through (5-109). Each of the following sets of six would then be independent:

$$(1)\ U_{SS}, U_{SV}, U_{VV}, U_{SN_A}, U_{VN_A}, U_{N_AN_A}$$

or

$$(2)\ U_{SS}, U_{SV}, U_{VV}, U_{SN_B}, U_{VN_B}, U_{N_BN_B}$$

or

$$(3)\ U_{SS}, U_{SN_A}, U_{SN_B}, U_{N_AN_A}, U_{N_AN_B}, U_{N_BN_B}$$

or

$$(4)\ U_{VV}, U_{VN_A}, U_{VN_B}, U_{N_AN_A}, U_{N_AN_B}, U_{N_BN_B}$$

Choosing the first set, the four dependent derivatives can be determined from Eqs. (5-106) through (5-109):

$$U_{SN_B} = \frac{1}{N_B}(-\underline{S}U_{SS} - \underline{V}U_{SV} - N_AU_{SN_A}) \tag{5-110}$$

$$U_{VN_B} = \frac{1}{N_B}(-\underline{S}U_{SV} - \underline{V}U_{VV} - N_AU_{VN_A}) \tag{5-111}$$

$$U_{N_AN_B} = \frac{1}{N_B}(-\underline{S}U_{SN_A} - \underline{V}U_{VN_A} - N_AU_{N_AN_A}) \tag{5-112}$$

$$\begin{aligned} U_{N_BN_B} = \frac{1}{N_B}\Big[&-\underline{S}\left(\frac{1}{N_B}\right)(-\underline{S}U_{SS} - \underline{V}U_{SV} - N_AU_{SN_A}) \\ &- \underline{V}\left(\frac{1}{N_B}\right)(-\underline{S}U_{SV} - \underline{V}U_{VV} - N_AU_{VN_A}) \\ &- N_A\left(\frac{1}{N_B}\right)(-\underline{S}U_{SN_A} - \underline{V}U_{VN_A} - N_AU_{N_AN_A})\Big] \end{aligned} \tag{5-113}$$

or

$$\begin{aligned} U_{N_BN_B} = \left(\frac{1}{N_B}\right)^2 [&\underline{S}^2U_{SS} + 2\underline{S}\underline{V}U_{SV} + \underline{V}^2U_{VV} + 2\underline{S}N_AU_{SN_A} \\ &+ 2\underline{V}N_AU_{VN_A} + N_A^2U_{N_AN_A}] \end{aligned} \tag{5-114}$$

When partial derivatives of one representation of the Fundamental Equation are available, but a problem can be more conveniently approached by using an alternative representation, it is possible to transform the available partial derivatives into the other representation. For example, we might want to determine A_{TT}, A_{TV}, or A_{VV} given G_{TT}, G_{TP}, and G_{PP}. To find these relationships, we can apply Legendre transform theory to a Taylor series expansion of the property of interest. Conceptually, the procedure is straightforward and the transformation rules are simply expressed, but the algebra involved in developing the results is cumbersome. We shall demonstrate the derivation for two relatively simple cases and then state the more general results.

First, let us consider the case with only one independent variable, namely,

$y = g(z)$. As described in Section 5.4, we shall follow the procedure outlined by Eqs. (5-35) through (5-37), except that we shall first express y as a Taylor series expansion in z. Thus, Eq. (5-35) becomes:

$$y = y^0 + y_z^0(z - z^0) + \left(\frac{1}{2}\right) y_{zz}^0(z - z^0)^2 + \ldots \tag{5-115a}$$

or

$$y = y^0 + y_z^0\, \delta z + \left(\frac{1}{2}\right) y_{zz}^0\, (\delta z)^2 + \ldots \tag{5-115b}$$

where

$$\delta z = z - z^0 \tag{5-115c}$$

$$y^0 = (y)_{z=z^0} \tag{5-115d}$$

$$y_z^0 = \left(\frac{dy}{dz}\right)_{z=z^0} \tag{5-115e}$$

$$y_{zz}^0 = \left(\frac{d^2y}{dz^2}\right)_{z=z^0} \tag{5-115f}$$

We would like to transform this expression for y as a function of z to ψ as a function of ξ where $\xi = (dy/dz)$. In Taylor series notation, we seek

$$\psi = \psi^0 + \psi_\xi^0(\xi - \xi^0) + \left(\frac{1}{2}\right) \psi_{\xi\xi}^0(\xi - \xi^0)^2 + \ldots \tag{5-116a}$$

or

$$\psi = \psi^0 + \psi_\xi^0\, \delta\xi + \left(\frac{1}{2}\right) \psi_{\xi\xi}^0\, (\delta\xi)^2 + \ldots \tag{5-116b}$$

where

$$\delta\xi = \xi - \xi^0 \tag{5-116c}$$

$$\psi^0 = (\psi)_{\xi=\xi^0} \tag{5-116d}$$

$$\psi_\xi^0 = \left(\frac{d\psi}{d\xi}\right)_{\xi=\xi^0} \tag{5-116e}$$

$$\psi_{\xi\xi}^0 = \left(\frac{d^2\psi}{d\xi^2}\right)_{\xi=\xi^0} \tag{5-116f}$$

As outlined in Section 5.4, we can differentiate Eq. (5-115a) to find an expression for ξ as a function of z, which is the analog of Eq. (5-36):

$$\xi = \frac{dy}{dz} = y_z^0 + y_{zz}^0(z - z^0) + \ldots \tag{5-117a}$$

or, since

$$y_z^0 = \left(\frac{dy}{dz}\right)_{z=z^0} = (\xi)_{z=z^0} = \xi^0 \tag{5-117b}$$

$$\xi = \xi^0 + y_{zz}^0(z - z^0) \tag{5-117c}$$

Thus, we obtain z as a function of ξ:

$$z = z^0 + \frac{\xi - \xi^0}{y^0_{zz}} \tag{5-118a}$$

or, using Eq. (5-116c),

$$z = z^0 + \frac{\delta\xi}{y^0_{zz}} \tag{5-118b}$$

and, using Eq. (5-115c),

$$\delta z = \frac{\delta\xi}{y^0_{zz}} \tag{5-118c}$$

The next step in the procedure is to solve for y as a function of ξ by substituting Eq. (5-118b) into Eq. (5-115a) or, alternatively, Eq. (5-118c) into Eq. (5-115b):

$$y = y^0 + \frac{y^0_z}{y^0_{zz}}(d\xi) + \frac{1}{2}\frac{1}{y^0_{zz}}(d\xi)^2 + \dots \tag{5-119a}$$

or

$$y = y^0 + \frac{\xi^0}{y^0_{zz}}(\delta\xi) + \frac{1}{2}\frac{1}{y^0_{zz}}(\delta\xi)^2 + \dots \tag{5-119b}$$

where we have made use of Eq. (5-117b) in Eq. (5-119b).

Now, let us substitute Eqs. (5-119b) and (5-118b) into the defining equation for ψ, Eq. (5-37):

$$\psi = y - \xi z$$

$$\psi = y^0 + \frac{\xi^0}{y^0_{zz}}(\delta\xi) + \frac{1}{2}\frac{1}{y^0_{zz}}(\delta\xi)^2 + \dots - \xi\left(z^0 + \frac{\delta\xi}{y^0_{zz}}\right) \tag{5-120a}$$

$$= y^0 - \xi z^0 - \frac{(\xi - \xi^0)}{y^0_{zz}}(\delta\xi) + \frac{1}{2}\frac{1}{y^0_{zz}}(\delta\xi)^2 + \dots \tag{5-120b}$$

$$= y^0 - \xi z^0 - \frac{1}{2}\frac{1}{y^0_{zz}}(\delta\xi)^2 + \dots \tag{5-120c}$$

where we have made use of Eq. (5-116c) in Eq. (5-120c). The variable ξ in the second term of Eq. (5-120c) can be eliminated by noting that

$$\begin{aligned} \xi z^0 &= \xi z^0 - \xi^0 z^0 + \xi^0 z^0 \\ &= z^0(\xi - \xi^0) + \xi^0 z^0 \\ &= z^0\,\delta\xi + \xi^0 z^0 \end{aligned} \tag{5-121}$$

Substituting Eq. (5-121) into Eq. (5-120c) and collecting terms in $\delta\xi$, we obtain the transformed Taylor series for ψ in terms of ξ:

$$\psi = (y^0 - \xi^0 z^0) + (-z^0)(\delta\xi) + \frac{1}{2}\left(\frac{-1}{y^0_{zz}}\right)(\delta\xi)^2 + \dots \tag{5-122}$$

Comparing Eqs. (5-122) and (5-116b) term by term,

$$\psi^0 = y^0 - \xi^0 z^0 \tag{5-123}$$

$$\psi_{\xi}^{0} = -z^{0} \tag{5-124}$$

$$\psi_{\xi\xi}^{0} = -\frac{1}{y_{zz}^{0}} \tag{5-125}$$

Eq. (5-123) is nothing more than Eq. (5-37), evaluated at z^0. Eq. (5-124) is the transform of the first-order derivative. It represents the inversion of Eq. (5-117b) and, for that reason, $-z$ is usually referred to as the *inverse transform of* ψ.

Eq. (5-125) is the transform of the second-order partial derivative, which is the relationship that we set out to determine. Note that we could have developed the transforms for third- and higher-order derivatives if we had retained higher-order terms in the Taylor series expansion.

We now consider the case of two independent variables, z_1 and z_2, wherein we desire the partial transform. That is,

$$y = f(z_1, z_2) \tag{5-126a}$$

is to be transformed into

$$\psi(\xi_1, z_2) \equiv y - \xi_1 z_1 \tag{5-126b}$$

where

$$\xi_1 = \left(\frac{\partial y}{\partial z_1}\right)_{z_2} \tag{5-126c}$$

The Taylor series expansion for y is

$$y = y^0 + (y_{z_1}^0\,\delta z_1 + y_{z_2}^0 \delta z_2) + \frac{1}{2}[y_{z_1z_1}^0\,(\delta z_1)^2 + 2y_{z_1z_2}^0\,(\delta z_1)\,(\delta z_2) + y_{z_2z_2}^0\,(\delta z_2)^2] + \dots \tag{5-126d}$$

where

$$y_{z_i}^0 = \left(\frac{\partial y}{\partial z_i}\right)_{z_1=z_1^0,\, z_2=z_2^0}$$

$$y_{z_iz_j}^0 = \left(\frac{\partial^2 y}{\partial z_i\,\partial z_j}\right)_{z_1=z_1^0,\, z_2=z_2^0}$$

$$\delta z_i = (z_i - z_i^0)$$

The desired transformation is the Taylor series expansion of ψ:

$$\psi = \psi^0 + (\psi_{\xi_1}^0\,\delta\xi_1 + \psi_{z_2}^0\,\delta z_2) + \frac{1}{2}[\psi_{\xi_1\xi_1}^0\,(\delta\xi_1)^2 + 2\psi_{\xi_1z_2}^0\,(\delta\xi_1)\,(\delta z_2) + \psi_{z_2z_2}^0\,(\delta z_2)^2] + \dots \tag{5-126e}$$

Noting that y^0, $y_{z_i}^0$ and $y_{z_iz_j}^0$ are constants, ξ_1 may be obtained by differentiating Eq. (5-126d) as indicated by Eq. (5-126c)

$$\xi_1 = y_{z_1}^0 + y_{z_1z_1}^0\,\delta z_1 + y_{z_1z_2}^0\,\delta z_2 \tag{5-126f}$$

Since

$$y_{z_1}^0 = \left(\frac{\partial y}{\partial z_1}\right)_{z_1=z_1^0,\, z_2=z_2^0} = \xi_1^0$$

and since

$$\delta\xi_1 = \xi_1 - \xi_1^0$$

Eq. (5-126f) can be solved for δz_1 in terms of $\delta\xi_1$ and δz_2:

$$\delta z_1 = \frac{\delta\xi_1 - y^0_{z_1 z_2}\,\delta z_2}{y^0_{z_1 z_1}} \tag{5-126g}$$

or

$$z_1 = z_1^0 + \frac{\delta\xi_1 - y^0_{z_1 z_2}\,\delta z_2}{y^0_{z_1 z_1}} \tag{5-126h}$$

Substituting Eq. (5-126g) into Eq. (5-126d) yields y as a function of ξ_1 and z_2; substituting this relation and Eq. (5-126h) into Eq. (5-126b) yields, after considerable algebraic manipulation:

$$\begin{aligned}\psi(\xi_1, z_2) = (y^0 - \xi_1 z_1^0 + y^0_{z_2}\,\delta z_2)&\\ + \frac{1}{2}\Bigg[-\frac{1}{y^0_{z_1 z_1}}(\delta\xi_1)^2 + 2\left(\frac{y^0_{z_1 z_2}}{y^0_{z_1 z_1}}\right)(\delta\xi_1)(\delta z_2)&\\ + \left\{y^0_{z_2 z_2} - \frac{(y^0_{z_1 z_2})^2}{y^0_{z_1 z_1}}\right\}(\delta z_2)^2\Bigg]&\end{aligned} \tag{5-126i}$$

The first bracket can be simplified as follows:

$$\begin{aligned}y^0 - \xi_1 z_1^0 + y^0_{z_2}\,\delta z_2 &= y^0 - \xi_1^0 z_1^0 - (\xi_1 z_1^0 - \xi_1^0 z_1^0) + y^0_{z_2}\,\delta z_2\\ &= (y^0 - \xi_1^0 z_1^0) + (-z_1^0\,\delta\xi_1 + y^0_{z_2}\,\delta z_2)\end{aligned} \tag{5-126j}$$

With Eq. (5-126j), Eq. (5-126i) becomes:

$$\begin{aligned}\psi(\xi_1, z_2) = (y^0 - \xi_1^0 z_1^0) + (-z_1^0\,\delta\xi_1 + y^0_{z_2}\,\delta z_2)&\\ + \frac{1}{2}\Bigg[-\frac{1}{y^0_{z_1 z_1}}(\delta\xi_1)^2 + 2\left(\frac{y^0_{z_1 z_2}}{y^0_{z_1 z_1}}\right)(\delta\xi_1)(\delta z_2)&\\ + \left\{y^0_{z_2 z_2} - \frac{(y^0_{z_1 z_2})^2}{y^0_{z_1 z_1}}\right\}(\delta z_2)^2\Bigg] + \dots&\end{aligned} \tag{5-127}$$

Comparing Eqs. (5-127) and (5-126e) term by term,

$$\psi = y - \xi_1 z_1 \tag{5-128}$$

$$\psi_{\xi_1} = -z_1 \tag{5-129}$$

$$\psi_{z_2} = y_{z_2} \tag{5-130}$$

$$\psi_{\xi_1\xi_1} = -\frac{1}{y_{z_1 z_1}} \tag{5-131a}$$

$$\psi_{\xi_1 z_2} = \frac{y_{z_1 z_2}}{y_{z_1 z_1}} \tag{5-131b}$$

$$\psi_{z_2 z_2} = y_{z_2 z_2} - \frac{(y_{z_1 z_2})^2}{y_{z_1 z_1}} \tag{5-131c}$$

Eqs. (5-128), (5-129), and (5-130) are not new results; they are nothing more than the partial transform results of Section 5.4, as given by Eqs. (5-40), (5-44), and (5-45).

Eqs. (5-131a), (5-131b), and (5-131c) are the desired results of this development. They are the second-order partial derivative transformations.

Although the transformation was performed for the case of two independent variables, it can be shown that the results are also valid for the transformation of $y(z_1, z_2, \ldots, z_n)$ into $\psi(\xi_1, z_2, \ldots, z_n)$. In this more general case, one more relationship is also obtained:

$$\psi_{z_i z_j} = y_{z_i z_j} - \frac{y_{z_1 z_i} y_{z_1 z_j}}{y_{z_1 z_1}} \qquad i, j \neq 1 \tag{5-131d}$$

The four equations, Eqs. (5-131a) through (5-131d), permit us to transform any second-order partial derivative in, for example, the $\underline{U}$-representation to functions of partials in either the $\underline{H}$- or $\underline{A}$-representation. Similarly, a partial of $\underline{A}$ or $\underline{H}$ could be transformed to functions of partials in $\underline{G}$.

Example 5.7

Express C_v of a pure material as a function of $C_p (= T(\partial S/\partial T)_P)$, $(\partial V/\partial T)_P$, and $(\partial V/\partial P)_T$.

Solution

C_v is related to a second-order derivative of $\underline{U}$ or $\underline{A}$, that is, $C_v = T/NU_{SS} = -TA_{TT}/N$. The set of three given partials contain P and T as independent variables and, thus, they should be related to second-order partials of $\underline{G}$. In fact,

$$\left(\frac{\partial S}{\partial T}\right)_P = -\left(\frac{\partial^2 \underline{G}}{\partial T^2}\right)_{P,N} = -\frac{G_{TT}}{N} \tag{5-132}$$

$$\left(\frac{\partial V}{\partial T}\right)_P = \frac{\partial}{\partial T}\left[\left(\frac{\partial \underline{G}}{\partial P}\right)_{T,N}\right]_{P,N} = \frac{G_{TP}}{N} \tag{5-133}$$

$$\left(\frac{\partial V}{\partial P}\right)_T = \left(\frac{\partial^2 \underline{G}}{\partial P^2}\right)_{T,N} = \frac{G_{PP}}{N} \tag{5-134}$$

Since these three derivatives are independent, the desired result can be obtained by transforming A_{TT} into functions of G_{TT}, G_{TP}, and G_{PP}. Associating y with $\underline{G}$ and ψ with $\underline{A}$, we have:

$$y = \underline{G}, \qquad z_1 = P, \qquad z_2 = T, \qquad z_3 = N$$

$$\psi = \underline{A}, \qquad \xi_1 = \underline{V}, \qquad z_2 = T, \qquad z_3 = N$$

or

$$A_{TT} = \psi_{z_2 z_2}$$

From Eq. (5-131c),

$$A_{TT} = \psi_{z_2 z_2} = y_{z_2 z_2} - \frac{(y_{z_1 z_2})^2}{y_{z_1 z_1}} = G_{TT} - \frac{(G_{PT})^2}{G_{PP}} \tag{5-135}$$

or

$$-\frac{N}{T}C_v = -\frac{N}{T}C_p - \frac{N^2\left(\frac{\partial V}{\partial T}\right)_P^2}{N\left(\frac{\partial V}{\partial P}\right)_T} \tag{5-136}$$

which simplifies to

$$C_v = C_p + \frac{T\left(\frac{\partial V}{\partial T}\right)_P^2}{\left(\frac{\partial V}{\partial P}\right)_T} \tag{5-137}$$

Example 5.8

As will be shown in Chapter 7, the criteria for the stable existence of a single phase of a pure material are:

$$U_{SS} > 0 \tag{5-138}$$

$$U_{SS}U_{VV} - U_{SV}^2 > 0 \tag{5-139}$$

Determine the equivalent criteria in terms of the Helmholtz free energy.

Solution

In this case we want to express partials of $\underline{U}$ in terms of partials of $\underline{A}$. Therefore, let

$$y = \underline{A}, \quad z_1 = T, \quad z_2 = \underline{V}, \quad z_3 = N$$

$$\psi = \underline{U}, \quad \xi_1 = -\underline{S}, \quad z_2 = \underline{V}, \quad z_3 = N$$

Using Eqs. (5-131a) through (5-131c),

$$U_{SS} = \psi_{-\xi_1 - \xi_1} = \psi_{\xi_1\xi_1} = -\frac{1}{y_{z_1z_1}} = -\frac{1}{A_{TT}} \tag{5-140}$$

$$U_{SV} = \psi_{-\xi_1 z_2} = -\psi_{\xi_1 z_2} = -\frac{y_{z_1z_2}}{y_{z_1z_1}} = -\frac{A_{TV}}{A_{TT}} \tag{5-141}$$

$$U_{VV} = \psi_{z_2z_2} = y_{z_2z_2} - \frac{(y_{z_1z_2})^2}{y_{z_1z_1}} = A_{VV} - \frac{(A_{TV})^2}{A_{TT}} \tag{5-142}$$

Therefore, the stability criteria become:

$$U_{SS} = -\frac{1}{A_{TT}} > 0 \tag{5-143}$$

or

$$A_{TT} < 0 \tag{5-144}$$

and

$$\begin{aligned} U_{SS}U_{VV} - U_{SV}^2 &= -\left(\frac{1}{A_{TT}}\right)\left[A_{VV} - \frac{(A_{TV})^2}{A_{TT}}\right] - \left(\frac{A_{TV}}{A_{TT}}\right)^2 \\ &= -\frac{A_{VV}}{A_{TT}} > 0 \end{aligned} \tag{5-145}$$

or, using Eq. (5-144),

$$A_{VV} > 0 \tag{5-146}$$

PROBLEMS

5.1. Given the function, $y = A(x - x_0)^2 + B$, what is the Legendre transform, $\psi = f(\xi)$?

5.2. Define ψ and express $d\psi$ in terms of the appropriate ξ_j and its conjugate variable for the following cases:
(a) $\psi = f(T, \underline{V}, \mu_1, \mu_2, \ldots, \mu_n)$
(b) $\psi = f\left[\left(\frac{1}{T}\right), \underline{V}, N_1, N_2, \ldots, N_n\right]$

5.3. Carry out the transformations indicated below:
(a) Express $(\partial \underline{S}/\partial P)_{T,N_1,\ldots,N_n}$ as a function of P, $\underline{V}$, T, and their derivatives.
(b) Express $(\partial \underline{H}/\partial P)_{T,N_1,\ldots,N_n}$ as a function of $\underline{G}$ and its derivatives and show how these may be given in terms of P, $\underline{V}$, T, N, and C_p.
(c) Express $[\partial(\underline{G}/T)/\partial(1/T)]_{P,N_1,\ldots,N_n}$ as a function of $\underline{H}$ and its derivatives.
(d) Express $(\partial \underline{H}/\partial \underline{V})_{T,N_1,\ldots,N_n}$ as a function of $\underline{G}$ and its derivatives and show how these may be given in terms of P, $\underline{V}$, T, N, and C_p.
(e) Express $(\partial T/\partial N_A)_{\underline{V},\underline{S},\mu_B}$ as a function of $\underline{U}$ and its *independent* derivatives.
(f) Express $(\partial T/\partial N_A)_{\underline{S},P,\mu_B}$ as a function of $\underline{U}$ and its *independent* derivatives.

5.4. For a one-component system, show:

(a) $\left(\frac{\partial \mu}{\partial N}\right)_{T,\underline{V}} = \left(-\frac{\underline{V}}{N}\right)\left(\frac{\partial \mu}{\partial \underline{V}}\right)_{T,N}$

(b) $\left(\frac{\partial P}{\partial N}\right)_{T,\underline{V}} = \left(-\frac{\underline{V}}{N}\right)\left(\frac{\partial P}{\partial \underline{V}}\right)_{T,N}$

5.5. Analyze and comment on the statement and proof given below:
"C_p is a function only of temperature irrespective of the material or phase."

Proof:

By definition, $C_p \equiv (\partial H/\partial T)_{P,N}$. Differentiating with respect to volume, at constant temperature,

$$\left(\frac{\partial C_p}{\partial V}\right)_T = \left(\frac{\partial}{\partial V}\right)\left[\left(\frac{\partial H}{\partial T}\right)_P\right]_T$$

but from the reciprocity relation, the right-hand side is,

$$= \left(\frac{\partial}{\partial T}\right)\left[\left(\frac{\partial H}{\partial V}\right)_T\right]_P$$

Since $H = U + PV$, $(\partial H/\partial V)_T = (\partial U/\partial V)_T + P + V(\partial P/\partial V)_T$.
Also, $dU = T\,dS - P\,dV$, so

$$\left(\frac{\partial U}{\partial V}\right)_T = T\left(\frac{\partial S}{\partial V}\right)_T - P = T\left(\frac{\partial P}{\partial T}\right)_V - P$$

Substituting,

$$\left(\frac{\partial H}{\partial V}\right)_T = T\left(\frac{\partial P}{\partial T}\right)_V - P + P + V\left(\frac{\partial P}{\partial V}\right)_T$$

and,

$$\begin{aligned}\left(\frac{\partial C_p}{\partial V}\right)_T &= \left(\frac{\partial}{\partial T}\right)\left[T\left(\frac{\partial P}{\partial T}\right)_V + V\left(\frac{\partial P}{\partial V}\right)_T\right]_P \\ &= T\left(\frac{\partial}{\partial T}\right)\left[\left(\frac{\partial P}{\partial T}\right)_V\right]_P + \left(\frac{\partial P}{\partial T}\right)_V \\ &\quad + V\left(\frac{\partial}{\partial T}\right)\left[\left(\frac{\partial P}{\partial V}\right)_T\right]_P + \left(\frac{\partial P}{\partial V}\right)_T\left(\frac{\partial V}{\partial T}\right)_P\end{aligned}$$

Reversing the order of differentiation on the first and third terms,

$$T\left(\frac{\partial}{\partial T}\right)\left[\left(\frac{\partial P}{\partial T}\right)_P\right]_V = V\left(\frac{\partial}{\partial V}\right)\left[\left(\frac{\partial P}{\partial T}\right)_P\right]_T = 0$$

and since $(\partial P/\partial V)_T(\partial V/\partial T)_P = -(\partial P/\partial T)_V$, then substituting,

$$\left(\frac{\partial C_p}{\partial V}\right)_T = 0, \qquad \text{Q.E.D.}$$

5.6. Express the following in terms of C_p, P, V, T, and derivatives of these variables:
(a) $(\partial \underline{S}/\partial T)_{G,N}$
(b) $(\partial A/\partial G)_{T,N}$

5.7. Evaluate each of the following partial derivatives for neon at 3000 atm and 70°F assuming that neon obeys the Clausius equation,

$$P(V - b) = RT, \qquad b = 3.0 \times 10^{-2}\ \text{ft}^3/\text{lb-mol}$$

Under these conditions, $C_p = 5.5$ Btu/lb-mol°R.
(a) $(\partial C_p/\partial P)_T$
(b) $(\partial U/\partial H)_P$
(c) Neon is expanded isothermally from 3000 atm to 2000 atm at 70°F in a turbine. Determine the heat load to the turbine under steady-state operation. Express your results in Btu/lb-mol.

5.8. Derive an expression to calculate the Gibbs free energy change for one mole of a nonideal gas from state P_1, T_1 to state P_2, T_2. List the experimental data required to carry out a numerical computation.

5.9. A reversible, isothermal pump is used to reduce the pressure in an isothermal test chamber from P_1 to P_2. Gas is rejected to a constant pressure P_0. If the temperatures of the pump and tank are both T_0, derive an expression for the work required. Express your answer, in so far as possible, in terms of free energy changes in the gas.

5.10. Given a thermodynamic property ψ, where:

$$\psi = \psi(\underline{S}, \underline{V}, \mu_1, \mu_2, \ldots, \mu_n)$$

express the derivative $(\partial\psi/\partial P)_{\underline{V},N}$ in the following set of variables and their derivatives (T, P, V, S, N). Describe the experimental data required to obtain a numerical value. What would the value of $(\partial \psi / \partial P)_{\underline{V},N}$ be for an ideal gas?

5.11. A spherical tank initially contains 1 g-mol of helium at 10 atm and 300°K (see Figure P5.11). We would like to plan an experiment in which helium is released from the sphere to the environment but at the same time maintain the total energy, $\underline{U}$, constant.

Assume no heat transfer to the walls during the process. Helium behaves as an ideal gas with a C_v of 3 and C_p of 5 cal/g-mol°K.

(a) Choose as a base state $H = 0$ at 300°K. What is the temperature of the residual helium when half the mass has been bled off? What is the pressure? What is the total energy input to the sphere?

(b) Determine $(\partial T/\partial P)_{\underline{U},\underline{V}}$ at the very instant the blow-down process starts.

(c) Repeat (a) and (b) if the base state was chosen so that $U = 0$ at 300°K.

(d) In (a) what is the temperature when 60% of the gas has been bled off?

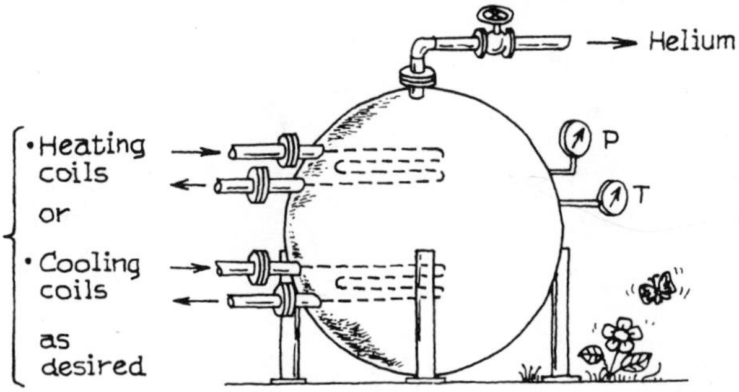

Figure P5.11

Single-Phase Simple Systems of Pure Materials

6

In Chapter 5 the theoretical basis for determining the information required to specify the thermodynamic properties of materials was described. In this chapter we shall continue this line of reasoning with reference to a specific type of system: the single-phase simple system of a pure material. Single-phase simple systems of mixtures are treated separately in Chapter 8.

In addition to discussing the Fundamental Equation of a pure material, we shall describe a few methods for obtaining the data required for applications, namely, the physical properties of single-phase pure materials. Specific correlations are covered in more detail in Appendix D.

6.1 Gibbs Free Energy Representation of the Fundamental Equation

For a single-phase simple system of a pure material, the independent variables for the energy representation are $\underset{\sim}{S}$, $\underset{\sim}{V}$, N, and for the Gibbs free energy representation, they are T, P, N. The latter set is usually more convenient because

T and P are the variables normally under the control of the experimentalist.

The Fundamental Equation for a single component in the Gibbs free energy form is

$$\underline{G} = f_G(T, P, N) \tag{6-1}$$

and, in differential form,

$$d\underline{G} = -\underline{S}\, dT + \underline{V}\, dP + \mu\, dN \tag{6-2}$$

The equations of state, obtained from the first-order partial derivatives of Eq. (6-1) are

$$-\underline{S} = g_1(T, P, N) = G_T \tag{6-3}$$

$$\underline{V} = g_2(T, P, N) = G_P \tag{6-4}$$

and

$$\mu = g_3(T, P, N) = G_N \qquad \text{(6-5)[1]}$$

The three independent second-order partial derivatives are

$$\frac{\partial^2 \underline{G}}{\partial T^2} = -\left(\frac{\partial \underline{S}}{\partial T}\right)_{P,N} = g_{11}(T, P, N) = G_{TT} \tag{6-6}$$

$$\frac{\partial^2 \underline{G}}{\partial P^2} = \left(\frac{\partial \underline{V}}{\partial P}\right)_{T,N} = g_{22}(T, P, N) = G_{PP} \tag{6-7}$$

$$\frac{\partial^2 \underline{G}}{\partial T\, \partial P} = -\left(\frac{\partial \underline{S}}{\partial P}\right)_{T,N} = \left(\frac{\partial \underline{V}}{\partial T}\right)_{P,N} = g_{12}(T, P, N) = G_{TP} \qquad \text{(6-8)[2]}$$

Eqs. (6-6) through (6-8) can be simplified because each of the g_{ij} functions are first order in mass. Thus,

$$-\left(\frac{\partial S}{\partial T}\right)_P = -\frac{1}{N}\left(\frac{\partial \underline{S}}{\partial T}\right)_{P,N} = g_{11}(T, P) \tag{6-9}$$

$$\left(\frac{\partial V}{\partial P}\right)_T = \frac{1}{N}\left(\frac{\partial \underline{V}}{\partial P}\right)_{T,N} = g_{22}(T, P) \tag{6-10}$$

[1] Note that the equations of state obtained from the $\underline{U}$, $\underline{H}$, $\underline{A}$, and $\underline{G}$ representations are equivalent. Given the three equations of state for any one representation, we could solve them simultaneously to obtain the three equations of state for any other representation. For example, compare Eqs. (6-3) to (6-5) with Eqs. (5-5) to (5-7).

[2] When expressing $\underline{G}$ in extensive form, there are three more partial derivatives in addition to Eqs. (6-6) through (6-8). These derivatives, however, are either zero or are redundant with Eqs. (6-3) through (6-5). Specifically,

$$\left(\frac{\partial^2 \underline{G}}{\partial N^2}\right) = \left(\frac{\partial \mu}{\partial N}\right)_{T,P} = 0 \qquad \text{(see Section 5.2)}$$

$$\left(\frac{\partial^2 \underline{G}}{\partial T\, \partial N}\right) = -\left(\frac{\partial \underline{S}}{\partial N}\right)_{T,P} = -S = \frac{g_1}{N} \qquad \text{[see Eq. (6-3)]}$$

and

$$\left(\frac{\partial^2 \underline{G}}{\partial P\, \partial N}\right) = \left(\frac{\partial \underline{V}}{\partial N}\right)_{T,P} = V = \frac{g_2}{N} \qquad \text{[see Eq. (6-4)]}$$

and

$$\left(\frac{\partial V}{\partial T}\right)_P = -\left(\frac{\partial S}{\partial P}\right)_T = \frac{1}{N}\left(\frac{\partial \underline{V}}{\partial T}\right)_{P,N} = g_{12}(T, P) \qquad (6\text{-}11)^3$$

These second-order partial derivatives are related to three widely used properties: *the heat capacity at constant pressure*, C_p; *the isothermal compressibility*, κ_T; and *the coefficient of thermal expansion*, α_p. By definition,

$$C_p \equiv T\left(\frac{\partial S}{\partial T}\right)_P \qquad (6\text{-}12)$$

$$\kappa_T \equiv -\frac{1}{V}\left(\frac{\partial V}{\partial P}\right)_T \qquad (6\text{-}13)$$

and

$$\alpha_p \equiv \frac{1}{V}\left(\frac{\partial V}{\partial T}\right)_P \qquad (6\text{-}14)$$

These three properties were among the first thermodynamic properties ever reported, principally because they can be measured by relatively simple experiments.

Let us now determine which data sets have the information equivalent to Eq. (6-1). We shall approach this problem once again by determining what data are required to reconstruct the Fundamental Equation.

Applying Euler's theorem to Eq. (6-2), we have the integrated form of the Fundamental Equation in the Gibbs free energy representation:

$$\underline{G} = \mu N \qquad (6\text{-}15)$$

Since Eq. (6-1) can be recovered by substituting $g_3(T, P)$ from Eq. (6-5) into Eq. (6-15), we need specify only one equation of state, namely, Eq. (6-5), to obtain the Fundamental Equation. This conclusion is not in contradiction to the discussion of Section 5.3 in which it was shown that two equations of state, namely, Eqs. (5-5) and (5-6), were necessary in order to reconstruct the Fundamental Equation to within an arbitrary constant. As shown below, we must in fact know Eqs. (6-3) and (6-4) in order to obtain Eq. (6-5).

Let us first express Eq. (6-5) in differential form. In Section 5.4 it was shown that the Gibbs-Duhem equation for a pure material is

$$d\mu = -S\,dT + V\,dP \qquad (6\text{-}16)$$

To evaluate g_3 of Eq. (6-5), we must express S and V as functions of T and P, substitute into Eq. (6-16), and integrate. The desired expressions for S and V are given by Eqs. (6-3) and (6-4) when written in intensive form. Thus,

$$d\mu = g_1(T, P)\,dT + g_2(T, P)\,dP \qquad (6\text{-}17)$$

and, by integration from a reference state at $T = T^0$ and $P = P^0$ where μ

[3] In the derivatives on the left-hand side of Eqs. (6-9) through (6-11), note that the subscript N has been omitted. Of course, $n + 1$ variables must always be held constant in partial differentiation. It is, however, common practice to omit the mole numbers when expressing partial derivatives of intensive variables.

is set equal to an arbitrary value of μ^0,

$$\mu = \mu^0 + \int_{T^0}^{T} [g_1(T, P)]_{P^0} \, dT + \int_{P^0}^{P} [g_2(T, P)]_T \, dP \tag{6-18}$$

Let us assume for the moment that Eqs. (6-3) and (6-4) are not known. In this case, we can evaluate g_1 and g_2 in the following manner.

Since S and V are properties, we can start with the exact differentials.

$$dS = \left(\frac{\partial S}{\partial T}\right)_P dT + \left(\frac{\partial S}{\partial P}\right)_T dP \tag{6-19}$$

and

$$dV = \left(\frac{\partial V}{\partial T}\right)_P dT + \left(\frac{\partial V}{\partial P}\right)_T dP \tag{6-20}$$

Of course, the partial derivatives in Eqs. (6-19) and (6-20) are related to the second-order partial derivatives of Eq. (6-1). Therefore,

$$dS = \frac{C_p}{T} dT - \alpha_p V \, dP \tag{6-21}$$

and

$$dV = \alpha_p V \, dT - \kappa_T V \, dP \tag{6-22}$$

Integrating from a reference state at T^0 and P^0 for which $S = S^0$ and $V = V^0$, we have

$$S = S^0 + \int_{T^0}^{T} \left(\frac{C_p}{T}\right)_{P^0} dT - \int_{P^0}^{P} (\alpha_p V)_T \, dP \tag{6-23}$$

and

$$V = V^0 + \int_{T^0}^{T} (\alpha_p V)_{P^0} \, dT - \int_{P^0}^{P} (\kappa_T V)_T \, dP \tag{6-24}$$

Eqs. (6-23) and (6-24) are equivalent to Eqs. (6-3) and (6-4), respectively. Substitution into Eq. (6-18) yields

$$\mu = \mu^0 - S^0(T - T^0) + V^0(P - P^0) - g'(T, P, T^0, P^0) + g''(T, P, T^0, P^0) \tag{6-25}$$

where

$$g'(T, P, T^0, P^0) = \int_{T^0}^{T} \left[\int_{T^0}^{T} \left(\frac{C_p}{T}\right)_{P^0} dT - \int_{P^0}^{P} (\alpha_p V)_T \, dP \right]_{P_0} dT \tag{6-26}$$

and

$$g''(T, P, T^0, P^0) = \int_{P^0}^{P} \left[\int_{T^0}^{T} (\alpha_p V)_{P^0} \, dT - \int_{P^0}^{P} (\kappa_T V)_T \, dP \right]_T dP \tag{6-27}$$

(Note that the second term in Eq. (6-26) vanishes because the upper limit of integration, the dummy variable P, is held constant at P^0 in the second integration for dT.) In Eq. (6-25), μ^0 and S^0 can be assigned arbitrary values; the absolute value of volume, however, has physical significance and, therefore, V^0 must be the actual volume corresponding to T^0 and P^0.

In summary, we have seen that the information content of the Fundamental Equation is contained in the data set consisting of the three second-order partial derivatives C_p, α_p, and κ_T in addition to one value of the specific volume in a reference state.

It should be clear that an equally valid data set consists of C_p and the *P-V-T* relationship, Eq. (6-4), since the latter contains the information content of α_p, κ_T, and the absolute value of volume. The *P-V-T* relationship is so commonly used that it is usually called *the* equation of state (as opposed to the equations of state, which is the general terminology for all the first-order derivatives of the Fundamental Equation). The data set of C_p and *P-V-T* is most commonly used to solve a wide variety of problems because this information is most readily available. In Section 6.3 we shall describe a method of obtaining any other derivatives involving thermodynamic properties directly from this data set.

Before concluding this discussion of the Gibbs free energy representation, it is interesting to compare the results with those for the energy in Section 5.3. From Eq. (6-25) for a change in state from T_1, P_1, to T_2, P_2,

$$\Delta G = \Delta\mu = -S^0(T_2 - T_1) + V^0(P_2 - P_1) - g'(T_2, P_2, T_1, P_1) + g''(T_2, P_2, T_1, P_1) \tag{6-28}$$

Whereas ΔU for the corresponding change in state would have a value independent of the reference values assigned to S^0 and μ^0, the value of ΔG, as given by Eq. (6-28) depends on the value of S^0. Since S^0 is an arbitrary constant, there can be no direct physical significance associated with ΔG for the general change in state under consideration.[4]

6.2 Fugacity

There is an additional thermodynamic property in common use that is closely related to the chemical potential: the fugacity, f. For a pure material, the fugacity is defined in Eq. (6-29):

$$\mu \equiv RT \ln f + \lambda(T) \qquad \text{(6-29)}^5$$

[4] Note that for isothermal processes, ΔG is no longer a function of S^0, and Eq. (6-16) reduces to $\Delta G = \int_{P_1}^{P_2} V\,dP$. In this case, we can attach physical significance to the value of ΔG. Specifically, it represents the negative of the reversible flow work for the isothermal process (see Section 4.11).

[5] The choice of the function defining the fugacity, namely, $RT \ln f$, was made by G. N. Lewis in 1901 [see *Proc. Am. Acad. Arts Sci.*, **37**, 49 (1901)]. The rationale behind this form was that the fugacity for a nonideal system would be analogous to the pressure for an ideal gas. That is, for an isothermal change in state for an ideal gas, $\Delta G = RT \ln (P_2/P_1)$, whereas for a nonideal system, $\Delta G = RT \ln (f_2/f_1)$. Although it is sometimes misleading to think of the fugacity simply as a "corrected pressure," the logarithmic relationship of fugacity to chemical potential is quite convenient in dealing with mixtures, phase equilibria, and chemical reactions.

We have seen that μ is a function of T, P, μ^0, and S^0. That portion of the chemical potential dependence on S^0 is included in the λ-functionality; thus, the value of λ depends on T and S^0. The value of the fugacity, as defined by Eq. (6-29), is dependent on T, P, and μ^0. In other words, the fugacity describes all of the pressure-dependence and a portion of the temperature-dependence (excluding the S^0-dependence) of the chemical potential.

The function $\lambda(T)$ in Eq. (6-29) can be eliminated by defining a reference state for the fugacity at some arbitrarily chosen pressure *but at the same temperature as the system.* Calling μ^0 the chemical potential of this reference state,

$$\mu^0 = RT \ln f^0 + \lambda(T) \tag{6-30}$$

Subtracting Eq. (6-30) from Eq. (6-29), we have

$$\mu - \mu^0 = RT \ln \left(\frac{f}{f^0}\right) \tag{6-31}$$

Special note must be made of the fact that the temperature of the reference state is a floating variable that changes as the temperature of the system of interest changes.

Eq. (6-31) indicates that the ratio of fugacity to the reference state fugacity is fixed by the defining equation, Eq. (6-29), but the absolute value of fugacity is not fixed. To fix the absolute value of fugacity, we include as part of the definition the requirement that the value of the fugacity approach the value of the pressure as the pressure approaches a small enough value such that the material behaves as an ideal gas. That is,

$$\operatorname*{limit}_{P \to P^*} \left(\frac{f}{P}\right) = 1 \tag{6-32}$$

or

$$\mu - \mu^* = RT \ln \left(\frac{f}{P^*}\right) \tag{6-33}$$

where P^* is in the range where ideal gas behavior is approached and μ^* is the chemical potential corresponding to the ideal gas state at T and P^*. Note that the use of Eq. (6-32) as part of the definition of fugacity removes the dependence of fugacity on μ^0. That is, fugacity is a function only of temperature and pressure.

By virtue of Eq. (6-32), the fugacity of an ideal gas is always equal in value to the pressure. The *fugacity coefficient*, ϕ, is defined as

$$\phi \equiv \left(\frac{f}{P}\right) \tag{6-34}$$

Hence, ϕ is a measure of the deviation of the system from ideal gas behavior. The fugacity coefficient, like the fugacity, is an intensive property of the system and is a function of any two other intensive variables for a single-phase pure material.

The variation of fugacity with temperature and pressure can be expressed as an exact differential. It is more convenient to consider $\ln f$ rather than f in this context. Thus,

$$d \ln f = \left(\frac{\partial \ln f}{\partial T}\right)_P dT + \left(\frac{\partial \ln f}{\partial P}\right)_T dP \tag{6-35}$$

The temperature variation can be evaluated from Eq. (6-29):

$$\left(\frac{\partial \ln f}{\partial T}\right)_P = \frac{1}{R}\left[\frac{\partial\left(\frac{\mu}{T}\right)}{\partial T}\right]_P - \frac{1}{R}\left[\frac{d\left(\frac{\lambda}{T}\right)}{dT}\right] \tag{6-36}$$

where the λ-term is written as a total differential because λ is not a function of P. The first term on the right-hand side of Eq. (6-36) reduces to

$$\left[\frac{\partial\left(\frac{\mu}{T}\right)}{\partial T}\right]_P = \frac{1}{T}\left(\frac{\partial \mu}{\partial T}\right)_P - \frac{\mu}{T^2} = -\frac{S}{T} - \frac{1}{T^2}(H - TS) = -\frac{H}{T^2} \tag{6-37}$$

wherein use was made of the Gibbs-Duhem equation for a pure material, Eq. (5-70), and Eq. (5-52). The second term on the right-hand side of Eq. (6-36) can be evaluated by applying Eq. (6-36) to the same system in an ideal gas reference state, namely,

$$\left(\frac{\partial \ln f^*}{\partial T}\right)_{P=P^*} = \frac{1}{R}\left[\frac{\partial\left(\frac{\mu^*}{T}\right)}{\partial T}\right]_{P=P^*} - \frac{1}{R}\frac{d\left(\frac{\lambda}{T}\right)}{dT} \tag{6-38}$$

Since $f^* = P^*$ by Eq. (6-32), the left-hand side of Eq. (6-38) vanishes. Thus,

$$\frac{1}{R}\frac{d\left(\frac{\lambda}{T}\right)}{dT} = \frac{1}{R}\left[\frac{\partial\left(\frac{\mu^*}{T}\right)}{\partial T}\right]_{P=P^*} = -\frac{H^*}{RT^2} \tag{6-39}$$

wherein we made use of Eq. (6-37) as applied to the ideal gas reference state at T and P^* with a corresponding enthalpy of H^*.

Substituting Eqs. (6-37) and (6-39) into Eq. (6-36),

$$\left(\frac{\partial \ln f}{\partial T}\right)_P = -\frac{(H - H^*)}{RT^2} \tag{6-40}$$

where $(H - H^*)$ is the difference in enthalpy of the real system at T and P and the enthalpy the material would have in an ideal gas state at T and P^*. Since the enthalpy of an ideal gas is independent of pressure, the value of P^* is immaterial provided that it is in the ideal gas range.

Note that the temperature-dependence of the fugacity is not related to any reference value assigned to the entropy, whereas the temperature-dependence of the chemical potential does depend on S^0.

The second differential in Eq. (6-35), the pressure-dependence of the fugacity, can be evaluated from Eq. (6-29) in conjunction with Eq. (5-70). Thus,

$$\left(\frac{\partial \ln f}{\partial P}\right)_T = \frac{1}{R}\left[\frac{\partial\left(\frac{\mu}{T}\right)}{\partial P}\right]_T = \frac{1}{RT}\left(\frac{\partial \mu}{\partial P}\right)_T = \frac{V}{RT} \tag{6-41}$$

Substituting Eqs. (6-40) and (6-41) into Eq. (6-35), we have for the total differential of $\ln f$,

$$d \ln f = -\left[\frac{(H - H^*)}{RT^2}\right] dT + \left(\frac{V}{RT}\right) dP \tag{6-42}$$

Example 6.1

Evaluate the total differential of the fugacity coefficient with respect to the properties T and P.

Solution

By analogy to the discussion above for fugacity, we might expect a simple solution if we work in terms of $\ln \phi$ rather than ϕ. Thus, from the differential of the logarithm of Eq. (6-34),

$$d \ln \phi = d \ln f - d \ln P \tag{6-43}$$

Substituting Eq. (6-42) into Eq. (6-43) and combining terms,

$$d \ln \phi = -\left[\frac{(H - H^*)}{RT^2}\right] dT + \left(\frac{V}{RT} - \frac{1}{P}\right) dP \tag{6-44}$$

If we define V^* as the volume an ideal gas would have at the system conditions of T and P (i.e., $V^* = RT/P$), Eq. (6-44) becomes

$$d \ln \phi = -\left[\frac{(H - H^*)}{RT^2}\right] dT + \left(\frac{V - V^*}{RT}\right) dP \tag{6-45}$$

6.3 Transformation of Partial Derivatives

We have seen in Section 6.1 that the three independent second-order partial derivatives of the Gibbs free energy representation are a convenient set because they can be measured readily. There are, nevertheless, many cases in which we must determine other partial derivatives. We shall now demonstrate that any partial derivative for a pure material can be expressed in terms of S, T, P, and V and also that any partial involving only these four variables can be expressed in terms of second-order partial derivatives of one of the four representations of the Fundamental Equation. Since we have already demonstrated in Section 5.6 that any second-order partial derivative of one representation of the Fundamental Equation can be expressed as a function of the three independent second-order partials of any other representation, it follows that any partial derivative can be expressed as a function of C_p, κ_T, and, α_p or their equivalent.

A partial derivative may involve intensive and extensive variables. For a *pure* material, only $n + 1 = 2$ intensive variables are independent; hence, a partial derivative involving only intensive variables can be expressed as

$(\partial b/\partial c)_d$ where it is implied that N is constant. That is,

$$\left(\frac{\partial b}{\partial c}\right)_{d,N} = \left(\frac{\partial b}{\partial c}\right)_d \tag{6-46}$$

We shall now show that for a pure material, any partial derivative involving extensive variables can always be reduced to expressions involving partial derivatives of entirely intensive variables.

Consider the derivative $(\partial b/\partial c)_{d,e}$ where one of the four variables is extensive.

1i. If c is extensive, then

$$\left(\frac{\partial b}{\partial \underline{c}}\right)_{d,e} = 0 \tag{6-47}$$

The proof follows from applying Euler's theorem to $b = f(\underline{c}, d, e)$:

$$0 = \left(\frac{\partial b}{\partial \underline{c}}\right)_{d,e} \underline{c}$$

1ii. If b is extensive, then

$$\left(\frac{\partial \underline{b}}{\partial c}\right)_{d,e} = \frac{1}{\left(\dfrac{\partial c}{\partial \underline{b}}\right)_{d,e}} = \frac{1}{0} = \infty \tag{6-48}$$

1iii. If d(or e) is extensive, then

$$\left(\frac{\partial b}{\partial c}\right)_{\underline{d},e} = \left(\frac{\partial b}{\partial c}\right)_e \tag{6-49}$$

Since

$$\left(\frac{\partial b}{\partial c}\right)_{\underline{d},e} = -\frac{\left(\dfrac{\partial \underline{d}}{\partial c}\right)_{b,e}}{\left(\dfrac{\partial \underline{d}}{\partial b}\right)_{c,e}} = -\frac{\left(\dfrac{\partial \underline{d}}{\partial N}\right)_{b,e}\left(\dfrac{\partial N}{\partial c}\right)_{b,e}}{\left(\dfrac{\partial \underline{d}}{\partial N}\right)_{c,e}\left(\dfrac{\partial N}{\partial b}\right)_{c,e}}$$

and

$$\left(\frac{\partial \underline{d}}{\partial N}\right)_{b,e} = \left(\frac{\partial \underline{d}}{\partial N}\right)_{c,e} = d$$

then,

$$\left(\frac{\partial b}{\partial c}\right)_{\underline{d},e} = -\frac{\left(\dfrac{\partial N}{\partial c}\right)_{b,e}}{\left(\dfrac{\partial N}{\partial b}\right)_{c,e}} = \left(\frac{\partial b}{\partial c}\right)_{e,N} = \left(\frac{\partial b}{\partial c}\right)_e$$

Thus, if only one variable is extensive, then the partial derivative is finite and nonzero only if d or e is extensive. The extensive variable may be deleted to yield a partial involving only intensive variables.

Now, consider the three cases in which two of the four variables are extensive.

2i. If b and c are extensive, then

$$\left(\frac{\partial \underline{b}}{\partial \underline{c}}\right)_{d,e} = \frac{b}{c} \tag{6-50}$$

which follows directly by applying Euler's theorem to $\underline{b} = f(\underline{c}, d, e)$:

$$\underline{b} = \left(\frac{\partial \underline{b}}{\partial \underline{c}}\right)_{d,e} \underline{c} \quad \text{and} \quad \underline{b} = Nb, \qquad \underline{c} = Nc$$

2ii. If b and d (or e) are extensive, then

$$\left(\frac{\partial \underline{b}}{\partial c}\right)_{\underline{d},e} = N\left[\left(\frac{\partial b}{\partial c}\right)_e - \frac{b}{d}\left(\frac{\partial d}{\partial c}\right)_e\right] \tag{6-51}$$

Since, expanding $\underline{b} = Nb$,

$$\left(\frac{\partial \underline{b}}{\partial c}\right)_{\underline{d},e} = N\left(\frac{\partial b}{\partial c}\right)_{\underline{d},e} + b\left(\frac{\partial N}{\partial c}\right)_{\underline{d},e} \tag{6-52}$$

Eq. (6-49) may be used to reduce $(\partial b/\partial c)_{\underline{d},e}$. The last term is reduced as follows:

$$\left(\frac{\partial N}{\partial c}\right)_{\underline{d},e} = -\frac{\left(\frac{\partial \underline{d}}{\partial c}\right)_{N,e}}{\left(\frac{\partial \underline{d}}{\partial N}\right)_{c,e}} = -\frac{N\left(\frac{\partial d}{\partial c}\right)_e}{d}$$

2iii. If d and e are extensive, then

$$\left(\frac{\partial b}{\partial c}\right)_{\underline{d},\underline{e}} = -\frac{\left(\frac{\partial \underline{d}}{\partial c}\right)_{b,\underline{e}}}{\left(\frac{\partial \underline{d}}{\partial b}\right)_{c,\underline{e}}} \tag{6-53}$$

The numerator and denominator can each be reduced by applying the results of case (2ii).

Any partial derivative involving three extensive variables can now be reduced to partials involving two extensive variables by using Eq. (6-52) followed by one or more of the steps illustrated above. Similarly, partials involving four extensive variables can be reduced to three, etc. The net result is that *any partial derivative for a pure material can be expressed in terms of partials involving only three intensive variables.*

Let us now consider derivatives in intensive variables only and show that they can be reduced to expressions involving only S, T, P, V.

In general, a partial derivative of intensive variables may involve U, H, A, G, S, T, P, V. We have excluded N because it is extensive; μ and f are also excluded because they are directly related to G for a pure material.

The first step in reducing the general derivative $(\partial b/\partial c)_d$ is to eliminate any of the four potential functions U, H, A, or G which may appear as b, c, or d. The three possibilities are treated as follows:

1. If b is U, H, A, or G, eliminate it by using one of Eqs. (5-53) through (5-56). For example,

$$\left(\frac{\partial G}{\partial S}\right)_T = -S\overbrace{\left(\frac{\partial T}{\partial S}\right)_T}^{0} + V\left(\frac{\partial P}{\partial S}\right)_T \tag{6-54}$$

2. If c is U, H, A, or G, invert the derivative by using the identity

$$\left(\frac{\partial c}{\partial b}\right)_d = \frac{1}{\left(\frac{\partial b}{\partial c}\right)_d} \tag{6-55}$$

and then proceed by step 1.

3. If d is U, H, A, or G, bring the potential function into the brackets by using the relation

$$\left(\frac{\partial b}{\partial c}\right)_d = -\frac{1}{\left(\frac{\partial c}{\partial d}\right)_b\left(\frac{\partial d}{\partial b}\right)_c} = -\frac{\left(\frac{\partial d}{\partial c}\right)_b}{\left(\frac{\partial d}{\partial b}\right)_c} \tag{6-56}$$

and then proceed by step 1.

Using these three steps, we can reduce any partial derivative to a function of partial derivatives involving only S, V, T, and P. It is readily shown that any such partial derivative is a second-order derivative of one of the four forms of the Fundamental Equation. Since two of the three variables in $(\partial b/\partial c)_d$ must be conjugate coordinates, there are three possibilities:

1. If b and c are conjugate coordinates, then c and d cannot be conjugate coordinates and, hence, c and d define the Legendre transform $\psi(c, d)$. Since b is the conjugate of c, it follows that $b \equiv \pm(\partial\psi/\partial c)_d$. Therefore,

$$\left(\frac{\partial b}{\partial c}\right)_d = \pm\left(\frac{\partial^2\psi}{\partial c^2}\right)_d = \pm\psi_{cc}(c, d) \tag{6-57}$$

For example, the appropriate representation for $(\partial P/\partial V)_T$ is $\psi(V, T)$ or A. Thus,

$$\left(\frac{\partial P}{\partial V}\right)_T = -\left(\frac{\partial^2 A}{\partial V^2}\right)_T = -A_{VV}$$

2. If b and d are conjugate coordinates, then $b \equiv \pm(\partial\psi/\partial d)_c$. Therefore,

$$\left(\frac{\partial b}{\partial c}\right)_d = \pm\frac{\partial^2\psi}{(\partial c)(\partial d)} = \pm\psi_{cd}(c, d) \tag{6-58}$$

For example, the appropriate representation for $(\partial P/\partial S)_V$ is $\psi(S, V)$ or U. Thus,

$$\left(\frac{\partial P}{\partial S}\right)_V = -\frac{\partial^2 U}{\partial S\,\partial V} = -U_{SV}$$

3. If c and d are conjugate coordinates, then

$$\left(\frac{\partial b}{\partial c}\right)_d = \frac{1}{\left(\frac{\partial c}{\partial b}\right)_d} = \pm[\psi_{bd}(b, d)]^{-1} \tag{6-59}$$

In summary, we have now demonstrated that any partial derivative for a pure material is expressible in terms of second-order derivatives of a representation of the Fundamental Equation. Using the Legendre transformation procedure described in Section 5.6, we can then convert from one representation to any other.

Example 6.2

Of the 24 partial derivatives involving S, T, P, V, 8 are given below. Choose a set of three that is independent and then express the others in terms of this set of three.

1	2	3	4	5	6
$\left(\frac{\partial S}{\partial T}\right)_P$,	$\left(\frac{\partial S}{\partial P}\right)_T$,	$\left(\frac{\partial S}{\partial T}\right)_V$,	$\left(\frac{\partial P}{\partial T}\right)_V$,	$\left(\frac{\partial S}{\partial P}\right)_V$,	$\left(\frac{\partial S}{\partial V}\right)_P$,
7	8				
$\left(\frac{\partial V}{\partial T}\right)_P$,	$\left(\frac{\partial V}{\partial P}\right)_T$				

Solution

Using Eqs. (6-57) through (6-59), we see that each derivative can be expressed in terms of a second-order partial derivative of a representation of the Fundamental Equation. They are, respectively,

1	2	3	4	5	6
$-G_{TT}$,	$-G_{TP}$,	$-A_{TT}$,	$-A_{TV}$,	$-(U_{VS})^{-1}$,	$(H_{PS})^{-1}$,
7	8				
G_{PT},	G_{PP}				

Since the second and seventh are clearly related by the Maxwell reciprocity theorem, the seventh can be eliminated. Of the seven remaining, the first (G_{TT}), the second (G_{TP}), and the eighth (G_{PP}) form an independent set since these are the three independent second-order derivatives of G. Using these as a base set, we see that the derivatives of A and H can be transformed directly by using Eqs. (5-131a) through (5-131d):

3. $$\left(\frac{\partial S}{\partial T}\right)_V = -A_{TT} = -\psi_{z_2z_2} = -y_{z_2z_2} + \frac{(y_{z_1z_2})^2}{y_{z_1z_1}} = -G_{TT} + \frac{(G_{TP})^2}{G_{PP}}$$

or

$$\left(\frac{\partial S}{\partial T}\right)_V = \left(\frac{\partial S}{\partial T}\right)_P + \frac{\left(\frac{\partial V}{\partial T}\right)_P^2}{\left(\frac{\partial V}{\partial P}\right)_T} \tag{6-59a}$$

4. $\left(\frac{\partial P}{\partial T}\right)_V = -A_{TV} = -\psi_{\xi_1 z_2} = -\frac{y_{z_1 z_2}}{y_{z_1 z_1}} = -\frac{G_{TP}}{G_{PP}}$

or

$$\left(\frac{\partial P}{\partial T}\right)_V = -\frac{\left(\frac{\partial V}{\partial T}\right)_P}{\left(\frac{\partial V}{\partial P}\right)_T} \tag{6-59b}$$

6. $\left(\frac{\partial S}{\partial V}\right)_P = (H_{SP})^{-1} = (\psi_{-\xi_1 z_2})^{-1} = -(\psi_{\xi_1 z_2})^{-1} = -\frac{y_{z_1 z_1}}{y_{z_1 z_2}} = -\frac{G_{TT}}{G_{TP}}$

or

$$\left(\frac{\partial S}{\partial V}\right)_P = \frac{\left(\frac{\partial S}{\partial T}\right)_P}{\left(\frac{\partial V}{\partial T}\right)_P} \tag{6-59c}$$

The fifth derivative requires a double transformation. First, let us transform it to the A-representation and thence to G.

5. $\left(\frac{\partial S}{\partial P}\right)_V = -(U_{VS})^{-1} = (\psi_{\xi_1 z_2})^{-1} = \frac{A_{TT}}{A_{TV}} = \frac{G_{TT} - (G_{TP})^2/G_{PP}}{G_{TP}/G_{PP}}$

$= -G_{TP} + \frac{G_{TT}G_{PP}}{G_{TP}}$

or

$$\left(\frac{\partial S}{\partial P}\right)_V = -\left(\frac{\partial V}{\partial T}\right)_P - \frac{\left(\frac{\partial S}{\partial T}\right)_P \left(\frac{\partial V}{\partial P}\right)_T}{\left(\frac{\partial V}{\partial T}\right)_P} \tag{6-59d}$$

Although the procedure described above is always valid and unequivocal, there are some shortcut methods which, when applicable, are less tedious. For example, Eq. (6-59b) could be obtained directly by applying Eq. (6-56) to $(\partial P/\partial T)_V$. This shortcut is applicable when we have a derivative $(\partial b/\partial c)_d$ in which b and c are the independent properties of the representation to which $(\partial b/\partial c)_d$ is to be transformed. That is, P and T in $(\partial P/\partial T)_V$ are the independent variables for G.

The second shortcut involves the use of the chain rule,

$$\left(\frac{\partial b}{\partial c}\right)_d = \left(\frac{\partial b}{\partial e}\right)_d \left(\frac{\partial e}{\partial c}\right)_d = \frac{\left(\frac{\partial b}{\partial e}\right)_d}{\left(\frac{\partial c}{\partial e}\right)_d} \tag{6-60}$$

If d is a property of the desired representation (e.g., T or P for G), but b and c are not, then e can be taken as the second independent property in the desired representation. Thus, Eq. (6-60) yields Eq. (6-59c) directly by substituting T for e.

In general, shortcuts should be sought first, but if none can be found, then recourse should be made to the general procedure. This point is illustrated in the following example.

Example 6.3

Evaluate the following partial derivatives as functions of P, V, T, their partial derivatives and C_p.

$$\text{(a)}\quad \left(\frac{\partial S}{\partial P}\right)_G \qquad \text{(b)}\quad \left(\frac{\partial A}{\partial G}\right)_T$$

Solution (a)

Using Eqs. (6-56) and (6-54), we find that

$$\left(\frac{\partial S}{\partial P}\right)_G = -\frac{\left(\frac{\partial G}{\partial P}\right)_S}{\left(\frac{\partial G}{\partial S}\right)_P} = \frac{-S\left(\frac{\partial T}{\partial P}\right)_S + V}{S\left(\frac{\partial T}{\partial S}\right)_P} \tag{6-61}$$

Using Eqs. (6-56) and (5-87) to eliminate $(\partial T/\partial P)_S$, we find that

$$\left(\frac{\partial T}{\partial P}\right)_S = -\frac{\left(\frac{\partial S}{\partial P}\right)_T}{\left(\frac{\partial S}{\partial T}\right)_P} = \frac{\left(\frac{\partial V}{\partial T}\right)_P}{\frac{C_p}{T}} \tag{6-62}$$

so that

$$\left(\frac{\partial S}{\partial P}\right)_G = \frac{VC_p}{TS} - \left(\frac{\partial V}{\partial T}\right)_P \tag{6-63}$$

The entropy, S, in Eq. (6-63) can be expressed as a function of the desired variables, as demonstrated previously in Section 6.1 [see Eq. (6-23)].

Solution (b)

This partial derivative can be reduced by the following shortcut procedure

$$\left(\frac{\partial A}{\partial G}\right)_T = \frac{\left(\frac{\partial A}{\partial V}\right)_T}{\left(\frac{\partial G}{\partial V}\right)_T} = -\frac{P}{V\left(\frac{\partial P}{\partial V}\right)_T} \tag{6-64}$$

in which Eq. (6-54) was employed to reduce the intermediate partial derivatives. The insertion of V was a convenient but not an arbitrary choice. Although P could have been used in place of V with equal simplicity, S would have been less convenient. The choice was guided by the fact that $A = f(T, V)$ and $G = f(T, P)$; since T was the constant in the differentiation, either T, V or T, P would be a convenient set of independent variables.

Example 6.4

A well-insulated vessel is divided into two compartments by a partition. The volume of each compartment is 1 ft³. One compartment initially contains

0.25 lb-mol of argon at 70°F, and the other compartment is initially evacuated. The partition is then removed, and the gas is allowed to equilibrate. What is the final temperature?

Note: Under these conditions, argon is not an ideal gas. Assume that the Van der Waals equation of state is valid:

$$\left(P + \frac{a}{V^2}\right)(V - b) = RT \tag{6-65}$$

where $a = 345$ atm-ft^6/(lb-mol)^2 and $b = 0.515$ ft^3/lb-mol. Assume that $C_v = 3.0$ Btu/lb-mol°R, independent of temperature or pressure. Assume that the mass of the walls can be neglected.

Solution

If we choose the gas as the system, the process occurs at constant energy. Furthermore, we know the initial and final volume of the system. Since we are concerned with a one-component simple system, we can evaluate the change in any property if the changes of two other properties are known. Let us disregard the actual path between the initial and final conditions because it was clearly irreversible. If the process were carried out reversibly, we can relate the temperature variation to the volume and energy by

$$dT = \left(\frac{\partial T}{\partial V}\right)_U dV + \left(\frac{\partial T}{\partial U}\right)_V dU \tag{6-66}$$

Eq. (6-66) is, of course, an exact differential. If we choose a reversible path in which U is constant, then Eq. (6-66) can be integrated:

$$T_f - T_i = \int_{V_i}^{V_f} \left(\frac{\partial T}{\partial V}\right)_U dV \tag{6-67}$$

The problem can be solved by expressing $(\partial T/\partial V)_U$ in terms of the available data. Thus,

$$\left(\frac{\partial T}{\partial V}\right)_U = -\frac{\left(\frac{\partial U}{\partial V}\right)_T}{\left(\frac{\partial U}{\partial T}\right)_V} = \frac{-T\left(\frac{\partial S}{\partial V}\right)_T + P}{T\left(\frac{\partial S}{\partial T}\right)_V} \tag{6-68}$$

We can evaluate $(\partial S/\partial V)_T$ in terms of P-V-T derivatives by applying the reciprocity relation to the Legendre transform $\psi(T, V)$ (see Section 5.4). Thus,

$$d\psi = -S\,dT - P\,dV \tag{6-69}$$

and

$$\left(\frac{\partial S}{\partial V}\right)_T = \left(\frac{\partial P}{\partial T}\right)_V \tag{6-70}$$

so that Eq. (6-68) becomes

$$\left(\frac{\partial T}{\partial V}\right)_U = \frac{-T\left(\frac{\partial P}{\partial T}\right)_V + P}{C_v} \tag{6-71}$$

From the Van der Waals equation, we find

$$\left(\frac{\partial P}{\partial T}\right)_V = \frac{R}{V - b} \tag{6-72}$$

so that Eq. (6-71) reduces to

$$\left(\frac{\partial T}{\partial V}\right)_U = -\frac{a}{C_v V^2} \tag{6-73}$$

Substituting Eq. (6-73) into Eq. (6-67), we obtain

$$T_f = T_i + \frac{a}{C_v}\left(\frac{1}{V_f} - \frac{1}{V_i}\right) \tag{6-74}$$

or

$$T_f = T_i - \frac{a}{2C_v V_i} \tag{6-75}$$

$$= 530 - \frac{345}{(2)(3)(0.3676)(4)} = 530 - 39$$

where $(0.3676)\text{ft}^3\text{atm} = 1\,\text{Btu}$. Thus,

$$T_f = 491°\text{R} = 31°\text{F}$$

6.4 Basis for Equation of State Correlations

In Example 6.4, it was suggested that, as an approximation, Van der Waals equation of state might be used to estimate the gas-phase P-V-T properties of argon. This suggestion was made more for computational convenience than to obtain a very accurate result. This famous relation is in fact quite crude but is representative of a large number of other similarly crude equations. Even the most recent and best equations of state can only predict approximately the effect of temperature and pressure on the specific volume of a pure substance.

It is of interest to explore further why we are so handicapped. In fact, we do have, for gases, a rigorous basis from which to start. From statistical mechanics it can be demonstrated[6] that pressure is related to the configurational integral Z^* as:

$$P = kT\left(\frac{\partial \ln Z^*}{\partial \underline{V}}\right)_{T,N} \tag{6-76}$$

where

$$Z^* = \int \underset{\underline{V}}{\cdots} \int \left\{\exp - \left(\frac{\Phi}{kT}\right)\right\} dx_1 \cdots dz_N \tag{6-77}$$

N is the number of molecules in the system and Φ is the total potential energy of interaction between these molecules. The spatial coordinates of each

[6] Hirschfelder, J. O., C. F. Curtiss, nd R. B. Bird, *Molecular Theory of Gases and Liquids* (New York: John Wiley & Sons, Inc., 1954).

molecule are given by x_j, y_j, z_j, and the integration is over the system volume $\underline{V}$. Evaluation of the configurational integral is, at present, impossible since we do not yet know how to relate Φ to x, y, and z with any real degree of confidence. In passing though, it is interesting to note that if there were *no* intermolecular potential energies (i.e., no forces of attraction or repulsion between molecules), then Z^* would simply be the system volume raised to the power N. From Eq. (6-76)

$$P = kT\left[\frac{\partial \ln (\underline{V}^N)}{\partial \underline{V}}\right]_{T,N} = \frac{NkT}{\underline{V}}$$

or the ideal gas equation results.

If the gas is not ideal, to make Eq. (6-77) more tractable, the assumption is usually invoked that the total potential energy of the system of N molecules can be equated to the sum of interaction energies between all possible binaries.

$$\Phi = \sum_{i=1}^{N} \sum_{j=i+1}^{N} u_{ij}(r_{ij}) \tag{6-78}$$

Here u_{ij} is the intermolecular potential energy just between molecules i and j and, in Eq. (6-78) u_{ij} is taken to be a function only of the molecular separation distance r_{ij}.[7] Thus, if one were examining a system of, say, three molecules, 1, 2, 3, the total intermolecular system energy would be taken to be $u_{12} + u_{13} + u_{23}$, with the energies varying with the spacial coordinates of the molecules.

Eq. (6-78) is probably not a bad approximation except for liquids or gases at very high pressures. It is more difficult to delineate the functional dependence of u_{ij} on r_{ij}. Many forms have been suggested (e.g., the famous Lennard-Jones 12-6 potential), but a detailed discussion of these relations is inappropriate here. Of more importance are generalizations that we can extract from the theory we have so briefly summarized above. For example, consider the general term $u_{ij}(r_{ij})$. We might inquire if it is possible to obtain a universal function u' if we nondimensionalize u_{ij}. That is, can we find a function of the form

$$u_{ij} = \epsilon u'\left(\frac{r_{ij}}{\sigma}\right) \tag{6-79}$$

where ϵ is some parameter with the units of energy and σ is a similar term with the units of length. For a pure material, ϵ and σ are constants. For a mixture, there would presumably be different ϵ and σ values for each type of binary interaction. The significant result of Eq. (6-79) is that if it were used in Eq. (6-78) which was, in turn, substituted into Eq. (6-77), then

[7] This particular assumption is not always made. Sometimes u_{ij} is assumed to be a function of the separation distance and some angle that is related to the orientations of the molecules.

$$Z^* = \sigma^{3N} \int \cdots \int \left\{ \exp - \Sigma \Sigma \left(\frac{\epsilon}{kT}\right) u'\left(\frac{r_{ij}}{\sigma}\right) \right\} d\left(\frac{x_1}{\sigma}\right) \cdots d\left(\frac{z_N}{\sigma}\right)$$

$$Z^* = \sigma^{3N} f\left(\frac{kT}{\epsilon}, \frac{V}{\sigma^3}\right) \tag{6-80}$$

Eq. (6-80) is both interesting and valuable. With the assumptions made, one concludes that Eq. (6-76) might be written in the nondimensional form,

$$\frac{P\sigma}{\epsilon} = f\left(\frac{kT}{\epsilon}, \frac{V}{\sigma^3}\right) \tag{6-81}$$

or

$$P^* = f(T^*, V^*) \tag{6-82}$$

Thus, for example, if the Lennard-Jones 12-6 potential were chosen for Eq. (6-79)

$$u_{ij} = 4\epsilon\left[\left(\frac{\sigma}{r_{ij}}\right)^{12} - \left(\frac{\sigma}{r_{ij}}\right)^{6}\right] \tag{6-83}$$

then the characteristic energy and distance parameters of this relation may be used to nondimensionalize P, T, and V in Eqs. (6-81) and (6-82) for a pure material. Alternatively, if one further assumes that there is a linear correlation between ϵ/σ and P_c, ϵ/k and T_c, and between σ^3 and V_c $(=Z_cRT_c/P_c)$, it is then obvious that with $Z \equiv PV/RT$

$$Z = f(Z_c, P_r, T_r) \tag{6-84}$$

This result is a familar relation expressing the compressibility factor as a function of $P/P_c = P_r$, $T/T_c = T_r$, and Z_c. In many cases Z_c is eliminated for simplicity and the famous two-parameter "law of corresponding states" results.

The key step in this very rapid development was Eq. (6-79); presumably, different "universal u_{ij} functions" that included other parameters might have been introduced. The dipole moment may, for example, be important when delineating intermolecular energies. Likewise, the polarizability, the quadrapole moment, a shape factor or other parameters could be introduced. If so, a similar development would lead to Eq. (6-84) where besides Z_c, P_r, T_r, there would be other dimensionless groups characteristic of the material.

Through statistical mechanics, one can even infer the functionality of Eq. (6-84): Z is closely approximated by a polynomial in inverse volume. Called the *virial equation*, it is written as:

$$Z = 1 + \frac{B}{V} + \frac{C}{V^2} + \ldots \tag{6-85}$$

If our development were sound, then the terms (B/V), (C/V^2), etc., should be expressible in terms of an intermolecular potential function u_{ij}. It can be shown that for the second virial term

$$B = 2\pi N_0 \int_0^{\infty} \left[1 - \left\{\exp - \left(\frac{u_{ij}}{kT}\right)\right\}\right] r_{ij}^2 \, dr_{ij} \tag{6-86}$$

where N_0 is Avogadro's number. Thus, with u_{ij} given say by Eq. (6-83), it is clear that (B/V) is a function only of (V/σ^3) and (kT/ϵ). It is only a short but obvious jump to obtain Eq. (6-82).

A similar treatment may be applied for the third and higher virials, although the mathematics are considerably more complicated.

We now have a choice of directions if we wish to pursue further the development of an equation of state. We may elect to concentrate on the theory-based virial equation or we may wish to emphasize analytical or graphical forms of the relation shown functionally in Eq. (6-84). Both paths have been extensively developed, and although the latter will be of more value to us in later portions of this chapter and in the remainder of the book, the former needs at least a few words to emphasize some of its more salient advantages and disadvantages. Let us, in our further discussion of the virial equation, truncate it after the second virial. This step will limit the usefulness of the equation to a region of moderate pressures, but inclusion of higher terms would merely overcomplicate matters and, in most cases, estimations of the third virial coefficient C can now only be made in a very approximate manner.

In the truncated form,

$$Z = 1 + \frac{B}{V} \tag{6-87}$$

The first point to be made is that B is a function only of temperature, as is obvious from Eq. (6-86). It is readily determinable if the binary intermolecular potential energy function $u_{ij}(r_{ij})$ is known. Though such functions are not available, we at least can approximate such relationships [by, for example, Eq. (6-83)]. Also, $B(T)$ turns out to be somewhat insensitive to the exact functional form of $u_{ij}(r_{ij})$. For any pure substance, we need to know the characteristic parameters [such as ϵ and σ in Eq. (6-83)] in order to perform any detailed calculations. Many authors have inserted particular $u_{ij}(r_{ij})$ functions, determined ϵ and σ from comparison of experimental B values with calculated ones, and tabulated ϵ, σ for a number of pure substances. Thus, even the theoretical, truncated form of the virial equation requires some specific constants before it may be employed to determine volumetric data or be operated upon to obtain thermodynamic properties.

Eq. (6-87) has a simple analytical form; in fact, if the substitution $V = ZRT/P$ is made and terms in P^2 are neglected, one obtains

$$Z = 1 + \frac{BP}{RT} \tag{6-88}$$

This form is even more widely used than Eq. (6-87). Z is predicted to be a linear function of P at constant T, which it certainly is at low pressures.

Example 6.5

A residual function α is often employed in thermodynamic analyses.

$$\alpha \equiv \frac{RT}{P} - V$$

Determine the limiting value of α as the pressure is reduced to a very low value.

Solution

At low pressures,

$$\alpha = \frac{RT}{P} - V = \frac{RT}{P} - \frac{ZRT}{P}$$

$$= \left(\frac{RT}{P}\right)(1 - Z) = \left(\frac{RT}{P}\right)\left(\frac{-BP}{RT}\right) = -B$$

Thus,

$$\lim_{P \to 0} \alpha = -B$$

Should we wish to obtain a numerical value of B for a specific material, the solution of Eq. (6-86) for a reasonable potential energy function is sought and a tabulation of the particular σ, ϵ values located. The Lennard-Jones 12-6 is probably the most widely used potential and many tabulations of σ and ϵ are available from this form.[8] (A word of caution is necessary. When using tabulations of ϵ and σ, even for the same potential function, never use ϵ from one tabulation and σ from another. These two parameters must be treated as a set. There exist multiple sets of ϵ, σ which are almost equally satisfactory in calculating virial coefficients, although values of ϵ and σ in different sets may differ considerably from one another.)

Leaving the virial equation, let us return to the more general functional form, Eq. (6-84). Since some liberties were taken in relating (ϵ/k) to T_c, (ϵ/σ) to P_c, and σ^3 to V_c, this relation is certainly not based on much definitive theory. The general concept, however, has been of immense importance to chemical engineers. The two-parameter "law," $Z = f(T_r, P_r)$, has been illustrated in many so-called reduced compressibility factor plots. Only in the last 15–20 years has there been much effort expended to improve these simple and easy-to-use diagrams. The major modifications came when it was realized that a *third parameter* [such as Z_c in Eq. (6-84)] should be introduced to set apart different classes of compounds. The detailed tables of Z as a

[8] See, for example, Hirschfelder *et al.*, *op. cit.*

function of T_r, P_r, and Z_c found in Hougen, Watson, and Ragatz[9] attest to the effort expended in developing this correlation.

An alternate approach that has been equally successful involves substituting for Z_c a parameter characteristic of the vapor pressure of a material. The most commonly used parameter is the acentric factor, ω,

$$\omega \equiv -\log P_{vp_r} \text{ (at } T_r = 0.7) - 1.000 \tag{6-89}$$

where P_{vp_r} is the reduced vapor pressure. As defined, ω for spherically symmetric molecules is essentially zero. For nonsymmetric molecules $\omega > 0$. Eq. (6-84) then becomes

$$Z = f(T_r, P_r, \omega) \tag{6-90}$$

As developed[10,11] for computational ease, the functionality is further expressed as

$$Z = Z^{(0)}(T_r, P_r) + \omega Z^{(1)}(T_r, P_r) \tag{6-91}$$

where $Z^{(0)}$ and $Z^{(1)}$ are functions simply of T_r and P_r. These functions are illustrated in Figures 6.1 and 6.2.

Figures 6.1 and 6.2 (or their table counterpart)[10,11] are excellent representations of most *P-V-T* data. Some recent analytical equations of state are but approximations to these figures. Should one accept this conclusion, then Eq. (6-91) becomes invaluable in obtaining other thermodynamic properties. This usage will be considered in detail later.

To conclude this section, we should at least note that in addition to the virial equation and the graphical or tabular representations of Z as a function of T_r, P_r, and possibly some third parameter, there have been hundreds, if not thousands, of analytical equations of state suggested. A review of these would be out of place here and those interested can find many excellent papers.[12-15] Those now in favor, however, usually will still fit in the categories under Eq. (6-84). An analytical form is certainly preferable for data storage purposes or for machine computation. On occasion throughout this book we will suggest certain analytical equations of state in the problems or in the examples. In general, however, to maintain a common thread, we will use Eq. (6-91) and relations derived from it.

[9] Hougen, O. A., K. M. Watson, and R. A. Ragatz, *Chemical Process Principles, Part II, Thermodynamics*, 2nd ed. (New York: John Wiley & Sons, Inc., 1959).

[10] Lewis, G. N., and M. Randall, *Thermodynamics*, 2nd ed., rev. by L. Brewer and K. S. Pitzer (New York: McGraw-Hill Book Company, 1961).

[11] Pitzer, K. S., D. Z. Lippman, R. F. Curl, C. M. Huggins, and D. E. Petersen, *J. Am. Chem. Soc.*, **77**, 3433 (1955).

[12] Martin, J. J., *Ind. Eng. Chem.*, **59** (12), 34 (1967).

[13] Tsonopoulos, P. L., and J. M. Prausnitz, *Cryogenics*, Oct. 1969, p. 315.

[14] Stiel, L. I., *Ind. Eng. Chem.*, **60** (5), 50 (1968).

[15] Leland, T. W., Jr., and P. S. Chappelear, *Ind. Eng. Chem.*, **60** (7), 15 (1968).

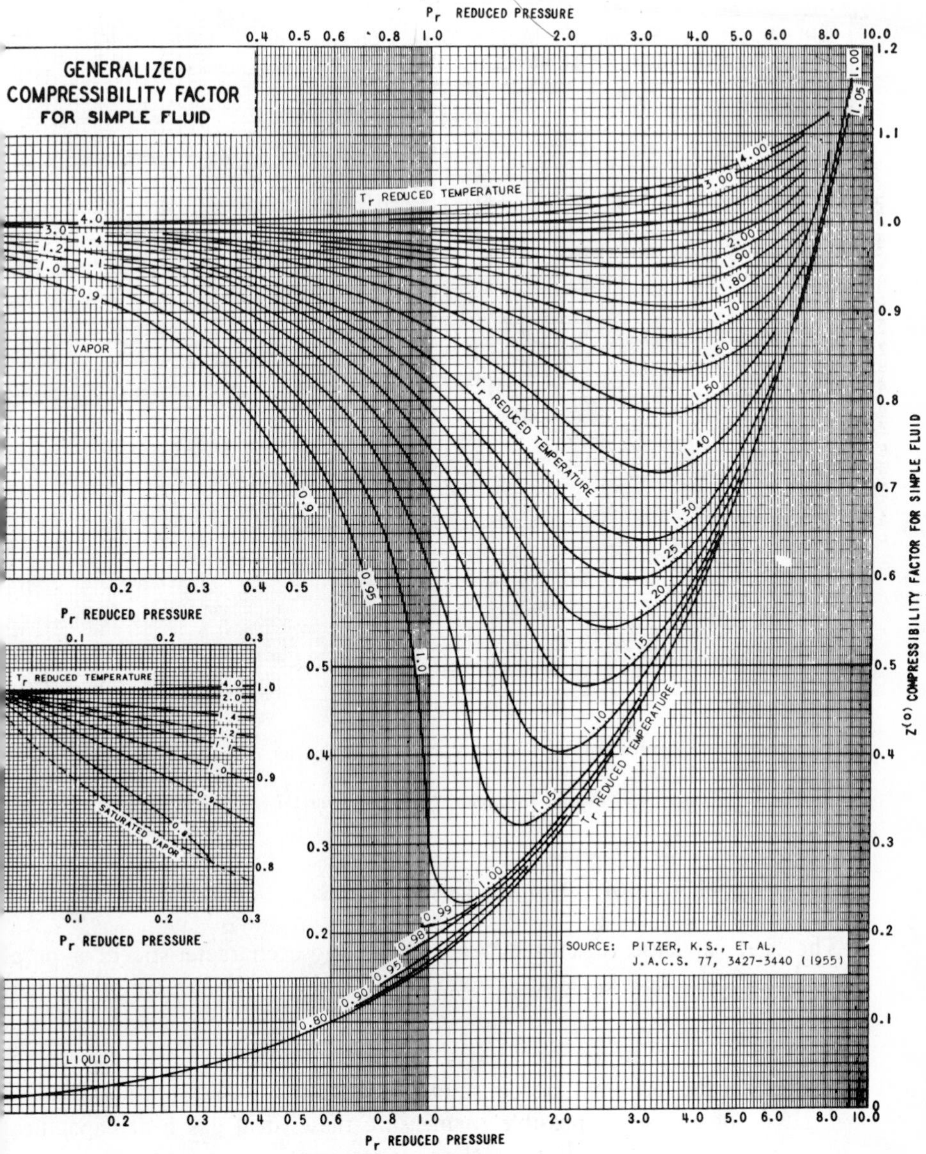

4 **Generalized compressibility factor for simple fluid.**

Figure 6.1. [W. C. Edmister, *Petrol. Refiner*, **37** (4), 173 (1958)]

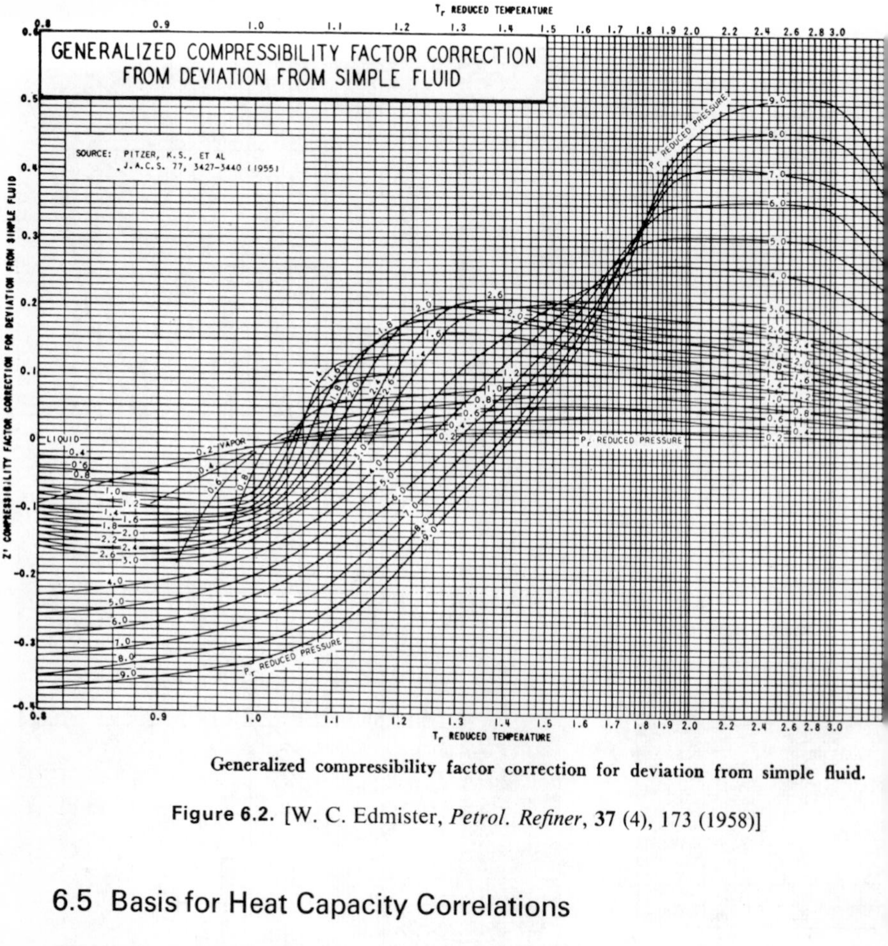

Generalized compressibility factor correction for deviation from simple fluid.

Figure 6.2. [W. C. Edmister, *Petrol. Refiner*, **37** (4), 173 (1958)]

6.5 Basis for Heat Capacity Correlations

The heat capacities discussed here will be those characteristic of a pure material in the ideal gas state. Two exist, C_p^* and C_v^*. They are simply related by:

$$C_p^* - C_v^* = R \tag{6-92}$$

where R is the universal gas constant and C_p^* and C_v^* are on a mole basis.

It is, of course, impossible to measure these ideal gas heat capacities calorimetrically although very close approximations are possible if the pressure in the system is maintained at a low value. Most C_p^* values are, however, calculated from C_v^* values [with Eq. (6-92)]; C_v^* is, in turn, determined from theoretical arguments and spectroscopic data. The essence of the method lies in expressing the *energy* of the ideal gas as a function of temperature; then, by differentiation, C_v^* is obtained. Without going into great depth, let us at least examine briefly the types of energies associated with molecules.

First, we rule out intermolecular energy since, by choosing an *ideal gas*, there are no interactions between molecules.[16] Next, energies associated with electron motion, or intranuclear movement, or even energy associated with mass by relativistic considerations may be discarded since these are either constants or temperature-dependent only at very high temperatures. For materials used normally in chemical engineering practice, none contribute to C_v^*.

Molecules do, however, move and have kinetic energy; in addition, these energies increase with temperature. By simple kinetic arguments it is easy to show that in *each* of the three possible directions of motion there is an energy of $(mv^2/2)$ or $(kT/2)$. k is Boltzmann's constant or R divided by Avogadro's number. This translational energy thus contributes $3(kT/2)$ per molecule or $3RT/2$ per mole. The translational contribution to C_v^* is then $(3R/2)$. Molecules having no other energy storage modes would have C_v^* values of about 3 cal/g-mol°K since R is close to 2 cal/g-mol°K. Such is the case for monatomic gases such as He, Ar, and Ne at low pressures. C_p^* is, of course, then 5 cal/g-mol°K.

More complex molecules can also store energy in other ways. Most are obvious from a mechanical visualization. Rotation of the molecule may occur both with the entire molecule and, to some degree, by rotation of certain segments in relation to one another. Individual atoms may also vibrate in relation to adjacent atoms. As the temperature increases, more energy can be stored in rotational and vibrational modes. To compute quantitatively the exact relation between energy and temperature is, however, not always a simple matter. Quantum mechanics enters the picture particularly with respect to vibrational energies since vibrational energy quanta turn out to be large. Nevertheless, with sufficient experimental spectroscopic data on the characteristic vibrational levels, as well as the moments of inertia, the barriers to internal rotation, and any rotational-vibrational interactions, it is possible to calculate accurately C_v^*.

Engineers, however, have taken the C_v^* values calculated from theory and have developed simple, approximate atomic or group additive techniques to allow rapid estimations of heat capacities as a function of temperature. One such method is detailed in Appendix D. For those who wish to pursue further either the theoretical or alternate approximation aspects, we can recommend several texts.[17-20]

[16] Note that with this elimination, C_v^* and C_p^* become pressure-independent.

[17] Janz, G. J., *Estimation of Thermodynamic Properties of Organic Compounds* (New York: Academic Press, Inc., 1958).

[18] Reid, R. C., and T. K. Sherwood, *Properties of Liquids and Gases*, 2nd ed. (New York: McGraw-Hill Book Company, 1966), Chap. 5.

[19] Glasstone, S., *Theoretical Chemistry* (Princeton, New Jersey: D. Van Nostrand Co., Inc., 1944).

[20] Slater, J. C., *Introduction to Chemical Physics* (New York: McGraw-Hill Book Company, 1939).

6.6 Effect of Pressure on Enthalpy and Internal Energy

Using the techniques shown in Section 6.3, it is easily shown that

$$\left(\frac{\partial H}{\partial P}\right)_T = V - T\left(\frac{\partial V}{\partial T}\right)_P \tag{6-93}$$

Introducting $Z \equiv PV/RT$,

$$\left(\frac{\partial H}{\partial P}\right)_T = \frac{-RT^2}{P}\left(\frac{\partial Z}{\partial T}\right)_P \tag{6-94}$$

Integrating,

$$H_2 - H_1 = -RT^2 \int_{P_1}^{P_2} \left(\frac{\partial Z}{\partial T}\right)_P d\ln P \tag{6-95}$$

Eq. (6-95) is convenient to use in calculating enthalpy changes from some finite pressure P to a state of infinite attenuation P^*. The integral does not diverge in this calculation since the integrand decreases more rapidly than $d\ln P$. If we rewrite, but use reduced variables $P_r = P/P_c$, $T_r = T/T_c$,

$$\frac{H^* - H}{RT_c} = T_r^2 \int_P^{P_r} \left(\frac{\partial Z}{\partial T_r}\right)_{P_r} d\ln P_r \tag{6-96}$$

In order to evaluate the integral, *P-V-T* data or some generalized *P-V-T* correlation is necessary. One type of integration is shown in Appendix D.

For internal energy, since $H = U + PV$,

$$\frac{U^* - U}{RT_c} = \frac{H^* - H}{RT_c} + T_r(Z - 1) \tag{6-97}$$

6.7 Variation of Entropy with Pressure

From Eq. (6-16) with the definition of the compressibility factor,

$$\left(\frac{\partial S}{\partial P}\right)_T = \frac{-R}{P}\left[Z + T\left(\frac{\partial Z}{\partial T}\right)_P\right] \tag{6-98}$$

In order to integrate Eq. (6-98), the base state of infinite attenuation is inconvenient because the entropy increases without limit as $P \longrightarrow 0$. A hypothetical reference state is chosen instead. By integrating Eq. (6-98) from P to a low pressure, P^*, where the gas behaves ideally, then integrating again from this low pressure state to some reference pressure P_s, *assuming in the latter integration that the material is an ideal gas*, and by adding, the entropy change for the two steps is:

$$\frac{S - S_s}{R} = \ln\frac{P_s}{P} + \int_{P^*}^{P}\left[1 - Z - T\left(\frac{\partial Z}{\partial T}\right)_P\right] d\ln P \tag{6-99}$$

Or, using reduced parameters, we find that

$$\frac{S - S_s}{R} = \ln \frac{P_s}{P} + \int_{P_r^*}^{P_r} \left[1 - Z - T_r \left(\frac{\partial Z}{\partial T_r} \right)_{P_r} \right] d \ln P_r \qquad (6\text{-}100)$$

The difference $(S - S_s)$ is the isothermal entropy change from P, real gas, to P_s, ideal gas. If the isothermal entropy change is desired between two real states at P_1 and P_2, Eq. (6-99) can be applied to both and the results subtracted whereby the pressures P_s and P^* are eliminated.

$$\frac{S_2 - S_1}{R} = -\ln \frac{P_2}{P_1} + \int_{P_1}^{P_2} \left[1 - Z - T \left(\frac{\partial Z}{\partial T} \right)_P \right] d \ln P \qquad (6\text{-}101)$$

For an ideal gas, the integral term vanishes.

Although Eq. (6-101) is useful in specific cases, generalized correlations of entropy with pressure are usually determined from Eq. (6-100). Two common reference states are used for P_s(i.e., either one atmosphere or the actual system pressure). In using any generalized correlation, it is necessary to know which reference pressure state has been chosen. This is illustrated in Appendix D in which we discuss the numerical evaluation of isothermal entropy changes.

6.8 Effect of Pressure on the Heat Capacity

The heat capacity at constant pressure, C_p, is defined in Eq. (6-12). Alternatively, it may be defined as $(\partial H/\partial T)_P$. Thus,

$$\left(\frac{\partial C_p}{\partial P} \right)_T = \frac{\partial^2 H}{\partial P \, \partial T} = \frac{\partial^2 H}{\partial T \, \partial P} = \frac{\partial}{\partial T} \left[\left(\frac{\partial H}{\partial P} \right)_T \right]_P$$

Integrating between P and P^* and using Eq. (6-94), we see that

$$\frac{C_p - C_p^*}{R} = \frac{\partial}{\partial T_r} \left(\frac{H - H^*}{RT_c} \right)_{P_r} \qquad (6\text{-}102)$$

In this case, the isothermal difference in heat capacity depends on the derivative of an integral [i.e., see Eq (6-96)].

6.9 Derivative Properties

As shown in the previous sections, many of the equations relating the variation of a property with pressure require some knowledge of derivatives such as $(\partial V/\partial P)_T$ or $(\partial V/\partial T)_P$. It is, therefore, convenient to define two new functions that are related to these derivatives:

$$Z_p = Z - P \left(\frac{\partial Z}{\partial P} \right)_T \qquad (6\text{-}103)$$

$$Z_T = Z + T \left(\frac{\partial Z}{\partial T} \right)_P \qquad (6\text{-}104)$$

where Z is the compressibility factor. Eqs. (6-103) and (6-104) may also be written in terms of the reduced pressure and temperature, P_r, T_r in place of P, T. The P-V-T derivatives of interest are:

$$\left(\frac{\partial V}{\partial T}\right)_P = \frac{RZ_T}{P} \qquad (6\text{-}105)$$

$$\left(\frac{\partial V}{\partial P}\right)_T = \frac{-RTZ_p}{P^2} \qquad (6\text{-}106)$$

$$\left(\frac{\partial P}{\partial T}\right)_V = \frac{PZ_T}{TZ_p} \qquad (6\text{-}107)$$

Both Z_p and Z_T are unity for an ideal gas; values for real fluids are correlated with reduced temperatures and pressures in Appendix D.

6.10 Bridgman Table of Derivatives

In this chapter only a few important thermodynamic derivatives were considered. A convenient reference table for finding any thermodynamic derivative is shown in Table 6.1. This representation is normally called a Bridg-

TABLE 6.1

BRIDGMAN TABLES IN TERMS OF Z_T AND Z_p

$$(\partial T)_P = -(\partial P)_T = 1$$

$$(\partial V)_P = -(\partial P)_V = \frac{RZ_T}{P}$$

$$(\partial S)_P = -(\partial P)_S = \frac{C_p}{T}$$

$$(\partial U)_P = -(\partial P)_U = C_p - RZ_T$$

$$(\partial H)_P = -(\partial P)_H = C_p$$

$$(\partial G)_P = -(\partial P)_G = -S$$

$$(\partial A)_P = -(\partial P)_A = -(S + RZ_T)$$

$$(\partial V)_T = -(\partial T)_V = \frac{Z_pRT}{P^2}$$

$$(\partial S)_T = -(\partial T)_S = \frac{RZ_T}{P}$$

$$(\partial U)_T = -(\partial T)_U = \frac{RZ_TT}{P} - \frac{RZ_pT}{P} = \frac{RT}{P}(Z_T - Z_p)$$

$$(\partial H)_T = -(\partial T)_H = -\frac{ZRT}{P} + \frac{Z_TRT}{P} = \frac{RT}{P}(Z_T - Z)$$

$$(\partial G)_T = -(\partial T)_G = -V = -\frac{ZRT}{P}$$

TABLE 6.1 (CONT.)

$$(\partial A)_T = -(\partial T)_A = -\frac{Z_p RT}{P}$$

$$(\partial S)_V = -(\partial V)_S = \frac{R}{P^2}(-C_p Z_p + RZ)_T^2$$

$$(\partial U)_V = -(\partial V)_U = -\frac{C_p Z_p RT}{P^2} + \frac{R^2 T Z_T^2}{P^2} = \frac{RT}{P^2}(-C_p Z_p + RZ)_T^2$$

$$(\partial H)_V = -(\partial V)_H = \frac{RT}{P^2}(-C_p Z_p + RZ_T^2 - RZZ_T)$$

$$(\partial G)_V = -(\partial V)_G = -\frac{RT}{P^2}(-SZ_p + RZZ_T)$$

$$(\partial A)_V = -(\partial V)_A = \frac{SZ_p RT}{P^2}$$

$$(\partial U)_S = -(\partial S)_U = \frac{R}{P}(-C_p Z_p + RZ_T^2)$$

$$(\partial H)_S = -(\partial S)_H = -\frac{ZRC_p}{P}$$

$$(\partial G)_S = -(\partial S)_G = \frac{R}{P}(SZ_T - C_p Z)$$

$$(\partial A)_S = -(\partial S)_A = \frac{R}{P}(-C_p Z_p + RZ_T^2 + SZ_T)$$

$$(\partial H)_U = -(\partial U)_H = \frac{RT}{P}(-ZC_p + ZRZ_T + C_p Z_p - RZ_T^2)$$

$$(\partial G)_U = -(\partial U)_G = \frac{RT}{P}(RZ_T Z - ZC_p + SZ_T - SZ_p)$$

$$(\partial A)_U = -(\partial U)_A = \frac{RT}{P}(-C_p Z_p + RZ_T^2 + SZ_T - SZ_p)$$

$$(\partial G)_H = -(\partial H)_G = \frac{RT}{P}(-SZ - C_p Z + SZ_T)$$

$$(\partial A)_H = -(\partial H)_A = \frac{RT}{P}[(S + RZ_T)(Z_T - Z) - C_p Z_p]$$

$$(\partial A)_G = -(\partial G)_A = -\frac{RT}{P}(ZS - Z_p S + RZZ_T)$$

nan table, and to use it, for any derivative $(\partial W/\partial X)_Y$, one locates $(\partial W)_Y$ and livides this entity by $(\partial X)_Y$. For example, to determine $(\partial H/\partial P)_S$, $(\partial H)_S = -ZRC_p/P$ and $(\partial P)_S = -C_p/T$. Thus, $(\partial H/\partial P)_S = ZRT/P = V$.

All derivatives are expressed in terms of Z, Z_p, Z_T, P, T, R, C_p, and S. Vith this formulation, one can obtain numerical values of the derivatives vhen techniques given in Appendix D are employed to determine Z, Z_p, Z_T, and C_p. For derivatives containing S, entropy must be related to some ase state (e.g., $S = 0$, $T = 0$, $P = 1$ atm) so that the numerical values of ertain derivatives depend on the choice of base state.

Example 6.6

For the isentropic compression or expansion of an ideal gas, the relations between P and V, V and T, and T and P are given as:

$$PV^{\kappa} = \text{constant}$$

$$TV^{\kappa-1} = \text{constant}$$

$$TP^{(1-\kappa)/\kappa} = \text{constant}$$

where $\kappa \equiv C_p/C_v$.

In the chemical industry, however, it is often necessary to compress or expand nonideal gases. Assuming these operations are still to be carried out isentropically, what are the comparable relations between P and V, etc?

Solution

To obtain the P, T relation, we desire first $(\partial P/\partial T)_S$.
But,

$$\left(\frac{\partial P}{\partial T}\right)_S = -\frac{\left(\frac{\partial S}{\partial T}\right)_P}{\left(\frac{\partial S}{\partial P}\right)_T} = \frac{\frac{-C_p}{T}}{-\left(\frac{\partial V}{\partial T}\right)_P} = \frac{\frac{C_p}{T}}{\frac{RZ_T}{P}} = \frac{C_p P}{RTZ_T}$$

$$\left(\frac{\partial \ln P}{\partial \ln T}\right)_S = \frac{C_p}{RZ_T}$$

thus,

$$\ln P = \int \frac{C_p}{RZ_T}\, d \ln T + C$$

Note that for ideal gases, $Z_T = 1$ and the relation becomes

$$\ln P = \frac{C_p}{R} \ln T + C = \frac{C_p}{C_p - C_v} \ln T + C$$

or

$$TP^{(1-\kappa)/\kappa} = \text{constant}$$

In a similar manner

$$\left(\frac{\partial \ln P}{\partial \ln V}\right)_S = \frac{C_p Z}{RZ_T^2 - C_p Z_p}$$

and

$$\left(\frac{\partial \ln V}{\partial \ln T}\right)_S = \frac{C_p Z_p - RZ_T^2}{RZZ_T}$$

Example 6.7

Common problems in petroleum production involve the estimation of liquid levels in gas wells, the location of hydrate freezes, and lost "pigs" or other obstructions in gas pipelines.[21] One way to obtain such information is to use acoustic reflection (i.e., one measures the time required between the initia ion of an acoustic wave and the receipt of the back-reflected wave). This ime and data on the acoustic velocity will then allow the path length to be letermined.

Acoustic velocities are given as:

$$\mathbf{V}^2 = \left(\frac{\partial P}{\partial \rho}\right)_S$$

where P is the pressure and ρ the mass density. The derivative is at constant entropy.

Suppose that in trying to locate a lost "pig" in a natural gas pipeline, a ound wave was initiated at Station A and a delay of 1.0 sec was measured before the back-reflected wave was noted. How far was the pig from Station A?

The methane is at 77°F and 2000 psia. At 77°F, the heat capacity of methane at zero pressure is 8.54 Btu/lb-mol°R. The critical properties are isted below:

$$T_c = 190°\text{K} \qquad V_c = 99.5 \text{ cc/g-mol} \qquad \omega = 0.013$$

$$P_c = 45.8 \text{ atm} \qquad Z_c = 0.290$$

Also calculate the distance assuming that methane is an ideal gas. Compare this result with the one obtained above.

The nonideal gas correlations of Appendix D may be used.

Solution

First, the definition of sonic velocity is rearranged into a more convenient form.

$$\mathbf{V}^2 = \left(\frac{\partial P}{\partial \rho}\right)_S = -V^2\left(\frac{\partial P}{\partial V}\right)_S$$

but,

$$dS = \left(\frac{\partial S}{\partial P}\right)_T dP + \left(\frac{\partial S}{\partial T}\right)_P dT$$

so

$$\left(\frac{\partial P}{\partial V}\right)_S = -\frac{\left(\frac{\partial S}{\partial T}\right)_P\left(\frac{\partial T}{\partial V}\right)_S}{\left(\frac{\partial S}{\partial P}\right)_T}$$

$$\left(\frac{\partial S}{\partial T}\right)_P = \frac{C_p}{T}$$

and

$$\left(\frac{\partial T}{\partial V}\right)_S\left(\frac{\partial S}{\partial T}\right)_V\left(\frac{\partial V}{\partial S}\right)_T = -1$$

[21] Thomas, L. K., R. W. Hankinson, and K. P. Phillips, *J. Petrol. Tech.*, July 1970, p. 889.

Therefore,

$$\left(\frac{\partial T}{\partial V}\right)_S = -\frac{\left(\frac{\partial S}{\partial V}\right)_T}{\left(\frac{\partial S}{\partial T}\right)_V}$$

$$\left(\frac{\partial S}{\partial T}\right)_V = \frac{C_v}{T}$$

substituting and noting that

$$\frac{\left(\frac{\partial S}{\partial V}\right)_T}{\left(\frac{\partial S}{\partial P}\right)_T} = \left(\frac{\partial P}{\partial V}\right)_T$$

then

$$\mathbf{V}^2 = -V^2\left(\frac{C_p}{C_v}\right)\left(\frac{\partial P}{\partial V}\right)_T$$

If the methane gas were assumed ideal,

$$\left(\frac{C_p}{C_v}\right) = \frac{C_p^*}{C_p^* - R}$$

$$\left(\frac{\partial P}{\partial V}\right)_T = -\frac{RT}{V^2}$$

$$\mathbf{V}^2 \text{ (ideal gas)} = \frac{RTC_p^*}{C_p^* - R}$$

The units of the gas constant R are obtained as follows:

$$R = 1.987\frac{\text{Btu}}{\text{lb-mol°R}} \times \frac{778 \text{ ft-lb}^F}{\text{Btu}} \times \frac{\frac{32.2 \text{ lb}^M\text{-ft}}{\text{sec}^2}}{\text{lb}^F} \times \frac{\text{lb-mol}}{M\,\text{lb}^M}$$

$$= \frac{4.978 \times 10^4}{M} \frac{\text{ft}^2}{\text{sec}^{2}\text{°R}}$$

$$\mathbf{V}^2 \text{ (ideal gas)} = \frac{(4.978)(10^4)(8.54)(537)}{(8.54 - 1.99)(16)} = 2.16 \times 10^6 \text{ ft}^2/\text{sec}^2$$

$$\mathbf{V} = 1470 \text{ ft/sec}$$

For a nonideal gas, using the Z_p and Z_T functions (discussed in Appendix D), we find that

$$V = \frac{ZRT}{P}$$

$$\left(\frac{\partial V}{\partial P}\right)_T = -\frac{RT}{P^2}\left[Z - P\left(\frac{\partial Z}{\partial P}\right)_T\right]$$

$$= -\frac{RTZ_p}{P^2}$$

$$V^2 = \left(\frac{ZRT}{P}\right)^2$$

$$C_p - C_v = \frac{RZ_T^2}{Z_p}$$

Thus,

$$\frac{C_p}{C_v} = \frac{C_p^* + \Delta C_p}{C_p^* + \Delta C_p - RZ_T^2/Z_p}$$

$$V^2 = \frac{RTZ^2(C_p^* + \Delta C_p)}{Z_p(C_p^* + \Delta C_p) - RZ_T^2}$$

At 77°F and 2000 psia, $T_r = (77 + 460)/(190)(1.8) = 1.57$; $P_r = (2000)/(45.8)(14.69) = 2.97$. From Figures 6.1, 6.2, D.7 through D.10, $Z^{(0)} = 0.85$, $Z^{(1)} = 0.21$; $Z_p^{(0)} = 0.88$; $Z_p^{(1)} = 0.16$; $Z_T^{(0)} = 1.56$; $Z_T^{(1)} = 0.25$; $\Delta C_p^{(0)} = 4.0$; $\Delta C_p^{(1)} = 0$. Thus,

$$Z = Z^{(0)} + \omega Z^{(1)} = 0.85 + (0.013)(0.21) = 0.85$$

$$Z_p = Z_p^{(0)} + \omega Z_p^{(1)} = 0.88 + (0.013)(0.16) = 0.88$$

$$Z_T = Z_T^{(0)} + \omega Z_T^{(1)} = 1.56 + (0.013)(0.25) = 1.56$$

$$\Delta C_p = \Delta C_p^{(0)} + \omega \Delta C_p^{(1)} = 4.0$$

$$V^2 = \frac{(4.978)(10^4)(537)(0.85)^2(8.54 + 4.0)}{(16)(0.88)[(8.54 + 4.0) - (1.987)(1.56)^2]}$$

$$= 2.42 \times 10^6 \text{ ft}^2/\text{sec}$$

$$V_S = 1550 \text{ ft/sec}$$

$$L = \frac{1550}{2} = 775 \text{ ft}$$

In this case, the sonic velocity for the nonideal gas does not differ appreciably from the ideal gas value. If, however, the pressure were higher or, more important, the temperature were lower, the true sonic velocity would be appreciably lower than that predicted from the ideal gas relation.

PROBLEMS

6.1. Calculate the change in entropy for one g-mol of normal hydrogen gas between the saturated vapor state, 1 atm, 20.4°K, and the state, 100 atm, 40°K. The heat capacities at 1 atm are tabulated below as a function of temperature. Gas phase compressibility data are shown in Figure P6.1.

T, °K	C_p, cal/g-mol°K at 1 atm
20.4	2.85
22	2.80
24	2.75
26	2.69
28	2.65
30	2.63
32	2.60
34	2.58
36	2.57
40	2.56

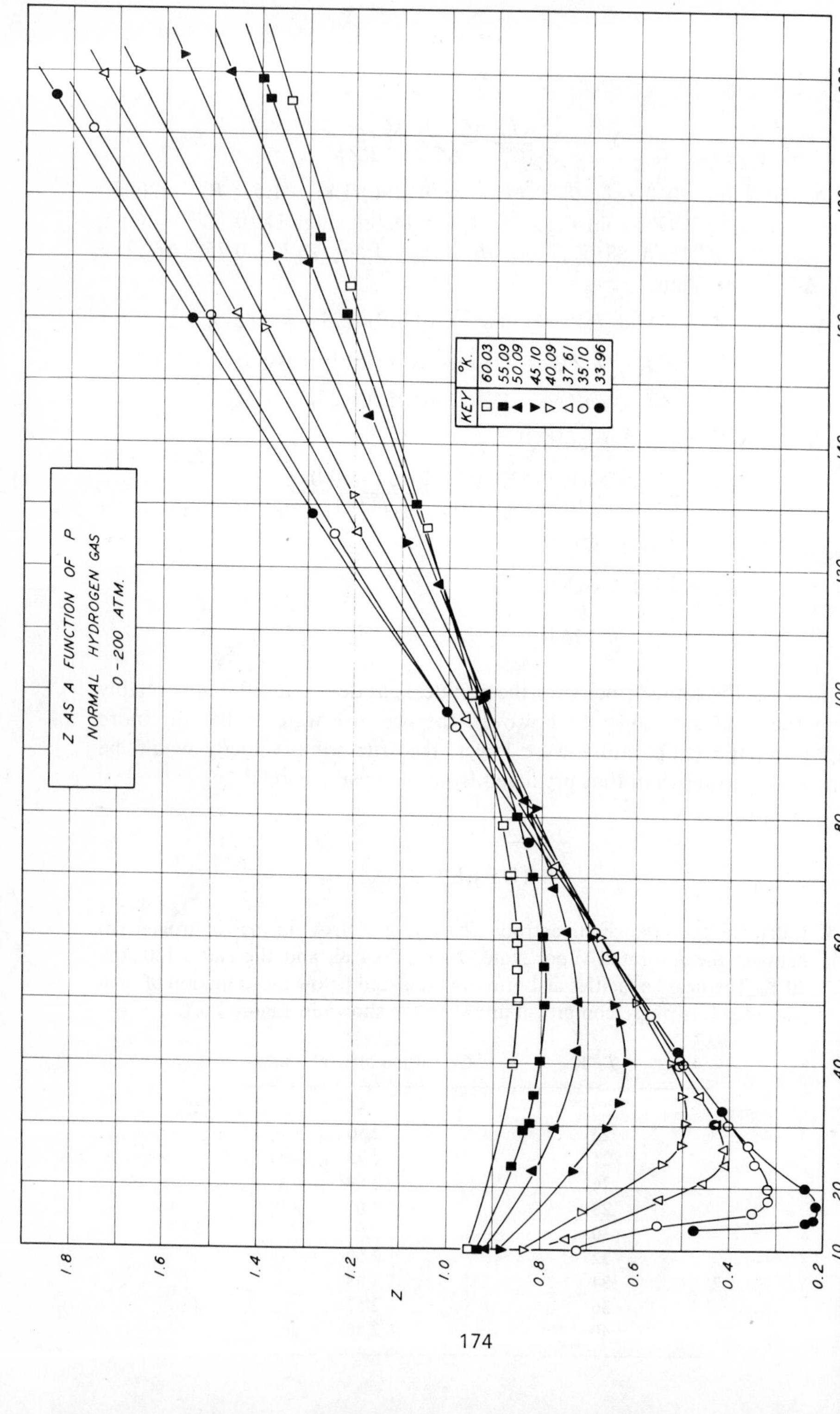
Z AS A FUNCTION OF P
NORMAL HYDROGEN GAS
0 - 200 ATM.
KEY °K.
□ 60.03
■ 55.09
▲ 50.09
▼ 45.10
▽ 40.09
△ 37.61
○ 35.10
● 33.96
Z
0.2
0.4
0.6
0.8
1.0
1.2
1.4
1.6
1.8
10
20
40
60
80
100
120
140
160
180
200

6.2. Estimate the value of C_v for propylene at 100°C, 0.0928 g/cm³ (41 atm) using only the data given below. Compare your result with the experimental value of 0.434 cal/g°K [*AIChE Journal*, **6**, 43 (1960)].

Data:

(a) C_p^*, ideal gas.

Ref.: *J. Research Nat. Bur. Standards*, **44**, 312 (1950).

T, °K	C_p^*, cal/g-mol°K
300	15.34
400	19.10
500	22.62

(b) *P-V-T* data.

Ref.: *Physica*, **19**, 287 (1953).

The data shown below express the gas density in terms of Amagat units. This unit is dimensionless and simply expresses the ratio of the true density to the density at 1 atm, 273°K, which for propylene is 4.544×10^{-5} g-mol/cm³.

Density, Amagats	75°C	100°C	125°C
		Pressure, atm	
6.4976	7.8365	8.4920	9.1415
8.3409	9.8430	10.7020	11.5514
10.1832	11.7558	12.8257	13.8818
12.1074	13.6579	14.9562	16.2349
14.3997	15.7973	17.3793	18.9330
16.6494	17.7674	19.6391	21.4729
18.3557	19.1775	21.2770	23.3297
19.5100	20.0937	22.3483	24.5522
23.0078	22.6744	25.4271	28.1038
27.6638	25.6809	29.1347	32.4745
32.1907	28.1611	32.3417	36.3613
36.9366	30.3249	35.3143	40.0840
41.6597	32.0613	37.9045	43.4519
42.4108	—	38.2855	43.9626
46.1992	—	40.0797	46.4052
47.3154	—	40.5730	47.0935
50.9531	—	42.0618	49.2283

6.3. The heat capacity at constant pressure at 1 atm for isobutane is tabulated below. Data showing the variation of the Joule-Thompson coefficient with temperature and pressure are also given. From these data, estimate C_p at 20 psia from 70 to 130°F.

Data:

(a) Heat capacities.

T, °F	C_p, 1 atm, Btu/lb°R
70	0.3896
100	0.3995
130	0.4096
160	0.4212
190	0.4332
220	0.4458
250	0.4594

(b) Joule-Thompson coefficients.

P, psia	Joule-Thompson coefficients, $(\partial T/\partial P)_H$, °F/psia						
	70°F	100°F	130°F	160°F	190°F	220°F	250°F
0	0.1500	0.1310	0.1119	0.0928	0.0739	0.0548	0.035
14.7	0.2088	0.1692	0.1384	0.1141	0.0928	0.0729	0.054
20	0.2303	0.1817	0.1478	0.1217	0.0991	0.0789	0.0599
40	0.3183	0.2307	0.1802	0.1478	0.1225	0.1003	0.0797
60	—	0.2785	0.2105	0.1724	0.1435	0.1188	0.0965

6.4. Estimate, from any generalized method, the value of C_v for isobutane gas at 300 psia and 220°F. The value of C_p at this temperature and pressure is 0.584 Btu/lb°R [*Ind. Eng. Chem.*, **30**, 673 (1938)].

6.5. Given the data below, estimate the fugacity of pure hydrogen gas at 80°K, 100 atm.

Data: (NBS Circular 564)

P, atm	T, °K	Enthalpy (Btu/lb)	Entropy (Btu/lb°R)
0.1	80	583	13.77
100	80	509	6.54

At 0.1 atm, 80°K, hydrogen may be assumed to be an ideal gas.

6.6. A turbine, which operates essentially adiabatically and reversibly, is being supplied with steam from a header some distance away. The connecting line, 3 in. in diameter, is well-insulated.

At the turbine exit the steam is dry and saturated at 14.7 psia. In the supply header the steam is also dry but saturated at 150 psia. Entering the connecting pipe, the steam velocity is 100 ft/sec.

What is the horsepower output of the turbine? Property data for steam may be found in the steam tables or other similar sources.*

6.7. A well-insulated, rigid, spherical storage tank has an internal volume of 25 ft³. An inlet line connects the tank to an ammonia gas header containing saturated ammonia vapor at 150 psia and 79°F.

* For example see, J. H. Keenan *et al.*, *Steam Tables—Thermodynamic Properties of Water Including Vapor, Liquid, and Solid Phases* (New York: John Wiley & Sons, Inc., 1969).

Initially, the tank is evacuated to a low pressure. When the ammonia inlet valve is opened, the tank pressure rapidly increases to 150 psia. Neglect any heat transfer between the walls and gas during filling.

(a) What is the gas temperature in the tank?

(b) Repeat (a) if the tank initially contained 5 lb of ammonia at 79°F.

A pressure-enthalpy diagram is given in Figure P6.7.

6.8. A well-insulated 100-gallon storage tank contains one gallon of liquid water at 90°F. The vapor space contains only water vapor (i.e., the vapor and liquid are in equilibrium at 90°F). Wet steam at 1 atm with a quality of 94% is added to the tank until the pressure rises to 1 atm.

How many pounds of steam will enter the tank, and how many pounds of liquid water are present at the completion of the process? Use the steam tables (see Problem 6.6 for reference) for the properties of water.

6.9. A new device has just been patented to measure temperatures between 10 and 40°K (see Figure P6.9). It consists essentially of a small rigid tube with a flexible bellows located inside. The bellows is welded to the right end and is closed from the environment except during filling. Different and nonideal gases are originally charged to A and B. The plate C forms a capacitor relative to bellows B. The bellows is quite flexible and tests indicate that $P_A \sim P_B$.

When this device is placed in a cold environment, the volumetric behavior of closed systems A and B is sufficiently different so that bellows B moves; the movement is easily related to the increase or decrease in the bellows volume. Thus, at any temperature, the volumes of both A and B can be determined because the sum is constant.

The device is originally charged to a pressure P_0 at T_0 with gas A in the tube and gas B in the bellows. The original mole ratio N_B/N_A is set at some value η. At this initial P_0, T_0, the gases A and B are essentially ideal. At lower temperatures, however, they become nonideal, but the P-V-T properties of both may be expressed by a simple virial equation of state:

$$Z = \left(\frac{PV}{RT}\right) = 1 + B'P$$

where $B'_A \neq B'_B$ are both functions of temperature but not of pressure. Assume $B'(T)$ values are known.

(a) Derive a relation to express $\phi_A = \underline{V}_A/(\underline{V}_A + \underline{V}_B)$ at any temperature in terms of B' values, T, and initial conditions.

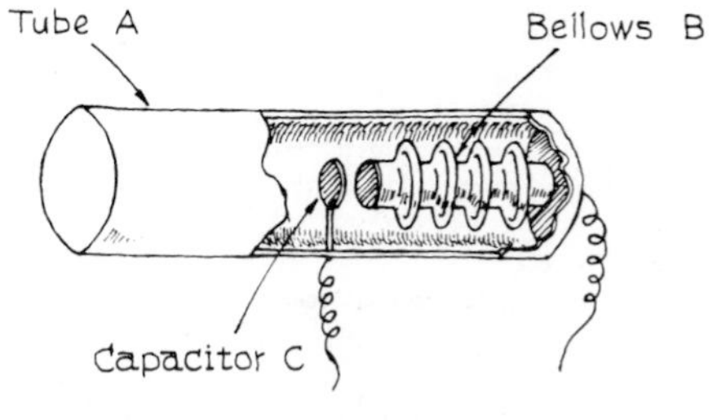

Figure P6.9

(Reprinted by permission from the ASHRAE THERMODYNAMIC PROPERTIES OF REFRIGERANTS)

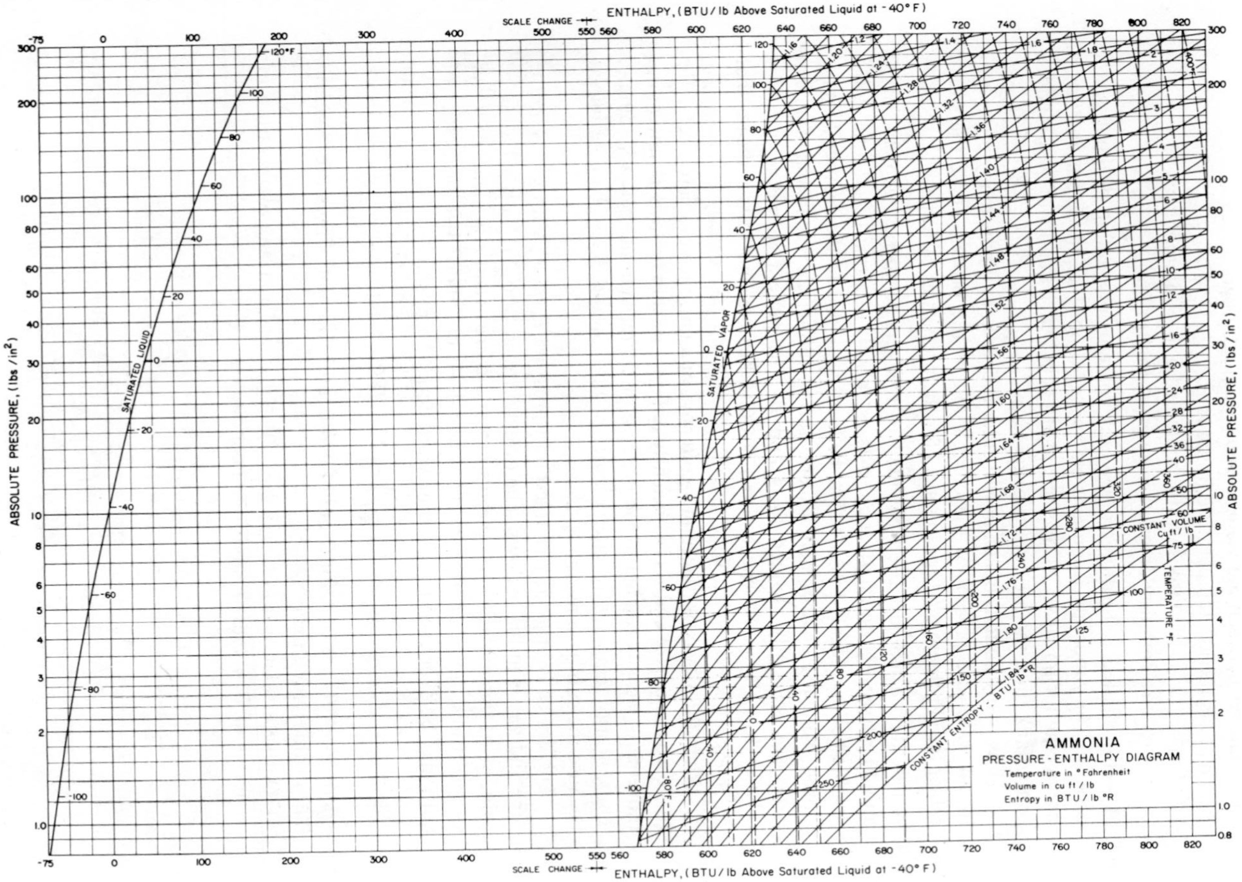

(b) Instead of measuring ϕ_A from our capacitance data, we could determine with other instrumentation the pressure in the cylinder A. Relate this pressure to B' and initial parameters (such as T_0, P_0, η, . . .) and indicate how the temperature could be inferred.

6.10. A typical phase diagram is shown in Figure P6.10 in which the specific volume

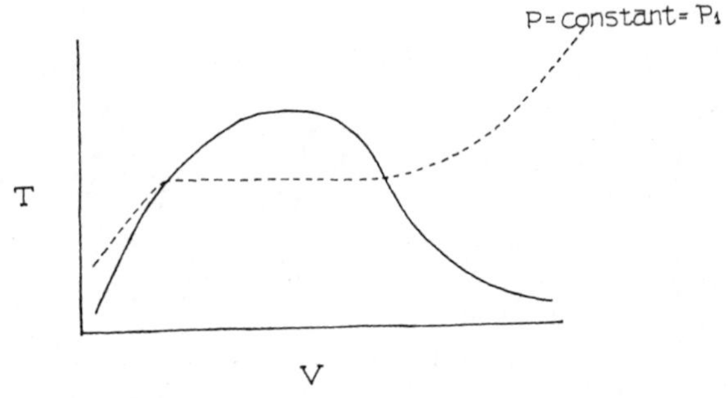

Figure P6.10

is plotted as a function of temperature; the saturated liquid and vapor envelope is shown as well as a single isobar at some pressure P_1.

For the material shown, choose some point along the P_1 isobar and assume this material to be expanded to a slightly lower pressure through a well-insulated cracked valve. Show clearly on the diagram those states where you would expect the temperature to decrease during expansion.

6.11. Gases are often cooled and even liquefied by expansion processes. Two common processes are those in which the gas expands isoenthalpically or isentropically. It is often necessary to formulate a rational basis in order to choose the preferable one. To aid in this decision, derive an expression for the ratio of $\mu_H/\mu_S = (\partial T/\partial P)_H/(\partial T/\partial P)_S$ in terms of the gas compressibility and derivative compressibility factors. Under what condition(s) is the ratio unity?

6.12. An isothermal, reciprocating compressor that operates essentially reversibly takes suction from a manifold containing CO_2 at 1 atm and 80°F. Gas is compressed and fed into a well-insulated 300-ft^3 storage tank originally containing CO_2 also at 1 atm and 80°F. What is the final temperature of the gas in the tank when the pressure reaches 700 psia?

It may be assumed that the exhaust pressure of the compressor is, at all times, equal to the tank pressure and also that the compressor displacement per stroke is small compared to the tank volume.

Plot the tank temperature and mass as a function of the pressure and on the same plot show the cumulative compressor work required.

Repeat the problem assuming CO_2 to be an ideal gas. CO_2 property data are shown in Figure P6.12.

6.13. A system contains 1 g-mol of gas at 10 atm and 0°C. By some series of irreversible operations, the pressure increases to 20 atm and the temperature to 15°C and 200 cal of work are done on the gas.

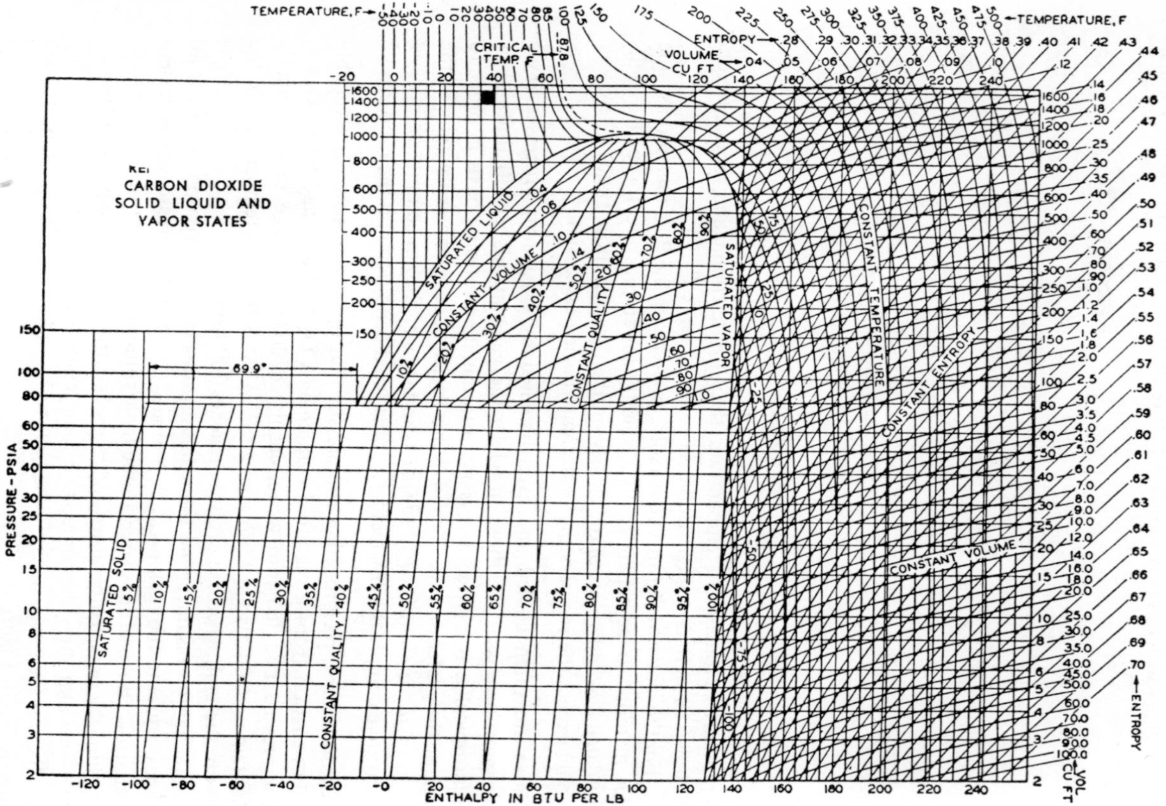

Plotted by M.B. Gelber for ASRE from data published in Industrial and Engineering Chemistry, Vol. 38, No. 2, pp. 185-200 (1946).

Figure P6.12. [Reprinted by permission from the ASHRAE Thermodynamic Properties of Refrigerants.]

What is the net heat flow into or out of the system? What is ΔH and ΔA?

Data:

$P(V - b) = RT$ with $b = 50$ cm^3/g-mol.

$C_v = 8.00$ cal/g-mol°K independent of temperature but a function of pressure.

6.14. Determine $(\partial T/\partial S)_H$ for NH_3 at 222 atm and 250°C.

Data:

$C_p = 10.2$ cal/g-mol°K at 222 atm, 250°C.

$T_c = 405$°K, $P_c = 111$ atm, $\omega = 0.250$.

The derivative compressibility plots in Appendix D may be used if desired.

6.15. Nitrogen gas at −178°F and 1230 psia fills one-half of a rigid, adiabatic cylinder. The other half is evacuated, and the two halves are separated by a membrane. If the membrane should rupture, what final temperature and pressure would result? Assume that the expansion is sufficiently rapid so that no heat transfer occurs between the gas and cylinder walls.

Data:

$C_p^* = 5$ cal/g-mol°K, independent of temperature.

$T_c = -233$°F, $P_c = 492$ psia.

6.16. Superheated ethanol vapor is flowing through a well-insulated pipeline. At two positions along this line the temperatures and pressures were found to be 450°F, 500 psia and 440°F, 455 psia, respectively. Neglect any velocity heads.

(a) Estimate the average value of C_p for ethanol in this pressure-temperature range.

(b) What is the change in entropy between the two positions.

(c) It has been reported that in some ethanol pipelines a temperature increase was found over a section of line where a pressure drop occurred. Explain when such a phenomenon would be observed.

Data:

Pressure, psia	Specific volume, ft^3/lb 450°F	440°F
500	0.324	0.309
400	0.443	0.427

6.17. If, on a plot of log pressure vs. enthalpy, the isentropic path were a straight line, determine how the compressibility factor would vary with temperature and pressure. Also, along such an isentropic path, if the material were expanded through a reversible, adiabatic turbine from P_1 to P_2, how much work could be obtained per mole of gas?

6.18. Saturated steam at a quality of unity fills a rigid cylinder at 390°F. It is desired to reduce the pressure and temperature of the steam simultaneously but to maintain always a saturated vapor. The cylinder is fitted with a frictionless piston to allow for changes in volume. Heat transfer is allowed. What is the direction of this heat flow initially and what is the magnitude when expressed as Btu/°R-lb of steam at the start of the pressure reduction?

6.19. For a given gas, the following property data are available at 100°F.

P, atm	V(ft³/lb)	C_p, Btu/lb°R	$(\partial T/\partial P)_H$, °R/atm
1.5	7.30	0.21	0.6395
2.0	5.25	0.23	0.6400
2.5	4.40	0.25	0.6406

What is the value of $(\partial T/\partial P)_S$ at 100°F and 2 atm? If the above data are insufficient, what other information would be required?

6.20. A material in the saturated vapor state is to be heated from T_1 to T_2 but always maintained as a saturated vapor in the process. What is the minimum work required to accomplish this operation assuming that a large sink is available at T_0? What is the total energy removed from this sink? Explain clearly what property data would be required to allow a numerical answer to be obtained.

6.21. Hydrogen is to be liquefied in the cycle illustrated in Figure P6.21. High-pressure gas is precooled to 40°K and passed into a heat exchanger that employs saturated hydrogen recycle gas as the coolant. The high-pressure cold vapor is then expanded to 1 atm across an insulated expansion valve. The liquid fraction is separated and the vapor recycled as shown. The heat transfer area of this final exchanger is very large.

(a) At what pressure should the compressor be operated to maximize the amount of liquid formed per pound of hydrogen flow?

(b) Sketch the state of hydrogen during the process on both T-S and H-P diagrams.

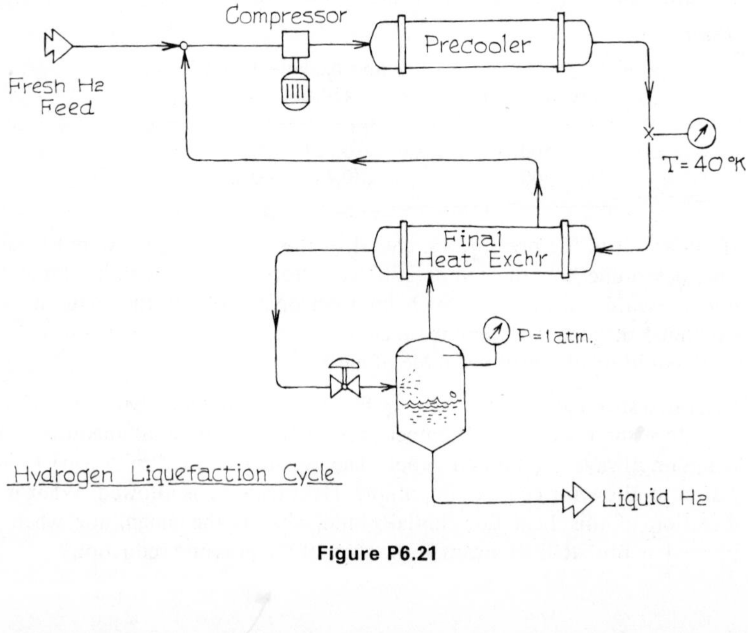

Figure P6.21

(c) How would you generalize your results for any material?

Data:

Normal boiling point of hydrogen = 20.6°K.

Enthalpy of vaporization at 20.6°K = 194 Btu/lb.

Compressibility factors for hydrogen are given in Problem 6.1.

6.22. Prepare a preliminary sketch for the shape of an adiabatic nozzle to expand a stream of saturated steam from 600 psia to 100 psia. Maintain the steam always in a saturated state. The steam enters the nozzle (where the inside diameter is 1 in.) at a rate of 1 lb/sec. Determine the area-pressure profile down the nozzle.

6.23.

MITY CORPORATION

Greedy, Massachusetts

Phone : 7-11711
Cable : Eureak

Mr. James Longthorne
c/o Faroutlake, Maine

Dear Jim:

Please forgive this intrusion upon your well-deserved vacation, but a problem has arisen in the office that demands immediate attention and all the rest of the design section is very busy except you and me.

I am, therefore, sending this problem to you in the hope that you can handle it as soon as possible.

A client has requested us to make rough calculations on his proposed process to liquefy monochlorodifluoromethane ($CHClF_2$) as per the attached flowsheet. The process gas enters at (1) at 1 atm and 20°C, is compressed adiabatically to (2). The compressor operates at about 95% of a reversible, adiabatic device and a similar efficiency is believed applicable for the electric drive motor. The hot, high-pressure gas at (2) is cooled to 400°K by heat transfer in exchanger *A* and in so doing makes saturated steam at 400°K. (The feed to *A* is water, liquid, saturated at 400°K.)

From (3) the gas is further cooled in exchanger *B* (which is essentially infinite in area), passed through a Joule-Thompson valve and into a liquid-vapor separator at 1 atm. Liquid is removed and saturated vapor is used as the cooling medium in exchanger *B*. The exit vapor at (*B*) is cooled and recycled, but do not worry about this part of the process.

Now, the customer demands (although we have our doubts) that there shall be a *maximum* amount of liquification per pass.

Would you please submit a flowsheet showing me your recommendations for:

(a) all line sizes
(b) motor horsepower for the compressor
(c) pounds of steam generated/day.
(d) liquid flow rate and the number of 50 gallon drums we can plan to ship per day

(e) all pressures and temperatures in the process streams

The flow rate at (1) is 10 lb/min and the customary flow velocities of 200 ft/sec (gas), 50 ft/sec (gas-liquid), 10 ft/sec (liquid) may be assumed.

Also attached are all the data on $CHClF_2$ that the client has available. Sorry it is not more; I have taken the liberty to ship you the text: R. C. Reid and T. K. Sherwood, *Properties of Gases and Liquids*, 2nd ed., McGraw-Hill, New York, 1966, with this letter. I'm sure you can estimate the desired properties.

Hope to hear from you soon. Have a pleasant vacation.

With regards,

Edward Col

Manager, Design Section

EC: jh

Attachment 1

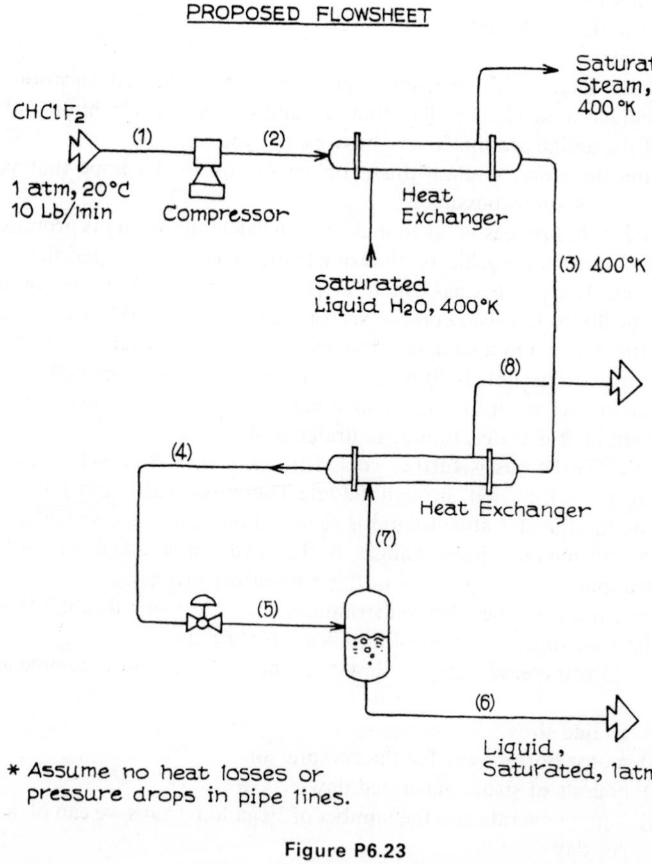

Figure P6.23

Attachment 2

Client Data on $CHClF_2$

Molecular weight = 86.48
Boiling point at 1 atm = −41.44°F
Melting point = −256°F
Color: Clear, water white
Odor: Ethereal
Toxicity: Group 5-A Classification of Underwriters' Laboratory Report MH-3134
Flammability: Nonflammable

6.24. While making a conceptual process design for a plant to be built in the near future, you are requested to analyze a small section of the overall design. In the section of concern to you, pure ethylidine chloride (1,1-dichloroethane) vapor is obtained from a still at about 194°F and 37 psia (see Figure P6.24). This vapor is to be compressed to 220 psia and then cooled and condensed to a saturated liquid. This hot liquid is to be stored as required for use later in the overall process. The estimated maximum flow rate is about 100 lb/min.

The compressor design has not yet been settled, but it is reasonable to assume that it operates adiabatically but with an efficiency of about 90% of theoretical. The pump drive unit is an electrical motor with an efficiency near 95%. The cooler-condenser is air-cooled. The storage tank is well-insulated and must be capable of storing a 12-hr flow of ethylidine chloride with a 10% ullage at the end of 12 hr. A schematic flow sheet is given in Figure P6.24.

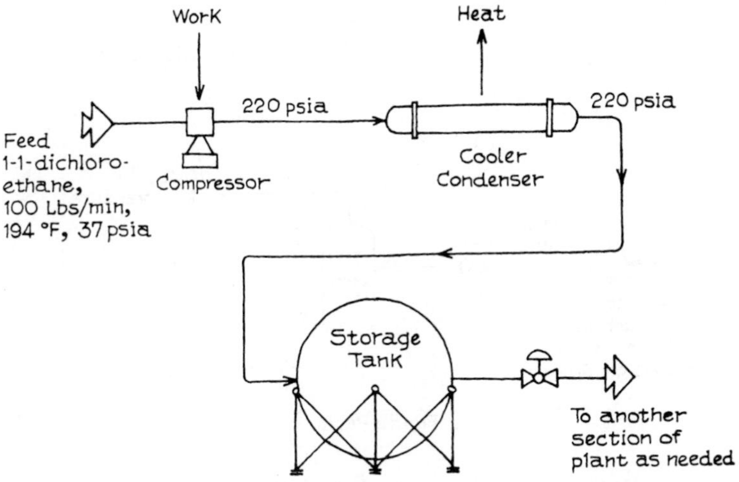

Figure P6.24

(a) What size horsepower motor do you recommend?
(b) What is the heat load in the cooler-condenser, in Btu/min?

(c) What size storage tank is needed, in gallons?
(d) What are the temperatures and specific volumes out of the compressor and cooler-condenser?
(e) If flow velocities are to be held near 20 ft/sec in the lines, what size piping do you recommend for all the connecting lines?

Data:

The Handbook of Chemistry and Physics lists a boiling point and liquid density at 20°C, and you may use these if desired.

Use any correlation methods you desire.

6.25. We are faced with handling a stream of pure carbon monoxide at 300 atm and 197°K. We would like to liquefy as great a fraction as possible, but there seems to be some disagreement about the process to be used. One suggestion has been to expand this high-pressure fluid across a *J-T* valve and take what liquid is formed (see Figure P6.25a).

Another suggestion is to expand the vapor in an adiabatic turbine to the saturation curve and then follow with a *J-T* expansion (see Figure P6.25b). The two-step operation has been suggested in order to avoid erosion of the turbine phase blades with a two-phase mixture.

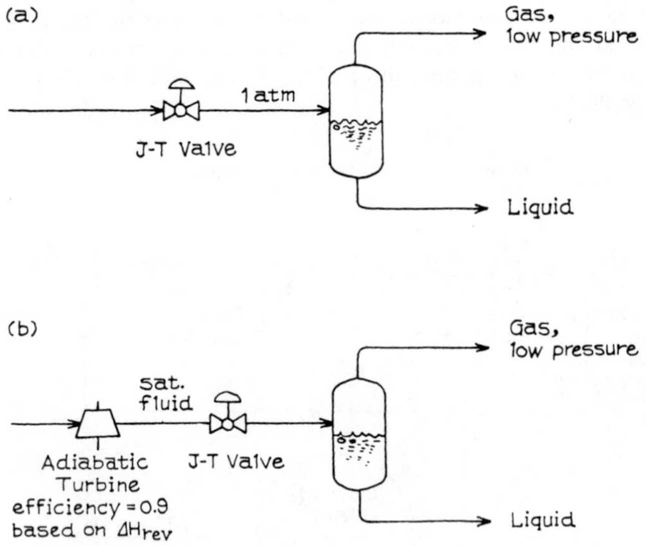

Figure P6.25

Evaluate these methods in order to indicate the fraction of initial gas one might be expected to liquefy in each. Also, suggest a better way to carry out the process to liquefy carbon monoxide that will give a larger fraction liquefied; draw a flow sheet, and calculate the fraction liquefied.

We wish we could supply you with thermodynamic property data but, unfortunately, we do not have access to any at the present time. Use any estimation methods you can find.

Equilibrium and Stability 7

In Postulate I it was stated that stable equilibrium states exist for all simple systems (see Section 2.7). In Postulate II it was stated that complex systems will approach an equilibrium condition (consistent with internal restraints) in which each simple system of the composite approaches a stable equilibrium state. In Chapter 4 we explored briefly some consequences of the stable equilibrium state. It was shown that the entropy of an isolated system (simple or complex) is a maximum at the equilibrium condition (see Section 4.7) and that the temperature, pressure, and all chemical potentials are uniform throughout a simple system in a stable equilibrium state (see Example 4.7).

In this chapter we shall examine in greater depth the criteria which can be applied to determine if a system has reached a stable equilibrium state. For example, we know from experience that uniformity of temperature, pressure, and chemical potential are necessary but not sufficient criteria to define a stable equilibrium state. Water vapor can be subccoled below 212°F at 1 atm and spontaneously form a liquid phase even though the original vapor may have been a homogeneous, uniform phase. Under other conditions, water vapor may be cooled to 211°F at 1 atm without condensation. Thus, we might wonder if water vapor at 211°F and 1 atm is in a stable equilibrium state.

Alternatively, we shall want to determine when a state is not a stable equilibrium state.

In pursuit of answers to these questions, we shall uncover a result that has not been alluded to in our prior developments. We shall find that there exist limits (e.g. , of temperature and pressure) beyond which some phases cannot exist in a stable equilibrium state regardless of the care taken to avoid a phase transition. That is, we cannot continue to subcool water vapor indefinitely, even if we could eliminate all foreign matter that usually nucleates liquid phase formation. Below some finite temperature, at a given pressure, water vapor is unstable and a transition of some kind is unavoidable. Such a state, which is characteristic of all materials, is called a state of *intrinsic instability.*

7.1 Classification of Equilibrium States

Up to this point we have used the term *equilibrium* only in connection with stable equilibrium states. In broader usage, any system that does not undergo a change with time is said to be in an equilibrium state. The adjective *stable* is reserved for those equilibrium states which, following a perturbation, will revert to the original equilibrium state. Alternatively, there are equilibrium states that are not stable (i.e., states may be permanently altered as a result of even a small perturbation).

There are four classes of equilibrium states. For ease of conceptualization, they can be described with the mechanical analogy of a ball on a solid surface in a gravitational field (see Figure 7.1). If the ball were pushed to the right or left and if it were to return to its original position, the state is stable [case (a)]. If the original position were metastable [case (b)], the ball would revert to the original position after a small perturbation, but there is the possibility that a large perturbation would displace the ball to a state of lower potential energy. If the original state were unstable [case (c)], then even a minor perturbation would displace it to a position of lower potential energy. A system in a state of neutral equilibrium [case (d)] would be altered by any perturbation, but the potential energy would remain unchanged.

In terms of these mechanical analogies, most real systems would be classified as metastable. All organic materials in the presence of oxygen could attain a more stable state by reacting to form CO_2 and H_2O; similarly, most metals are metastable in relation to their oxides. In many such cases, the barriers to transition to the more stable state may be large enough to prevent the change from occurring within the time span of interest. Thus, for all practical purposes, these states can be considered stable if the barriers are large in relation to the magnitude of the perturbation.

Since almost all real systems are metastable with respect to some pertur-

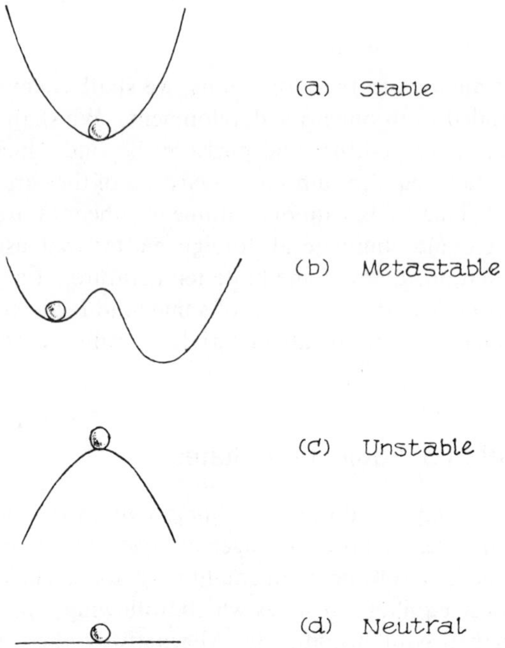

Figure 7.1. Classification of equilibrium states.

bation yet stable with respect to minor perturbations, it is of greater utility to approach equilibrium more pragmatically than that suggested by the simple mechanical analogies of Figure 7.1. We shall consider only minor perturbations to a given system. If minor perturbations leave the system unchanged, then we shall define the original state as a stable equilibrium state [see Figure 7.2(a)].

In real systems, the extent to which a barrier exists is dictated not by thermodynamic reasoning but by rate or kinetic considerations. If the rate of a possible transition is too slow to be significant within the time span of interest, we shall consider the barrier as an impenetrable internal restraint. For simple systems, such barriers may be visualized as activation energies that prevent chemical reactions or as the lack of solid nuclei needed to initiate phase transitions. Complex systems may contain additional barriers such as adiabatic, rigid, or impermeable internal walls.

For practical purposes, metastable systems are treated as stable if the barriers to transition are large in relation to minor perturbations. If one of the barriers is relatively small or nonexistent, then the original state is treated as unstable with respect to that particular kind of transition. It should be emphasized that the delineation of internal barriers forms part of the definition of a system. Alteration or elimination of an internal barrier can change a stable system to an unstable system.

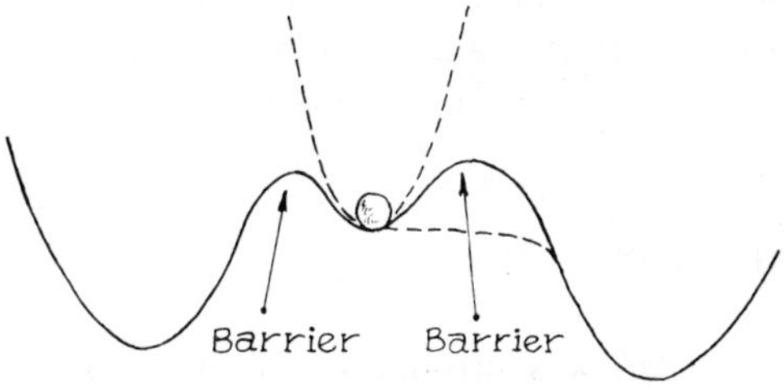

(a) A state stable with respect to minor perturbations.

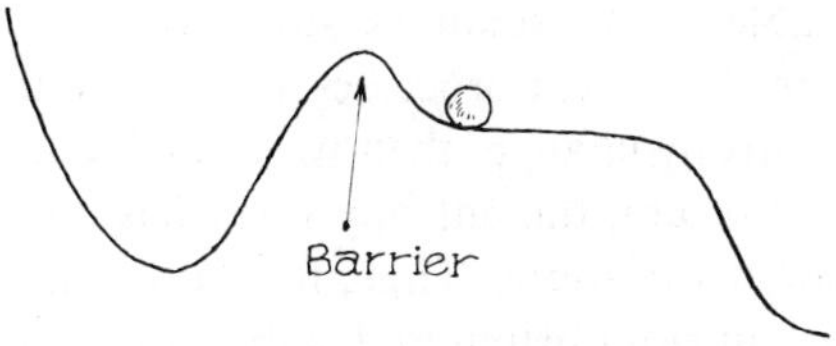

(b) The state in (a) with the right barrier removed. This state is unstable with respect to perturbations which move the ball to the right.

Figure 7.2

An interesting state is depicted by the mechanical analogy in Figure 7.2b. A minor perturbation to the right leads to an unstable state whereas a push to the left indicates that the system is stable with respect to this kind of variation. Although the analysis may not reveal what the final state of the system might be, it will indicate that the original system is intrinsically unstable and some transition must result from a minor perturbation. Although it may be difficult experimentally to force systems to their *limit of intrinsic stability*, there is sufficient evidence to suggest that such limits exist.

In general, potential transitions in real systems are more complicated than the horizontal motion of a ball on a surface, which has only two kinds of perturbations (i.e., displacement to either the left or right). Instead of testing a system for all possible transitions, we usually test for only those that we suspect may occur. Hence, we classify the stability of a system with respect to particular kinds of perturbations.

The process by which we test for stability is to perform thought experiments in which we envision a transition occurring as a result of a small, finite change in one or more properties of the system. We then determine if the proposed variation leads to a more favored state. This kind of thought experiment is referred to as a *virtual process.*

7.2 Extrema Principles

The *stable equilibrium state* was introduced in Postulate II as that state which all isolated systems approach and eventually attain. We shall now develop criteria for testing whether or not a given system has attained a stable equilibrium state. The results of Example 4.2 provide the basis for our quantitative treatment. There, it was proved that, for an adiabatic process occurring in a closed system, the total change in entropy must be either zero if the process is reversible or positive if the process is spontaneous or irreversible. This result is also applicable to any isolated system, which is a special case of a closed, adiabatic system. If all real processes in an isolated system occur with a zero or positive entropy change, then we can conclude that if an isolated system were at equilibrium, the entropy must have a maximum value with respect to any allowed variations. Thus, to test whether or not a given isolated system is, in fact, in equilibrium and stable, we propose virtual processes to test certain variations. If for such variations we can show that the total entropy decreases, then the proposed variation was impossible and the

original state was an equilibrium and stable state—at least with respect to the proposed variation.

Almost all systems will be stable with respect to some variations and unstable with respect to others. Thus, the entropy maximum principle is relative and, in the last analysis, we will only be able to propose and test a finite number of variations for a given system. Therefore, we will be able to make only a qualified statement about the stability of an equilibrium state.

The method we shall employ to test for an entropy maximum is illustrated in Figure 7.3 for the simple case of a single allowed variation in parameter

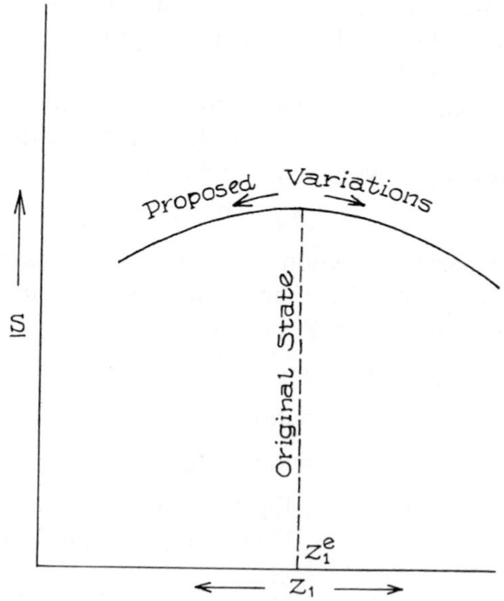

Figure 7.3 Variation in the system entropy with a single variable z_1.

z_1, where $\underset{\sim}{S}$ is a function of variables $z_1, z_2, \ldots, z_{n+2}$. As z_1 is varied, there is a value $z_1 = z_1^e$, at which the system entropy is maximized. At this point, $(\partial \underset{\sim}{S}/\partial z_1) = 0$, $(\partial^2 \underset{\sim}{S}/\partial z_1^2) < 0$, and the system is in a stable equilibrium state with respect to variations of z_1. If we looked at a system originally at $z_1 = z_1^e$ and proposed a virtual process in which z_1 was varied by $\pm\delta z_1$, then the resulting change in $\underset{\sim}{S}$, or $\Delta \underset{\sim}{S}$, can be calculated by expanding $\underset{\sim}{S}$ in a Taylor series about *the conditions of the original state provided that* δz_1 *is a small perturbation.* Thus,

$$\Delta \underset{\sim}{S} = \delta \underset{\sim}{S} + \left(\frac{1}{2!}\right) \delta^2 \underset{\sim}{S} + \left(\frac{1}{3!}\right) \delta^3 \underset{\sim}{S} + \cdots \left(\frac{1}{m!}\right) \delta^m \underset{\sim}{S} + \cdots \qquad (7\text{-}1)$$

where $\Delta \underset{\sim}{S}$ is the resultant change in $\underset{\sim}{S}$ due to the small perturbation, $\delta \underset{\sim}{S}$ is the *first-order variation* of $\underset{\sim}{S}$, and $\delta^m \underset{\sim}{S}$ is the mth*-order variation* of $\underset{\sim}{S}$. By definition,

$$\delta \underline{S} \equiv \sum_{i=1}^{n+2} \left(\frac{\partial \underline{S}}{\partial z_i}\right) \delta z_i = \sum_{i=1}^{n+2} S_{z_i}\, \delta z_i \tag{7-2}$$

$$\delta^2 \underline{S} \equiv \sum_{i=1}^{n+2} \sum_{j=1}^{n+2} \left(\frac{\partial^2 \underline{S}}{\partial z_i\, \partial z_j}\right) \delta z_i\, \delta z_j = \sum_{i=1}^{n+2} \sum_{j=1}^{n+2} S_{z_i z_j}\, \delta z_i\, \delta z_j \tag{7-3}$$

$$\delta^3 \underline{S} \equiv \sum_{i=1}^{n+2} \sum_{j=1}^{n+2} \sum_{k=1}^{n+2} \left(\frac{\partial^3 \underline{S}}{\partial z_i\, \partial z_j\, \partial z_k}\right) \delta z_i\, \delta z_j\, \delta z_k = \sum_{i=1}^{n+2} \sum_{j=1}^{n+2} \sum_{k=1}^{n+2} S_{z_i z_j z_k}\, \delta z_i\, \delta z_j\, \delta z_k \tag{7-4}$$

where $\delta z_i = z_i - z_i^0$ (superscript 0 denotes the value in the original state) and each of the partial derivatives is evaluated at the conditions prevailing in the original state. The shorthand notation for the partial derivatives will be used throughout this chapter.

If $\Delta \underline{S}$ represents the entropy change from the original state to the perturbed state, and if $\underline{S}$ is a maximum in the former state, then the mathematical equivalent of the entropy maximum principle is

$$\Delta \underline{S} < 0 \tag{7-5}$$

If $\underline{S}$ is a smoothly varying function of z_i, then it is necessary and sufficient for a maximum in $\underline{S}$ that $\delta \underline{S} = 0$ and $\delta^m \underline{S} < 0$ where $\delta^m \underline{S}$ is the lowest order, nonvanishing variation of $\underline{S}$. That is,

$$\delta \underline{S} = 0 \qquad \text{and} \tag{7-6}$$

$$\left.\begin{aligned} \delta^2 \underline{S} &\leqslant 0 \qquad \text{but, if } = 0\text{, then} \\ \delta^3 \underline{S} &\leqslant 0 \qquad \text{but, if } = 0\text{, then—etc.} \end{aligned}\right\} \tag{7-7}$$

The appropriate inequality in Eq. (7-7) forms the *criterion of stability* and the equality in Eq. (7-6), $\delta \underline{S} = 0$, the *criterion of equilibrium* in the entropy representation.

These criteria apply only to isolated systems. Since the parameters z_1, $z_2, \ldots, z_{n+2}$ are related to $\underline{U}, \underline{V}, N_1, \ldots, N_n$, the isolation requirement places restraints upon the allowed variations of these independent parameters. In particular, if the system in question is a composite of two subsystems, then any proposed virtual process must be consistent with the restraining equations of isolation:

$$\delta \underline{U} = \delta \underline{U}^{(1)} + \delta \underline{U}^{(2)} = 0 \tag{7-8}$$

$$\delta \underline{V} = \delta \underline{V}^{(1)} + \delta \underline{V}^{(2)} = 0 \tag{7-9}$$

$$\delta M = \delta M^{(1)} + \delta M^{(2)} = 0 \tag{7-10}$$

The proposed virtual processes must also be consistent with any internal restraints (e.g., rigid walls), as discussed in more depth in Section 7.3.

We shall now consider alternate criteria of equilibrium and stability that are applicable for systems that are not isolated. We often encounter systems that interact with external heat, work, and mass reservoirs, and we must also be able to treat these cases. During such interactions, different restraints will be imposed (e.g., constant $\underline{S}$, $\underline{V}$, M; constant T, $\underline{V}$, M; etc.). The insight gained in Chapter 5 with alternate forms of the Fundamental Equation should, however, lead us to expect that alternate extremum principles could also be developed with the potential functions, $\underline{U}$, $\underline{A}$, $\underline{H}$, $\underline{G}$.

The duality of the entropy and energy representation of the Fundamental Equation can readily be applied to prove that for a system at constant $\underline{S}$, $\underline{V}$, M, the total internal energy must be a minimum. To demonstrate this we shall employ a simple logical proof and follow this by an example to show that the energy-minimization and entropy-maximization do indeed yield identical criteria of equilibrium and intrinsic stability.

Consider a system that is supposed to be in a state of stable equilibrium. In proving this statement to be true, we isolated the system and showed that *all* allowable variations led to a decrease in the total entropy; that is, for every variation that we considered at constant $\underline{U}$, $\underline{V}$, M, we found $\underline{S}_f$ to be less than $\underline{S}_i$. Now, however, we wish to consider variations at constant $\underline{S}$, $\underline{V}$, M, but to allow $\underline{U}$ to vary. We now take the system at $\underline{S}_f$ and allow it to react reversibly with an external system to return the value of entropy to $\underline{S}_i$, maintaining $\underline{V}$, M constant. Since $\underline{S}_f - \underline{S}_i < 0$, we must transfer energy *into* the system to return to $\underline{S}_i$. In so doing, we increase $\underline{U}$ in this two-step process. Thus, one concludes that for any variation in a system at constant $\underline{S}$, $\underline{V}$, M, the internal energy will increase—if the system were initially at equilibrium. The converse is also true; that is, if a variation within a system held at constant $\underline{S}$, $\underline{V}$, M leads to a decrease in internal energy, then the system was not initially in a state of equilibrium.

In an analogous form to Eqs. (7-5), (7-1), (7-6), and (7-7), at constant $\underline{S}$, $\underline{V}$, M, we obtain

$$\Delta\underline{U} > 0 \quad \text{(for variations originating from a stable equilibrium state)} \tag{7-11}$$

where

$$\Delta\underline{U} = \delta\underline{U} + \left(\frac{1}{2!}\right)\delta^2\underline{U} + \left(\frac{1}{3!}\right)\delta^3\underline{U} + \cdots \tag{7-12}$$

$$\delta\underline{U} = \sum_{i=1}^{n+2}\left(\frac{\partial\underline{U}}{\partial z_i}\right)\delta z_i \text{ (equilibrium)} = 0 \tag{7-13}$$

$$\delta^2\underline{U} \geqslant 0 \text{ and } \delta^m\underline{U} > 0 \tag{7-14}$$

where $\delta^m\underline{U}$ is the lowest nonvanishing variation

The restraining equations of isolation are Eqs. (7-9), (7-10), and

$$\delta\underline{S} = \delta\underline{S}^{(1)} + \delta\underline{S}^{(2)} = 0 \tag{7-15}$$

which replaces Eq. (7-8).

For other alternate extrema, let us consider a system that may be simple or complex, with no restrictions on the number of components or the number of phases, but with no significant body force fields. We also have available large thermal and work reservoirs (R_T and R_P) to which the system may be connected, if desired, to hold the temperature and/or pressure constant (see Figure 7.4). We will still maintain our system at constant total mass during any variation, not because we could not also have used external mass reservoirs, but because the inclusion of mass variations is treated later.

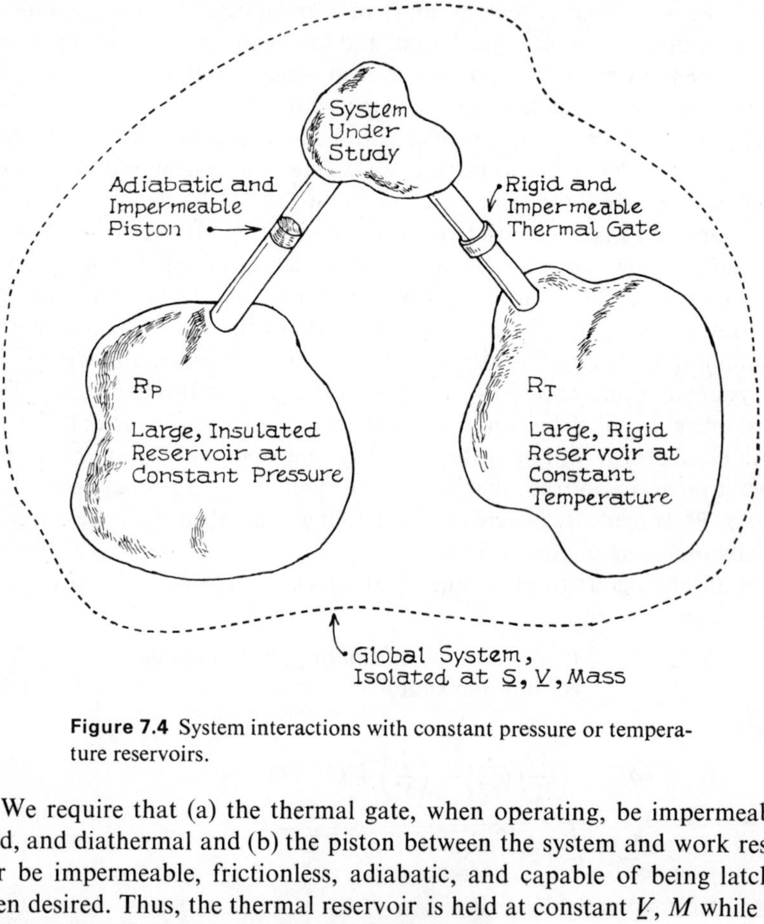

Figure 7.4 System interactions with constant pressure or temperature reservoirs.

We require that (a) the thermal gate, when operating, be impermeable, rigid, and diathermal and (b) the piston between the system and work reservoir be impermeable, frictionless, adiabatic, and capable of being latched when desired. Thus, the thermal reservoir is held at constant $\underline{V}$, M while the work reservoir is held at constant $\underline{S}$, M. The system restraints may be varied depending on our desire to have interactions with one or both of the adjacent revervoirs. Finally, we will assume that the reservoirs are large compared to the system so that small variations in the system as a result of movements of the piston connecting the system to R_P, or heat transfer with R_T, will not change the pressure level in R_P or the temperature level in R_T by any significant degree.

Our global system then consists of the original system together with R_P and R_T. These three subsystems are placed in an environment such that the *total* $\underline{S}$, $\underline{V}$, and mass are held constant. We shall investigate some possible variations of this global system and determine the consequences as we apply the energy-minimization criteria.

In general, if the global system is originally in a stable equilibrium state, then for any proposed virtual process,

$$\Delta\underline{U}^{\Sigma} = \Delta(\underline{U} + \underline{U}^{R_T} + \underline{U}^{R_P}) > 0 \tag{7-16}$$

$$\Delta\underline{S}^{\Sigma} = \Delta(\underline{S} + \underline{S}^{R_T}) = \Delta\underline{S}^{R_P} = 0 \tag{7-17}$$

$$\Delta\underline{V}^{\Sigma} = \Delta(\underline{V} + \underline{V}^{R_P}) = \Delta\underline{V}^{R_T} = 0 \tag{7-18}$$

$$\Delta M^{\Sigma} = \Delta M = \Delta M^{R_T} = \Delta M^{R_P} = 0 \tag{7-19}$$

where the superscript Σ refers to the global system, R_T and R_P to the thermal and work reservoirs, and the original system under study is not superscripted.

Case (*a*). The thermal gate is inoperative. The piston is unlatched in order to allow an interaction between the system and R_P in order to keep the system at constant pressure. Since the movement of the piston is also assumed to be frictionless and thus will operate in a reversible manner, its displacement will vary neither the entropy of the system nor the entropy of R_P. The results of any such variation will then apply to a system at constant $\underline{S}$, P, and M.

For the pressure reservoir, the First Law yields

$$\Delta\underline{U}^{R_P} = -P\,\Delta\underline{V}^{R_P} \tag{7-20}$$

or, using Eq. (7-18),

$$\Delta\underline{U}^{R_P} = P\,\Delta\underline{V} \tag{7-21}$$

Substituting Eq. (7-21) into Eq. (7-16),

$$\Delta\underline{U}^{\Sigma} = \Delta\underline{U} + P\,\Delta\underline{V} > 0 \tag{7-22}$$

or

$$\Delta\underline{H} > 0 \qquad (\underline{S}, P, M \text{ constant}) \tag{7-23}$$

That is, for a system maintained at constant $\underline{S}$, P, M, the enthalpy is a minimum for a stable equilibrium state. For such a system, the criteria of equilibrium and stability are, respectively,

$$\delta\underline{H} = 0 \qquad (\underline{S}, P, M \text{ constant}) \tag{7-24}$$

$$\delta^m\underline{H} > 0 \qquad (\underline{S}, P, M \text{ constant}) \tag{7-25}$$

where $\delta^m\underline{H}$ is the lowest-order nonvanishing variation of $\underline{H}$.

Case (*b*). Let us lock the piston and allow no interaction of the original system with R_P. Now open the thermal gate. The small system is held at constant $\underline{V}$, T, mass. A possible variation is to allow heat transfer between

R_T and the small system. Then, since $\Delta \underline{V} = \Delta \underline{V}^{R_T} = 0$,

$$\Delta \underline{U}^{R_T} = T\,\Delta \underline{S}^{R_T} = -T\,\Delta \underline{S} \tag{7-26}$$

Hence, Eq. (7-16) becomes

$$\Delta \underline{U}^{\Sigma} = \Delta \underline{U} - T\,\Delta \underline{S} > 0 \tag{7-27}$$

or

$$\Delta \underline{A} > 0 \qquad (T,\ \underline{V},\ M \text{ constant}) \tag{7-28}$$

or

$$\delta \underline{A} = 0 \qquad (T,\ \underline{V},\ M \text{ constant}) \tag{7-29}$$

$$\delta^m \underline{A} > 0 \qquad (T,\ \underline{V},\ M \text{ constant}) \tag{7-30}$$

Thus, our extremum principle for systems held at constant total volume, temperature, and mass is that the Helmholtz free energy should be a minimum with respect to all allowable variations.

Case (*c*). If we allow simultaneous interactions of our small system with both R_P and R_T, by an approach similar to that used in (a) and (b),

$$\Delta \underline{G} > 0 \qquad (T,\ P,\ M \text{ constant}) \tag{7-31}$$

or

$$\delta \underline{G} = 0 \qquad (T,\ P,\ M \text{ constant}) \tag{7-32}$$

$$\delta^m \underline{G} > 0 \qquad (T,\ P,\ M \text{ constant}) \tag{7-33}$$

Thus, the Gibbs free energy appears as the potential function to be minimized for systems at constant temperature, pressure, and mass.

7.3 Criteria of Equilibrium

For an isolated system, it was shown in the preceding section that a necessary (but not sufficient) condition of a system in a stable equilibrium state is that $\delta \underline{S} = 0$ for any virtual process. In this section we shall examine the consequences of this condition, first for the general case of a complex system and then for multiphase and chemically reacting simple systems.

Consider an isolated, complex system containing two subsystems, each of which contains a nonreacting binary mixture of components A and B (see Figure 7.5). The criterion of equilibrium for the isolated composite, Eq. (7-6), can be expanded in terms of the properties of the two subsystems:

$$\begin{aligned}\delta \underline{S} &= \delta \underline{S}^{(1)} + \delta \underline{S}^{(2)} \\ &= \frac{1}{T^{(1)}}\,\delta \underline{U}^{(1)} + \frac{1}{T^{(2)}}\,\delta \underline{U}^{(2)} + \frac{P^{(1)}}{T^{(1)}}\,\delta \underline{V}^{(1)} + \frac{P^{(2)}}{T^{(2)}}\,\delta \underline{V}^{(2)} \\ &\quad - \frac{\mu_A^{(1)}}{T^{(1)}}\,\delta N_A^{(1)} - \frac{\mu_A^{(2)}}{T^{(2)}}\,\delta N_A^{(2)} - \frac{\mu_B^{(1)}}{T^{(1)}}\,\delta N_B^{(1)} - \frac{\mu_B^{(2)}}{T^{(2)}}\,\delta N_B^{(2)} = 0\end{aligned} \tag{7-34}$$

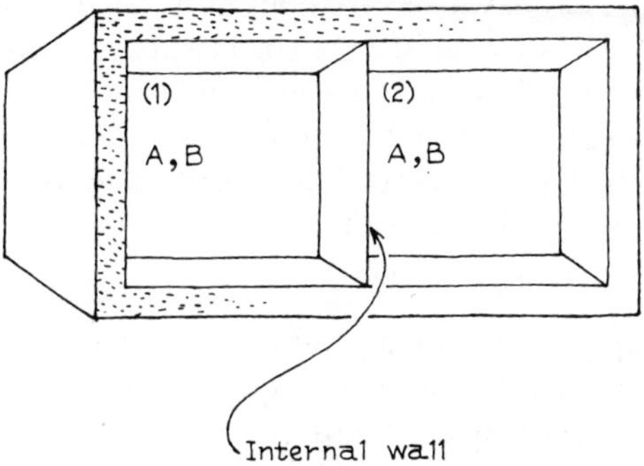

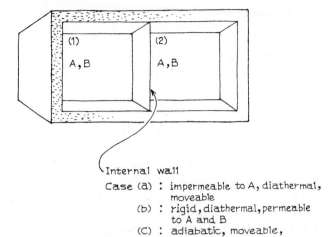

Figure 7.5. Equilibrium in a complex system.

We note that the restraining equations of isolation, Eqs. (7-8) through (7-10), place restrictions on the allowable variations in any virtual process that is considered. That is, $\delta \underline{U}^{(1)}$ and $\delta \underline{U}^{(2)}$ are not independently variable. Substituting the restraining equations of isolation into Eq. (7-34) and simplifying, one obtains:

$$\delta \underline{S} = \left(\frac{1}{T^{(1)}} - \frac{1}{T^{(2)}}\right) \delta \underline{U}^{(1)} + \left(\frac{P^{(1)}}{T^{(1)}} - \frac{P^{(2)}}{T^{(2)}}\right) \delta \underline{V}^{(1)} - \left(\frac{\mu_A^{(1)}}{T^{(1)}} - \frac{\mu_A^{(2)}}{T^{(2)}}\right) \delta N_A^{(1)} - \left(\frac{\mu_B^{(1)}}{T^{(1)}} - \frac{\mu_B^{(2)}}{T^{(2)}}\right) \delta N_B^{(1)} = 0 \tag{7-35}$$

The variations $\delta \underline{U}^{(1)}$, $\delta \underline{V}^{(1)}$, $\delta N_A^{(1)}$, $\delta N_B^{(1)}$ are independently variable only if the composite is a simple system. By definition, a complex system contains some additional internal restraints that must be recognized in any virtual process imposed upon the system. Although there is no universal result applicable to all complex systems, we illustrate three cases that demonstrate the method of obtaining the appropriate criteria of equilibrium (see Figure 7.5).

1. The internal boundary is semipermeable to B, diathermal and movable. The additional restraining equation is then:

$$\delta N_A^{(1)} = \delta N_A^{(2)} = 0 \tag{7-36}$$

and Eq. (7-35) reduces to:

$$\delta \underline{S} = \left(\frac{1}{T^{(1)}} - \frac{1}{T^{(2)}}\right) \delta \underline{U}^{(1)} + \left(\frac{P^{(1)}}{T^{(1)}} - \frac{P^{(2)}}{T^{(2)}}\right) \delta \underline{V}^{(1)} - \left(\frac{\mu_B^{(1)}}{T^{(1)}} - \frac{\mu_B^{(2)}}{T^{(2)}}\right) \delta N_B^{(1)} = 0 \tag{7-37}$$

For all virtual processes in which variations in $\delta \underline{U}^{(1)}$, $\delta \underline{V}^{(1)}$, and $\delta N_B^{(1)}$ are considered, if $\delta \underline{S}$ is to vanish, the coefficients of each variation must be zero. Therefore, the criteria of equilibrium are $T^{(1)} = T^{(2)}$, $P^{(1)} = P^{(2)}$ and $\mu_B^{(1)} = \mu_B^{(2)}$. Note that there is no restriction on $\mu_A^{(1)}$ or $\mu_A^{(2)}$.

2. The internal wall is rigid, diathermal, and permeable to both A and B. With the same approach, but with

$$\delta \underline{V}^{(1)} = \delta \underline{V}^{(2)} = 0 \tag{7-38}$$

used instead of Eq. (7-36), the temperatures and chemical potentials of both components are found to be equal in both phases. Note that there is no restriction on $P^{(1)}$ and $P^{(2)}$.

3. The internal wall is adiabatic, movable, and permeable. By analogy to cases (1) and (2), it might be thought that $\delta \underline{U}^{(1)}$ and $\delta \underline{U}^{(2)}$ were zero. Mass interchange between the subsystems, however, can also vary the energy of each compartment; thus we have, in reality, no additional restraints and, at equilibrium, the temperature, pressure, and component chemical potentials are equal in each subsystem. A similar result would be found even if the boundary were rigid except, as in (2), there would have been no restriction on the subsystem pressures. Thus, the adiabatic-permeable case is very similar to the diathermal-permeable case.

In contrast to the lack of universal results for complex systems, there are broad generalities that apply to simple systems. Consider an isolated, simple, multicomponent system containing π phases. The equilibrium criterion, Eq. (7-6), then becomes:

$$\begin{aligned}\delta \underline{S} &= \sum_{s=1}^{\pi} \delta \underline{S}^{(s)} \\ &= \sum_{s=1}^{\pi} S_U^{(s)}\, \delta \underline{U}^{(s)} + \sum_{s=1}^{\pi} S_V^{(s)}\, \delta \underline{V}^{(s)} + \sum_{s=1}^{\pi} \sum_{j=1}^{n} S_{N_j}^{(s)}\, \delta N_j^{(s)} \\ &= 0\end{aligned} \tag{7-39}$$

where superscript (s) is a dummy variable denoting the phase. The constraints resulting from isolation, with no chemical reaction, are:

$$\delta \underline{U} = \sum_{s=1}^{\pi} \delta \underline{U}^{(s)} = 0 \tag{7-40}$$

$$\delta \underline{V} = \sum_{s=1}^{\pi} \delta \underline{V}^{(s)} = 0 \tag{7-41}$$

$$\delta N_j = \sum_{s=1}^{\pi} \delta N_j^{(s)} = 0 \tag{7-42}$$

In Eq. (7-39), as indicated before, $S_U^{(s)} = (\partial \underline{S}^{(s)}/\partial \underline{U}^{(s)})_{\underline{V},N} = (1/T^{(s)})$, $S_V^{(s)} = P^{(s)}/T^{(s)}$ and $S_{N_j}^{(s)} = -\mu_j^{(s)}/T^{(s)}$.

To include the $n + 2$ restraining equations in order to eliminate $(n + 2)$

dependent variables, it is convenient to use the method of Lagrange undetermined multipliers. Let us define the arbitrary multipliers to Eqs. (7-40), (7-41), and (7-42) as $(1/T^{(1)})$, $(P^{(1)}/T^{(1)})$, and $(\mu_j^{(1)}/T^{(1)})$, respectively. Then, multiplying each constraint equation by its respective multiplier and subtracting all from Eq. (7-39), there results:

$$\delta \underline{S} = \sum_{s=2}^{\pi} \left(\frac{1}{T^{(s)}} - \frac{1}{T^{(1)}}\right) \delta \underline{U}^{(s)} + \sum_{s=2}^{\pi} \left(\frac{P^{(s)}}{T^{(s)}} - \frac{P^{(1)}}{T^{(1)}}\right) \delta \underline{V}^{(s)} - \sum_{s=2}^{\pi} \sum_{j=1}^{n} \left(\frac{\mu_j^{(s)}}{T^{(s)}} - \frac{\mu_j^{(1)}}{T^{(1)}}\right) \delta N_j^{(s)} = 0 \tag{7-43}$$

Since each variation in $\delta \underline{U}^{(s)}$, $\delta \underline{V}^{(s)}$, and $\delta N_j^{(s)}$, is now independent, it is immediately obvious that:

$$T^{(1)} = T^{(2)} = \cdots T^{(\pi)} \tag{7-44}$$

$$P^{(1)} = P^{(2)} = \cdots P^{(\pi)} \tag{7-45}$$

$$\mu_j^{(1)} = \mu_j^{(2)} = \cdots \mu_j^{(\pi)} \tag{7-46}$$

These temperature-, pressure-, and chemical potential-equalities between phases hold for all simple multiphase systems.

Finally, the treatment for multiphase systems can be extended in the following manner to cases in which chemical reactions occur. The procedure shown is applicable except that the constraint on mole numbers, Eq. (7-42), must be modified for those components that react chemically. If a reaction involving components 1 through i occurs, the general form may be expressed as:

$$\nu_1 C_1 + \nu_2 C_2 + \cdots \nu_i C_i = 0 \tag{7-47}$$

or

$$\sum_{j=1}^{i} \nu_j C_j = 0 \tag{7-48}$$

Eqs. (7-47) and (7-48) are in reality atom balances; C_j could be considered the chemical formula for j and ν_j the molar stoichiometric multiplier or coefficient. *For products, ν_j is always defined as a positive number and, for reactants, a negative number.*

If the reaction is stoichiometrically balanced, then there is no net change in mass, and,

$$\sum_{j=1}^{i} m_j \nu_j = 0 \tag{7-49}$$

where m_j is the molecular weight of j. The stoichiometric coefficient defines the ratio of mole changes of reacting components. That is, for a single reaction,

$$\frac{\delta N_1}{\nu_1} = \frac{\delta N_2}{\nu_2} = \cdots \frac{\delta N_i}{\nu_i} \tag{7-50}$$

Since the δN_j variations are related, only one may be varied independently. To simplify bookkeeping, a new variable, the *extent of reaction*, ξ, is introduced,

$$\delta\xi = \frac{\delta N_j}{\nu_j}, \qquad j = 1, 2, \ldots, i \tag{7-51}$$

so that all δN_j for reacting species may be expressed in terms of $\delta\xi$. Eq. (7-42) is then modified, for reacting components, to:

$$\left.\begin{aligned} \delta N_j &= \nu_j\,\delta\xi \qquad &(7\text{-}52)\\ \sum_{s=1}^{\pi} \delta N_j^{(s)} &= \nu_j\,\delta\xi \qquad &(7\text{-}53) \end{aligned}\right\} j = 1, 2, \ldots, i$$

For inert species, $\delta N_j = 0$ and

$$\sum_{s=1}^{\pi} \delta N_j^{(s)} = 0 \qquad j = i + 1, \ldots, n \tag{7-54}$$

Again employing the same Lagrange undetermined multipliers as used in Eq. (7-43), with Eqs. (7-53) and (7-54) in place of (7-42), Eq. (7-43) is again obtained except that there is one additional term, i.e.,

$$-\left(\frac{1}{T^{(1)}}\right) \sum_{j=1}^{i} (\nu_j \mu_j)\,\delta\xi$$

Since ξ can be varied independently from $\underline{U}^{(s)}$, $\underline{V}^{(s)}$, and $N_j^{(s)}$, we have, at equilibrium, in addition to the equalities of temperature, pressure, and chemical potential between phases, the requirement that

$$\sum_{j=1}^{i} \nu_j \mu_j = 0 \tag{7-55}$$

The superscript on μ_j is deleted since the chemical potential of component j is equal in all phases.

For cases in which multiple reactions occur, it is readily shown by an identical treatment that an equation of the form of Eq. (7-55) applies for each reaction in which there is chemical equilibrium. This general criterion will be developed further in Chapter 10.

Example 7.1

At low temperatures, a mixture of water and excess ferric chloride forms a solid phase of $FeCl_3 \cdot 6H_2O$. If equilibrium existed in an isolated system of water and ferric chloride such that there was a gas phase consisting only of water vapor, a liquid phase with only H_2O and $FeCl_3$ and a solid phase of the hexahydrate, clearly specify all equilibrium criteria that apply to this system. (Neglect any ionization of the $FeCl_3$.)

Solution

Eq. (7-39) as written for this three-phase system is:

$$\delta \underline{S} = \left(\frac{1}{T^V}\right)\delta \underline{U}^V + \left(\frac{P^V}{T^V}\right)\delta \underline{V}^V - \left(\frac{\mu_w^V}{T^V}\right)\delta N_w^V + \left(\frac{1}{T^L}\right)\delta \underline{U}^L + \left(\frac{P^L}{T^L}\right)\delta \underline{V}^L$$
$$-\left(\frac{\mu_{\mathrm{FeCl_3}}^L}{T^L}\right)\delta N_{\mathrm{FeCl_3}}^L - \left(\frac{\mu_w^L}{T^L}\right)\delta N_w^L + \left(\frac{1}{T^S}\right)\delta \underline{U}^S + \left(\frac{P^S}{T^S}\right)\delta \underline{V}^S$$
$$-\left(\frac{\mu_{\mathrm{hyd}}^S}{T^S}\right)\delta N_{\mathrm{hyd}}^S = 0$$

The constraints placed on the system are:

$$\delta \underline{U}^V + \delta \underline{U}^L + \delta \underline{U}^S = 0 \tag{7-56}$$

$$\delta \underline{V}^V + \delta \underline{V}^L + \delta \underline{V}^S = 0 \tag{7-57}$$

and there is also the reaction,

$$6H_2O + FeCl_3 = FeCl_3 \cdot 6H_2O$$

so that,

$$\delta \xi = \frac{\delta N_w}{-6} = \frac{\delta N_{\mathrm{FeCl_3}}}{-1} = \frac{\delta N_{\mathrm{hyd}}}{1}$$

and

$$\delta N_w = \delta N_w^L + \delta N_w^V = -6\,\delta\xi \tag{7-58}$$

$$\delta N_{\mathrm{FeCl_3}} = \delta N_{\mathrm{FeCl_3}}^L = -\delta\xi \tag{7-59}$$

$$\delta N_{\mathrm{hyd}} = \delta N_{\mathrm{hyd}}^S = \delta\xi \tag{7-60}$$

Using Lagrange multipliers $(1/T^V)$ to Eq. (7-56), (P^V/T^V) to Eq. (7-57), (μ_w^V/T^V) to Eq. (7-58), $(\mu_{\mathrm{FeCl_3}}^L/T^V)$ to Eq. (7-59), and $(\mu_{\mathrm{hyd}}^S/T^V)$ to Eq. (7-60), adding to the $\delta \underline{S}$ expression, we see immediately, since all variations are now independent, that:

$$T = T^V = T^L = T^S$$
$$P = P^V = P^L = P^S$$
$$\mu_w^V = \mu_w^L$$
$$6\mu_w^L + \mu_{\mathrm{FeCl_3}}^L = \mu_{\mathrm{hyd}}^S$$

These are the equilibrium criteria.

Example 7.2

Three reactors in an isolation chamber are initially charged as follows: *A* has pure methane, *B* pure hydrogen, and *C* pure ammonia. They are interconnected by rigid, semipermeable membranes as shown in Figure 7.6. All

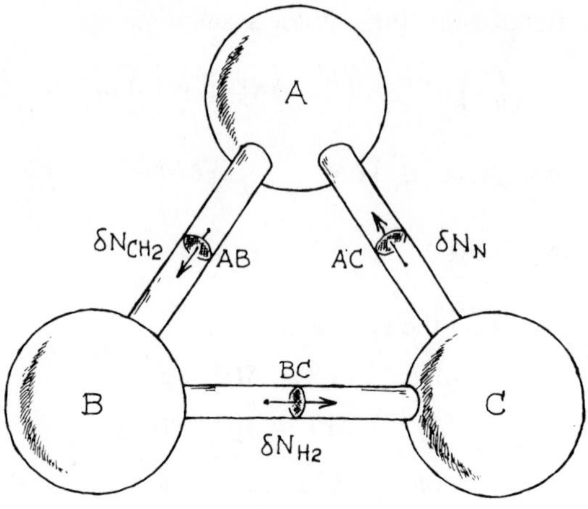

Figure 7.6 Sketch for Example 7.2.

reactors have suitable catalysts so that the reactions:

$$CH_4 = CH_2 + H_2 \tag{1}$$

$$NH_3 = N + \left(\frac{3}{2}\right)H_2 \tag{2}$$

are always in equilibrium. Membrane AB is permeable only to CH_2, BC to H_2, and AC to nitrogen atoms. What are the equilibrium criteria for this system?

Solution

Reactions (1) and (2) can occur in A; call the extent of reactions for these $\delta\xi_1^A$ and $\delta\xi_2^A$, respectively. In B only (1) occurs with $\delta\xi_1^B$ and in C only (2) with $\delta\xi_2^C$. With $\delta \underline{V}^A = \delta \underline{V}^B = \delta \underline{V}^C = 0$, the entropy variation of the total system is:

$$\begin{aligned}
\delta \underline{S} &= \delta \underline{S}^A + \delta \underline{S}^B + \delta \underline{S}^C \\
&= \left(\frac{1}{T^A}\right)\delta \underline{U}^A - \left(\frac{\mu_{CH_4}^A}{T^A}\right)\delta N_{CH_4}^A - \left(\frac{\mu_{CH_2}^A}{T^A}\right)\delta N_{CH_2}^A - \left(\frac{\mu_{H_2}^A}{T^A}\right)\delta N_{H_2}^A \\
&\quad - \left(\frac{\mu_N^A}{T^A}\right)\delta N_N^A - \left(\frac{\mu_{NH_3}^A}{T^A}\right)\delta N_{NH_3}^A + \left(\frac{1}{T^B}\right)\delta \underline{U}^B - \left(\frac{\mu_{CH_4}^B}{T^B}\right)\delta N_{CH_4}^B \\
&\quad - \left(\frac{\mu_{CH_2}^B}{T^B}\right)\delta N_{CH_2}^B - \left(\frac{\mu_{H_2}^B}{T^B}\right)\delta N_{H_2}^B + \left(\frac{1}{T^C}\right)\delta \underline{U}^C - \left(\frac{\mu_{H_2}^C}{T^C}\right)\delta N_{H_2}^C \\
&\quad - \left(\frac{\mu_N^C}{T^C}\right)\delta N_N^C - \left(\frac{\mu_{NH_3}^C}{T^C}\right)\delta N_{NH_3}^C = 0
\end{aligned}$$

The conditions of restraint are:

$$(1)\ \delta \underline{U}^A + \delta \underline{U}^B + \delta \underline{U}^C = 0$$

$$(2)\ \delta N^A_{CH_2} = -\delta N_{CH_2} + \delta \xi^A_1$$

$$(3)\ \delta N^A_{H_2} = \delta \xi^A_1 + \left(\frac{3}{2}\right) \delta \xi^A_2$$

$$(4)\ \delta N^A_{CH_4} = \delta \xi^A_1$$

$$(5)\ \delta N^A_N = \delta N_N + \delta \xi^A_2$$

$$(6)\ \delta N^A_{NH_3} = -\delta \xi^A_2$$

$$(7)\ \delta N^B_{H_2} = -\delta N_{H_2} + \delta \xi^B_1$$

$$(8)\ \delta N^B_{CH_4} = -\delta \xi^B_1$$

$$(9)\ \delta N^B_{CH_2} = \delta N_{CH_2} + \delta \xi^B_1$$

$$(10)\ \delta N^C_{H_2} = \delta N_{H_2} + \left(\frac{3}{2}\right) \delta \xi^C_2$$

$$(11)\ \delta N^C_{NH_3} = -\delta \xi^C_2$$

$$(12)\ \delta N^C_N = -\delta N_N + \delta \xi^C_2$$

Multiplying (1) by $1/T'$, adding to $\delta \underline{S}$, and also substituting (2) through (12) into the expression for $\delta \underline{S}$, there results:

$$\left(\frac{1}{T^A} - \frac{1}{T'}\right) \delta \underline{U}^A + \left(\frac{1}{T^B} - \frac{1}{T'}\right) \delta \underline{U}^B + \left(\frac{1}{T^C} - \frac{1}{T'}\right) \delta \underline{U}^C + \left(\frac{\mu^A_{CH_2}}{T^A} - \frac{\mu^B_{CH_2}}{T^B}\right) \delta N_{CH_2}$$
$$+ \left(\frac{\mu^C_N}{T^C} - \frac{\mu^A_N}{T^A}\right) \delta N_N + \left(\frac{\mu^B_{H_2}}{T^B} - \frac{\mu^C_{H_2}}{T^C}\right) \delta N_{H_2} - \frac{1}{T^A}(\mu^A_{CH_2} + \mu^A_{H_2} - \mu^A_{CH_4})\, \delta \xi^A_1$$
$$- \frac{1}{T^A}\left(\mu^A_N + \left(\frac{3}{2}\right) \mu^A_{H_2} - \mu^A_{NH_3}\right) \delta \xi^A_2 - \frac{1}{T^B}(\mu^B_{CH_2} + \mu^B_{H_2} - \mu^B_{CH_4})\, \delta \xi^B_1$$
$$- \frac{1}{T^C}\left(\mu^C_N + \left(\frac{3}{2}\right) \mu^C_{H_2} - \mu^C_{NH_3}\right) \delta \xi^C_2 = 0$$

Therefore, the criteria of equilibrium are:

$$T^A = T^B = T^C$$

$$\mu^A_{CH_2} = \mu^B_{CH_2} = \mu^A_{CH_4} - \mu^A_{H_2} = \mu^B_{CH_4} - \mu^B_{H_2}$$

$$\mu^B_{H_2} = \mu^C_{H_2}$$

$$\mu^A_N = \mu^C_N = \mu^A_{NH_3} - \left(\frac{3}{2}\right) \mu^A_{H_2} = \mu^C_{NH_3} - \left(\frac{3}{2}\right) \mu^C_{H_2}$$

Note that no statement was made on whether the semipermeable membranes were diathermal or adiabatic. Identical equilibrium criteria would result in either case.

7.4 Criteria of Stability

Let us examine a simple, isolated system with but a single phase and allow no chemical reactions to occur. Let us further suppose that the system satisfies the criteria of equilibrium; that is, T, P, and μ_j are uniform throughout the system and, hence, $\delta \underline{S} = 0$. To determine if $\underline{S}$ is a maximum, we must show that the lowest-order nonvanishing variation in $\underline{S}$ is negative (see Section 7.2).

It might appear, at first glance, that no internal variations are possible for an isolated, single-phase system. Nevertheless, we can restructure our homogeneous system into one large portion α and a second small portion β (see Figure 7.7). That is, we conceptually insert a membrane enclosing some

Figure 7.7 Conceptual visualization of a subsystem β within a homogeneous system α.

finite element β inside the system so that we may distinguish this element from the remainder of the system, α. We must, however, allow this membrane to be diathermal, nonrigid, and permeable to all components so that the composite system is still a simple system. In the original state both subsystems have properties identical to those of the α-phase.

Employing the stability criteria of Eq. (7-7), *for small variations*, we should examine second-order inequalities first. Of course, if $\delta^2 \underline{S}$ should be zero, then we must examine higher-order terms. Let us agree to write second-order derivatives in a shorthand notation such as $S^{\alpha}_{UV} \equiv (\partial^2 \underline{S}^{\alpha}/\partial \underline{U}^{\alpha} \partial \underline{V}^{\alpha})$, etc. Then, if the original system (i.e., the α-phase) is stable,

$$\begin{aligned}
\delta^2 \underline{S} &= \delta^2(\underline{S}^{\alpha} + \underline{S}^{\beta}) = \delta^2 \underline{S}^{\alpha} + \delta^2 \underline{S}^{\beta} \\
&= S^{\alpha}_{UU}(\delta \underline{U}^{\alpha})^2 + 2S^{\alpha}_{UV}\,\delta \underline{U}^{\alpha}\,\delta \underline{V}^{\alpha} + S^{\alpha}_{VV}(\delta \underline{V}^{\alpha})^2 + \sum_{j=1}^{n}\sum_{k=1}^{n} S^{\alpha}_{N_jN_k}\,\delta N^{\alpha}_j\,\delta N^{\alpha}_k \\
&\quad + 2\sum_{j=1}^{n} S^{\alpha}_{UN_j}\,\delta \underline{U}^{\alpha}\,\delta N^{\alpha}_j + 2\sum_{j=1}^{n} S^{\alpha}_{VN_j}\,\delta \underline{V}^{\alpha}\,\delta N^{\alpha}_j \qquad (7\text{-}61) \\
&\quad + \text{similar terms for subsystem } \beta < 0
\end{aligned}$$

Noting that in the Taylor expansion the second-order partial derivatives are evaluated at the initial conditions, and since the α-phase is identical to the β-phase at the outset of the perturbation, it follows that $N^{\alpha} S^{\alpha}_{XY} = N^{\beta} S^{\beta}_{XY}$, where X and Y may be $\underline{U}$, $\underline{V}$, or N_j. Furthermore, the α- and β-phase variations ~~are related by the equations of isolation~~: may be adjusted so that:

$$(\delta \underline{U}^{\alpha})^2 = (\delta \underline{U}^{\beta})^2 \qquad (7\text{-}62)$$

$$(\delta \underline{V}^{\alpha})^2 = (\delta \underline{V}^{\beta})^2 \qquad (7\text{-}63)$$

$$(\delta N^{\alpha}_j)^2 = (\delta N^{\beta}_j)^2 \qquad (7\text{-}64)$$

With these substitutions, Eq. (7-61) simplifies to:

$$\begin{aligned}
\delta^2 \underline{S} &= \frac{N}{N^{\beta}}\Big[S^{\alpha}_{UU}(\delta \underline{U}^{\alpha})^2 + 2S^{\alpha}_{UV}(\delta \underline{U}^{\alpha})(\delta \underline{V}^{\alpha}) + S^{\alpha}_{VV}(\delta \underline{V}^{\alpha})^2 \\
&\quad + 2\sum_{j=1}^{n}(S^{\alpha}_{UN_j}\,\delta \underline{U}^{\alpha} + S^{\alpha}_{VN_j}\,\delta \underline{V}^{\alpha})\,\delta N^{\alpha}_j + \sum_{j=1}^{n}\sum_{k=1}^{n} S^{\alpha}_{N_jN_k}(\delta N^{\alpha}_j)(\delta N^{\alpha}_k)\Big] \qquad (7\text{-}65) \\
&= 2(\delta^2 \underline{S}^{\alpha}) < 0
\end{aligned}$$

Note that Eq. (7-65) contains only derivatives and variations for the α-phase. Although the β-phase was introduced in order to allow us to vary the parameters of the α-phase, we see that the stability of the composite reduces to determining the stability of the original α-phase. Thus, our stability analysis, when completed, will tell us whether or not the original system is stable. If it is unstable, the analysis will not tell us what other phase may form in its place, but it will show that some transformation, leading to a more stable condition, would occur.

Note that we could have eliminated the α terms instead of the β terms, in which case we would have obtained Eq. (7-65) with β superscripts instead of α superscripts. Since the properties of the α-and β-phases are indentical in the initial state, and since the Taylor series exansion is for deviations from that initial state, it is immaterial which superscript we retain.

It is more convenient to explore the consequences of stability in the energy representation instead of the entropy representation because the $\underline{U}$ criteria can be readily modified to the $\underline{H}$, $\underline{A}$, or $\underline{G}$ forms using the Legendre transform technique discussed in Section 5.6. In the internal energy representation, the analog of Eq. (7-65) is

$$\delta^2\underline{U} = U_{SS}(\delta\underline{S})^2 + 2U_{SV}(\delta\underline{S})(\delta\underline{V}) + U_{VV}(\delta\underline{V})^2 + 2\sum_{j=1}^{n}(U_{SN_j}\,\delta\underline{S} + U_{VN_j}\,\delta\underline{V})\,\delta N_j + \sum_{j=1}^{n}\sum_{k=1}^{n} U_{N_jN_k}\,\delta N_j\,\delta N_k > 0 \tag{7-66}$$

where we have dropped the superscript α and the factor N/N^B of Eq. (7-65).

The conditions under which $\delta^2\underline{U} > 0$ or $\delta^2\underline{S} < 0$ are called the criteria of *intrinsic stability* since they relate to a single phase. The conditions under which these criteria are first violated, namely $\delta^2\underline{U} = 0$ or $\delta^2\underline{S} = 0$, are called the *limits of intrinsic stability*. These systems may be stable or unstable, as will be discussed later.

Before applying the variation procedure to Eq. (7-66), it is convenient to develop an equivalent form in terms of a sum of squares. By so doing, we can more easily test if $\delta^2\underline{U}$ is greater or less than zero since all variations, being squared, are positive and only the sign of the coefficients are important. Since the algebra is somewhat tedious, we shall illustrate first the form for a pure material and then extrapolate to a multicomponent system.

For a pure material, Eq. (7-66) can be viewed as a polynomial in $\delta\underline{S}$, $\delta\underline{V}$, and δN. The sum of squares procedure then yields:[1]

$$\delta^2\underline{U} = U_{SS}(\delta Z_1)^2 + \left(U_{VV} - \frac{U_{SV}^2}{U_{SS}}\right)(\delta Z_2)^2 + \left\{\left(U_{NN} - \frac{U_{SN}^2}{U_{SS}}\right) - \frac{\left(U_{VN} - \dfrac{U_{SV}U_{SN}}{U_{SS}}\right)^2}{\left(U_{VV} - \dfrac{U_{SV}^2}{U_{SS}}\right)}\right\}(\delta Z_3)^2 \tag{7-67}$$

where

$$\delta Z_1 = \delta\underline{S} + \frac{U_{SV}}{U_{SS}}\,\delta\underline{V} + \frac{U_{SN}}{U_{SS}}\,\delta N \tag{7-68}$$

$$\delta Z_2 = \delta\underline{V} + \frac{(U_{SS}U_{VN} - U_{SV}U_{SN})}{(U_{SS}U_{VV} - U_{SV}^2)}\,\delta N \tag{7-69}$$

$$\delta Z_3 = \delta N \tag{7-70}$$

The coefficients of the squared terms can be related to the following determinants:

$$\mathcal{A}_1 = |\, U_{SS} \,| \tag{7-71}$$

$$\mathcal{A}_2 = \begin{vmatrix} U_{SS} & U_{SV} \\ U_{SV} & U_{VV} \end{vmatrix} \tag{7-72}$$

[1] Obviously, the sum of squares shown in Eq. (7-67) is only one of a number of possible permutations since equally valid forms could be written interchanging $\underline{S}$ with $\underline{V}$, $\underline{S}$ with N, etc.

$$\mathcal{A}_3 = \begin{vmatrix} U_{SS} & U_{SV} & U_{SN} \\ U_{SV} & U_{VV} & U_{VN} \\ U_{SN} & U_{VN} & U_{NN} \end{vmatrix} \tag{7-73}$$

That is, Eq. (7-67) becomes:

$$\delta^2 \underline{U} = \mathcal{A}_1 (\delta Z_1)^2 + \frac{\mathcal{A}_2}{\mathcal{A}_1} (\delta Z_2)^2 + \frac{\mathcal{A}_3}{\mathcal{A}_2} (\delta Z_3)^2 \tag{7-74}$$

This result can be readily extrapolated to a multicomponent system. If we define $\mathcal{A}_0 = 1$ and

$$\mathcal{A}_i = \begin{vmatrix} U_{SS} & U_{SV} & U_{SN_1} & \cdots & U_{SN_{i-2}} \\ U_{SV} & U_{VV} & U_{VN_1} & \cdots & U_{VN_{i-2}} \\ U_{SN_1} & U_{VN_1} & U_{N_1N_1} & \cdots & U_{N_1N_{i-2}} \\ \cdot & \cdot & \cdot & & \cdot \\ \cdot & \cdot & \cdot & & \cdot \\ \cdot & \cdot & \cdot & & \cdot \\ U_{SN_{i-2}} & U_{VN_{i-2}} & U_{N_1N_{i-2}} & \cdots & U_{N_{i-2}N_{i-2}} \end{vmatrix} \tag{7-75}$$

then Eq. (7-74) extrapolates to: (so sayeth the Lord.)

$$\delta^2 \underline{U} = \sum_{k=1}^{n+2} \frac{\mathcal{A}_k}{\mathcal{A}_{k-1}} (\delta Z_k)^2 \tag{7-76}$$

The variation, δZ_k, that results when $\delta^2 \underline{U}$ is expressed as a sum of squares is normally a rather complex expression [see Eqs. (7-68) and (7-69)]. The actual form is, however, not important to us since we are never interested in evaluating $\delta^2 \underline{U}$ but only in determining if it is positive, negative, or zero. The $(\delta Z_k)^2$ term is always positive or zero.

We shall now show that the last term in the sum of Eq. (7-76) vanishes because $\mathcal{A}_{n+2}$ must always be zero. The $\mathcal{A}_{n+2}$ determinant contains $(n+3)(n+2)/2$ second-order partial derivatives, of which $(n+2)$ are dependent (see Section 5.6). As illustrated in Example 5.6, any row or column of $\mathcal{A}_{n+2}$ can be expressed as a linear combination of the other rows or columns and, thus, $\mathcal{A}_{n+2}$ must equal zero. Therefore,

$$\delta^2 \underline{U} = \sum_{k=1}^{n+1} \frac{\mathcal{A}_k}{\mathcal{A}_{k-1}} (\delta Z_k)^2 \tag{7-77}$$

Eq. (7-77) can now be used to identify the criteria of intrinsic stability. If $\delta^2 \underline{U} > 0$ (i.e., positive definite), then the coefficient of each squared term must be positive.* Therefore, the criteria of intrinsic stability are:

$$\mathcal{A}_k > 0 \qquad (k = 1, \ldots, n+1) \tag{7-78}$$

For a pure material, we require:

$$\mathcal{A}_1 = U_{SS} > 0 \tag{7-79}$$

$$\mathcal{A}_2 = (U_{SS}U_{VV} - U_{SV}^2) > 0 \tag{7-80}$$

* this assumes that the δZ_k's are independently variable.
The whole thing assumes $\mathcal{A}_k \neq 0$ for $k = 1, \ldots, n+1$

Note that if Eqs. (7-79) and (7-80) are obeyed, it follows that U_{VV} must be greater than zero.

For a binary mixture of A and B, in addition to Eqs. (7-79) and (7-80), we require:

$$\mathcal{A}_3 = \frac{(U_{SS}U_{VV} - U_{SV}^2)(U_{SS}U_{N_AN_A} - U_{SN_A}^2) - (U_{SS}U_{VN_A} - U_{SV}U_{SN_A})^2}{U_{SS}} > 0 \tag{7-81}$$

Thus, there appear to be $n + 1$ inequalities required to establish intrinsic stability. We shall now prove that it is necessary and sufficient that only $\mathcal{A}_{n+1}$ be positive in order to establish stability.

To introduce the general idea, let us return to Eqs. (7-79) and (7-80) and ask if either implies the other? We know in a stable phase both U_{SS} and U_{VV} are greater than zero. Now let U_{SS} become smaller and smaller. Unless U_{VV} simultaneously increases without limit, $\mathcal{A}_2$ will become negative *before* U_{SS} can become zero (or negative). Thus, for a pure material the statement that $\mathcal{A}_2 > 0$ implies that $\mathcal{A}_1 > 0$. U_{SS} or $\mathcal{A}_1$ will always be positive for a stable phase and this criterion will never be the first to be violated (i.e., as shown above, $\mathcal{A}_2$ became negative before $\mathcal{A}_1$). Similarly, it will be shown below that $\mathcal{A}_3$ is violated before $\mathcal{A}_2$, etc.

In more general terms, the limit of intrinsic stability is the point at which any one of the $(n + 1)$ positive determinants becomes zero. The first determinant to become nonpositive (i.e., zero) must be $\mathcal{A}_{n+1}$ by the following reasoning. In general, it is true that the $\mathcal{A}_i$ determinant contains within it a product of $\mathcal{A}_{i-1}$. That is, $\mathcal{A}_i$ can be written as:[2]

$$\mathcal{A}_i = \frac{a\mathcal{A}_{i-1} - b^2}{\mathcal{A}_1} \tag{7-82}$$

Thus, recalling that $\mathcal{A}_1$ must be positive, $\mathcal{A}_i$ becomes zero when

$$\mathcal{A}_{i-1} = \frac{b^2}{a} \tag{7-83}$$

Alternatively, if $\mathcal{A}_{i-1}$ is zero, then $\mathcal{A}_i$ is negative and, hence, $\mathcal{A}_i$ has already violated the stability condition. Thus, $\mathcal{A}_{n+1}$ will be the first determinant to vanish and, hence, the limit of intrinsic stability is simply

$$\mathcal{A}_{n+1} = 0 \tag{7-84}$$

[2] For example, in Eq. (7-81) where $\mathcal{A}_i = \mathcal{A}_3$

$$a = (U_{SS}U_{N_AN_A} - U_{SN_A}^2)$$
$$b = (U_{SS}U_{VN_A} - U_{SV}U_{SN_A})$$
$$\mathcal{A}_1 = U_{SS}$$

Since $\mathcal{A}_n$ is positive when $\mathcal{A}_{n+1}$ vanishes, an equivalent criterion is

$$\frac{\mathcal{A}_{n+1}}{\mathcal{A}_n} = 0 \tag{7-85}$$

where this ratio is the coefficient of the last variation in Eq. (7-77).

In Sections 7.5 and 7.6 we shall evaluate these determinants in terms of partial derivatives. To simplify that task, we shall now show that any i^{th}-order determinant, $\mathcal{A}_i$, can be reduced to a single second-order partial derivative of some transformed representation of the Fundamental Equation. We shall illustrate the technique by transforming the second-order variation of $\underline{U}$ into the $\underline{A}$-representation.

Consider a ternary system. $\underline{U}$ is a function of $\underline{S}$, $\underline{V}$, N_A, N_B, N_C, and $\underline{A}$ is a function of T, $\underline{V}$, N_A, N_B, N_C. Let us first transform the second-order partial derivatives in $\underline{U}$ to derivatives in $\underline{A}$ and then express the determinants $\mathcal{A}_i$ in terms of these derivatives in $\underline{A}$.

Using the Legendre transformation technique developed in Section 5-6 and Example 5-7, we have

$$y = \underline{A}, \quad z_1 = T, \quad z_2 = \underline{V}, \quad z_3 = N_A, \quad z_4 = N_B, \quad z_5 = N_C$$

$$\psi = \underline{U}, \quad \xi_1 = -\underline{S}, \quad z_2 = \underline{V}, \quad z_3 = N_A, \quad z_4 = N_B, \quad z_5 = N_C$$

Thus,

$$U_{SS} = \psi_{\xi_1\xi_1} = -\frac{1}{y_{z_1z_1}} = -\frac{1}{A_{TT}} \tag{7-86}$$

$$U_{VV} = \psi_{z_2z_2} = y_{z_2z_2} - \frac{y_{z_1z_2}^2}{y_{z_1z_1}} = A_{VV} - \frac{A_{TV}^2}{A_{TT}} \tag{7-87}$$

$$U_{SV} = -\psi_{\xi_1z_2} = -\frac{y_{z_1z_2}}{y_{z_1z_1}} = -\frac{A_{TV}}{A_{TT}} \tag{7-88}$$

$$U_{SN_i} = -\psi_{\xi_1z_3} = -\frac{y_{z_1z_3}}{y_{z_1z_1}} = -\frac{A_{TN_i}}{A_{TT}} \tag{7-89}$$

$$U_{VN_i} = \psi_{z_2z_3} = y_{z_2z_3} - \frac{y_{z_1z_3}y_{z_1z_2}}{y_{z_1z_1}} = A_{VN_i} - \frac{A_{TN_i}A_{TV}}{A_{TT}} \tag{7-90}$$

$$U_{N_iN_i} = \psi_{z_3z_3} = y_{z_3z_3} - \frac{y_{z_1z_3}^2}{y_{z_1z_1}} = A_{N_iN_i} - \frac{A_{TN_i}^2}{A_{TT}} \tag{7-91}$$

$$U_{N_iN_j} = \psi_{z_3z_4} = y_{z_3z_4} - \frac{y_{z_1z_3}y_{z_1z_4}}{y_{z_1z_1}} = A_{N_iN_j} - \frac{A_{TN_i}A_{TN_j}}{A_{TT}} \tag{7-92}$$

For the ternary system, $\mathcal{A}_{n+1} = \mathcal{A}_4$ is the highest-order nontrivial determinant. Using the transformations given above, we have:

$$\mathcal{A}_4 = \begin{vmatrix} U_{SS} & U_{SV} & U_{SN_A} & U_{SN_B} \\ U_{SV} & U_{VV} & U_{VN_A} & U_{VN_B} \\ U_{SN_A} & U_{VN_A} & U_{N_AN_A} & U_{N_AN_B} \\ U_{SN_B} & U_{VN_B} & U_{N_AN_B} & U_{N_BN_B} \end{vmatrix}$$

$$\mathfrak{a}_4 = \frac{1}{(A_{TT})^4}\begin{vmatrix} -1 & -A_{TV} & -A_{TN_A} & -A_{TN_B} \\ -A_{TV} & (A_{TT}A_{VV} - A_{TV}^2) & (A_{TT}A_{VN_A} - A_{TN_A}A_{TV}) & (A_{TT}A_{VN_B} - A_{TN_B}A_{TV}) \\ -A_{TN_A} & (A_{TT}A_{VN_A} - A_{TN_A}A_{TV}) & (A_{TT}A_{N_AN_A} - A_{TN_A}^2) & (A_{TT}A_{N_AN_B} - A_{TN_A}A_{TN_B}) \\ -A_{TN_B} & (A_{TT}A_{VN_B} - A_{TN_B}A_{TV}) & (A_{TT}A_{N_AN_B} - A_{TN_A}A_{TN_B}) & (A_{TT}A_{N_BN_B} - A_{TN_B}^2) \end{vmatrix} \tag{7-93}$$

The $\mathfrak{a}_4$-determinant can be simplified by multiplying column one by $-A_{TV}$ and adding the result to column 2, and similarly multiplying column 1 by $-A_{TN_A}$ and $-A_{TN_B}$, respectively, and adding to columns 3 and 4, respectively. Thus,

$$\mathfrak{a}_4 = \frac{1}{A_{TT}^4}\begin{vmatrix} -1 & 0 & 0 & 0 \\ -A_{TV} & A_{TT}A_{VV} & A_{TT}A_{VN_A} & A_{TT}A_{VN_B} \\ -A_{TN_A} & A_{TT}A_{VN_A} & A_{TT}A_{N_AN_A} & A_{TT}A_{N_AN_B} \\ -A_{TN_B} & A_{TT}A_{VN_B} & A_{TT}A_{N_AN_B} & A_{TT}A_{N_BN_B} \end{vmatrix}$$

$$= -\frac{1}{A_{TT}}\begin{vmatrix} A_{VV} & A_{VN_A} & A_{VN_B} \\ A_{VN_A} & A_{N_AN_A} & A_{N_AN_B} \\ A_{VN_B} & A_{N_AN_B} & A_{N_BN_B} \end{vmatrix} \tag{7-94}$$

Thus, we have reduced the fourth-order determinant down to a third-order determinant. Note also that $\mathfrak{a}_3$ (which is a subset of $\mathfrak{a}_4$) can be reduced to a second-order determinant and $\mathfrak{a}_2$ to a first-order determinant.

If we repeat the process by transforming $\underline{A}$-derivatives to $\underline{G}$-derivatives, we find that the determinants are again reduced in rank. The results are given in Table 7.1.

We can utilize these results to simplify the criterion of stability by choosing an appropriate representation depending on the number of components in the system. If we have a pure material (see Table 7.1), then the limit of intrinsic stability as given by Eq. (7-85) is $(\mathfrak{a}_2/\mathfrak{a}_1)_{\underline{U}} = 0$ or

$$\left(\frac{\mathfrak{a}_2}{\mathfrak{a}_1}\right)_{\underline{U}} = \frac{U_{SS}U_{VV} - U_{SV}^2}{U_{SS}} = A_{VV} = \left(\frac{\partial^2 \underline{A}}{\partial \underline{V}^2}\right)_{T,N} = \left(\frac{\mathfrak{a}_1}{\mathfrak{a}_0}\right)_{\underline{A}} \tag{7-95}$$

or, simply

$$-\left(\frac{\partial P}{\partial \underline{V}}\right)_{T,N} = 0 \tag{7-96}$$

Similarly, if we have a binary mixture, the limit of intrinsic stability is $(\mathfrak{a}_3/\mathfrak{a}_2)_{\underline{U}} = 0$, or

$$\left(\frac{\mathfrak{a}_3}{\mathfrak{a}_2}\right)_{\underline{U}} = G_{N_AN_A} = \left(\frac{\partial^2 \underline{G}}{\partial N_A^2}\right)_{T,P,N_B} = \left(\frac{\partial \mu_A}{\partial N_A}\right)_{T,P,N_B} = \left(\frac{\mathfrak{a}_1}{\mathfrak{a}_0}\right)_{\underline{G}} = 0 \tag{7-97}$$

Table 7.1

Summary of $\mathfrak{A}$ Functions for a Ternary System

$\underline{U}(\underline{S}, \underline{V}, N_A, N_B, N_C)$	$\underline{A}(T, \underline{V}, N_A, N_B, N_C)$	$\underline{G}(T, P, N_A, N_B, N_C)$	$\underline{G}'(T, P, \mu_A, N_B, N_C)$
$\boxed{\left(\frac{\mathfrak{A}_1}{\mathfrak{A}_0}\right)_{\underline{U}} = U_{SS}}$	$\left(\frac{\mathfrak{A}_1}{\mathfrak{A}_0}\right)_{\underline{U}} = -\frac{1}{A_{TT}}$	$\left(\frac{\mathfrak{A}_1}{\mathfrak{A}_0}\right)_{\underline{U}} = \frac{-G_{PP}}{\begin{vmatrix} G_{TT} & G_{TP} \\ G_{TP} & G_{PP} \end{vmatrix}}$	$\left(\frac{\mathfrak{A}_1}{\mathfrak{A}_0}\right)_{\underline{U}} = \frac{-\begin{vmatrix} G'_{PP} & G'_{PN_A} \\ G'_{PN_A} & G'_{N_AN_A} \end{vmatrix}}{\begin{vmatrix} G'_{TT} & G'_{TP} & G'_{TN_A} \\ G'_{TP} & G'_{PP} & G'_{PN_A} \\ G'_{TN_A} & G'_{PN_A} & G'_{N_AN_A} \end{vmatrix}}$
$\left(\frac{\mathfrak{A}_2}{\mathfrak{A}_1}\right)_{\underline{U}} = \frac{\begin{vmatrix} U_{SS} & U_{SV} \\ U_{SV} & U_{VV} \end{vmatrix}}{U_{SS}}$	$\boxed{\left(\frac{\mathfrak{A}_2}{\mathfrak{A}_1}\right)_{\underline{U}} = A_{VV} = \left(\frac{\mathfrak{A}_1}{\mathfrak{A}_0}\right)_{\underline{A}}}$	$\left(\frac{\mathfrak{A}_2}{\mathfrak{A}_1}\right)_{\underline{U}} = \frac{-1}{G_{PP}}$	$\left(\frac{\mathfrak{A}_2}{\mathfrak{A}_1}\right)_{\underline{U}} = \frac{-G'_{N_AN_A}}{\begin{vmatrix} G'_{PP} & G'_{PN_A} \\ G'_{PN_A} & G'_{N_AN_A} \end{vmatrix}}$
$\left(\frac{\mathfrak{A}_3}{\mathfrak{A}_2}\right)_{\underline{U}} = \frac{\begin{vmatrix} U_{SS} & U_{SV} & U_{SN_A} \\ U_{SV} & U_{VV} & U_{VN_A} \\ U_{SN_A} & U_{VN_A} & U_{N_AN_A} \end{vmatrix}}{\begin{vmatrix} U_{SS} & U_{SV} \\ U_{SV} & U_{VV} \end{vmatrix}}$	$\left(\frac{\mathfrak{A}_3}{\mathfrak{A}_2}\right)_{\underline{U}} = \frac{\begin{vmatrix} A_{VV} & A_{VN_A} \\ A_{VN_A} & A_{N_AN_A} \end{vmatrix}}{A_{VV}} = \left(\frac{\mathfrak{A}_2}{\mathfrak{A}_1}\right)_{\underline{A}}$	$\boxed{\left(\frac{\mathfrak{A}_3}{\mathfrak{A}_2}\right)_{\underline{U}} = G_{N_AN_A} = \left(\frac{\mathfrak{A}_1}{\mathfrak{A}_0}\right)_{\underline{G}}}$	$\left(\frac{\mathfrak{A}_3}{\mathfrak{A}_2}\right)_{\underline{U}} = -\frac{1}{G'_{N_AN_A}}$
$\left(\frac{\mathfrak{A}_4}{\mathfrak{A}_3}\right)_{\underline{U}} = \frac{\begin{vmatrix} U_{SS} & U_{SV} & U_{SN_A} & U_{SN_B} \\ U_{SV} & U_{VV} & U_{VN_A} & U_{VN_B} \\ U_{SN_A} & U_{VN_A} & U_{N_AN_A} & U_{N_AN_B} \\ U_{SN_B} & U_{VN_B} & U_{N_AN_B} & U_{N_BN_B} \end{vmatrix}}{\begin{vmatrix} U_{SS} & U_{SV} & U_{SN_A} \\ U_{SV} & U_{VV} & U_{VN_A} \\ U_{SN_A} & U_{VN_A} & U_{N_AN_A} \end{vmatrix}}$	$\left(\frac{\mathfrak{A}_4}{\mathfrak{A}_3}\right)_{\underline{U}} = \left(\frac{\mathfrak{A}_3}{\mathfrak{A}_2}\right)_{\underline{A}} = \frac{\begin{vmatrix} A_{VV} & A_{VN_A} & A_{VN_B} \\ A_{VN_A} & A_{N_AN_A} & A_{N_AN_B} \\ A_{VN_B} & A_{N_AN_B} & A_{N_BN_B} \end{vmatrix}}{\begin{vmatrix} A_{VV} & A_{VN_A} \\ A_{VN_A} & A_{N_AN_A} \end{vmatrix}}$	$\left(\frac{\mathfrak{A}_4}{\mathfrak{A}_3}\right)_{\underline{U}} = \frac{\begin{vmatrix} G_{N_AN_A} & G_{N_AN_B} \\ G_{N_AN_B} & G_{N_BN_B} \end{vmatrix}}{G_{N_AN_A}} = \left(\frac{\mathfrak{A}_2}{\mathfrak{A}_1}\right)_{\underline{G}}$	$\boxed{\left(\frac{\mathfrak{A}_4}{\mathfrak{A}_3}\right)_{\underline{U}} = G'_{N_BN_B} = \left(\frac{\mathfrak{A}_1}{\mathfrak{A}_0}\right)_{\underline{G}'}}$

Thus, for a n-component mixture, the stability criterion of $(\mathcal{A}_{n+1}/\mathcal{A}_n)_{\underline{U}} = 0$ can be written as a single second-order derivative by transforming the first n variables of the set $\underline{S}$, $\underline{V}$, $N_1, \ldots, N_n$ into their conjugate coordinates, or $T, P, \mu_1, \ldots, \mu_{n-2}, N_{n-1}, N_n$. For a ternary mixture, this set is T, P, μ_A, N_B, N_C, which we shall call the $\underline{G}'$-representation (see Table 7.1). The limit of intrinsic stability is then simply

$$\left(\frac{\mathcal{A}_4}{\mathcal{A}_3}\right)_{\underline{U}} = G'_{N_B N_B} = \left(\frac{\partial^2 \underline{G}'}{\partial N_B^2}\right)_{T,P,\mu_A,N_C} = \left(\frac{\mathcal{A}_1}{\mathcal{A}_0}\right)_{\underline{G}'} = 0 \tag{7-98}$$

Since $\underline{G}'$ is a function of T, P, μ_A, N_B, N_C, then

$$\underline{G}' = \underline{U} - T\underline{S} + P\underline{V} - \mu_A N_A \tag{7-99}$$

and

$$d\underline{G}' = -\underline{S}\, dT + \underline{V}\, dP - N_A\, d\mu_A + \mu_B\, dN_B + \mu_C\, dN_C \tag{7-100}$$

Thus,

$$\left(\frac{\partial \underline{G}'}{\partial N_B}\right)_{T,P,\mu_A,N_C} = \mu_B \tag{7-101}$$

and Eq. (7-98) simplifies to:

$$G'_{N_B N_B} = \left(\frac{\partial \mu_B}{\partial N_B}\right)_{T,P,\mu_A,N_C} = 0 \tag{7-102}$$

For a quaternary system, a $\underline{G}''$-representation in terms of $T, P, \mu_A, \mu_B, N_C, N_D$ would yield the limit of intrinsic stability as simply:

$$G''_{N_C N_C} = \left(\frac{\partial \mu_C}{\partial N_C}\right)_{T,P,\mu_A,\mu_B,N_D} = 0 \tag{7-103}$$

In summary, we have seen that once again the Legendre transformations have allowed us to reduce a cumbersome set of relations to a very simple form. Having convinced ourselves that the results are rigorously correct, we need not be discouraged by the tedious algebra because we can immediately turn to the simplified results and apply them when needed.

7.5 Stability of Pure Materials

For a *pure* material, we have seen in the preceding section that if $\delta^2\underline{U}$ is to be positive, then $(\mathcal{A}_1)_{\underline{U}}$ and $(\mathcal{A}_2)_{\underline{U}}$ must be greater than zero. It was also shown that $(\mathcal{A}_2/\mathcal{A}_1)_{\underline{U}} > 0$ is both necessary and sufficient criterion of stability for pure materials ($n = 1$).

The determinant $(\mathcal{A}_1)_{\underline{U}}$ is related to the specific heat at constant volume:

$$(\mathcal{A}_1)_{\underline{U}} = U_{SS} = \left(\frac{\partial^2 \underline{U}}{\partial \underline{S}^2}\right)_{\underline{V},N} = \left(\frac{\partial T}{\partial \underline{S}}\right)_{\underline{V},N} = \frac{1}{N}\left(\frac{\partial T}{\partial S}\right)_V = \frac{T}{NC_v} \tag{7-104}$$

In a stable equilibrium state, $(\mathcal{A}_1)_{\underline{U}} > 0$. Since T and N are always positive, it follows that

$$C_v > 0 \tag{7-105}$$

Eq. (7-105) is referred to as the *criterion of thermal stability*. It must always be satisfied for a stable equilibrium state. Nevertheless, it is not a sufficient criterion because C_v can be positive even when $(\mathfrak{a}_2/\mathfrak{a}_1)_{\underline{V}}$ is negative.

Using Eqs. (7-95) and (7-96), we see that the more stringent test of stability is

$$\left(\frac{\mathfrak{a}_2}{\mathfrak{a}_1}\right)_{\underline{V}} = -\left(\frac{\partial P}{\partial \underline{V}}\right)_{T,N} = -\frac{1}{N}\left(\frac{\partial P}{\partial V}\right)_T > 0 \tag{7-106}$$

or

$$\left(\frac{\partial P}{\partial V}\right)_T < 0 \tag{7-107}$$

Eq. (7-107) is called the *criterion of mechanical stability*. It is sometimes written in the equivalent form:

$$\kappa_T = -\frac{1}{V}\left(\frac{\partial V}{\partial P}\right)_T > 0 \tag{7-108}$$

From the criteria of thermal and mechanical stability, limitations on other second-order partials can also be derived. From Eqs. (5-137), (6-13) and (6-14), one can show that

$$C_p = C_v + \frac{\alpha_p^2 VT}{\kappa_T}$$

Since $\alpha_p^2 VT$ is always positive and κ_T is positive for mechanical stability, it follows that

$$C_p > C_v \quad \text{and, thus,} \quad C_p > 0 \tag{7-109}$$

The coefficient of isentropic expansion, κ_S, can also be expanded in terms of C_p, C_v and κ_T as follows:

$$\kappa_S = -\frac{1}{V}\left(\frac{\partial V}{\partial P}\right)_S = \frac{1}{V}\frac{\left(\frac{\partial S}{\partial P}\right)_V}{\left(\frac{\partial S}{\partial V}\right)_P} = \frac{1}{V}\left(\frac{C_v}{C_p}\right)\frac{\left(\frac{\partial T}{\partial P}\right)_V}{\left(\frac{\partial T}{\partial V}\right)_P} = -\frac{1}{V}\left(\frac{C_v}{C_p}\right)\left(\frac{\partial V}{\partial P}\right)_T \tag{7-110}$$

or

$$\kappa_S = \frac{C_v \kappa_T}{C_p} > 0 \tag{7-111}$$

Example 7.4

Compressibility factor plots of Z vs. pressure for various temperatures (such as the one shown for hydrogen in Problem 6.1) show some unusual curvature, especially in the vicinity of the critical point. Show that any tangent to an isotherm always must intersect the Z-axis with a positive value.

Solution

As shown in Figure 7.8, at any T' and P', the value of $\epsilon = Z - P(\partial Z/\partial P)_T$.

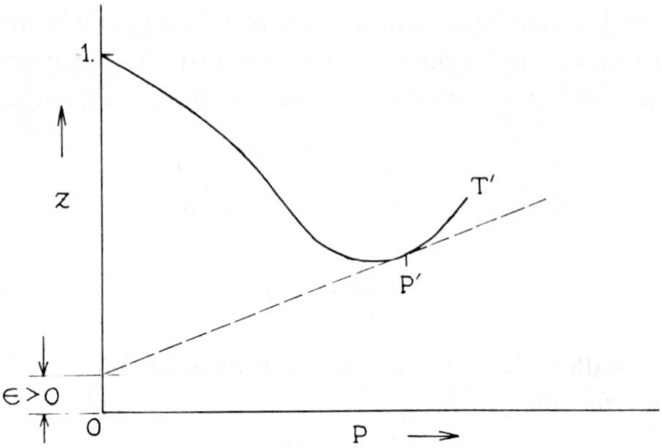

Figure 7.8

But from Eq. (6-103), $\epsilon = Z_p$ which by Eq. (6-106),

$$\epsilon = Z_p = \frac{\left(\dfrac{-P^2}{RT}\right)}{\left(\dfrac{\partial P}{\partial V}\right)_T}$$

But by Eq. (7-107), $(\partial P/\partial V)_T < 0$; therefore, $\epsilon > 0$.

The necessary and sufficient criterion of stability for a pure material, Eq. (7-107), can be visualized with the aid of Figure 7.9. This is a typical P-V plot for the liquid and gaseous regions of a pure substance; also shown is an isotherm at temperature T. Reducing the pressure at constant temperature one follows the isotherm to point A where the liquid is called a saturated liquid and would be in equilibrium with saturated vapor at C. The isotherm for the vapor phase thus normally emerges at C. At both A and C, $(\partial P/\partial V)_T < 0$ and both phases are stable.

Suppose, however, we propose an experiment in which a liquid under conditions at A is further expanded isothermally to some lower pressure noted by B. The criterion $(\partial P/\partial V)_T < 0$ is still satisfied and B is a stable equilibrium state, at least with respect to *small* variations in volume. We stress the fact that at B the variations must be small since it is not difficult to show that there are possible variations that could lead to a state of higher entropy; that is, with regard to some variations, B is not a stable equilibrium state. One obvious variation fitting this description would be to allow some vapor to form.

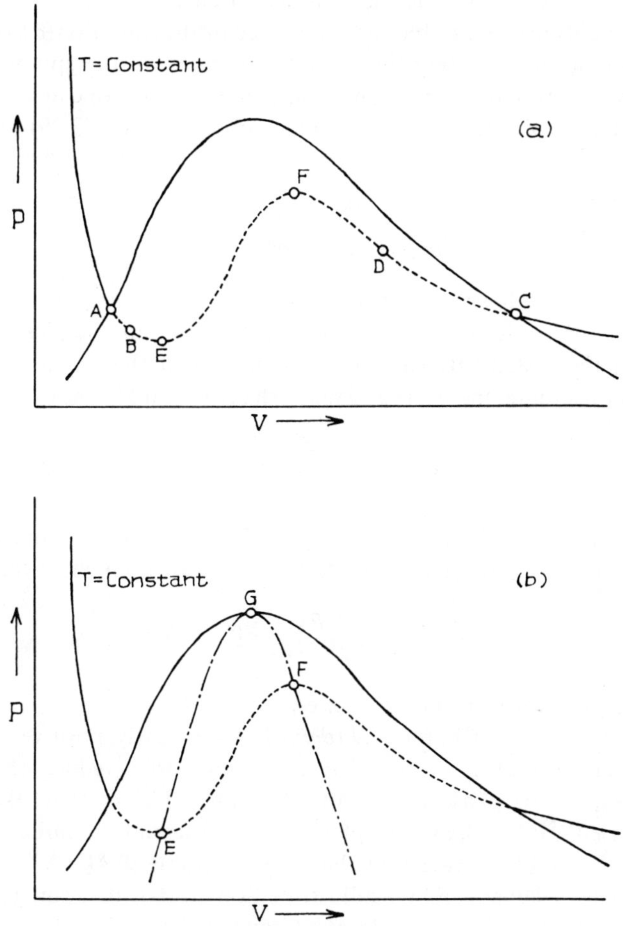

Figure 7.9 Metastable equilibrium.

State B is, therefore, metastable. To ever attain this state we would need to start with a very clean, degassed liquid, free of nucleation motes in order to prevent vapor from forming.

In a similar manner, we could have begun with the saturated vapor at C and compressed isothermally to D. Again, this latter state is metastable.

It is generally assumed that A, B, C, and D lie on a continuous curve. If this is true, the curve would have the general shape of A-B-E-F-D-C in Figure 7.9(a). At E and F it is obvious that $(\partial P/\partial V)_T = 0$, which is the limit of intrinsic stability for the temperature in question. These points fall on the *spinoidal* curve shown by the dot-dashed curve in Figure 7.9(b). States on the spinoidal curve are stable and are defined by the condition $(\partial P/\partial V)_T = 0$.

To examine more closely the stability of states on the spinoidal curve, since we are interested in variations along an isotherm, we can conveniently apply the Helmholtz free energy extrema principle. Eqs. (7-29) and (7-30), written in the form:

$$\delta \underline{A} = 0$$

$$\delta^2 \underline{A} \geqslant 0, \qquad \delta^m \underline{A} > 0$$

where $\delta^m \underline{A}$ is the lowest-order nonvanishing variation. The significance of $\delta \underline{A})_{T,\underline{V},N} = 0$ leads to the equality of pressures and chemical potentials throughout the system. By similar reasoning, when there is a variation in volume at constant temperature and mass, the criterion $\delta^2 \underline{A}$ may be reduced to

$$\delta^2 \underline{A} = -\left(\frac{\partial P}{\partial \underline{V}}\right)_{T,N} \delta \underline{V}^2$$

As we have indicated, for points E and F, $(\partial P/\partial V)_T = 0$ and, thus, the third-order variation of $\underline{A}$ must be considered. We must then show that:

$$\delta^3 \underline{A} = -\left(\frac{\partial^2 P}{\partial \underline{V}^2}\right)_{T,N} \delta \underline{V}^3 > 0$$

or, if this equals zero, then $\delta^4 \underline{A} > 0$, etc.

Now, at point E on Figure 7.9 $(\partial^2 P/\partial V^2)_T$ is clearly positive; thus, one would conclude that for variations in volume that reduce the volume (i.e., $\delta \underline{V}^3 < 0$), the phase is intrinsically stable. If, however, the volume were increased, $\delta \underline{V}^3 > 0$, and the phase would be unstable. Similar reasoning applied to F would indicate that the phase is stable if $\delta \underline{V}^3 > 0$ but not if $\delta \underline{V}^3 < 0$. The arguments may be illustrated on an $\underline{A}$-$\underline{V}$ diagram as shown in Figure 7.10. Points A and C represent saturated liquid and vapor phases at the given temperature. These points have a common tangent, the slope of which is $-P$. E and F represent states in which $\partial^2 \underline{A}/\partial \underline{V}^2 = 0$ and they correspond to points E and F respectively in Figure 7.9. For any system between E and F, $\partial^2 \underline{A}/\partial \underline{V}^2 < 0$ and this system is unstable. Points E and F represent stable states only for particular variations (i.e., for E, $\delta \underline{V} < 0$ and for F, $\delta \underline{V} > 0$).

At higher temperatures, states comparable to E and F on the spinoidal curve begin to approach each other and, at *the critical point*, they coincide. That is, the two phases become indistinguishable (see Figure 7.9, point G) as the intensive properties of each phase (e.g., density, specific enthalpy, etc.) become identical. For a pure substance, if a liquid in equilibrium with its vapor is heated until the meniscus disappears, we denote the vapor pressure at this *critical temperature* as the *critical pressure*. At this unique

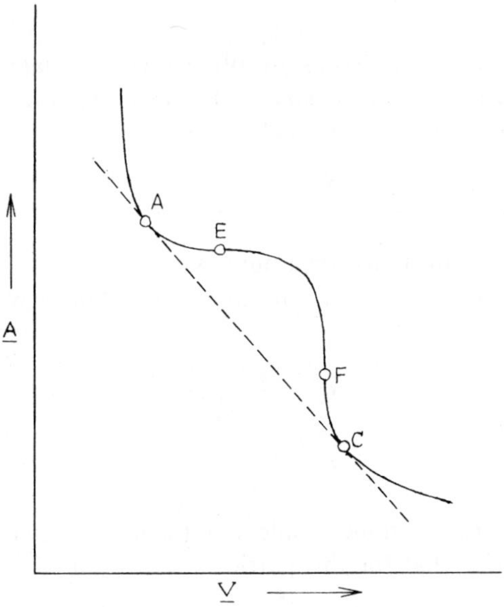

Figure 7.10 Helmholtz free energy-volume diagram.

point, from the arguments shown above, it is clear that:

$$\left.\begin{aligned} \delta \underline{A} &= 0 \\ \delta^2 \underline{A} &= 0 \\ \delta^3 \underline{A} &= 0 \\ \delta^4 \underline{A} &\geqslant 0 \end{aligned}\right\} \tag{7-112}$$

or

$$\left.\begin{aligned} \left(\frac{\partial P}{\partial \underline{V}}\right)_{T,N} &= 0 \\ \left(\frac{\partial^2 P}{\partial \underline{V}^2}\right)_{T,N} &= 0 \\ \left(\frac{\partial^3 P}{\partial \underline{V}^3}\right)_{T,N} &\leqslant 0 \end{aligned}\right\} \tag{7-113}$$

Above the critical point, $(\partial P/\partial V)_T$ is everywhere negative, and all such systems are intrinsically stable.

This treatment has only briefly touched on the extraordinary behavior of matter at the critical point between liquids and gases. Yet, the study of matter in this region yields a number of interesting and unusual results.[3] Other related critical phenomena, for mixtures, are discussed in Section 7.6.

[3] See, for example, the excellent review article, "The Critical Region," by J. V. Sengers and A. L. Sengers, *Chem. Eng. News*, **46**, No. 25 (1968) 104–18.

Example 7.5

The stability criteria, Eq. (7-113) are often used to relate constants in an equation of state to critical constants. To illustrate, the well-known two-constant equation of state of Redlich-Kwong[4]

$$P = \frac{RT}{V - b} - \frac{a}{T^{1/2}V(V + b)}$$

employs a and b as constants for a pure material.

To nondimensionalize these constants, we define new parameters, $\boldsymbol{\Omega}_a$ and $\boldsymbol{\Omega}_b$:

$$a = \frac{\Omega_a R^2 T_c^{5/2}}{P_c}$$

$$b = \frac{\Omega_b R T_c}{P_c}$$

The terms $\boldsymbol{\Omega}_a$ and $\boldsymbol{\Omega}_b$ are dimensionless and can be evaluated from the stability criteria. Show that for this particular equation of state they are pure numbers.

Solution

One approach is to differentiate the Redlich-Kwong equation of state directly and substitute in Eq. (7-113),

$$\left(\frac{\partial P}{\partial V}\right)_{T_c} = 0$$

$$\left(\frac{\partial^2 P}{\partial V^2}\right)_{T_c} = 0$$

These equations could be solved simultaneously for $\boldsymbol{\Omega}_a$ and $\boldsymbol{\Omega}_b$ when $P = P_c$, $T = T_c$, and $V = V_c$. This technique is, however, algebraically complex and a simpler, alternate scheme is demonstrated.

We define two new parameters A and B as:

$$A = \frac{aP}{R^2 T^{5/2}} = \frac{\Omega_a P_r}{T_r^{2 \cdot 5}}$$

$$B = \frac{bP}{RT} = \frac{\Omega_b P_r}{T_r}$$

where $P_r = P/P_c$ and $T_r = T/T_c$. Employing the compressibility factor $Z = PV/RT$, with a and b in terms of A and B, the Redlich-Kwong equation

[4] Otto Redlich and J. N. S. Kwong, "On the Thermodynamics of Solutions, V," *Chem. Rev.*, **44**, (1949), 233.

becomes:

$$Z^3 - Z^2 + (A - B^2 - B)Z - AB = 0$$

(This is Eq. (E-11) in Appendix E.)

At the critical point, $P_r = 1$, $T_r = 1$ and $A = \Omega_a$, $B = \Omega_b$, $Z = Z_c$.

$$Z_c^3 - Z_c^2 + (\Omega_a - \Omega_b^2 - \Omega_b)Z_c - \Omega_a\Omega_b = 0$$

To develop the stability criteria, we note that at $T < T_c$ there are three real roots for V (see Figure 7.9); for $T > T_c$, there is only one real root for V; and at $T = T_c$, there are three equal, real roots for V. Thus at $T = T_c$ there are also three equal, real roots for Z. Let us call these α. Then,

$$(Z_c - \alpha)^3 = 0 = Z_c^3 - 3\alpha Z_c^2 + 3\alpha^2 Z_c - \alpha^3 = 0$$

Comparing coefficients to the Redlich-Kwong equation at the critical point,

$$-3\alpha = -1$$

$$\Omega_a - \Omega_b^2 - \Omega_b = 3\alpha^2$$

$$-\Omega_a\Omega_b = -\alpha^3$$

Thus,

$$\alpha = \frac{1}{3} = Z_c$$

$$\Omega_a = [9(2^{1/3} - 1)]^{-1} = 0.427480\ldots$$

$$\Omega_b = \frac{2^{1/3} - 1}{3} = 0.0866403\ldots$$

The last two relations are those given in Eqs. (E-14) and (E-15).

It is common practice to proceed by choosing equal roots for V_c (or Z_c) at the critical. In some cases, where the assumed equation of state is of even higher order in volume, more than three equal roots have been chosen, e.g., Barner and Alder[5] in their fifth-order equation assume five equal roots at the critical to aid them in formulating a generalized equation of state.

7.6 Stability of Multicomponent Systems

In Section 7.4 the stability criteria for mixtures were delineated. For a binary system, $n + 1$ is 3 and, thus $(\mathcal{A}_3/\mathcal{A}_2)_{\underline{V}} > 0$ is necessary and sufficient to identify a stable equilibrium state provided $\delta^2\underline{U}$ does not vanish. When $(\mathcal{A}_3/\mathcal{A}_2)_{\underline{V}} = 0$, the system is at the limit of intrinsic stability or, in other words, it lies on the spinoidal curve. To determine if this system is stable, we must examine higher-order variations in $\underline{U}$.

[5] H. E. Barner and S. B. Adler, "Three-Parameter Formulation of the Joffe Equation of State," *Ind. Eng. Chem. Fundamentals*, **9**, (1970), 521.

Using Table 7.1, we see that it is clear that the stability criteria for a binary mixture are simplified by working in the $\underline{G}$-representation. Denoting the components by subscripts 1 and 2, then

$$\left(\frac{\mathcal{A}_3}{\mathcal{A}_2}\right)_U = G_{11} = \left(\frac{\partial \mu_1}{\partial N_1}\right)_{T,P,N_2} \geqslant 0 \tag{7-114}$$

where the equality applies to the spinoidal curve. Note that the inequalities representing thermal and mechanical stability, Eqs. (7-105) and (7-107), hold even on the spinoidal curve; they are necessary criteria but not sufficient since Eq. (7-114) supercedes them. Eq. (7-114) is commonly called the *criterion of diffusional stability*, although there is no conceptual connection with the process of eddy or molecular diffusion.

The stability criterion for a binary system can be expressed in terms of mole fractions instead of moles. This alternate form is often more convenient in practice. From Eq. (7-114),

$$\begin{aligned} G_{11} &= \left(\frac{\partial^2 \underline{G}}{\partial N_1^2}\right)_{T,P,N_2} = \left(\frac{\partial \mu_1}{\partial N_1}\right)_{T,P,N_2} \\ &= \left(\frac{\partial \mu_1}{\partial x_1}\right)_{T,P} \left(\frac{\partial x_1}{\partial N_1}\right)_{T,P,N_2} = \left(\frac{x_2}{N}\right)\left(\frac{\partial \mu_1}{\partial x_1}\right)_{T,P} \geqslant 0 \end{aligned} \tag{7-115}$$

Therefore, an alternate criterion for intrinsic stability in a binary system at constant temperature and pressure is:

$$\left(\frac{\partial \mu_1}{\partial x_1}\right)_{T,P} \geqslant 0 \tag{7-116}$$

To visualize this inequality, examine Figure 7.11 in which, for a hypothetical binary system at constant pressure, μ_1 is plotted against x_1 at several different temperatures. QKR defines the locus of all phases that are in equilibrium (i.e., a phase at A is in phase equilibrium with a phase at C at T_1). At these points $(\partial \mu_1/\partial x_1)$ is positive and the phase is stable. A continuation of the T_1 isotherm from A to E or from C to F would define metastable equilibrium states that could be attained experimentally only if the phase transitions were inhibited. Such curves are quite similar to those in Figures 7.9 and 7.10 which were developed earlier to treat mechanical stability. States between E and F are unstable and at E and F

$$\left(\frac{\partial \mu_1}{\partial x_1}\right)_{P,T} = 0 \tag{7-117}$$

Eq. (7-117) defines states on the spinoidal curve $Q'KR'$. Both the equilibrium phase envelope (QKR) and the spinoidal curve become tangent at K, the critical point for this binary. The critical temperature is at T_2 and x_1^c for this particular pressure. It is obvious that Eq. (7-117) would still apply at point K.

To determine if the critical point is stable, we must examine variations of

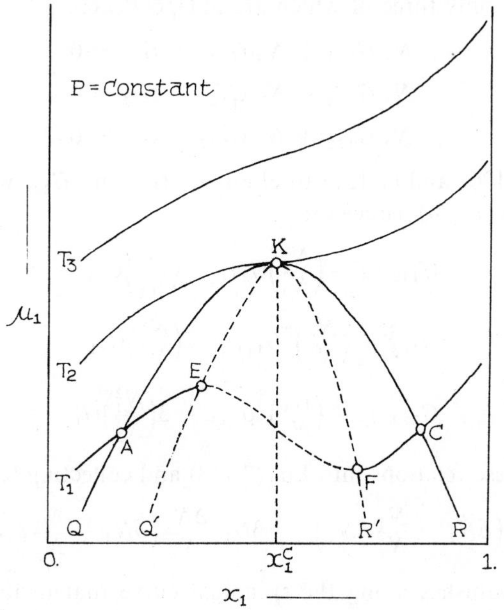

Figure 7.11 Chemical potential as a function of composition for a binary system.

order greater than two. For the binary mixture, the third-order variation of $\underline{G}$ is

$$\begin{aligned} \partial^3 \underline{G} = {} & G_{111}\, \delta N_1^3 + 3G_{112}\, \delta N_1^2\, \delta N_2 + 3G_{122}\, \delta N_1\, \delta N_2^2 + G_{222}\, \partial N_2^3 \\ & + \text{terms in } \delta T^3,\quad \delta T^2\, \delta P,\quad \delta T\, \delta P^2,\quad \delta P^3,\quad \delta T^2\, \delta N_i, \\ & \delta T\, \delta N_i^2,\quad \delta P^2\, \delta N_i,\quad \delta P\, \delta N_i^2 \end{aligned} \tag{7-118}$$

In the $\underline{G}$-representation, the extremum principle applies to a system held at constant T and P. Therefore, all terms in δT and δP in Eq. (7-118) are zero. The remaining terms in δN_1 and δN_2 can be simplified because only one of the third-order partial deriavtives, G_{ijk}, is independent. G_{112}, G_{122}, and G_{222} can be expressed as a function of G_{111} by twice differentiating the Gibbs-Duhem equation of the binary. That is,

$$N_1\, dG_1 + N_2\, dG_2 = 0 \quad (\text{constant } T, P) \tag{7-119}$$

Hence,

$$N_1\, G_{11} + N_2\, G_{12} = 0 \tag{7-120}$$

and

$$N_1\, G_{12} + N_2\, G_{22} = 0 \tag{7-121}$$

Differentiating Eqs. (7-120) and (7-121) with respect to N_1 and N_2 yields four equations, only three of which are independent:

$$N_1 G_{111} + N_2 G_{112} + G_{11} = 0 \tag{7-122}$$

$$N_1 G_{112} + N_2 G_{122} + G_{12} = 0 \tag{7-123}$$

$$N_1 G_{122} + N_2 G_{222} + G_{22} = 0 \tag{7-124}$$

Using Eqs. (7-120) and (7-121) to eliminate G_{12} and G_{22}, we see that Eqs. (7-122) through (7-124) reduce to:

$$G_{112} = -\left(\frac{N_1}{N_2}\right)G_{111} - \left(\frac{1}{N_2}\right)G_{11} \tag{7-125}$$

$$G_{122} = \left(\frac{N_1}{N_2}\right)^2 G_{111} + 2\left(\frac{N_1}{N_2^2}\right)G_{11} \tag{7-126}$$

$$G_{222} = -\left(\frac{N_1}{N_2}\right)^3 G_{111} - 3\left(\frac{N_1^2}{N_2^3}\right)G_{11} \tag{7-127}$$

Substituting these relations into Eq. (7-118) and collecting terms,

$$\delta^3\underline{G} = G_{111}\left(\delta N_1 - \frac{N_1}{N_2}\delta N_2\right)^3 - 3G_{11}\frac{\delta N_2}{N_2}\left(\delta N_1 - \frac{N_1}{N_2}\delta N_2\right)^2 \tag{7-128}$$

Since G_{11} vanishes along the spinoidal curve that includes the critical point, and since the coefficient of G_{111} can be either positive or negative, then, if the critical point is a stable equilibrium state,

$$G_{111} = 0 \tag{7-129}$$

and

$$\delta^3\underline{G} = 0 \tag{7-130}$$

To continue our search for whether or not the critical point is stable, we must determine if

$$\delta^4\underline{G} \geqslant 0 \tag{7-131}$$

It can be shown by a similar approach that

$$\delta^4\underline{G} = G_{1111}\left[\delta N_1 - \left(\frac{N_1}{N_2}\right)\delta N_2\right]^4 + \text{terms in } G_{111} \text{ and } G_{11} \geqslant 0 \tag{7-132}$$

If

$$G_{1111} > 0 \tag{7-133}$$

the critical point is stable.

The significance of Eq. (7-129) is readily shown, i.e.,

$$G_{111} = \left(\frac{\partial G_{11}}{\partial N_1}\right)_{T,P,N_2} = \frac{\partial}{\partial N_1}\left[\left(\frac{\partial \mu_1}{\partial N_1}\right)_{T,P}\right]_{T,P,N_2} = \frac{\partial}{\partial N_1}\left[\frac{x_2}{N}\left(\frac{\partial \mu_1}{\partial x_1}\right)_{T,P}\right]_{T,P,N_2}$$
$$= \frac{x_2}{N}\frac{\partial}{\partial x_1}\left[\frac{x_2}{N}\left(\frac{\partial \mu_1}{\partial x_1}\right)_{T,P}\right]_{T,P} = -\frac{x_2}{N^2}\left(\frac{\partial \mu_1}{\partial x_1}\right)_{T,P} + \left(\frac{x_2}{N}\right)^2\left(\frac{\partial^2 \mu_1}{\partial x_1^2}\right)_{T,P} \tag{7-134}$$

Since $(\partial \mu_1/\partial x_1)$ is zero on the spinoidal curve, Eq. (7-134) also states that at the critical point,

$$\left(\frac{\partial^2 \mu_1}{\partial x_1^2}\right)_{T,P} = 0 \tag{7-135}$$

Similarly, from Eq. (7-133),

$$\left(\frac{\partial^3 \mu_1}{\partial x_1^3}\right)_{T,P} > 0 \tag{7-136}$$

It can be seen by inspection of Figure 7.11 that Eqs. (7-135) and (7-136) are satisfied at the critical point and, hence, it is a stable equilibrium state.

Example 7.6

As will be shown in Chapter 8, the chemical potential of a component in a liquid mixture, μ_i, can be expressed as a function of the chemical potential of pure liquid i at the same T and P of the mixture, μ_{i_0}, and a property of the mixture, γ_i, called the activity coefficient:

$$\mu_i = \mu_i^0 + RT \ln (\gamma_i x_i) \tag{7-137}$$

where

$$\mu_i^0 = f(T, P)$$

and

$$\gamma_i = f(T, P, x_1, \ldots, x_{n-1}) \tag{7-138}$$

An ideal solution is defined as one for which

$$\gamma_i = 1 \text{ for all } i. \tag{7-139}$$

Just as there are nonideal gas equations that attempt to correlate the deviation from the ideal gas law, there are correlating equations available in the form of Eq. (7-138) for liquid mixtures.

Typical of activity coefficient correlating functions for binary systems at constant T and P is the van Laar equation:

$$\ln \gamma_1 = \frac{A}{\left(1 + \dfrac{A}{B}\dfrac{x_1}{x_2}\right)^2} \tag{7-140}$$

$$\ln \gamma_2 = \frac{A}{\left(1 + \dfrac{B}{A}\dfrac{x_2}{x_1}\right)^2} \tag{7-141}$$

where A and B are independent of concentration.

You are faced with the problem of modelling the chemical potential of a binary liquid mixture that exhibits partial miscibility over the range of conditions of interest. Determine which of the two correlating equations—(1) ideal or (2) van Laar—can be eliminated from considerations based on stability criteria.

Solution

From the preceding discussion, we have seen that if a phase of a binary mixture is stable for all compositions (i.e., miscible in all proportions), then

$$\left(\frac{\partial \mu_1}{\partial x_1}\right)_{T,P} > 0 \tag{7-142}$$

If the system exhibits partial miscibility, then

$$\left(\frac{\partial \mu_1}{\partial x_1}\right)_{T,P} \leqslant 0 \tag{7-143}$$

for some range of compositions within the immiscibility gap.

At the critical point for incipient immiscibility,

$$\left(\frac{\partial \mu_1}{\partial x_1}\right)_{T,P} = 0 \tag{7-144}$$

$$\left(\frac{\partial^2 \mu_1}{\partial x_1^2}\right)_{T,P} = 0 \tag{7-145}$$

and

$$\left(\frac{\partial^3 \mu_1}{\partial x_1^3}\right)_{T,P} > 0 \tag{7-146}$$

Since any correlation that successfully models a partially miscible system must also describe a critical point for incipient immiscibility, Eqs. (7-144) through (7-146) can be used to test any proposed correlating equation. We can restate the criteria of the critical condition in terms of activity coefficients since, in general,

$$\mu_i = \mu_{i_0} + RT \ln \gamma_i x_i \tag{7-147}$$

Thus, the corresponding criteria are:

$$\left(\frac{\partial \mu_1}{\partial x_1}\right)_{T,P} = RT\left[\left(\frac{\partial \ln \gamma_1}{\partial x_1}\right)_{T,P} + \frac{1}{x_1}\right] = 0 \tag{7-148}$$

$$\left(\frac{\partial^2 \mu_1}{\partial x_1^2}\right)_{T,P} = RT\left[\left(\frac{\partial^2 \ln \gamma_1}{\partial x_1^2}\right)_{T,P} - \frac{1}{x_1^2}\right] = 0 \tag{7-149}$$

and

$$\left(\frac{\partial^3 \mu_1}{\partial x_1^3}\right)_{T,P} = RT\left[\left(\frac{\partial^3 \ln \gamma_1}{\partial x_1^3}\right)_{T,P} + \frac{2}{x_1^3}\right] > 0 \tag{7-150}$$

1. *Ideal solution*: $\gamma_1 = 1$. By inspecting Eq. (7-148), we see that the first-order partial derivative of μ_1 is always positive and, hence, the ideal solution is miscible in all proportions.

2. *van Laar*: γ_1 given by Eq. (7-140). Defining $Z = A/B$, we see that Eq. (7-148) becomes, after differentiation and simplification,

$$\frac{2AZ}{x_{c_2}^2\left[1 + Z\frac{x_{c_1}}{x_{c_2}}\right]^3} + \frac{1}{x_{c_1}} = 0 \tag{7-151}$$

where x_c denotes the critical solution composition. Similarly, Eq. (7-149) becomes:

$$-\frac{2AZ\left(2-\dfrac{3Z}{x_{c_2}+Zx_{c_2}}\right)}{x_{c_2}^3\left(1+Z\dfrac{x_{c_1}}{x_{c_2}}\right)^3}-\frac{1}{x_{c_2}^2}=0 \tag{7-152}$$

Eqs. (7-151) and (7-152) contain three independent variables: x_{c_1}, A, and Z. Thus, we can solve for x_{c_1} as a function of A or Z. Choosing the latter,

$$x_{c_1}=\frac{(Z^2+1-Z)^{1/2}-Z}{1-Z} \tag{7-153}$$

The third criterion, Eq. (7-150), reduces to:

$$\frac{1}{x_{c_1}x_{c_2}^2}\left[\frac{18Z}{Dx_{c_2}}-\frac{12Z^2}{D^2x_{c_2}^2}-6\right]+\frac{2}{x_{c_1}^3}>0 \tag{7-154}$$

where

$$D=1+Z\frac{x_{c_1}}{x_{c_2}}$$

If the van Laar correlation is to describe a critical solution condition, then there must be values of Z for which x_{c_1} [from Eq. (7-153)] lies between 0

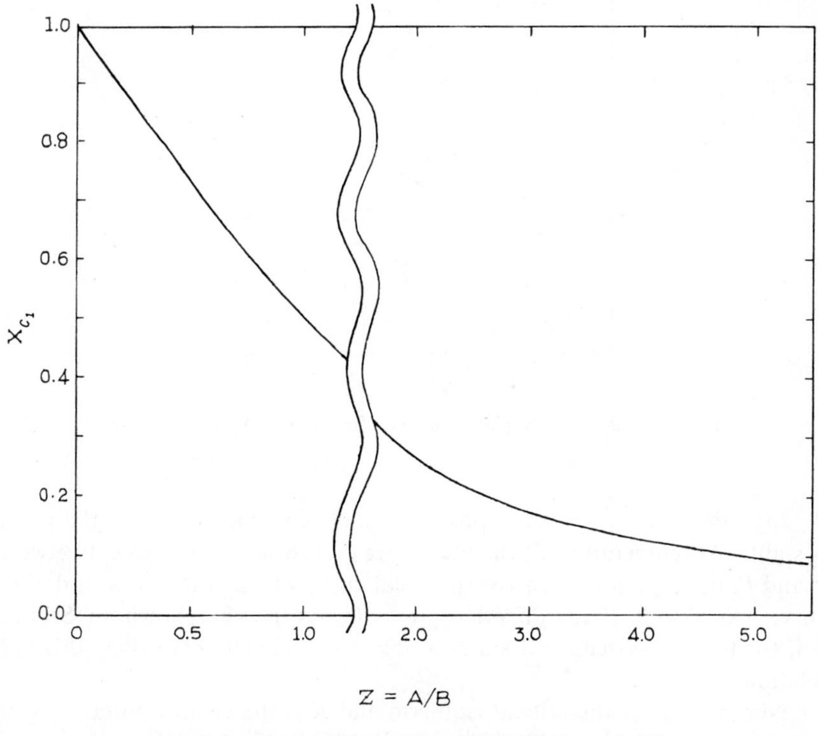

Figure 7.12 Test of van Laar equation in Example 7.6.

and 1 *and* for which the inequality of Eq. (7-154) is satisfied. It can be shown that these conditions are satisfied for all values of Z between 0 and ∞ and, thus, the coefficients A and B must be of the same sign. The relationship of x_{c_1} to Z is shown in Figure 7.12.

Now let us briefly examine systems that are more complex than a binary. For a ternary system, the necessary and sufficient criterion of a stable equilibrium state is $(\mathcal{Q}_4/\mathcal{Q}_3)_{\underline{V}} \geqslant 0$, which can be expressed in terms of $\underline{G}'(T, P, \mu_1, N_2, N_3)$ as simply:

$$G'_{22} = \left(\frac{\partial \mu_2}{\partial N_2}\right)_{T,P,\mu_1,N_3} \geqslant 0 \tag{7-155}$$

Eq. (7-155) can be visualized with the aid of Figure 7.13. At temperature

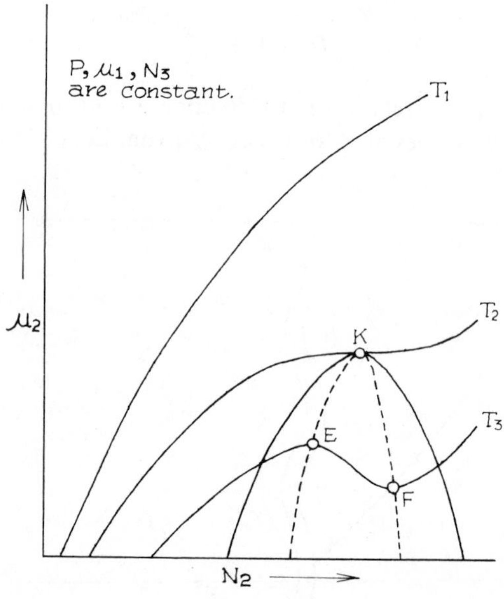

Figure 7.13 Stability plot of μ_2 as a function of N_2 for a ternary system.

T_1, $(\partial \mu_2/\partial N_2)_{T,P,\mu_1,N_3}$ is always positive, which would indicate that the phase is stable. At temperature T_3 the phases are everywhere stable except between E and F; these points lie on the spinoidal curve which is shown as a dashed curve. At E and F $(\partial \mu_2/\partial N_2)_{T,P,\mu_1,N_3} = 0$. As it was shown in Section 7.4, for ternary systems, all states along the spinoidal curve obey this null relation.

Note that T_2 is the critical isotherm and K is the critical point. K is on the spinoidal curve so that

$$G'_{22} = \left(\frac{\partial \mu_2}{\partial N_2}\right)_{T,P,\mu_1,N_3} = 0 \tag{7-156}$$

In addition, the isotherm must have a point of inflection at K so that

$$G'_{222} = \left(\frac{\partial^2 \mu_2}{\partial N_2^2}\right)_{T,P,\mu_1,N_3} = 0 \tag{7-157}$$

and

$$G'_{2222} = \left(\frac{\partial^3 \mu_2}{\partial N_2^3}\right)_{T,P,\mu_1,N_3} > 0 \tag{7-158}$$

Thus, the critical point is a stable equilibrium state.

In many instances for nonideal mixtures, the chemical potential is a complex polynomial and/or transcendental function of mole fractions (e.g., see Example 7.6) and, consequently, it is difficult if not impossible to transform $\underline{G}$ into a closed form of $\underline{G}'$. When calculations are to be done by hand instead of by machine, it is usually necessary to treat the mixture in the $\underline{G}$-representation. As shown in Table 7.1, for a ternary system, the appropriate analog to Eq. (7-114) is

$$G'_{22} = \left(\frac{\mathcal{C}_4}{\mathcal{C}_3}\right)_{U} = \frac{G_{11}G_{22} - G_{12}^2}{G_{11}} \geqslant 0 \tag{7-159}$$

or simply

$$G_{11}G_{22} - G_{12}^2 \geqslant 0 \tag{7-160}$$

At the critical point the equality applies and we must examine the analog to Eq. (7-157) in the $\underline{G}$-representation. If we apply the Legendre transformation to the third-order partial derivatives in the same way that second-order derivatives were transformed in Section 5.6, it can be shown that

$$G'_{222} = \left(\frac{\partial^2 \mu_2}{\partial N_2^2}\right)_{T,P,\mu_1,N_3} = \frac{1}{G_{11}^2}[G_{222}G_{11}^2 - G_{12}(3G_{122}G_{11} + G_{111}G_{22}) + G_{112}(2G_{12}^2 + G_{11}G_{22})] = 0 \tag{7-161}$$

The bracket in Eq. (7-162) is sometimes written as the determinant, $\underline{M}$:

$$\underline{M} = \begin{vmatrix} \left(\frac{\partial \mathcal{C}_{2G}}{\partial N_1}\right)_{N_2,N_3} & \left(\frac{\partial \mathcal{C}_{2G}}{\partial N_2}\right)_{N_1,N_3} \\ G_{12} & G_{22} \end{vmatrix} \tag{7-162}$$

where $\mathcal{C}_{2G}$ is the 2×2 matrix resulting from the transformation of $\mathcal{C}_4$ from the $\underline{U}$- to the $\underline{G}$-representation (see Table 7.1).

$$\mathcal{C}_{2G} = \begin{vmatrix} G_{11} & G_{12} \\ G_{12} & G_{22} \end{vmatrix} \tag{7-163}$$

The general case for an n-component mixture can be treated in either the $\underline{G}^{(n-2)'}$ or the $\underline{G}$-representation. For example, the critical point is defined by

$$\mathcal{A}_{n+1} = \mathcal{A}_{(n-1)G} = \begin{vmatrix} G_{11} & G_{12} & \cdots & G_{1(n-1)} \\ G_{12} & G_{22} & \cdots & G_{2(n-1)} \\ \cdot & \cdot & & \cdot \\ \cdot & \cdot & & \cdot \\ \cdot & \cdot & & \cdot \\ G_{1(n-1)} & G_{2(n-1)} & & G_{(n-1)(n-1)} \end{vmatrix} = 0 \qquad (7\text{-}164)$$

$$\underline{M} = \begin{vmatrix} \dfrac{\partial \mathcal{A}_{(n-1)G}}{\partial N_1} & \dfrac{\partial \mathcal{A}_{(n-1)G}}{\partial N_2} & \cdots & \dfrac{\partial \mathcal{A}_{(n-1)G}}{\partial N_{n-1}} \\ G_{12} & G_{22} & & G_{2(n-1)} \\ \cdot & \cdot & & \cdot \\ \cdot & \cdot & & \cdot \\ \cdot & \cdot & & \cdot \\ G_{1(n-1)} & G_{2(n-1)} & & G_{(n-1)(n-1)} \end{vmatrix} = 0 \qquad (7\text{-}165)$$

or, respectively,

$$\left(\frac{\partial \mu_{n-1}}{\partial N_{n-1}}\right)_{T, P, \mu_1, \cdots, \mu_{n-2}, N_n} = 0 \qquad (7\text{-}166)$$

and

$$\left(\frac{\partial^2 \mu_{n-1}}{\partial N_{n-1}^2}\right)_{T, P, \mu_1, \cdots, \mu_{n-2}, N_n} = 0 \qquad (7\text{-}167)$$

The criteria for the critical state, Eqs. (7-164) and (7-165), are the same as those stated by Gibbs in his famous paper,[6] and which have been used to estimate the critical properties of mixtures.[7]

PROBLEMS

7.1. A well-insulated steel bomb initially contains liquid and vapor of compound A (see Figure P7.1). The vapor is known to decompose under the conditions of the experiment to B and C by the reaction

$$A = 2\,B + C$$

Compound C is soluble in liquid A, but compound B does not dissolve to any measurable extent. Also, the reaction does not occur to any appreciable extent in the liquid.

Derive the criteria of equilibrium for the system of liquid plus vapor.

[6] "On the Equilibrium of Heterogeneous Substances," *The Collected Works of J. Willard Gibbs, Vol. I, Thermodynamics* (New Haven: Yale University Press, 1957), p. 132.

[7] R. R. Spear, R. L. Robinson, and K.-C. Chao, "Critical States of Ternary Mixtures and Equations of State," *Ind. Eng. Chem. Fund.*, **10**, (1971), 588.

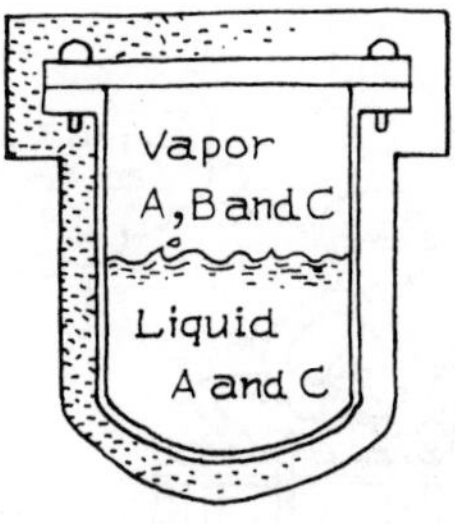

Figure P7.1

7.2. A closed vessel of volume, $\underline{V}$, contains liquid and vapor phases of a pure material. The vessel is immersed in a constant temperature bath at temperature T^* (see Figure P7.2).

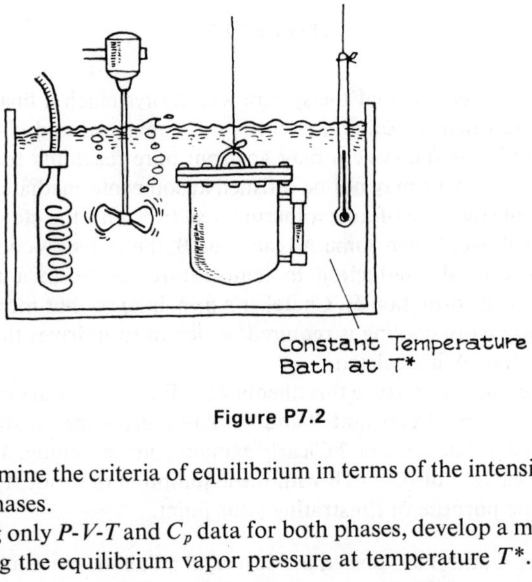

Figure P7.2

(a) Determine the criteria of equilibrium in terms of the intensive properties of the phases.

(b) Using only P-V-T and C_p data for both phases, develop a method for determining the equilibrium vapor pressure at temperature T^*.

7.3. The criteria for chemical equilibrium in a multicomponent, single-phase system,

$$\sum \nu_i \mu_i = 0 \tag{A}$$

is derived from systems at constant ($\underline{S}$, $\underline{V}$), ($\underline{S}$, P), (T, $\underline{V}$), or (T, P), by applying the minimization principle from $\underline{U}$, $\underline{H}$, $\underline{A}$, or $\underline{G}$, respectively. There is some question about the criteria of equilibrium for a system at constant P and $\underline{V}$. For example, a constant volume bomb may initially contain compound A and an anticatalyst to prevent the reaction

$$A \rightleftharpoons 2B \tag{B}$$

The anticatalyst is removed, and as the reaction proceeds, heat is added or removed in order to maintain a constant pressure (see Figure P7.3).

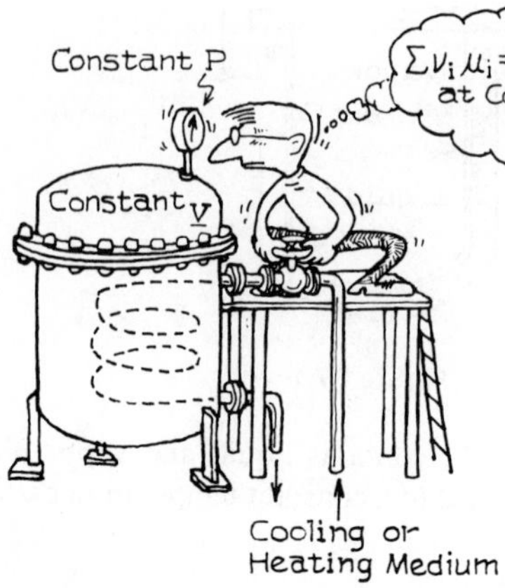

Figure P7.3

One student claims that the system will always reach a final state in which Eq. (A) is satisfied. Another student is not convinced and claims that under certain conditions the system may go from pure reactants to pure products, while Eq. (A) may or may not be satisfied at some intermediate point. He gives as an example the case of an exothermic reaction with the stoichiometric relation of (B) above. When some A reacts to 2B, the system is cooled to maintain constant P, but the reduction in temperature results in a more favorable condition for B formation (L'Chatelier's principle) so that more A reacts, and consequently more cooling is required, which in turn drives the reaction to B, etc., until all of A is depleted.

You are asked to resolve this dilemma: Is Eq. (A) a valid criteria of equilibrium for a system at constant P and $\underline{V}$? Are there some conditions for which Eq. (A) need not be satisfied? Clearly explain your reasoning. You may use the gas phase reaction of $A \rightleftharpoons 2B$ with the assumption of an ideal mixture of ideal gases for the purpose of illustrating your point.

7.4. Subsystems A and B are placed inside a rigid adiabatic container but separated by an internal wall that is restrictive both to volume and matter but not to energy (i.e., it is rigid, impermeable, and diathermal).

Subsystem A has 9 cm^3 and contains 3 moles, while B has 4 cm^3 and 2 moles. The total energy of A and B is 64 calories.

The fundamental equations for the subsystems are:

$$\underline{S}_A = C(N_A \underline{V}_A \underline{U}_A)^{1/3}$$
$$\underline{S}_B = C(N_B \underline{V}_B \underline{U}_B)^{1/3}$$

where C is a constant.

When thermal equilibrium is attained, what are the values of $\underline{U}_A$ and $\underline{U}_B$?

7.5. In Figure P7.5, the tube is well-insulated and the initial conditions are:

Section 1 $\underline{V} = 89.6$ liters
$T = 0°C$
$P = 1$ atm
gas, pure A

Section 2 $\underline{V} = 22.4$ liters
$T = 0°C$
$P = 1$ atm
gas, pure B

Piston L is rigid, diathermal, and initially impermeable. Suddenly, however, it is made permeable to gas A, but not to B. Piston K is simultaneously unlatched and connected to a large pressure reservoir at 2 atm, and piston M is unlatched and connected to a pressure reservoir at 3 atm. There is no friction in the pistons, but the pressure changes are effected so that oscillations are reduced. Assume ideal gases.

After equilibrium is attained,

(a) What are the criteria of equilibrium?

(b) Describe quantitatively the states of the gases on either side of the piston.

(c) Repeat (a) and (b) assuming that the piston L is permeable to both gases.

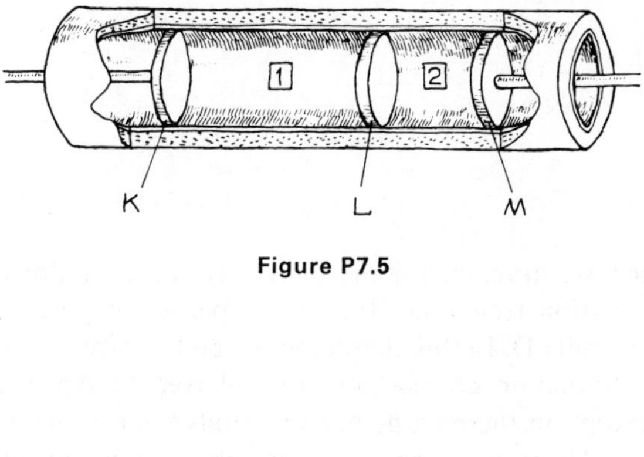

Figure P7.5

Single-Phase, Simple Systems of Mixtures 8

In Chapter 6 we developed many thermodynamic relationships for pure materials. Estimation techniques for the properties of pure substances are included in Appendix D. In this chapter we extend our treatment to mixtures, and property estimation correlations are deferred to Appendix E.

With one exception, there are no new thermodynamic principles introduced in this chapter.[1] There are a few new concepts such as partial molal quantities and mixing functions, but these involve only special definitions of properties in order to facilitate the discussion and analysis of multicomponent data.

In spite of the fact that there are few new concepts or principles, many relationships may appear unduly complex because the notation is often formidable. Yet, in the development of equations to describe mixtures, one must resort to detailed subscripting in order to denote the component involved or the conditions of restraint for partial differentiation.

To avoid repetitive derivations in our initial treatment, a generalized prop-

[1] The one exception involves the ideal entropy or Gibbs free energy of mixing; this is discussed in Section 8.4.

erty $\underline{B}$ (extensive) or B (intensive) is used. Relationships obtained for $\underline{B}$ and B are then applicable for properties such as energy, volume, entropy, enthalpy, or free energy. The treatment is limited to single-phase,[2] multicomponent, simple systems. The mixtures may involve solid solutions (provided that they are in a stable equilibrium state), liquids (including aqueous solutions of electrolytes), and gases. The components are assumed to be nonreacting[3] and nondissociating. Thus, we treat a salt in water as if we were unaware of dissociation and, thereby, define heats of solution, activities, fugacities, etc. for the undissociated salt "molecule." These derived quantities can be related to solubility, osmotic pressure, etc., provided that we are not interested in the properties of the dissociated ions (e.g., ion activities, half-cell potentials, etc.).

8.1 Extensive and Intensive Differentials

For many, this section may simply serve as a review of some elements of calculus.

Any extensive variable, $\underline{B}$, can be expanded in differential form as a function of $(n + 2)$ other variables and, for single-phase, simple systems there are no restrictions on the choice of variables. In Table 8.1, columns (b) and (c), we illustrate two variable sets that are commonly employed for mixtures. For reference, the corresponding relations for pure materials are shown in column (a).

In column (b), the independent variables are n mole numbers plus two other properties. In column (c), we have $(n - 1)$ mole fractions, the total number of moles, and two other properties. In this latter set, we have eliminated one mole fraction (i.e., x_n) so that any of the other $(n - 1)$ mole fractions can be varied independently. Thus, when $\underline{B}$ is expressed as a partial derivative with respect to x_i, x_n must vary so that $dx_n = -dx_i$.

Eq. (8-2a–c) represents chain-rule expansions of $\underline{B}$ in terms of the independent variables. One should note that the subscript N_i in column (b) denotes that *all* mole numbers are held constant. In column (c) the subscript N indicates that the total moles are constant. Of course, the combined x and N subscripts in column (c) are equivalent to the N_i subscript in column (b). Also, subscripts such as $N_j[i]$ denote that all N_j are held constant except N_i. Similarly, the subscript $x[i, n]$ means that all x_j are constant except x_i and x_n.

Eq. (8-2a–c) may be integrated by using Euler's theorem as shown in Appendix C. If Y_1 and Y_2 are extensive, Eq. (8-3a–c) is obtained; if they are intensive, the equations simplify to the form of Eq. (8-4a–c).

If Y_1 and Y_2 are intensive variables, the form of Eq. (8-4b) indicates that

[2] Multiphase systems are covered in Chapter 9.

[3] Reacting systems are covered in Chapter 10.

TABLE 8.1
THE GENERAL EXTENSIVE PROPERTY, $\underline{B}$

PURE MATERIAL (a)	MIXTURE (b)	MIXTURE (c)
Independent variables Y_1, Y_2, N	Independent variables $Y_1, Y_2, N_1, \ldots, N_n$	Independent variables $Y_1, Y_2, x_1, \ldots, x_{n-1}, N$
$\underline{B} = f(Y_1, Y_2, N)$ (8-1a)	$\underline{B} = f(Y_1, Y_2, N_1, \ldots, N_n)$ (8-1b)	$\underline{B} = f(Y_1, Y_2, x_1, \ldots, x_{n-1}, N)$ (8-1c)
$d\underline{B} = \left(\frac{\partial \underline{B}}{\partial Y_1}\right)_{Y_2,N} dY_1 + \left(\frac{\partial \underline{B}}{\partial Y_2}\right)_{Y_1,N} dY_2 + \left(\frac{\partial \underline{B}}{\partial N}\right)_{Y_1,Y_2} dN$ (8-2a)	$d\underline{B} = \left(\frac{\partial \underline{B}}{\partial Y_1}\right)_{Y_2,N_i} dY_1 + \left(\frac{\partial \underline{B}}{\partial Y_2}\right)_{Y_1,N_i} dY_2 + \sum_{i=1}^{n} \left(\frac{\partial \underline{B}}{\partial N_i}\right)_{Y_1,Y_2,N_j[i]} dN_i$ (8-2b)	$d\underline{B} = \left(\frac{\partial \underline{B}}{\partial Y_1}\right)_{Y_2,x,N} dY_1 + \left(\frac{\partial \underline{B}}{\partial Y_2}\right)_{Y_1,x,N} dY_2 + \sum_{i=1}^{n-1} \left(\frac{\partial \underline{B}}{\partial x_i}\right)_{Y_1,Y_2,x[i,n],N} dx_i + \left(\frac{\partial \underline{B}}{\partial N}\right)_{Y_1,Y_2,x} dN$ (8-2c)
INTEGRATION BY EULER'S THEOREM ASSUMING Y_1 AND Y_2 ARE EXTENSIVE PROPERTIES		
$\underline{B} = \left(\frac{\partial \underline{B}}{\partial Y_1}\right)_{Y_2,N} Y_1 + \left(\frac{\partial \underline{B}}{\partial Y_2}\right)_{Y_1,N} Y_2 + \left(\frac{\partial \underline{B}}{\partial N}\right)_{Y_1,Y_2} N$ (8-3a)	$\underline{B} = \left(\frac{\partial \underline{B}}{\partial Y_1}\right)_{Y_2,N_i} Y_1 + \left(\frac{\partial \underline{B}}{\partial Y_2}\right)_{Y_1,N_i} Y_2 + \sum_{i=1}^{n} \left(\frac{\partial \underline{B}}{\partial N_i}\right)_{Y_1,Y_2,N_j[i]} N_i$ (8-3b)	$\underline{B} = \left(\frac{\partial \underline{B}}{\partial Y_1}\right)_{Y_2,x,N} Y_1 + \left(\frac{\partial \underline{B}}{\partial Y_2}\right)_{Y_1,x,N} Y_2 + \left(\frac{\partial \underline{B}}{\partial N}\right)_{Y_1,Y_2,x} N$ (8-3c)
SIMPLIFICATIONS IF Y_1 AND Y_2 ARE INTENSIVE PROPERTIES		
$\underline{B} = \left(\frac{\partial \underline{B}}{\partial N}\right)_{Y_1,Y_2} N$ (8-4a)	$\underline{B} = \sum_{i=1}^{n} \left(\frac{\partial \underline{B}}{\partial N_i}\right)_{Y_1,Y_2,N_j[i]} N_i$ (8-4b)	$\underline{B} = \left(\frac{\partial \underline{B}}{\partial N}\right)_{Y_1,Y_2,x} N$ (8-4c)
$\left(\frac{\partial \underline{B}}{\partial N}\right)_{Y_1,Y_2} = \frac{\underline{B}}{N} = B$ (8-5a)	$\left(\frac{\partial \underline{B}}{\partial N_i}\right)_{Y_1,Y_2,N_j[i]} \neq \frac{\underline{B}}{N_i}$ (8-5b)	$\left(\frac{\partial \underline{B}}{\partial N}\right)_{Y_1,Y_2,x} = \frac{\underline{B}}{N} = B$ (8-5c)
WITH $Y_1 = T$ AND $Y_2 = P$		
$\underline{B} = \left(\frac{\partial \underline{B}}{\partial N}\right)_{T,P} N = BN$ (8-6a)	$\underline{B} = \sum_{i=1}^{n} \left(\frac{\partial \underline{B}}{\partial N_i}\right)_{T,P,N_j[i]} N_i = \sum_{i=1}^{n} \bar{B}_i N_i$ (8-6b)	$\underline{B} = \left(\frac{\partial \underline{B}}{\partial N}\right)_{T,P,x} N = BN$ (8-6c)
Eq. (8-2a) becomes: $d\underline{B} = \left(\frac{\partial \underline{B}}{\partial T}\right)_{P,N} dT + \left(\frac{\partial \underline{B}}{\partial P}\right)_{T,N} dP + B\,dN$ (8-7a)	Eq. (8-2b) becomes: $d\underline{B} = \left(\frac{\partial \underline{B}}{\partial T}\right)_{P,N_i} dT + \left(\frac{d\underline{B}}{\partial P}\right)_{T,N_i} dP + \sum^{n} \bar{B}_i\,dN_i$ (8-7b)	Eq. (8-2c) becomes: $d\underline{B} = \left(\frac{\partial \underline{B}}{\partial T}\right)_{P,x,N} dT + \left(\frac{\partial \underline{B}}{\partial P}\right)_{T,x,N} dP + \sum_{i=1}^{n-1} \left(\frac{\partial \underline{B}}{\partial x_i}\right)_{T,P,x[i,n],N} dx_i + B\,dN$ (8-7c)

the extensive property $\underline{B}$ may be expressed simply as a weighted average of the partial derivatives $(\partial \underline{B}/\partial N_i)_{Y_1,Y_2,N_j[i]}$. In such cases, it has been found that T and P form a convenient set of Y_1 and Y_2, and we define the derivative as a *partial molal property*, $\bar{B}_i$:

$$\bar{B}_i \equiv \left(\frac{\partial \underline{B}}{\partial N_i}\right)_{T,P,N_j[i]} \tag{8-8}$$

Note that partial molal properties are intensive and depend on the temperature, pressure, and composition of the system. The corresponding property for a pure material is, of course, the specific property, B.

With the definition of $\bar{B}_i$, Eqs. (8-2b) and (8-4b) are transformed to (8-7b) and (8-6b), as shown in Table 8.1. Although not shown, many of the derivatives in column (c) can also be rewritten to employ partial molal properties. For example, let us express the derivative $(\partial \underline{B}/\partial x_i)_{T,P,x[i,n],N}$ of Eq. (8-7c) in terms of partial molal quantities. Starting with Eq. (8-7b), divide all differentials by dx_j and impose upon these derivatives the restraints that $T, P, x[j, n], N$ be constant. The first two terms on the right-hand side vanish; thus,

$$\left(\frac{\partial \underline{B}}{\partial x_j}\right)_{T\ P,x[j,n],N} = \sum_{i=1}^{n} \bar{B}_i \left(\frac{\partial N_i}{\partial x_j}\right)_{T,P,x[j,n],N} \tag{8-9}$$

Since

$$N_i = x_i N \tag{8-10}$$

$$\left(\frac{\partial N_i}{\partial x_j}\right)_{T,P,x[j,n],N} = \begin{Bmatrix} 0 & i \neq j, n \\ 1 & i = j \\ -1 & i = n \end{Bmatrix} N \tag{8-11}$$

so that Eq. (8-9) becomes,

$$\left(\frac{\partial \underline{B}}{\partial x_j}\right)_{T,P,x[j,n],N} = N(\bar{B}_j - \bar{B}_n) \tag{8-12}$$

Example 8.1

A ternary mixture of 50 mol-% *n*-propanol, 25 mol-% *n*-pentanol, and 25 mol-% *n*-heptane is prepared by a two-step process in which pentanol and heptane are mixed in vat No.1 and this mixture is then added to vat No. 2 containing the propanol. Each vat is equipped with stirrer and heating coils; the addition is slow enough so that the process can be considered isothermal at 25°C. What is the cooling load required for each vat, expressed in Btu/1b of product?

Solution

First consider vat No. 1; choosing the contents as the system and employing a basis of 1 g-mol of product, at constant pressure

DATA

Mole fractions		Partial molal enthalpies, (−J/g-mol)		
Heptane	Propanol	Heptane	Propanol	Pentanol
0.00	0.00	—	—	0.0
0.00	0.25	—	67.1	6.5
0.00	0.50	—	46.6	16.0
0.00	0.75	—	18.2	54.9
0.00	1.00	—	0.0	—
0.25	0.00	1153.4	—	47.5
0.25	0.25	1155.5	167.4	53.5
0.25	0.50	1165.5	136.2	74.5
0.25	0.75	1200.8	106.4	—
0.50	0.00	864.8	—	237.1
0.50	0.25	884.1	335.7	229.2
0.50	0.50	919.2	280.7	—
0.75	0.00	361.3	—	1155.8
0.75	0.25	425.6	1203.1	—
1.00	0.00	0.0	—	—

$$đQ = d\underline{H} - h_e dn_e$$

$$Q = \underline{H}_{\text{final}} - \underline{H}_{\text{initial}} - h_{a5}n_{a5} - h_h n_h \quad = \Delta H \text{ mixing}$$

where the subscripts a5 and h denote n-pentanol and heptane, respectively. $\underline{H}_{\text{initial}}$ is zero since the vat was initially empty.

$$\underline{H}_{\text{final}} = \bar{H}_{a5}n_{a5} + \bar{H}_h n_h$$

Since $n_{a5} = n_h = 0.5$ g-mol and the enthalpy base for the pure components is zero (i.e., $h_{a5} = h_h = 0$), then

$$-Q = (0.5)(864.8) + (0.5)(237.1) = 551 \text{ J/g-mol}$$

$$= \frac{(551)(1.8)}{(4.19)(94)} = 2.9 \text{ Btu/lb of product from vat No. 1}$$

where 94 is the molecular weight of the mixture.

For vat No. 2, using the same approach,

$$\underline{H}_{\text{final}} = n_{a3}\bar{H}_{a3} + n_{a5}\bar{H}_{a5} + n_h\bar{H}_h$$

$$\underline{H}_{\text{initial}} = 0, \text{ pure } a3 \text{ (}n\text{-propanol)}$$

$$h_e = 551 \text{ J/g-mol (from above)}$$

$$n_e = 0.5$$

$$-Q = (0.5)(136.2) + (0.25)(74.5) + (0.25)(1165.5) - (0.5)(551)$$

$$= 103.1 \text{ J/g-mol} = \frac{(103.1)(1.8)}{(4.19)(77)} = 0.57 \text{ Btu/lb of product}$$

The general relations for an intensive property are given in Table 8.2;

columns (b) and (c) refer to a mixture and column (a) to a pure material. Note that when Y_1 and Y_2 are intensive variables, the sets in columns (a) and (c) involve $(n+1)$ intensive and one extensive variable. Since any intensive variable can be expressed as a function of any other $(n+1)$ *intensive properties*, it follows that $(\partial B/\partial N)$ in Eqs. (8-16a) and (8-16c) must vanish.

Eqs. (8-18b) and (8-18c) are the differential forms of the two most commonly used sets for expressing an intensive property of a mixture. They contain the terms $(\partial B/\partial N_i)_{T,P,N_j[i]}$ and $(\partial B/\partial x_i)_{T,P,x[i,n]}$, each of which can be expressed in terms of partial molal quantities. Since $(\partial B/\partial N_i)_{T,P,N_j[i]}$ already contains the partial molal set of properties $T, P, N_1, \ldots, N_n$, the transformation follows directly from the substitution of $B = \underline{B}/N$:

$$\left(\frac{\partial B}{\partial N_i}\right)_{T,P,N_j[i]} = \left(\frac{\partial \frac{\underline{B}}{N}}{\partial N_i}\right)_{T,P,N_j[i]} = \frac{\bar{B}_i}{N} - \frac{\underline{B}}{N^2}\left(\frac{\partial N}{\partial N_i}\right)_{T,P,N_j[i]} = \frac{1}{N}(\bar{B}_i - B) \quad (8\text{-}19)$$

In a similar manner, $(\partial B/\partial x_i)_{T,P,x[i,n]}$ can be related to $(\partial \underline{B}/\partial x_i)_{T,P,x[i,n]}$:

$$\left(\frac{\partial B}{\partial x_i}\right)_{T,P,x[i,n]} = \left(\frac{\partial \frac{\underline{B}}{N}}{\partial x_i}\right)_{T,P,x[i,n]} = \frac{1}{N}\left(\frac{\partial \underline{B}}{\partial x_i}\right)_{T,P,x[i,n]} = \bar{B}_i - \bar{B}_n \quad (8\text{-}20)^4$$

With Eqs. (8-19) and (8-20), Eqs. (8-18b) and (8-18c) of Table 8.2 become:

$$dB = \left(\frac{\partial B}{\partial T}\right)_{P,N_i} dT + \left(\frac{\partial B}{\partial P}\right)_{T,N_i} dP + \frac{1}{N}\sum_{i=1}^{n} \bar{B}_i\, dN_i - \frac{B}{N}\, dN \quad (8\text{-}21)$$

$$dB = \left(\frac{\partial B}{\partial T}\right)_{P,x} dT + \left(\frac{\partial B}{\partial P}\right)_{T,x} dP + \sum_{i=1}^{n-1} (\bar{B}_i - \bar{B}_n)\, dx_i \quad (8\text{-}22)$$

Expanding the sum in Eq. (8-22),

$$dB = \left(\frac{\partial B}{\partial T}\right)_{P,x} dT + \left(\frac{\partial B}{\partial P}\right)_{T,x} dP + \sum_{i=1}^{n} \bar{B}_i\, dx_i \quad (8\text{-}23)$$

Example 8.2

Relate the partial derivative $(\partial B/\partial x_i)_{T,P,x[i,k]}$ to $(\partial B/\partial x_i)_{T,P,x[i,n]}$.

Solution

From Eq. (8-20),

$$\left(\frac{\partial B}{\partial x_i}\right)_{T,P,x[i,n]} = \bar{B}_i - \bar{B}_n$$

$$\left(\frac{\partial B}{\partial x_i}\right)_{T,P,x[i,k]} = \bar{B}_i - \bar{B}_k$$

[4] Note that N is held constant during differentiation because $(\partial/\partial x_i)_{T,P,x[i,n]}$ implies that N is constant. See note at the bottom of Table 8.2.

Table 8.2
The General Intensive Property, B

PURE MATERIAL (a)	MIXTURE (b)	MIXTURE (c)
Independent variables Y_1, Y_2, N	Independent variables $Y_1, Y_2, N_1, \ldots, N_n$	Independent variables $Y_1, Y_2, x_1, \ldots, x_{n-1}, N$
$B = f(Y_1, Y_2, N)$ (8-13a)	$B = f(Y_1, Y_2, N_1, \ldots, N_n)$ (8-13b)	$B = f(Y_1, Y_2, x_1, \ldots, x_{n-1}, N)$ (8-13c)
$dB = \left(\frac{\partial B}{\partial Y_1}\right)_{Y_2,N} dY_1 + \left(\frac{\partial B}{\partial Y_2}\right)_{Y_1,N} dY_2 + \left(\frac{\partial B}{\partial N}\right)_{Y_1,Y_2} dN$ (8-14a)	$dB = \left(\frac{\partial B}{\partial Y_1}\right)_{Y_2,N_i} dY_1 + \left(\frac{\partial B}{\partial Y_2}\right)_{Y_1,N_i} dY_2 + \sum_{i=1}^{n} \left(\frac{\partial B}{\partial N_i}\right)_{Y_1,Y_2,N_j[i]} dN_i$ (8-14b)	$dB = \left(\frac{\partial B}{\partial Y_1}\right)_{Y_2,x,N} dY_1 + \left(\frac{\partial B}{\partial Y_2}\right)_{Y_1,x,N} dY_2 + \sum_{i=1}^{n-1} \left(\frac{\partial B}{\partial x_i}\right)_{Y_1,Y_2,x[i,n],N} dx_i + \left(\frac{\partial B}{\partial N}\right)_{Y_1,Y_2,x} dN$ (8-14c)
INTEGRATION BY EULER'S THEOREM ASSUMING Y_1 AND Y_2 ARE EXTENSIVE PROPERTIES		
$0 = \left(\frac{\partial B}{\partial Y_1}\right)_{Y_2,N} Y_1 + \left(\frac{\partial B}{\partial Y_2}\right)_{Y_1,N} Y_2 + \left(\frac{\partial B}{\partial N}\right)_{Y_1,Y_2} N$ (8-15a)	$0 = \left(\frac{\partial B}{\partial Y_1}\right)_{Y_2,N} Y_1 + \left(\frac{\partial B}{\partial Y_2}\right)_{Y_1,N} Y_2 + \sum_{i=1}^{n} \left(\frac{\partial B}{\partial N_i}\right)_{Y_1,Y_2,N_j[i]} N_i$ (8-15b)	$0 = \left(\frac{\partial B}{\partial Y_1}\right)_{Y_2,x,N} Y_1 + \left(\frac{\partial B}{\partial Y_2}\right)_{Y_1,x,N} Y_2 + \left(\frac{\partial B}{\partial N}\right)_{Y_1,Y_2,N} N$ (8-15c)
SIMPLIFICATIONS IF Y_1 AND Y_2 ARE INTENSIVE PROPERTIES		
$0 = \left(\frac{\partial B}{\partial N}\right)_{Y_1,Y_2}$ (8-16a)	$0 = \sum_{i=1}^{n} \left(\frac{\partial B}{\partial N_i}\right)_{Y_1,Y_2,N_j[i]} N_i$ $\left(\text{but} \left(\frac{\partial B}{\partial N_i}\right)_{Y_1,Y_2,N_j[i]} \neq 0\right)$ (8-16b)	$0 = \left(\frac{\partial B}{\partial N}\right)_{Y_1,Y_2,x}$ (8-16c)

WITH $Y_1 = T$ AND $Y_2 = P$

$$0 = \left(\frac{\partial B}{\partial N}\right)_{T,P} \qquad \text{(8-17a)}$$

Eq. (8-14a) becomes:

$$dB = \left(\frac{\partial B}{\partial T}\right)_P dT + \left(\frac{\partial B}{\partial P}\right)_T dP \qquad \text{(8-18a)}$$

$$0 = \sum_{i=1}^{n} \left(\frac{\partial B}{\partial N_i}\right)_{T,P,N_j[i]} N_i \qquad \text{(8-17b)}$$

Eq. (8-14b) becomes:

$$dB = \left(\frac{\partial B}{\partial T}\right)_{P,N_i} dT + \left(\frac{\partial B}{\partial P}\right)_{T,N_i} dP + \sum_{i=1}^{n} \left(\frac{\partial B}{\partial N_i}\right)_{T,P,N_j[i]} dN_i \qquad \text{(8-18b)}$$

$$0 = \left(\frac{\partial B}{\partial N}\right)_{T,P,x} \qquad \text{(8-17c)}$$

Eq. (8-14c) becomes:

$$dB = \left(\frac{\partial B}{\partial T}\right)_{P,x} dT + \left(\frac{\partial B}{\partial P}\right)_{T,x} dP + \sum_{i=1}^{n-1} \left(\frac{\partial B}{\partial x_i}\right)_{T,P,x[i,n]} dx_i \qquad \text{(8-18c)}$$

Note that we could have expressed B as a function of $(n + 1)$ intensive variables plus one extensive variable, e.g., $(T, P, x_1, \ldots, x_{n-1}, N)$ or $(T, P, x_1, \ldots, x_{n-1}, \underline{Y})$ where $\underline{Y}$ is extensive. In differential form, then

$$dB = \left(\frac{\partial B}{\partial T}\right)_{P,x,\underline{Y}} dT + \left(\frac{\partial B}{\partial P}\right)_{T,x,\underline{Y}} dP + \sum_{i=1}^{n-1} \left(\frac{\partial B}{\partial x_i}\right)_{T,P,x[i,n],\underline{Y}} dx_i + \left(\frac{\partial B}{\partial \underline{Y}}\right)_{T,P,x} d\underline{Y}$$

Applying Euler's theorem, $(\partial B/\partial \underline{Y})_{T,P,x} = 0$, regardless of what $\underline{Y}$ is chosen. It then follows that all derivations such as $(\partial B/\partial T)_{P,x,\underline{Y}}$ can be written as $(\partial B/\partial T)_{P,x}$ where it is then implied that some extensive variable, $\underline{Y}$, is held constant.

then

$$\left(\frac{\partial B}{\partial x_i}\right)_{T,P,x[i,k]} - \left(\frac{\partial B}{\partial x_i}\right)_{T,P,x[i,n]} = \bar{B}_n - \bar{B}_k = \left(\frac{\partial B}{\partial x_n}\right)_{T,P,x[n,k]}$$

8.2 Partial Molal Properties

A partial molal property was defined in Eq. (8-8) and is related to a mixture property by Eq. (8-6b). It follows from the definition, that if $\underline{W}$ and $\underline{Z}$ are any two extensive properties, and Y is either temperature or pressure, and if $\underline{W} = Y\underline{Z}$, then $\bar{W}_i = Y\bar{Z}_i$. Thus, the conjugate sets $T\underline{S}$ and $P\underline{V}$, when differentiated as indicated in Eq. (8-8), become $T\bar{S}_i$ and $P\bar{V}_i$. It also follows that any derivative of the form $(\partial \underline{Z}/\partial Y_1)_{Y_2,N_i}$ where Y_1 and Y_2 are T and P, when operated on by $(\partial/\partial N_i)_{T,P,N_j[i]}$, becomes $(\partial \bar{Z}_i/\partial Y_1)_{Y_2,N_i}$ because the order of differentiation is immaterial when the independent variables are consistent. Thus, for example,

$$\frac{\partial}{\partial N_i}\left[\left(\frac{\partial \underline{H}}{\partial P}\right)_{T,N_i}\right]_{T,P,N_j[i]} = \frac{\partial}{\partial P}\left[\left(\frac{\partial \underline{H}}{\partial N_i}\right)_{T,P,N_j[i]}\right]_{T,N_i} = \left(\frac{\partial \bar{H}_i}{\partial P}\right)_{T,N_i} \quad (8\text{-}24)$$

Hence, from $\underline{H} = \underline{U} + P\underline{V}$, we could immediately write:

$$\bar{H}_i = \bar{U}_i + P\bar{V}_i \quad (8\text{-}25)$$

Also, from $C_p = T(\partial S/\partial T)_{P,x}$, we could define $\underline{C}_p = NC_p = T(\partial \underline{S}/\partial T)_{P,N_i}$, and it then follows that:

$$\bar{C}_{p_i} = \frac{\partial}{\partial N_i}[NC_p]_{T,P,N_j[i]} = T\left(\frac{\partial \bar{S}_i}{\partial T}\right)_{P,N_i} \quad (8\text{-}26)$$

Transformations of potential functions require more effort. Let us apply the partial molal operator to

$$d\underline{A} = -\underline{S}\,dT - P\,d\underline{V} + \sum_{m=1}^{n} \mu_m\,dN_m$$

Then,

$$d\bar{A}_i = -\bar{S}_i\,dT - P\,d\bar{V}_i + \sum_{m=1}^{n}\left(\frac{\partial \mu_m}{\partial N_i}\right)_{T,P,N_j[i]} dN_m + \sum_{m=1}^{n} \mu_m d\left[\left(\frac{\partial N_m}{\partial N_i}\right)_{T,P,N_j[i]}\right] \quad (8\text{-}27)$$

Since

$$\left(\frac{\partial N_m}{\partial N_i}\right)_{T,P,N_j[i]} = \begin{Bmatrix} 0 & m \neq i \\ 1 & m = i \end{Bmatrix} \quad (8\text{-}28)$$

the last term reduces to $d(1) = 0$. The third term on the right-hand side can be left unchanged or transformed to a mole fraction derivative.

Partial molal quantities can be evaluated by several methods. If the property $\underline{B}$ can be conveniently measured, directly or indirectly, $\bar{B}_i$ can be found by measuring the change in $\underline{B}$ upon addition of a small amount of component i to a mixture while holding the temperature and pressure constant. From the definition, Eq. (8-8),

$$\bar{B}_i \equiv \left(\frac{\partial \underline{B}}{\partial N_i}\right)_{T,P,N_j[i]} = \lim_{\Delta N_i \to 0} \left(\frac{\Delta \underline{B}}{\Delta N_i}\right)_{T,P,N_j[i]} \tag{8-29}$$

It is essential to keep in mind the fact that $\bar{B}_i$ is a property of the mixture and not simply a property of component i. Thus, $\bar{B}_i$ will generally vary with mixture composition. To emphasize this point, we note that $\bar{B}_i$ is an intensive mixture property and, as such, we could apply any of the equations in Table 8.2 to this property. For example, expressing $\bar{B}_i = f(T, P, x_1, \ldots, x_{n-1})$, Eq. (8-18c) becomes:

$$d\bar{B}_i = \left(\frac{\partial \bar{B}_i}{\partial T}\right)_{P,x} dT + \left(\frac{\partial \bar{B}_i}{\partial P}\right)_{T,x} dP + \sum_{j=1}^{n-1} \left(\frac{\partial \bar{B}_i}{\partial x_j}\right)_{T,P,x[j,n]} dx_j \tag{8-30}$$

If experimental data or analytical expressions for $\underline{B}$ or B are available, $\bar{B}_i$ can be evaluated directly. Consider three possible cases:

(1) $\underline{B} = f(T, P, N_1, \ldots, N_n)$

(2) $B = f(T, P, N_1, \ldots, N_n)$

(3) $B = f(T, P, x_1, \ldots, x_{i-1}, x_{i+1}, \ldots, x_n)$

For case (1) $\bar{B}_i$ can be evaluated by differentiation using Eq. (8-8). For case (2) we can obtain $(\partial B/\partial N_i)_{T,P,N_j[i]}$ directly from the data and use Eq. (8-19) to solve for $\bar{B}_i$:

$$\bar{B}_i = B + N\left(\frac{\partial B}{\partial N_i}\right)_{T,P,N_j[i]} \tag{8-31}$$

For case (3) we can obtain $(\partial B/\partial x_j)_{T,P,x[j,i]}$ directly from the data and then relate this partial to one of the forms above. For example, let us express $(\partial B/\partial N_i)_{T,P,N_j[i]}$ in Eq. (8-31) as a function of $(\partial B/\partial x_j)_{T,P,x[j,i]}$. We first express B as a differential in terms of $T, P, x_1, \ldots, x_{i-1}, x_{i+1}, \ldots, x_n$:

$$dB = \left(\frac{\partial B}{\partial T}\right)_{P,x} dT + \left(\frac{\partial B}{\partial P}\right)_{T,x} dP + \sum_{j \neq i} \left(\frac{\partial B}{\partial x_j}\right)_{T,P,x[j,i]} dx_j \tag{8-32}$$

Dividing by dN_i and imposing the restraint of constant T, P, and $N_j[i]$,

$$\left(\frac{\partial B}{\partial N_i}\right)_{T,P,N_j[i]} = -\frac{1}{N} \sum_{j \neq i} x_j \left(\frac{\partial B}{\partial x_j}\right)_{T,P,x[j,i]} \tag{8-33}$$

Substituting into Eq. (8-31),

$$\bar{B}_i = B - \sum_{j \neq i} x_j \left(\frac{\partial B}{\partial x_j}\right)_{T,P,x[j,i]} \tag{8-34}$$

In Eq. (8-34) we have obtained $\bar{B}_i$ from a data set in which x_i was eliminated. This equation cannot be used to obtain $\bar{B}_k$ unless we transform the given data to a set in which x_k is eliminated. Alternatively, we can use Eq. (8-20) (with k substituted for n) to solve for $\bar{B}_k$ from the data set of case (3). Thus,

$$\bar{B}_k = \bar{B}_i + \left(\frac{\partial B}{\partial x_k}\right)_{T,P,x[k,i]} \tag{8-35}$$

or, using Eq. (8-34),

$$\bar{B}_k = B + \left(\frac{\partial B}{\partial x_k}\right)_{T,P,x[k,i]} - \sum_{j \neq i} x_j \left(\frac{\partial B}{\partial x_j}\right)_{T,P,x[j,i]} \tag{8-36}$$

For a binary system of A and C, Eq. (8-36) reduces to:

$$\left.\begin{aligned} \bar{B}_A &= B - y_C\left(\frac{\partial B}{\partial y_C}\right)_{T,P} \\ \bar{B}_C &= B - y_A\left(\frac{\partial B}{\partial y_A}\right)_{T,P} \end{aligned}\right\} \tag{8-37}$$

Equation set (8-37) can be easily visualized as shown in Figure 8.1 in which B is plotted as a function of y_C at constant T and P. At any y_C, a tangent to the curve, when extrapolated, intersects the $y_C = 0$ axis at $\bar{B}_A$ and the $y_C = 1$ axis at $\bar{B}_C$.[5] The application of Eq. (8-36) to a ternary system is shown in Example 8.3. Even in this relatively simple system, the data required to calculate partial molal properties are considerable—and only rarely available.

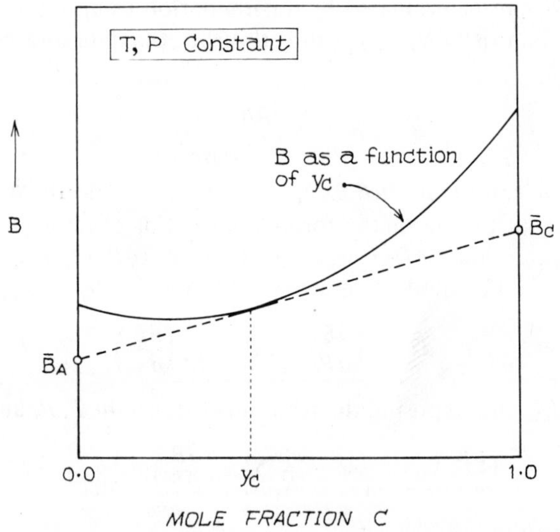

Figure 8.1 Tangent intercept rule for the binary A-C.

[5] There are occasions when, in determining partial molal quantities, they become either very large or very small at low concentrations (e.g., partial molal entropy or free energy), and this intercept technique is not particularly accurate. An alternate technique for handling such cases is found in H. C. van Ness and R. V. Mrazek, "Treatment of Thermodynamic Data for Homogeneous Binary Systems," *AIChE J.*, **5**, (1959), 209.

Example 8.3

Apply Eq. (8-36) to a ternary system of A, D, and C to obtain $\bar{B}_A$.

Solution

For a ternary system of A, D, and C, there are several ways to express $\bar{B}_A$. Suppose that C is the component "eliminated"; then from Eq. (8-36):

$$\bar{B}_A = B + \left(\frac{\partial B}{\partial y_A}\right)_{T,P,y_D} - y_A\left(\frac{\partial B}{\partial y_A}\right)_{T,P,y_D} - y_D\left(\frac{\partial B}{\partial y_D}\right)_{T,P,y_A}$$

$$= B + (1 - y_A)\left(\frac{\partial B}{\partial y_A}\right)_{T,P,y_D} - y_D\left(\frac{\partial B}{\partial y_D}\right)_{T,P,y_A}$$

On the other hand, if D were eliminated,

$$\bar{B}_A = B + (1 - y_A)\left(\frac{\partial B}{\partial y_A}\right)_{T,P,y_C} - y_C\left(\frac{\partial B}{\partial y_C}\right)_{T,P,y_A}$$

The two expressions for $\bar{B}_A$ can be shown to be equivalent by equating them, using Eq. (8-20) while noting that $(1 - y_A) = y_D + y_C$. The choice between the two ways of expressing $\bar{B}_A$ depends on the data available; in each case, derivatives are required with the composition of different components held constant.

Example 8.4

Show how Eq. (8-31) may be used to estimate the partial molal volume of a component i in a gas mixture assuming that the volumetric properties of the mixture may be determined from $V_m = Z_m RT/P$ and the compressibility factor Z_m can be approximated in the same manner as a pure component [i.e., as a function of T_{r_m}, P_{r_m}, and ω_m (see Eq. (6-91)]. To define T_{c_m}, P_{c_m}, ω_m, assume that a mole fraction rule is applicable [i.e., Eqs. (E-4), (E-5) and (E-6) of Appendix E].

Solution

Writing Eq. (8-31) for volume,

$$\bar{V}_i = V + N\left(\frac{\partial V}{\partial N_i}\right)_{T,P,N_j[i]} = \frac{RT}{P}\left[Z_m + N\left(\frac{\partial Z_m}{\partial N_i}\right)_{T,P,N_j[i]}\right]$$

But, since $Z_m = f[(T/T_{c_m}), (P/P_{c_m}), \omega_m]$, the derivative of Z_m with respect to N_i may be expanded to give:

$$\bar{V}_i = \frac{RT}{P}\left\{Z + \left[\left(\frac{\partial Z}{\partial T_r}\right)_{P_r,\omega} g(T_r) + \left(\frac{\partial Z}{\partial P_r}\right)_{T_r,\omega} g(P_r) + \left(\frac{\partial Z}{\partial \omega}\right)_{T_r,P_r} g(\omega)\right]\right\}$$

where the subscript m is understood to apply to Z, T_r, P_r, and ω, and the

$g(\ )$ functions are given as:

$$g(T_r) = N\left(\frac{\partial T_r}{\partial N_i}\right)_{T,P,N_j[i]}, \text{ etc.}$$

Expanding the $g(\ )$ functions with Eqs. (E-3), (E-4), and (E-5) of Appendix E, and introducing the derivative compressibility functions of Section 6.9, we see that

$$\bar{V}_i = \frac{RT}{P}\left\{Z + \left[\left(\frac{Z_T - Z}{T_r}\right)g(T_r) + \left(\frac{Z - Z_P}{P_r}\right)g(P_r) + Z^{(1)}g(\omega)\right]\right\}$$

$$g(T_r) = \left(-\frac{T}{T_{c_m}^2}\right)(T_{c_i} - T_{c_m})$$

$$g(P_r) = \left(-\frac{P}{P_{c_m}^2}\right)(P_{c_i} - P_{c_m})$$

$$g(\omega) = \omega_i - \omega_m$$

Thus, to estimate $\bar{V}_i$ from the pure component critical constants and acentric factors, T_{c_m}, P_{c_m}, and ω_m are calculated from Eqs. (E-3), (E-4), and (E-5). T_r and P_r for the mixture are then determined. Z and $Z^{(1)}$ are found from Eq. (6-91) and Figures 6.1 and 6.2; Z_p and Z_T can be calculated from Eqs. (D.15) and (D.16) with Figures D.7 through D.10 of Appendix D.

This is but one example of a more general technique to estimate partial molal properties from generalized correlations.[6]

8.3 Duhem Relations for Partial Molal Quantities

For an n-component system, there are n partial molal quantities representing any extensive property. The set of $n + 2$ variables, $T, P, \bar{B}_1, \ldots, \bar{B}_n$ are all intensive and, therefore, only $n + 1$ variables from this set are independent: one can be expressed as a function of the other $n + 1$ variables.

The dependency can be expressed in a form analogous to the Gibbs-Duhem equation for chemical potential [see Eq. (5-69)]. Any property, $\underline{B}$, for which $\bar{B}_i$ is employed as in Eq. (8-6b), can be expanded as a function of $T, P, N_1, \ldots, N_n$:

$$d\underline{B} = \left(\frac{\partial \underline{B}}{\partial T}\right)_{P,N_i} dT + \left(\frac{\partial \underline{B}}{\partial P}\right)_{T,N_i} dP + \sum_{i=1}^{n} \bar{B}_i\, dN_i \tag{8-38}$$

But $d\underline{B}$ can also be expressed by differentiating Eq. (8-6b):

$$d\underline{B} = \sum_{i=1}^{n} N_i\, d\bar{B}_i + \sum_{i=1}^{n} \bar{B}_i\, dN_i \tag{8-39}$$

[6] A more complete treatment is available in M. S. Morgan and R. C. Reid "Generalized Partial Quantities from Pitzer's Expansion and Pseudocritical Rules," *AIChE J*, **16**, (1970), 889.

Subtracting Eq. (8-39) from (8-38) yields:

$$\sum_{i=1}^{n} N_i \, d\bar{B}_i = \left(\frac{\partial B}{\partial T}\right)_{P,N_i} dT + \left(\frac{\partial B}{\partial P}\right)_{T,N_i} dP \tag{8-40}$$

or, dividing by total moles, N,

$$\sum_{i=1}^{n} x_i \, d\bar{B}_i = \left(\frac{\partial B}{\partial T}\right)_{P,x} dT + \left(\frac{\partial B}{\partial P}\right)_{T,x} dP \tag{8-41}$$

Eqs. (8-40) and (8-41) are the most common forms of the *Duhem relation* for partial molal quantities. When integrated, they permit evaluation of any $\bar{B}_i$ in terms of the other $(n - 1)$ values of $\bar{B}_j$.

An alternate form of the Duhem relation can be obtained as follows. Noting that

$$\left(\frac{\partial B}{\partial T}\right)_{P,x} = \left[\frac{\partial\left(\sum_{i=1}^{n} \bar{B}_i x_i\right)}{\partial T}\right]_{P,x} = \sum_{i=1}^{n} x_i \left(\frac{\partial \bar{B}_i}{\partial T}\right)_{P,x} \tag{8-42}$$

and

$$\left(\frac{\partial B}{\partial P}\right)_{T,x} = \sum_{i=1}^{n} x_i \left(\frac{\partial \bar{B}_i}{\partial P}\right)_{T,x} \tag{8-43}$$

Eq. (8-41) can be rewritten as:

$$\sum_{i=1}^{n} x_i \left[d\bar{B}_i - \left(\frac{\partial \bar{B}_i}{\partial T}\right)_{P,x} dT - \left(\frac{\partial \bar{B}_i}{\partial P}\right)_{T,x} dP\right] = 0 \tag{8-44}$$

Substituting Eq. (8-30) for $d\bar{B}_i$ in Eq. (8-44) yields:

$$\sum_{i=1}^{n} x_i \left[\sum_{j \neq k} \left(\frac{\partial \bar{B}_i}{\partial x_j}\right)_{T,P,x[j,k]} dx_j\right] = 0 \tag{8-45}$$

or, changing the order of the summations,

$$\sum_{j \neq k} \left[\sum_{i=1}^{n} x_i \left(\frac{\partial \bar{B}_i}{\partial x_j}\right)_{T,P,x[j,k]}\right] dx_j = 0 \tag{8-46}$$

Since the brackets are coefficients of $(n - 1)$ terms in dx_j, all of which are independent (because x_k has been eliminated), it follows that each term in brackets must vanish. Thus, an equivalent Duhem relation is

$$\sum_{i=1}^{n} x_i \left(\frac{\partial \bar{B}_i}{\partial x_j}\right)_{T,P,x[j,k]} = 0 \tag{8-47}$$

Eq. (8-47) could also have been obtained by dividing Eq. (8-41) by dx_j at constant $T, P, x[j, k]$.

We shall have occasion to use this form in treating fugacities of phases in equilibrium, as described in Chapter 9.

Example 8.5

For a binary of components 1 and 2, if the partial molal enthalpy $\bar{H}_1$ were available as a function of mole fraction x_1, show how $\bar{H}_2$ and the mixture enthalpy H can be determined. The data $\bar{H}_1 = f(x_1)$ are at constant T and P.

Solution

From Eq. (8-47),

$$x_1 \left(\frac{\partial \bar{H}_1}{\partial x_1}\right)_{T,P} + x_2 \left(\frac{\partial \bar{H}_2}{\partial x_1}\right)_{T,P} = 0$$

Separating and integrating between x_1^0 and x_1,

$$(\bar{H}_2)_{x_1} - (\bar{H}_2)_{x_1^0} = -\int_{x_1^0}^{x_1} \left(\frac{x_1}{x_2}\right)\left(\frac{\partial \bar{H}_1}{\partial x_1}\right) dx_1$$

Let x_1^0 be 0 (i.e., pure component 2). Then $(\bar{H}_2)_{x_1^0} = H_2$. If this is substituted in the above, the right-hand side may be found from the $\bar{H}_1 = f(x_1)$ data and $\bar{H}_2$ is then related to the pure component enthaply H_2. To obtain H, Eq. (8-6b) divided by N yields:

$$H = x_1\bar{H}_1 + x_2\bar{H}_2$$

We have already seen that all n $(\bar{B}_i)$s can be found from a data set of the form $B = f(T, P, x_1, \ldots, x_{n-1})$. We have also shown that any $\bar{B}_i$ can be found—to within an arbitrary constant—from the other $(n - 1)(\bar{B}_j)$s. It is then possible to reconstruct B from these $(\bar{B}_i)s$ by dividing Eq. (8-6b) by N, i.e.,

$$B = \sum_{i=1}^{n} \bar{B}_i x_i \tag{8-48}$$

Thus, $(n - 1)(\bar{B}_i)$s have an information content equivalent to B.

Frequently, data are reported in the literature for all $(n)(\bar{B}_i)$s (most commonly for binary systems in which $\bar{B}_1$ and $\bar{B}_2$ are measured independently). The redundant information can be used to check the consistency of the data. At any given T, P, and concentration, we can calculate the partial derivative $(\partial \bar{B}_i/\partial x_j)_{T,P,x[j,k]}$ for each component, and then use Eq. (8-47) to verify that the sum does indeed vanish.

In some cases, reported data will not give a satisfactory consistency check, but these may be the only available data. In that case, it is always possible to smooth the data in order to obtain a set of partial molal quantities that are consistent. The procedure is simply to reconstruct B from Eq. (8-48) and then apply an equation of the form of Eq. (8-34) or (8-36) to obtain the partial molal quantities. As shown in the following example, this set will always be consistent.

Example 8.6

Using Eqs. (8-34) and (8-36), prove that a set of partial molal properties obtained only from data of the form, $B = f(T, P, x_1, \ldots, x_{i-1}, x_{i+1}, \ldots, x_n)$ will always form a consistent set.

Solution

Multiply Eq. (8-36) by x_k and sum over all k except $k = i$.

$$\sum_{k \neq i} x_k \bar{B}_k = \sum_{k \neq i} x_k \left[B + \left(\frac{\partial B}{\partial x_k}\right)_{T,P,x[i,k]} - \sum_{j \neq i} x_j \left(\frac{\partial B}{\partial x_j}\right)_{T,P,x[i,j]} \right]$$

Multiply Eq. (8-34) by x_i and add to the above.

$$\sum_{k=1}^{n} x_k \bar{B}_k = B\left(x_i + \sum_{k \neq i} x_k\right) + \sum_{k \neq i} x_k \left(\frac{\partial B}{\partial x_k}\right)_{T,P,x[i,k]}$$
$$- \left(\sum_{k \neq i} x_k + x_i\right)\left[\sum_{j \neq i} x_j \left(\frac{\partial B}{\partial x_j}\right)_{T,P,x[i,j]}\right]$$

The second and third terms on the right-hand side cancel, and the result reduces to:

$$\sum_{k=1}^{n} x_k \bar{B}_k = B$$

Thus, the set of $\bar{B}_k$ obtained from $B = f(T, P, x_1, \ldots, x_{i-1}, x_{i+1}, \ldots, x_n)$ will always form a consistent set.

8.4 The Gibbs Free Energy of a Mixture

In Section 6.1 we reconstructed the Gibbs free energy representation of the Fundamental Equation of a pure material. The exercise was useful, not because we were interested in the Fundamental Equation, *per se*, but because it helped us to clarify its information content. Similarly, let us explore the reconstruction of the Fundamental Equation of a mixture in order to clarify the minimum information necessary to define all thermodynamic properties of a single-phase, multicomponent system. We shall restrict this discussion to the Gibbs free energy representation in order to take advantage of the convenience of the $T, P, N_1, \ldots, N_n$ set.

In Section 5.3 it was shown that the Fundamental Equation could be reconstructed to within an arbitrary constant if we knew $n + 1$ equations of state which are the first-order partial derivatives of the Fundamental Equation. For the Gibbs free energy representation of a mixture, the differential form of the Fundamental Equation is

$$d\underline{G} = -\underline{S}\, dT + \underline{V}\, dP + \sum_{i=1}^{n} \mu_i\, dN_i \tag{8-49}$$

and the equations of state are

$$-\underline{S} = \left(\frac{\partial \underline{G}}{\partial T}\right)_{P,N_i} = g_1(T, P, N_1, \ldots, N_n) \tag{8-50}$$

$$\underline{V} = \left(\frac{\partial \underline{G}}{\partial P}\right)_{T,N_i} = g_2(T, P, N_1, \ldots, N_n) \tag{8-51}$$

$$\mu_i = \left(\frac{\partial G}{\partial N_i}\right)_{T, P, N_j[i]} = g_i(T, P, N_1, \ldots, N_n) \qquad \text{(8-52)}^6$$

Of these $n + 2$ partial derivatives, we could solve for any one—to within an arbitrary constant—as a function of the other $n + 1$ by integrating the Gibbs-Duhem equation:

$$N_i \, d\mu_i = -\underline{S} \, dT + \underline{V} \, dP - \sum_{j \neq i} N_j \, d\mu_j \qquad \text{(8-53)}$$

Thus, the Fundamental Equation can be reconstructed to within an arbitrary constant from $n + 1$ first-order partial derivatives.

Alternatively, the Fundamental Equation can be reconstructed from second-order partial derivatives. As shown in Section 6.1, the Fundamental Equation for a pure material could be obtained from the three second-order partial derivatives. For a mixture, $(n + 2)(n + 1)/2$ independent second-order partial derivatives are required (see Section 5.6). For example, for a binary system of A and B, we would need to know G_{TT}, G_{TP}, G_{PP}, G_{TA}, G_{PA}, G_{AA}.

Only in rare cases do we have this information available. We shall, however, show shortly that there are mixtures (which we shall call *ideal mixtures*) for which the independent second-order derivatives can be expressed as functions of only the mole fractions and second-order derivatives of the *pure components*. The idealized condition is always approached for gases at low pressure, that is, for an *ideal gas mixture*. If we know the values of the second-order derivatives at low pressure, then we could determine these derivatives at any other pressure of interest provided that we know how each varies with pressure. That is, we must know the $(n + 2)(n + 1)/2$ third-order partial derivatives formed by differentiating the second-order derivatives with respect to pressure. As shown in Example 8.7 below, the necessary third-order derivatives can be obtained from the equation of state. Thus, the net result is that the Fundamental Equation of a mixture can be expressed in terms of the Fundamental Equation of the pure components if we use the ideal gas mixture concept and the equation of state of the mixture, $V = f(T, P, y_1, \ldots, y_{n-1})$.

Example 8.7

Prove that for a binary mixture the six independent second-order partial derivatives can be determined, given the equation of state $V = f(T, P, y_A)$ and the following derivatives at low pressure: G_{TT}^*, G_{TP}^*, G_{PP}^*, G_{TA}^*, G_{PA}^*, G_{AA}^*.

Solution

The equation of state can be viewed as

$$G_P = \underline{V} = f(T, P, N_A)$$

[6] Note that although

$$\mu_i = \left(\frac{\partial \underline{U}}{\partial N_i}\right)_{\underline{S}, \underline{V}, N_j[i]} = \left(\frac{\partial \underline{H}}{\partial N_i}\right)_{\underline{S}, P, N_j[i]} = \left(\frac{\partial \underline{A}}{\partial N_i}\right)_{T, \underline{V}, N_j[i]} = \left(\frac{\partial \underline{G}}{\partial N_i}\right)_{T, P, N_j[i]}$$

μ_i is the partial molal *Gibbs free energy* and not the partial molal energy, etc.

Differentiating successively with respect to T, P, and N_A, one obtains G_{TP}, G_{PP}, and G_{PA}. Differentiating each of these three derivatives with respect to T, P, and N_A, one obtains six independent third-order derivatives: G_{TTP}, G_{TPP}, G_{PPP}, G_{TPA}, G_{PPA}, G_{PAA}. These can be written as:

$$\frac{\partial}{\partial P}[G_{TT}, G_{TP}, G_{PP}, G_{TA}, G_{PA}, G_{AA}]_{T,N}$$

Thus, we have obtained the pressure variation of each of the independent second-order derivatives. At any given T, P, N_A of interest, any second-order derivative can be found by integration from P^* to P:

$$G_{z_1z_2} = G^*_{z_1z_2} + \int_{P^*}^{P} G_{z_1z_2P}\, dP$$

Thus, given the derivative at low pressure, $G^*_{z_1z_2}$, the derivative at higher pressures can be obtained by using the equation of state.

Let us now show how the Fundamental Equation is reconstructed from the equation of state and pure component properties. We shall develop expressions for the chemical potential of a component in a mixture, $\bar{G}_i$, and then reconstruct $G = \sum \bar{G}_i y_i$. Since there are no methods for measuring $\bar{G}_i$ directly, we must relate the chemical potential of a component in a mixture to the chemical potential of the pure component by analyzing an isothermal, reversible mixing process. Consider, for example, the process of Figure 8.2 in which pure j is added continuously to a mixture containing j. Let us start with mixture and pure j at the same temperature and pressure.[7] The flow rate of pure j in relation to the mixture is $\delta \dot{N}_j/\dot{N}$, where $\delta \dot{N}_j \ll \dot{N}$. The entire process is isothermal. To ensure reversibility, the pure component is pressurized to P_1 so that the chemical potential of j is the same on both sides of the rigid membrane, M, which is permeable only to j. For the system enclosed by the dashed lines, the work per unit mole of pure j processed is

$$W = -\int_{P}^{P_1} V_j\, dP \tag{8-54}$$

But for a steady-state, isothermal, reversible process, the work is the negative of the change in Gibbs free energy:

$$-W = \frac{\underline{G}(T, P, N_1, \ldots, N_j + \delta N_j, \ldots, N_n) - \underline{G}(T, P, N_1, \ldots, N_j, \ldots, N_n)}{\delta N_j} - G_j(T, P) \tag{8-55}$$

[7] Note that the state of aggregation of pure j is the normal or most stable state at T and P. Thus, the initial states of pure j and the mixture may be different.

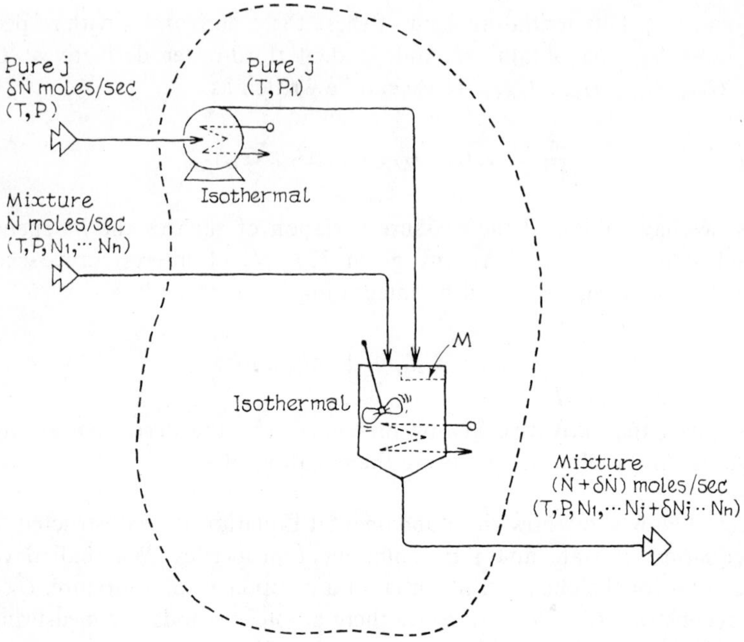

Figure 8.2 Simple mixing.

By definition,

$$\frac{\underline{G}(T, P, N_1, \ldots, N_j + \delta N_j, \ldots, N_n) - \underline{G}(T, P, N_1, \ldots, N_j, \ldots, N_n)}{\delta N_j}$$

$$= \left(\frac{\delta \underline{G}}{\delta N_j}\right)_{T, P, N_i[j]} = \bar{G}_j(T, P, N_1, \ldots, N_n) \tag{8-56}$$

as δN_j becomes infinitesimal in relation to N. Substituting into Eq. (8-55), equating with Eq. (8-54), and rearranging,

$$\bar{G}_j(T, P, N_1, \ldots, N_n) = G_j(T, P) + \int_P^{P_1} V_j \, dP \tag{8-57}$$

Since $\mu_j = \bar{G}_j$, we could evaluate the chemical potential of j from Eq. (8-57). Note that, with the exception of P_1, the right-hand side contains properties of only pure j and, hence, could be evaluated in relation to some base state if we knew the C_p and P-V-T behavior of pure j. Without conducting at least one experiment on the mixture, however, we cannot determine P_1. For each composition of interest, we must determine, from experimental information or otherwise, the pressure for which the mixture and pure j are at equilibrium; that is, the pressure P_1 for which

$$\mu_{\text{pure } j}(T, P_1) = \mu_j(T, P, N_1, \ldots, N_n) \tag{8-58}$$

It has been found experimentally that there is a region of conditions wherein P_1 can be related uniquely to the mixture pressure. At low pressure and moderately high temperature, where the volumetric behavior approaches that of an ideal gas, the partial pressures of j are equal, at equilibrium, across any membrane semipermeable to j. For a membrane separating pure j from a mixture, under such conditions,

$$p_j^* \equiv y_j P^* = P_1^* \tag{8-59}$$

where the superscript asterisk indicates that both mixture and pure component behave as ideal gases.[8]

Let us now use this fact to modify the process of Figure 8.2 so that we can mix ideal gases. The revised process is shown in Figure 8.3. First, the mixture is expanded isothermally and reversibly to P^* and pure j likewise to $y_j P^*$ where y_j is the mole fraction of j in the mixture. Eqs. (8-55) and (8-56) are unchanged while Eq. (8-54) becomes:

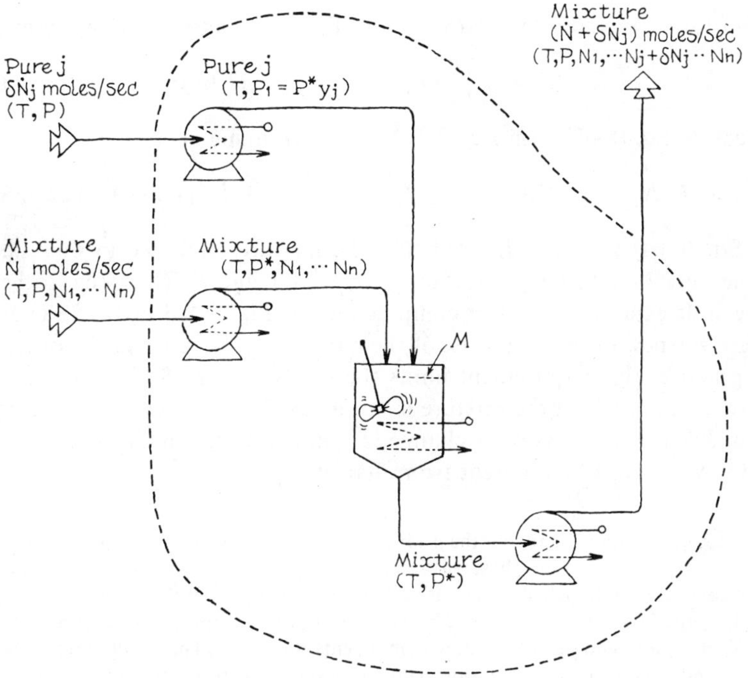

Figure 8.3 Mixing at low pressures.

[8] On one hand, the equality of partial pressures under ideal gas conditions might be viewed as a "fifth postulate" because the principle cannot be proved from any corollaries of the previous postulates. On the other hand, it can be shown that this principle follows from the statistical thermodynamic definition of entropy.

$$W = -\int_P^{y_jP^*} V_j\,dP - \left[\left[\int_P^{P^*} V(T, P, N_1, \ldots, N_j, \ldots, N_n)\,dP\right]\frac{\dot{N}}{\delta\dot{N}_j}\right.$$
$$\left. - \left[\int_{P^*}^{P} V(T, P, N_1, \ldots, N_j + \delta N_j, \ldots, N_n)\,dP\right]\frac{\dot{N} + \delta\dot{N}_j}{\delta\dot{N}_j}\right] \tag{8-60}$$

The second and third terms can be simplified in the limit of infinitesimal $\delta\dot{N}_j$ in relation to $\dot{N}$ because

$$\bar{V}_j(T, P, N_1, \ldots, N_n)$$
$$= \lim_{\delta N_j \to dN_j}\left[\frac{\underline{V}(T, P, N_1, \ldots, N_j + \delta N_j, \ldots, N_n) - \underline{V}(T, P, N_1, \ldots, N_j, \ldots, N_n)}{\delta N_j}\right] \tag{8-61}$$

The first term on the right-hand side of Eq. (8-60) can also be rearranged as follows, if we recognize that in the ideal gas region $V_j = RT/P$:

$$\int_P^{y_jP^*} V_j\,dP = \int_P^{P^*} V_j\,dP + \int_{P^*}^{y_jP^*} \frac{RT}{P}\,dP = \int_P^{P^*} V_j\,dP + RT\ln y_j \tag{8-62}$$

Substituting Eqs. (8-61) and (8-62) into Eq. (8-60) and collecting terms,

$$W = -\int_{P^*}^{P} (\bar{V}_j - V_j)\,dP - RT\ln y_j \tag{8-63}$$

Equating Eq. (8-63) with Eq. (8-55) and rearranging yields

$$\bar{G}_j^{\alpha}(T, P, N_1, \ldots, N_n) = G_j^{\beta}(T, P) + RT\ln y_j + \int_{P^*}^{P} (\bar{V}_j^{\alpha} - V_j^{\beta})\,dP \tag{8-64}$$

In Eq. (8-64) we have denoted that the mixture might be in one state, α, while pure j might be in another state, β, at T and P. Thus, the isothermal, reversible expansions and/or compressions in Figure 8.3 involve a change in phase whenever the initial state of the mixture and/or pure j are not gaseous. In practice, this requirement limits the utility of Eq. (8-64), particularly in those cases in which the mixture is not a gas. In such a case, the integration from P^* to P will involve a change in phase, during which the composition of the vapor will be different from that of liquid.[9]

[9] It might be noted here that the problem of having a mixture of one phase (α) and pure components of a different phase (β) is normally overcome by insisting that phase β be identical to α even though, in some cases, this would require that the pure component exist in a hypothetical state at the given T and P. For example, suppose that the T and P were at 110°C and 1 atm and one of the pure components was water. The stable state for water at this T and P is, of course, a vapor. Thus, if a vapor mixture containing water is to be synthesized, no problem results because the pure component and mixture are in the same state. If, however, the mixture were liquid, one could either use liquid or vapor water at 110°C and 1 atm as the pure material state. If the pure liquid state is used, then the properties of pure liquid at 110°C and 1 atm must be estimated because they cannot be readily measured for superheated liquids.

To return to our initial objective of synthesizing the Fundamental Equation of the mixture, we can write Eq. (8-64) for each component. Since,

$$G^{\alpha}(T, P, N_1, \ldots, N_n) = \sum_{j=1}^{n} y_j \bar{G}_j^{\alpha} \tag{8-65}$$

it follows that:

$$G^{\alpha} = \sum_{j=1}^{n} y_j G_j^{\beta_j} + RT \sum_{j=1}^{n} y_j \ln y_j + \int_{P^*}^{P} \left(V^{\alpha} - \sum_{j=1}^{n} y_j V_j^{\beta_j}\right) dP \tag{8-66}$$

where we have made use of the fact that

$$V^{\alpha}(T, P, N_1, \ldots, N_n) = \sum_{j=1}^{n} y_j \bar{V}_j^{\alpha}$$

In conclusion, then, we see from Eq. (8-66) that the information equivalent of the Fundamental Equation of a mixture is knowledge of the Gibbs free energy of each pure material (obtainable from C_p and P-V-T data) and P-V-T data for the mixture. The P-V-T data must cover the entire range of pressures from the ideal gas state to the pressure in question, including any phase changes that may occur. As shown in Appendix E, this information may be obtained by a variety of means when the mixture is a gas, but it is unlikely to be available when the mixture is a condensed phase.

The significance of Eqs. (8-64) and (8-66) extends far beyond their use as a method for reconstructing the Fundamental Equation, as we shall presently attempt to demonstrate. First, let us note that the partial molal and specific entropies and enthalpies of a mixture can also be obtained from these relations because

$$S = -\left(\frac{\partial G}{\partial T}\right)_{P, N_i} \quad \text{or} \quad \bar{S}_j = -\left(\frac{\partial \bar{G}_j}{\partial T}\right)_{P, N_i} \tag{8-67}$$

and

$$H = \left[\frac{\partial\left(\frac{G}{T}\right)}{\partial\left(\frac{1}{T}\right)}\right]_{P, N_i} \quad \text{or} \quad \bar{H}_j = \left[\frac{\partial\left(\frac{\bar{G}_j}{T}\right)}{\partial\left(\frac{1}{T}\right)}\right]_{P, N_i} \tag{8-68}$$

Thus,[10]

$$\bar{S}_j = S_j - R \ln y_j - \int_{P^*}^{P} \frac{\partial}{\partial T}(\bar{V}_j - V_j)_{P, N_i}\, dP \tag{8-69}$$

$$S = \sum_{j=1}^{n} y_j S_j - R \sum_{j=1}^{n} y_j \ln y_j - \int_{P^*}^{P} \frac{\partial}{\partial T}\left(V - \sum_{j=1}^{n} y_j V_j\right)_{P, N_i} dP \tag{8-70}$$

[10] As noted in footnote 9, properties such as S_j or H_j may be evaluated from pure component data in the stable state at T, P, but normally they are evaluated at T, P, pure component *in the same state of aggregation* as that of the mixture, even though this may introduce hypothetical states.

$$\bar{H}_j = H_j - T^2 \int_{P^*}^{P} \frac{\partial}{\partial T}\left(\frac{\bar{V}_j - V_j}{T}\right)_{P,N_i} dP \tag{8-71}$$

$$H = \sum_{j=1}^{n} y_j H_j - T^2 \int_{P^*}^{P} \frac{\partial}{\partial T}\left(\frac{V - \sum_{j=1}^{n} y_j V_j}{T}\right)_{P,N_i} dP \tag{8-72}$$

Now, if we inspect Eqs. (8-69) through (8-72), we note that *if each of the integrals should vanish, the properties of the mixture can be synthesized directly from the pure component properties.* This is the basis of an idealization for mixtures which is as significant as is the ideal gas law for the vapor state. That is, *we define an ideal mixture as one for which every component obeys the following equations:*

$$\bar{V}_j = V_j \text{ (from } P^* \text{ to } P) \tag{8-73}$$

and

$$\left(\frac{\partial \bar{V}_j}{\partial T}\right)_{P,N_i} = \left(\frac{\partial V_j}{\partial T}\right)_{P,N_i} \text{ (from } P^* \text{ to } P) \tag{8-74}$$

Under these conditions,

$$\bar{S}_j^{ID} = S_j - R \ln y_j \tag{8-75}$$

$$S^{ID} = \sum_{j=1}^{n} y_j S_j - R \sum_{j=1}^{n} y_j \ln y_j \tag{8-76}$$

$$\bar{H}_j^{ID} = H_j \tag{8-77}$$

$$H^{ID} = \sum_{j=1}^{n} y_j H_j \tag{8-78}$$

$$\bar{G}_j^{ID} = \mu_{j(\text{in mixture})}^{ID} = \mu_{j(\text{pure component})} + RT \ln y_j \tag{8-79}$$

$$G^{ID} = \sum_{j=1}^{n} y_j \mu_{j(\text{pure component})} + RT \sum_{j=1}^{n} y_j \ln y_j \tag{8-80}$$

The ideal mixture or ideal solution is used in much the same manner as the ideal gas law. As a first approximation, the mixture is considered ideal and then the deviations from ideality are expressed by a variety of methods.

Example 8.8

A new process calls for an aqueous feed of ammonium nitrate at high pressure at 25°C. The pressure currently being considered is 10 kilobars. (1 bar = 0.987 atm = 10^5 N/m^2.) To complete our design, the solubility of NH_4NO_3 in the feed stream is desired. A literature search of this system has provided the data shown below. What is your best estimate of the solubility at 10 kilobars?

Data. At 25°C and 1 atm, the solubility of NH_4NO_3 in water is 67.63 weight percent salt. At a pressure of about 1 bar (or 1 atm), the chemical potentials of both components have been determined at 25°C as a function of composition. These data are shown in Figure 8.4. Note that the chemical

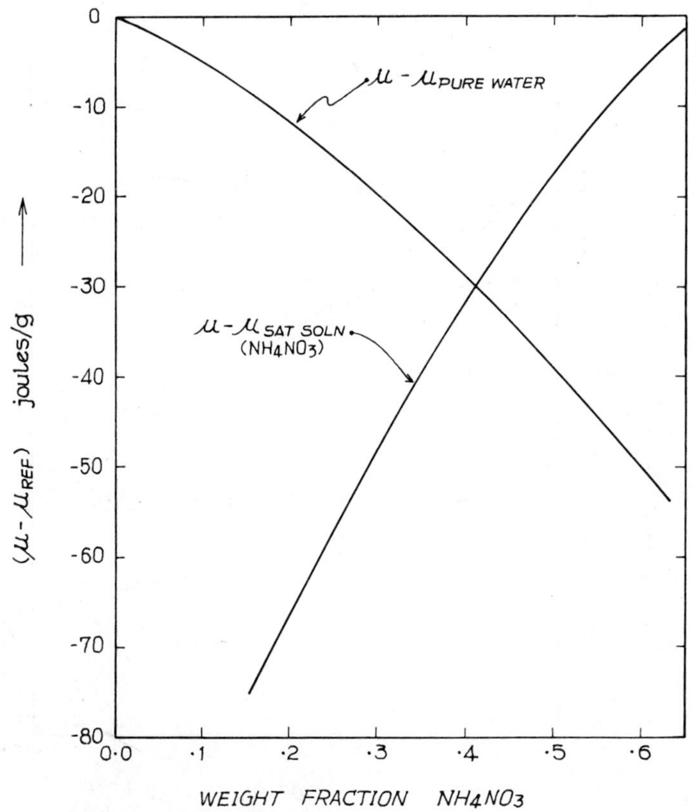

Figure 8.4 Chemical potentials for water and NH_4NO_3 in solution at 25°C, 1 bar.

potential of NH_4NO_3 is referenced to a saturated solution whereas, for water, the reference is pure water.

Volumetric data for the NH_4NO_3—H_2O system at 25°C are given in Figures 8.5 and 8.6.

Solution

We recall from Chapter 7 that under conditions of phase equilibrium, the temperature, pressure, and component chemical potentials are equal in all phases. Let x be the solubility (weight fraction) of NH_4NO_3 in the high-pressure solution. Now visualize an isothermal cyclic process beginning with a solution at x, at 1 bar. The change in chemical potential of NH_4NO_3 from this solution to the solid phase at 1 bar can be found from Figure 8.4 once

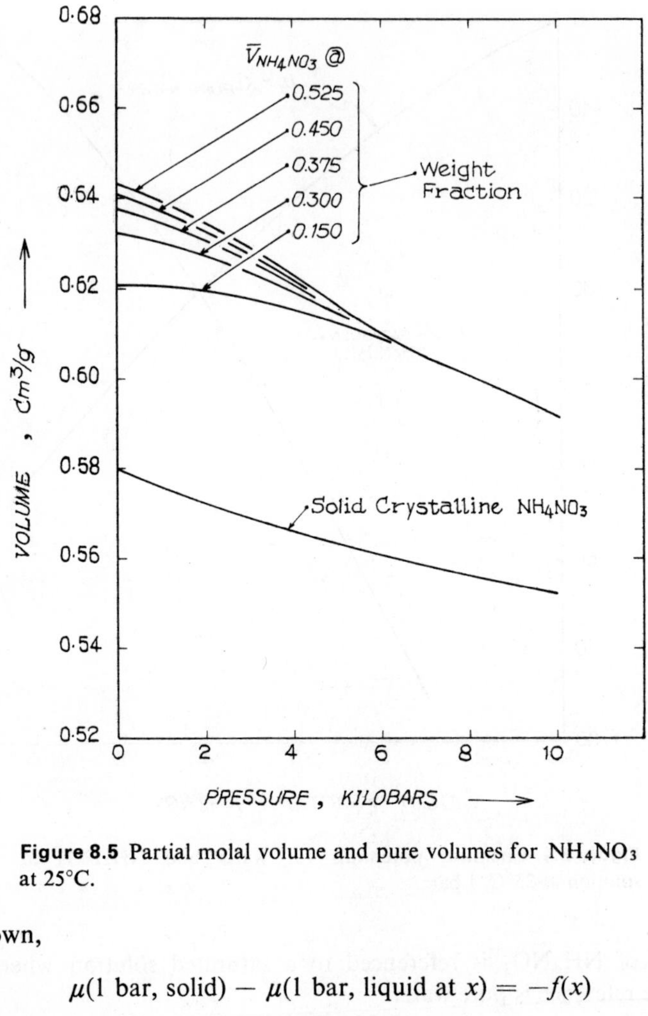

Figure 8.5 Partial molal volume and pure volumes for NH_4NO_3 at 25°C.

x is known,

$$\mu(1 \text{ bar, solid}) - \mu(1 \text{ bar, liquid at } x) = -f(x)$$

Next, compress the pure solid to 10 kb.

$$\mu(10 \text{ kb, solid}) - \mu(1 \text{ bar, solid}) = \int_{1\text{ b}}^{10\text{ kb}} V^S_{NH_4NO_3}\, dP$$

This solid is in equilibrium with liquid at x, i.e.,

$$\mu(10 \text{ kb, liquid at } x) - \mu(10 \text{ kb, solid}) = 0$$

Finally expand this liquid back to 1 bar,

$$\mu(1 \text{ bar, liquid at } x) - \mu(10 \text{ kb, liquid at } x) = \int_{10\text{ kb}}^{1\text{ b}} \bar{V}^L_{NH_4NO_3}\, dP$$

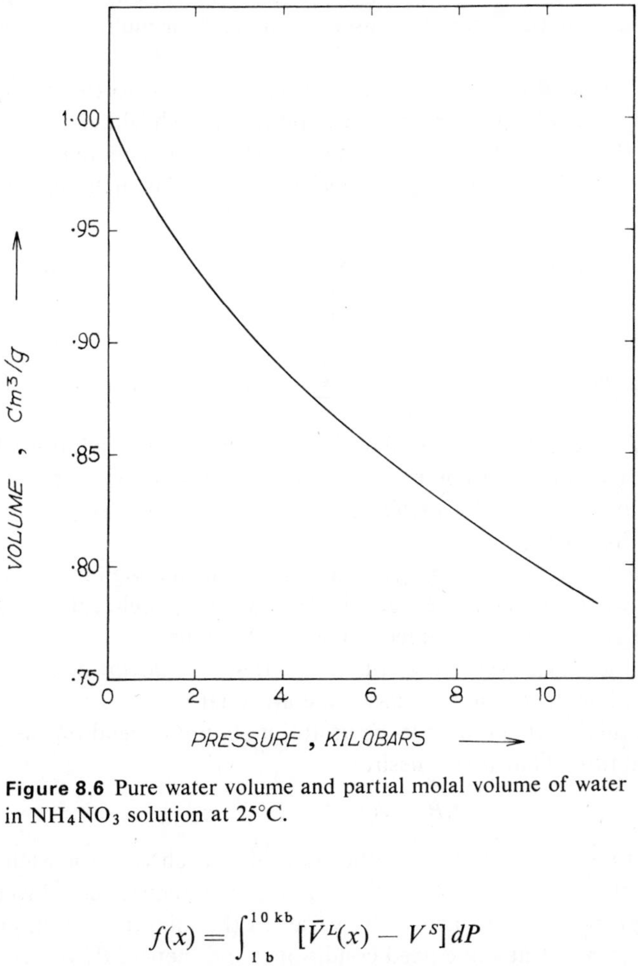

Figure 8.6 Pure water volume and partial molal volume of water in NH_4NO_3 solution at 25°C.

Adding,

$$f(x) = \int_{1\ \mathrm{b}}^{10\ \mathrm{kb}} [\bar{V}^L(x) - V^S]\, dP$$

By trial and error, with Figure 8.4 to obtain $f(x)$ and Figures 8.5 and 8.6 to obtain $\bar{V}^L(x)$ and V^S as a function of pressure, x is found to be 0.295. Thus, at 10 kb (about 9,900 atm) the solubility of NH_4NO_3 is only about 29.5 weight percent.

8.5 Mixing and Excess Functions

The experimental data required to calculate the Gibbs free energy of a nonideal mixture (i.e., volumetric data for the mixture and pure components from P^* to P) are not usually available. Consequently, the deviation concept, in which real properties are obtained from deviations from idealized or hypothetical mixtures, has been used extensively for multicomponent, single-phase systems.

In the general case, a mixture property is related to the properties of a reference state, which can be real or hypothetical. The difference between the value of the actual and reference state properties is denoted by the symbol Δ, which is called the *mixing or solution function.* The defining equation for Δ is:

$$\Delta \underline{B} \equiv \underline{B}(T, P, N_1, \ldots, N_n) - \sum_{j=1}^{n} N_j \bar{B}_j^0(T^0, P^0, x_1^0, \ldots, x_{n-1}^0) \qquad (8\text{-}81)$$

or

$$\Delta B \equiv B(T, P, x_1, \ldots, x_{n-1}) - \sum_{j=1}^{n} x_j \bar{B}_j^0(T^0, P^0, x_1^0, \ldots, x_{n-1}^0) \qquad (8\text{-}82)$$

where $\Delta \underline{B}$ and ΔB are the total and specific mixing (or solution) functions. For reasons that will become apparent later, the N_j or x_j (not x_j^0) within the summation is taken as the *actual mole number or mole fraction of the mixture*, and not that of the reference state.

It follows from the defining equations that the mixing functions are specified only when the reference states have been clearly delineated. (Note that a reference state must be defined for each component.) Reference states are chosen either for convenience (the properties are known) or for practicality (the deviations from the reference state are small).

If the mixing function is to be useful, it should depend on the properties of the mixture. That is, we desire:

$$\Delta B = f(T, P, x_1, \ldots, x_{n-1}) \qquad (8\text{-}83)$$

This requirement limits us to either one of two choices for each reference state variable; namely, $T^0, P^0, x_1^0, \ldots, x_{n-1}^0$ are either set equal to the actual mixture properties (and, hence, $\bar{B}_j^0$ varies as the mixture conditions change) or they can be set at some fixed conditions (and, hence, $\bar{B}_j^0$ is a constant).

In practice, the most common reference state is the pure component at the same temperature, pressure, and state of aggregation of the mixture. That is, for component j, $T^0 = T$, $P^0 = P$, $x_j^0 = 1$, $x_i^0 = 0$ $(i \neq j)$ and, therefore, $\bar{B}_j^0 = B_j(T, P)$. In this case, the temperature and pressure vary as the state of the mixture changes, but the mole fractions are fixed. Note that this definition of the reference state satisfies Eq. (8-83). Whenever the pure components exist in the same state of aggregation of the mixture at T and P, this reference state is *real.* There are, however, a number of cases in which the stable state of the pure material is different from that of the mixture and, hence, the reference state as defined above is *hypothetical* (e.g., an inorganic salt in an aqueous solution or a liquid mixture above the critical temperature of one of the components). In these cases, it is not uncommon to select other reference states (e.g., the saturated salt solution or an infinitely dilute solution).

The use of the mixture mole numbers in the summation of Eq. (8-81) results in a convenient definition of the partial molal mixing function. Substituting Eq. (8-6b)

$$\underline{B} = \sum_{j=1}^{n} N_j \bar{B}_j$$

into Eq. (8-81) yields:

$$\Delta\underline{B} = \sum_{j=1}^{n} N_j(\bar{B}_j - \bar{B}_j^0) \tag{8-84}$$

Applying the partial molal operator,

$$\overline{\Delta B}_j = \left[\frac{\partial(\Delta\underline{B})}{\partial N_j}\right]_{T,P,N_i[j]} = (\bar{B}_j - \bar{B}_j^0) \tag{8-85}$$

or, rewriting Eq. (8-84),

$$\Delta\underline{B} = \sum_{j=1}^{n} N_j\,\overline{\Delta B}_j \tag{8-86}$$

Dividing both sides of Eq. (8-86) by N,

$$\Delta B = \sum_{j=1}^{n} x_j\,\overline{\Delta B}_j \tag{8-87}$$

It also follows from Eq. (8-83) that

$$d(\Delta B) = \left(\frac{\partial(\Delta B)}{\partial T}\right)_{P,x} dT + \left(\frac{\partial(\Delta B)}{\partial P}\right)_{T,x} dP + \sum_{j\neq i}\left(\frac{\partial(\Delta B)}{\partial x_j}\right)_{T,P,x[i,j]} dx_j \tag{8-88}$$

Note that Eqs. (8-85), (8-87), and (8-88) are completely analogous to Eqs. (8-8), (8-48), and (8-30), with B replaced by ΔB. Since these three equations formed the basis for all the partial molal quantity relationships developed in Sections 8.2 and 8.3, it follows that all relationships developed for B also apply to ΔB. Thus, the analog to Eq. (8-34) is

$$\overline{\Delta B}_i = \Delta B - \sum_{j\neq i} x_j\left(\frac{\partial(\Delta B)}{\partial x_j}\right)_{T,P,x[j,i]} \tag{8-89}$$

and, hence, the slope-intercept method can be used to evaluate partial molal mixing functions from specific mixing functions. Similarly, the analogs to Eqs. (8-41) and (8-47) are

$$\sum_{i=1}^{n} x_i\,d(\overline{\Delta B}_i) = \left(\frac{\partial(\Delta B)}{\partial T}\right)_{P,x} dT + \left(\frac{\partial(\Delta B)}{\partial P}\right)_{T,x} dP \tag{8-90}$$

and

$$\sum_{i=1}^{n} x_i\left(\frac{\partial(\overline{\Delta B}_i)}{\partial x_j}\right)_{T,P,x[j,k]} = 0 \tag{8-91}$$

and, thus, the Duhem relation can be used to check consistency of data or to find one partial molal mixing function from a set of $(n - 1)$ others.

Example 8.9

If one g-mol of pure sulfuric acid is diluted with N_w g-mol of water, the heat evolved is about equal to:

$$Q\,(\text{kcal}) = \frac{17.86 N_w}{N_w + 1.7983} \qquad (18°\text{C}, N_w < 20)$$

What is the differential enthalpy of solution for water and acid for a solution containing 40 mole percent acid?

Solution

The heat evolved for one mole of acid is given in the problem statement. If we wished to obtain Q for N_A moles of acid and N_w moles of water, in cal,

$$Q = \frac{N_A\left(\dfrac{17{,}860\, N_w}{N_A}\right)}{\dfrac{N_w}{N_A} + 1.7983}$$

Then the integral heat of solution, $\Delta\underline{H}$, is $-Q$ and therefore,

$$\begin{aligned}\overline{\Delta H}_w &= \left(\frac{\partial \Delta \underline{H}}{\partial N_w}\right)_{T,P,N_A} \\ &= \frac{-32{,}100\, x_A^2}{(1 + 0.798 x_A)^2} \\ &= -2{,}950 \text{ cal/g-mol water}\end{aligned}$$

Similarly,

$$\begin{aligned}\overline{\Delta H}_A &= \left(\frac{\partial \Delta \underline{H}}{\partial N_A}\right)_{T,P,N_w} \\ &= \frac{-17{,}860\, x_w^2}{(1 + 0.798\, x_A)^2} \\ &= -3{,}700 \text{ cal/g-mol acid}\end{aligned}$$

Mixing functions can also be defined for the ideal solution, introduced in Section 8.4. In this case, the reference state is invariably taken as *the pure components at the same temperature, pressure, and state of aggregation as that of the mixture*. Thus, the superscripts in Eq. (8-82) can be deleted:

$$\Delta B = B - \sum_{j=1}^{n} x_j B_j \tag{8-92}$$

For an ideal solution,

$$\Delta B^{ID} = B^{ID} - \sum_{j=1}^{n} x_j B_j \tag{8-93}$$

For the Gibbs free energy, entropy, and enthalpy of an ideal solution, substituting Eqs. (8-80), (8-76), and (8-78), respectively, into Eq. (8-93) yields:

$$\Delta G^{ID} = RT \sum_{j=1}^{n} x_j \ln x_j \tag{8-94}$$

$$\Delta S^{ID} = -R \sum_{j=1}^{n} x_j \ln x_j \tag{8-95}$$

$$\Delta H^{ID} = 0 \tag{8-96}$$

or

$$\overline{\Delta G}_j^{ID} = RT \ln x_j \tag{8-97}$$

$$\overline{\Delta S}_j^{ID} = -R \ln x_j \tag{8-98}$$

$$\overline{\Delta H}_j^{ID} = 0 \tag{8-99}$$

The deviation of a mixture from ideal solution behavior is commonly denoted by the *excess function*, $\underline{B}^{EX}$ or B^{EX}:

$$\underline{B}^{EX} \equiv \underline{B} - \underline{B}^{ID} \quad \text{or} \quad B^{EX} = B - B^{ID} \tag{8-100}$$

From Eq. (8-6b) for $\underline{B}$ and $\underline{B}^{ID}$,

$$\underline{B} = \sum_{j=1}^{n} N_j \bar{B}_j$$

$$\underline{B}^{ID} = \sum_{j=1}^{n} N_j \bar{B}_j^{ID} \tag{8-101}$$

it follows that

$$\underline{B}^{EX} = \sum_{j=1}^{n} N_j (\bar{B}_j - \bar{B}_j^{ID}) \tag{8-102}$$

Applying the partial molal operator to Eq. (8-100):

$$\bar{B}_j^{EX} = \bar{B}_j - \bar{B}_j^{ID} \tag{8-103}$$

or

$$\underline{B}^{EX} = \sum_{j=1}^{n} N_j \bar{B}_j^{EX} \tag{8-104}$$

Thus, B^{EX} and $\bar{B}_j^{EX}$ are completely analogous to B and $\bar{B}_j$, and we could calculate $\bar{B}_j^{EX}$ from B^{EX} using the equations developed in Section 8.2 or we could use the Duhem equation as developed in Section 8.3.

By analogy to Eq. (8-100), excess mixing functions are commonly defined:

$$\Delta \underline{B}^{EX} = \Delta \underline{B} - \Delta \underline{B}^{ID} \tag{8-105}$$

or

$$\overline{\Delta B}_j^{EX} = \overline{\Delta B}_j - \overline{\Delta B}_j^{ID} \tag{8-106}$$

But, unlike $\Delta \underline{B}$ or $\Delta \underline{B}^{ID}$, an excess function does not depend directly on the choice of the reference state. This can be readily appreciated by expanding $\overline{\Delta B}_j$ and $\overline{\Delta B}_j^{ID}$ in Eq. (8-106):

$$\overline{\Delta B}_j = \bar{B}_j - \bar{B}_j^{0} \tag{8-106a}$$

$$\overline{\Delta B}_j^{ID} = \bar{B}_j^{ID} - \bar{B}_j^{0} \tag{8-106b}$$

Thus,

$$\overline{\Delta B}_j^{EX} = \bar{B}_j - \bar{B}_j^{ID} = \bar{B}_j^{EX} \tag{8-107}$$

Therefore, ΔB^{EX} and B^{EX} are identical, provided that the same reference state is used for ΔB and ΔB^{ID}. As mentioned above, the conventional reference state for ΔB^{ID} is taken as pure components at the same temperature, pressure, and state of aggregation of the mixture. Unless otherwise specified, the same reference state is used for ΔB whenever the excess function is applied.

Because the temperature and pressure of the ideal mixture are taken equal to those of the real mixture, it follows that if

$$W = YZ \tag{8-108}$$

where Y is T or P, then

$$W^{EX} = YZ^{EX} \tag{8-109}$$

Thus, it can be shown that

$$H^{EX} = U^{EX} + PV^{EX} \tag{8-110}$$

and

$$G^{EX} = H^{EX} - TS^{EX} \tag{8-111}$$

The three excess functions, G^{EX}, H^{EX}, and S^{EX} can be evaluated, by difference, from the functions for these properties derived in Section 8.4 and the corresponding ideal functions as defined earlier in this section. Thus,

$$\bar{G}_j^{EX} = \int_{P^*}^{P} \overline{\Delta V}_j \, dP; \quad G^{EX} = \int_{P^*}^{P} \Delta V \, dP \tag{8-112}$$

$$\bar{H}_j^{EX} = -T^2 \int_{P^*}^{P} \left[\frac{\partial \left(\dfrac{\overline{\Delta V}_j}{T} \right)}{\partial T} \right]_{P,x} dP; \quad H^{EX} = -T^2 \int_{P^*}^{P} \left[\frac{\partial \left(\dfrac{\Delta V}{T} \right)}{\partial T} \right]_{P,x} dP \tag{8-113}$$

$$\bar{S}_j^{EX} = -\int_{P^*}^{P} \left[\frac{\partial (\overline{\Delta V}_j)}{\partial T} \right]_{P,x} dP; \quad S^{EX} = -\int_{P^*}^{P} \left[\frac{\partial (\Delta V)}{\partial T} \right]_{P,x} dP \tag{8-114}$$

For an ideal solution, the excess functions are zero, by definition. In general, if the components of a mixture have similar force fields and are not significantly different in size and symmetry, ideal solution behavior is a good first approximation. For mixtures that are free of strong associative forces such as hydrogen bonding, solvation, or complexes, *regular solution behavior* is sometimes a more accurate approximation than ideal behavior. A regular solution is defined as one for which $\bar{S}_j^{EX}$ is zero for all components. Thus,

$$G^{EX} = H^{EX} = \Delta H = \Delta H^{EX} \tag{8-115}$$

or

$$\bar{G}_j^{EX} = \bar{H}_j^{EX} = \overline{\Delta H}_j^{EX} = \overline{\Delta H}_j \tag{8-116}$$

That is, the free energy of mixing can be synthesized *from knowledge of only the enthalpy of mixing.*

We mention in passing a third model that is less commonly employed; an *athermal* solution is defined as one for which $\bar{H}_j^{EX} \equiv 0$ for all j. In this case, $G^{EX} = -TS^{EX}$, and estimations of G^{EX} are made from liquid models that allow an estimate of the excess entropy of mixing.

8.6 Fugacity and Fugacity Coefficient

In Section 6.2, the fugacity of a pure substance, f_j, was introduced as a property that expressed all of the pressure dependence and a portion of the temperature dependence (excluding the reference state entropy) of the Gibbs free energy or chemical potential. Similarly, the fugacity of a mixture, f, and the fugacity of a component in a mixture, $\hat{f}_j$, are defined to reflect the pressure- and part of the temperature- and composition-dependency of G and $\bar{G}_j$, respectively.[11] The defining equations are given in Table 8.3.

The limiting condition required to fix the absolute values of f and $\hat{f}_j$ cannot be chosen arbitrarily; they must be consistent with the limiting condition already specified for the pure component fugacity [i.e., Eq. (6-32), which is repeated in Table 8.3 as Eq. (8-118a)]. In other words, as y_j approaches unity and as P approaches a pressure, P^*, in the ideal gas range, f must approach f_j which, in turn, approaches P^*. Therefore, the limiting condition for f is:

$$\lim_{P \to P^*} \left(\frac{f}{P} \right) = 1 \tag{8-118b}$$

The limiting condition for $\hat{f}_j$ can be defined in the following manner. First, we note that when y_j approaches unity, $\hat{f}_j$ must approach f_j and, consequently, $\lambda_j(T)$ in Eq. (8-117c) must be the same $\lambda_j(T)$ in Eq. (8-117a). Thus, subtracting Eq. (8-117a) from Eq. (8-117c) yields:

$$\bar{G}_j - G_j = RT \ln \frac{\hat{f}_j}{f_j} \tag{8-121}$$

Now consider a mixture containing j in equilibrium with pure j across a membrane permeable only to j. At equilibrium, $\bar{G}_j = G_j$ and, therefore, $\hat{f}_j = f_j$. These equalities are valid for all pressures. As P approaches P^*, f_j approaches the pressure of pure j which, in turn, must equal the partial pressure of j in the mixture (see Section 8.4). Thus, $\hat{f}_j$ must approach y_jP, or, as shown in Eq. (8-118c)

$$\lim_{P \to P^*} \left(\frac{\hat{f}_j}{y_jP} \right) = 1$$

[11] The symbol ^ above f is used to denote the fugacity of a component in a mixture. As shown later in this section, $\hat{f}_j$ is not a partial molal property.

TABLE 8.3
SUMMARY OF FUGACITY RELATIONS

Pure component (a)		Mixture (b)		Component in a mixture (c)	
$G_j \equiv RT \ln f_j + \lambda_j(T)$	(8-117a)	$G \equiv RT \ln f + \Lambda(T, y_1, \ldots, y_n)$	(8-117b)	$\bar{G}_j \equiv RT \ln \hat{f}_j + \lambda_j(T)$	(8-117c)
$\frac{f_j}{P} \longrightarrow 1 \text{ as } P \longrightarrow P^*$	(8-118a)	$\frac{f}{P} \longrightarrow 1 \text{ as } P \longrightarrow P^*$	(8-118b)	$\frac{\hat{f}_j}{y_j P} \longrightarrow 1 \text{ as } P \longrightarrow P^*$	(8-118c)
$\left(\frac{\partial \ln f_j}{\partial P}\right)_T = \frac{V_j}{RT}$	(8-119a)	$\left(\frac{\partial \ln f}{\partial P}\right)_{T,y} = \frac{V}{RT}$	(8-119b)	$\left(\frac{\partial \ln \hat{f}_j}{\partial P}\right)_{T,y} = \frac{\bar{V}_j}{RT}$	(8-119c)
$\left(\frac{\partial \ln f_j}{\partial T}\right)_P = \frac{-(H_j - H_j^*)}{RT^2}$	(8-120a)	$\left(\frac{\partial \ln f}{\partial T}\right)_{P,y} = \frac{-(H - \sum y_j H_j^*)}{RT^2}$	(8-120b)	$\left(\frac{\partial \ln \hat{f}_j}{\partial T}\right)_{P,y} = \frac{-(\bar{H}_j - H_j^*)}{RT^2}$	(8-120c)

The Λ-functionality in Eq. (8-117b) must also be consistent with that of Eq. (8-117c). Substituting Eqs. (8-117b) and (8-117c) into Eq. (8-122)

$$G = \sum_{j=1}^{n} y_j \bar{G}_j \tag{8-122}$$

yields

$$RT \ln f + \Lambda(T, y) = RT \sum_{j=1}^{n} y_j \ln \hat{f}_j + \sum_{j=1}^{n} y_j \lambda_j(T) \tag{8-123}$$

This relationship must be valid for all pressures. As P approaches P^*, it reduces to:

$$RT \ln P^* + \Lambda(T, y) = RT \sum_{j=1}^{n} y_j \ln (y_j P^*) + \sum_{j=1}^{n} y_j \lambda_j(T) \tag{8-124}$$

which then can be simplified to:

$$\Lambda(T, y) = RT \sum_{j=1}^{n} y_j \ln y_j + \sum_{j=1}^{n} y_j \lambda_j(T) \tag{8-125}$$

Note that as y_j approaches unity, Λ approaches λ_j, as required.

The pressure- and temperature-dependence of f and $\hat{f}_j$ can be determined in a manner analogous to that used in Section 6.2. The pressure-dependency, as given in Eqs. (8-119b) and (8-119c) is:

$$\left(\frac{\partial \ln f}{\partial P}\right)_{T,y} = \frac{1}{RT}\left(\frac{\partial G}{\partial P}\right)_{T,y} = \frac{V}{RT}$$

and

$$\left(\frac{\partial \ln \hat{f}_j}{\partial P}\right)_{T,y} = \frac{1}{RT}\left(\frac{\partial \bar{G}_j}{\partial P}\right)_{T,y} = \frac{\bar{V}_j}{RT}$$

The temperature-dependencies follow directly from the defining equations, Eq. (8-117b) and (8-117c), and from the λ_j-derivative, as given in Eq. (6-39):

$$\left(\frac{d\left(\frac{\lambda_j}{T}\right)}{dT}\right) = -\frac{H_j^*}{T^2}$$

where H_j^* is the enthalpy of pure j in the ideal gas state. Thus,

$$\left(\frac{\partial \ln f}{\partial T}\right)_{P,y} = \frac{1}{R}\left(\frac{\partial\left(\frac{G}{T}\right)}{\partial T}\right)_{P,y} - \frac{1}{R}\sum_{j=1}^{n} y_j \left(\frac{d\left(\frac{\lambda_j}{T}\right)}{dT}\right) = -\frac{1}{RT^2}\left(H - \sum_{j=1}^{n} y_j H_j^*\right) \tag{8-126}$$

and

$$\left(\frac{\partial \ln \hat{f}_j}{\partial T}\right)_{P,y} = \frac{1}{R}\left(\frac{\partial\left(\frac{\bar{G}_j}{T}\right)}{\partial T}\right)_{P,y} - \frac{1}{R}\left(\frac{d\left(\frac{\lambda_j}{T}\right)}{dT}\right) = -\frac{(\bar{H}_j - H_j^*)}{RT^2} \tag{8-127}$$

We noted above that $\hat{f}_j$ is not a partial molal quantity. That is, neither f nor $\hat{f}_j$ obey a relation of the form of Eq. (8-48). The relationship between these variables can, however, be found by substituting Eq. (8-125) into Eq. (8-123), which after simplification yields

$$\ln f = \sum_{j=1}^{n} y_j \ln\left(\frac{\hat{f}_j}{y_j}\right) \tag{8-128}$$

By analogy to Eq. (8-48), we might consider $\ln(\hat{f}_j/y_j)$ as the partial molal property of $\ln f$. But if so, then we must be able to show that:

$$\ln\left(\frac{\hat{f}_j}{y_j}\right) = \left[\frac{\partial(N \ln f)}{\partial N_j}\right]_{T,P,N_k[j]} \tag{8-129}$$

This relation can be proved in the following manner. Substituting Eq. (8-125) into Eq. (8-117b), multiplying by N, and differentiating with respect to N_j gives:

$$\begin{aligned}\bar{G}_j = \left(\frac{\partial G}{\partial N_j}\right)_{T,P,N_k[j]} &= RT\left[\frac{\partial(N \ln f)}{\partial N_j}\right]_{T,P,N_k[j]} + RT \ln y_j \\ &+ RT \sum_{i=1}^{n} N_i \left(\frac{\partial \ln y_i}{\partial N_j}\right)_{T,P,N_k[j]} + \lambda_j(T)\end{aligned} \tag{8-130}$$

Note that the summation vanishes. Equating the remaining terms to $\bar{G}_j$ as given by Eq. (8-117c), after simplification, one finds the desired result, Eq. (8-129). Thus, $\ln(\hat{f}_j/y_j)$ is truly the partial molal property of $\ln f$ and, therefore, all of the general relationships derived previously for partial molal quantities can be applied by substituting $\ln(\hat{f}_j/y_j)$ for $\bar{B}_j$ and $\ln f$ for B. In particular, it follows from Eq. (8-23) that:

$$d\ln f = \left(\frac{\partial \ln f}{\partial T}\right)_{P,y} dT + \left(\frac{\partial \ln f}{\partial P}\right)_{T,y} dP + \sum_{j=1}^{n} \ln\left(\frac{\hat{f}_j}{y_j}\right) dy_j \tag{8-131}$$

or, with Eqs. (8-119b) and (8-120b),

$$d\ln f = -\left(\frac{H - \sum y_j H_j^*}{RT^2}\right) dT + \left(\frac{V}{RT}\right) dP + \sum_{j=1}^{n} \ln\left(\frac{\hat{f}_j}{y_j}\right) dy_j \tag{8-132}$$

Also, the Duhem relation corresponding to Eq. (8-41) is

$$\sum_{j=1}^{n} y_j\, d\ln\left(\frac{\hat{f}_j}{y_j}\right) = \left(\frac{\partial \ln f}{\partial T}\right)_{P,y} dT + \left(\frac{\partial \ln f}{\partial P}\right)_{T,y} dP \tag{8-133}$$

which can be simplified, because $\sum_{j=1}^{n} y_j\, d\ln y_j = 0$, to

$$\sum_{j=1}^{n} y_j\, d\ln \hat{f}_j = \left(\frac{\partial \ln f}{\partial T}\right)_{P,y} dT + \left(\frac{\partial \ln f}{\partial P}\right)_{T,y} dP \tag{8-134}$$

Expanding the partial derivatives:

$$\left(\frac{\partial \ln f}{\partial T}\right)_{P,y} = -\frac{1}{RT^2}\left(H - \sum_{j=1}^{n} y_j H_j^*\right) = \sum_{j=1}^{n} y_j\left[-\left(\frac{\bar{H}_j - H_j^*}{RT^2}\right)\right] \tag{8-135}$$

and

$$\left(\frac{\partial \ln f}{\partial P}\right)_{T,y} = \frac{V}{RT} = \sum_{j=1}^{n} y_j \left(\frac{\bar{V}_j}{RT}\right) \tag{8-136}$$

Substituting into Eq. (8-134), and collecting terms yields:

$$\sum_{j=1}^{n} y_j \left[d \ln \hat{f}_j + \left(\frac{\bar{H}_j - H_j^*}{RT^2}\right) dT - \left(\frac{\bar{V}_j}{RT}\right) dP \right] = 0 \tag{8-137}$$

This equation can be simplified by noting that

$$d \ln \hat{f}_j = \left(\frac{\partial \ln \hat{f}_j}{\partial T}\right)_{P,y} dT + \left(\frac{\partial \ln \hat{f}_j}{\partial P}\right)_{T,y} dP + \sum_{i \neq k} \left(\frac{\partial \ln \hat{f}_j}{\partial y_i}\right)_{T,P,y[i,k]} dy_i \tag{8-138}$$

which is Eq. (8-18c) applied to $\ln \hat{f}_j$. Substituting Eq. (8-138) into Eq. (8-137), we obtain

$$\sum_{j=1}^{n} y_j \left[\sum_{i \neq k} \left(\frac{\partial \ln \hat{f}_j}{\partial y_i}\right)_{T,P,y[i,k]} dy_i \right] = 0 \tag{8-139}$$

or, inverting the order of summation,

$$\sum_{i \neq k} \left[\sum_{j=1}^{n} y_j \left(\frac{\partial \ln \hat{f}_j}{\partial y_i}\right)_{T,P,y[i,k]} \right] dy_i = 0 \tag{8-140}$$

Since the brackets are coefficients of $n - 1$ terms in dy_i, all of which are independent because y_k has been eliminated, it follows that each term in brackets must vanish. Therefore, an equivalent Duhem relation for fugacity is

$$\sum_{j=1}^{n} y_j \left(\frac{\partial \ln \hat{f}_j}{\partial y_i}\right)_{T,P,y[i,k]} = 0 \tag{8-141}$$

Finally, we can apply the generalizations developed previously for obtaining $\ln (\hat{f}_j/y_j)$ from data or correlations for $\ln f$. From Eq. (8-36), it follows that:

$$\ln\left(\frac{\hat{f}_k}{y_k}\right) = \ln f + \left(\frac{\partial \ln f}{\partial y_k}\right)_{T,P,y[i,k]} - \sum_{j \neq i} y_j \left(\frac{\partial \ln f}{\partial y_j}\right)_{T,P,y[i,j]} \tag{8-142}$$

which, it can be readily shown, is equivalent to the more commonly used form,

$$\ln\left(\frac{\hat{f}_k}{y_k P}\right) = \ln\left(\frac{f}{P}\right) + \left[\frac{\partial \ln\left(\frac{f}{P}\right)}{\partial y_k}\right]_{T,P,y[i,k]} - \sum_{j \neq i} y_j \left[\frac{\partial \ln\left(\frac{f}{P}\right)}{\partial y_j}\right]_{T,P,y[i,j]} \tag{8-143}$$

For a binary mixture of components A and B, Eq. (8-143) becomes

$$\ln\left(\frac{\hat{f}_B}{y_B P}\right) = \ln\left(\frac{f}{P}\right) - y_A \left[\frac{\partial \ln\left(\frac{f}{P}\right)}{\partial y_A}\right]_{T,P} \tag{8-144}$$

If f/P for the mixture can be obtained as a function of y, temperature, and pressure, a graph of $\ln (f/P)$ vs. y_A can then be constructed, as shown in

Figure 8.7. The slope-intercept method can be applied, as indicated in the figure to yield $\ln(\hat{f}_B/y_BP)$ and $\ln(\hat{f}_A/y_AP)$.

Alternatively, if P-V-T data are available for the mixture from P^* to the pressure in question, the fugacity of a component can be evaluated by integrating Eq. (8-119c):

$$RT \ln \frac{\hat{f}_j^\alpha}{\hat{f}_j^*} = \int_{P^*}^{P} \bar{V}_j \, dP \tag{8-145}$$

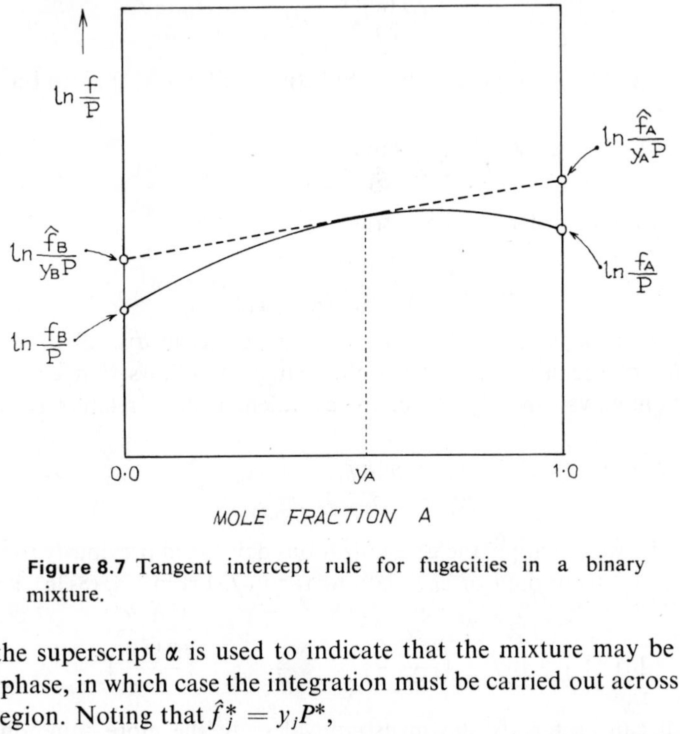

Figure 8.7 Tangent intercept rule for fugacities in a binary mixture.

where the superscript α is used to indicate that the mixture may be a condensed phase, in which case the integration must be carried out across a two-phase region. Noting that $\hat{f}_j^* = y_jP^*$,

$$RT \ln \frac{\hat{f}_j^\alpha}{y_jP^*} = \int_{P^*}^{P} \bar{V}_j \, dP \tag{8-146}$$

The fugacity of a component in a mixture can also be related to the fugacity of the pure material. Integrating Eq. (8-119a) from P^* to P,

$$RT \ln \frac{f_j^\alpha}{P^*} = \int_{P^*}^{P} V_j \, dP \tag{8-147}$$

and subtracting this result from Eq. (8-146) yields

$$RT \ln \frac{\hat{f}_j^\alpha}{y_jf_j^\alpha} = \int_{P^*}^{P} (\bar{V}_j - V_j) \, dP = \int_{P^*}^{P} \overline{\Delta V}_j \, dP \tag{8-148}$$

If the mixture is ideal, then by definition, the integral vanishes. Thus, for an ideal solution,

$$\hat{f}_j^\alpha = y_j f_j^\alpha \qquad \text{(ideal solution)} \tag{8-149}$$

which is known as the Lewis-Randall rule. Note that this relation applies to an ideal solution even if the pure material does not obey ideal gas behavior. If the mixture is a vapor and if the pure materials forming an ideal solution also follow ideal gas behavior, then $\hat{f}_j = y_j P = p_j$.

For vapor mixtures a commonly used relation for $\hat{f}_j$ is obtained by adding and subtracting (RT/P) within the integral in Eq. (8-146). After rearranging, and letting $P^* \longrightarrow 0$,

$$RT \ln \frac{\hat{f}_j^V}{y_j P} = \int_0^P \left(\bar{V}_j^V - \frac{RT}{P} \right) dP \tag{8-150}$$

The group $(\hat{f}_j^V / y_j P)$ is called the *fugacity coefficient of j in the mixture*, ϕ_j. To obtain ϕ_j, an equation of state applicable to the mixture must be used to evaluate $\bar{V}_j$ at constant composition and temperature from a low pressure to the system pressure.

By definition,

$$\hat{f}_j^V = \phi_j y_j P \tag{8-151}$$

If the mixture formed an ideal solution, then from Eq. (8-149)

$$\phi_j = \frac{f_j^V}{P} \qquad \text{(ideal solution)} \tag{8-152}$$

whereas, for an ideal gas mixture,

$$\phi_j = 1 \qquad \text{(ideal gas mixture)} \tag{8-153}$$

Eq. (8-150) is not in a particularly convenient form because almost all mixture equations of state are explicit in pressure. In Example 8.10 an alternate formulation for ϕ_j is given and the calculational procedure is illustrated with the virial equation of state.

Example 8.10

Derive an expression for ϕ_i suitable for a pressure-explicit equation of state. Demonstrate the utility of this relation using the mixture virial equation of state:

$$Z = \frac{P\underline{V}}{NRT} = 1 + \frac{BN}{\underline{V}} + \frac{CN^2}{\underline{V}^2} + \cdots \tag{8-154}$$

where B and C are the mixture second and third virial coefficients. Expressed as a function of composition, they are:

$$B = \sum_{i=1}^{n} \sum_{j=1}^{n} B_{ij} y_i y_j \tag{8-155}$$

$$C = \sum_{i=1}^{n} \sum_{j=1}^{n} \sum_{k=1}^{n} C_{ijk} y_i y_j y_k \tag{8-156}$$

B_{ij} and C_{ijk} are functions only of temperature and the specific species; B_{ii} reflects binary interactions between i and i, B_{ij} between i and j, C_{iii} ternary interactions between i, i, and i, C_{ijk} between i, j, and k, etc. There are many ways to estimate B_{ij} and C_{ijk}.[12] These techniques, however, are not discussed here.

Reduce the final equation to obtain a relation for ϕ_A in a binary of A and B and illustrate its use to calculate the fugacity coefficient of neopentane in a mixture with methane at 90°C and 10 atm. The mole fraction of neopentane is 0.608. At 90°C, the value of the interaction virial is reported to be -106 cm³/g-mol[13] and the values of the pure component second virials are -24.2 cm³/g-mol (methane) and -566 cm³/g-mol (neopentane).[14] Neglect the third virial terms in this calculation.

Solution

From Eq. (5-55),

$$d\underline{A} = -\underline{S}\,dT - P\,d\underline{V} + \sum_{i=1}^{n} \mu_i\,dN_i$$

$$\left(\frac{\partial \underline{A}}{\partial \underline{V}}\right)_{T,N} = -P \tag{8-157}$$

$$\underline{A}(T, \underline{V}, N) - \underline{A}(T, \underline{V}^*, N) = -\int_{\underline{V}^*}^{\underline{V}} P\,d\underline{V} \tag{8-158}$$

where $\underline{V}^*$ represents the (very large) volume as $P \longrightarrow P^* \longrightarrow 0$. Adding and subtracting $(NRT/\underline{V})$ within the integral,

$$\underline{A}(T, \underline{V}, N) - \underline{A}(T, \underline{V}^*, N) = \int_{\underline{V}^*}^{\underline{V}} \left(\frac{NRT}{\underline{V}} - P\right) d\underline{V} - \int_{\underline{V}^*}^{\underline{V}} \left(\frac{NRT}{\underline{V}}\right) d\underline{V} \tag{8-159}$$

Next, we differentiate Eq. (8-159) with respect to N_i at constant T, $\underline{V}$, $N_j[i]$ and use Eq.(5-61)

$$\bar{G}_i(T, \underline{V}, N) - \bar{G}_i(T, \underline{V}^*, N)$$

$$= \int_{\underline{V}^*}^{\underline{V}} \left[\left(\frac{RT}{\underline{V}}\right) - \left(\frac{\partial P}{\partial N_i}\right)_{T,\underline{V},N_j[i]}\right] d\underline{V} - RT \ln \frac{\underline{V}}{\underline{V}^*} \tag{8-160}$$

$$= RT \ln \frac{\hat{f}_i}{\hat{f}_i^*} \tag{8-161}$$

$$= RT \ln \frac{\hat{f}_i}{P^* y_i} \tag{8-162}$$

where Eqs. (8-117c) and (8-118c) have been used. Noting further that:

[12] See, for example, J. M. Prausnitz, *Molecular Thermodynamics of Fluid-Phase Equilibria* (Englewood Cliffs, N. J.: Prentice-Hall, Inc., 1969), Chap. 5.

[13] S. D. Hamann, J. A. Lambert, and R. B. Thomas, "The Second Virial Coefficients of Some Gas Mixtures," *Australian J. Chem.*, **8**, (1955), 149.

[14] J. A. Huff and T. M. Reed III, "Second Virial Coefficients of Mixtures of Nonpolar Molecules from Correlations on Pure Components," *J. Chem. Eng. Data*, **8**, (1963), 306.

$$RT \ln \frac{\underline{V}}{\underline{V}^*} = RT \ln \left[\frac{\left(\frac{ZNRT}{P}\right)}{\left(\frac{NRT}{P^*}\right)} \right] = RT \ln Z + RT \ln \frac{P^*}{P}$$

and substituting into Eqs. (8-160) and (8-162)

$$RT \ln \phi_i = RT \ln \left(\frac{\hat{f}_i}{Py_i}\right) = \int_{\underline{V}^* \to \infty}^{\underline{V}} \left[\left(\frac{RT}{\underline{V}}\right) - \left(\frac{\partial P}{\partial N_i}\right)_{T,\underline{V},N_j[i]} \right] d\underline{V} - RT \ln Z \tag{8-163}$$

This is the desired result. To employ the virial equation of state, solving Eq. (8-154) for P,

$$P = \frac{NRT}{\underline{V}} + \frac{N^2BRT}{\underline{V}^2} + \frac{N^3CRT}{\underline{V}^3} + \cdots \tag{8-164}$$

whence, with Eqs. (8-155) and (8-156),

$$\left(\frac{\partial P}{\partial N_i}\right)_{T,\underline{V},N_j[i]} = \frac{RT}{\underline{V}} + \frac{2NRT}{\underline{V}^2} \sum_{j=1}^{n} y_j B_{ij} + \frac{3N^2RT}{\underline{V}^3} \sum_{j=1}^{n} \sum_{k=1}^{n} y_j y_k C_{ijk} + \cdots \tag{8-165}$$

Substituting Eq. (8-165) into Eq. (8-163) and integrating

$$\ln \phi_i = \frac{2}{V} \sum_{j=1}^{n} y_j B_{ij} + \frac{3}{2V^2} \sum_{j=1}^{n} \sum_{k=1}^{n} y_j y_k C_{ijk} - \ln Z \tag{8-166}$$

where Z is given by Eq. (8-154) and $V = \underline{V}/N$.

For a binary of A and B,

$$\ln \phi_A = \frac{2}{V}(y_A B_{AA} + y_B B_{AB}) + \frac{3}{2V^2}(y_A^2 C_{AAA} + y_B^2 C_{ABB} + 2y_A y_B C_{AAB}) - \ln Z \tag{8-167}$$

For the system 39.2% methane (M) and 60.8% neopentane (P) at 90°C and 10 atm, neglecting the third virial terms, from Eq. (8-155), we see that

$$\begin{aligned} B &= y_M^2 B_{MM} + 2y_M y_P B_{MP} + y_P^2 B_{PP} \\ &= (0.392)^2(-24.2) + (2)(0.392)(0.608)(-106) + (0.608)^2(-566) \\ &= -265 \text{ cm}^3/\text{g-mol} \end{aligned}$$

$$Z = 1 + \frac{B}{V} \sim 1 + \frac{BP}{RT} = 1 + \frac{(-265)(10)}{(82.07)(363)} = 0.91$$

$$V = \frac{ZRT}{P} = \frac{(0.91)(82.07)(363)}{(10)} = 2710 \text{ cm}^3/\text{g-mol}$$

$$\begin{aligned} \ln \phi_P &= \frac{2}{V}(y_P B_{PP} + y_M B_{MP}) - \ln Z \\ &= \frac{2}{2710}[(0.608)(-566) + (0.392)(-106)] - \ln(0.91) \\ &= -0.142 - (-0.095) = -0.047 \end{aligned}$$

$$\phi_P = 0.956$$

8.7 Activity and Activity Coefficient

To calculate the fugacity of a component in a *gas mixture*, either Eq. (8-150) or Eq. (8-163) may be used. Both require that an equation of state for the mixture be available. The calculational procedure was illustrated in Example 8.10 using the virial equation.

Either Eq. (8-150) or Eq. (8-163) may also be used to calculate component fugacities in the liquid phase, although in such cases the isothermal integrations must be carried out from the ideal gas state up to the dew point and through the two-phase region to the liquid phase. In such a procedure there is an uncertainty about the physical significance of the integrands $\{[\bar{V}_j - (RT/P)]$ in Eq. (8-150) and $[(RT/V) - (\partial P/\partial N_j)]$ in Eq. (8-163)$\}$ between the limits of intrinsic stability of the gas and liquid phases (see Chapter 7). Besides such conceptual problems, very few, if any, equations of state apply equally well both to the gas and liquid phase.[15]

Therefore, an alternative method is commonly used to describe the Gibbs free energy of a component in a condensed phase mixture. Recall that fugacities were referenced to an ideal gas state [i.e., from Eqs. (8-121) and (8-118c)],

$$\bar{G}_j^\alpha(T, P, x_1, \ldots, x_{n-1}) - \bar{G}_j^*(T, P^*, x_1, \ldots, x_{n-1}) = RT \ln \frac{\hat{f}_j^\alpha}{x_j P^*} \qquad (8\text{-}168)$$

By replacing the ideal gas reference state with a reference state that represents a condensed phase, the integration across the two-phase region is avoided. Let us denote this new reference or *standard state* by a superscript zero and define:

$$\bar{G}_j^\alpha(T, P, x_1, \ldots, x_{n-1}) - \bar{G}_j^0(T, P^0, x_1^0, \ldots, x_{n-1}^0) \equiv RT \ln a_j \qquad (8\text{-}169)$$

where a_j is called the *activity*. Note that the standard-state temperature is equal to the system temperature, but the other standard-state conditions, $P^0, x_1^0, \ldots, x_{n-1}^0$, can be chosen arbitrarily. We can still, however, use Eq. (8-117c) to define a fugacity in the standard state,

$$\bar{G}_j^0(T, P^0, x_1^0, \ldots, x_{n-1}^0) = RT \ln \hat{f}_j^0 + \lambda_j(T) \qquad (8\text{-}170)$$

Thus, it follows that:

$$\bar{G}_j^\alpha - \bar{G}_j^0 = RT \ln \frac{\hat{f}_j^\alpha}{\hat{f}_j^0} \qquad (8\text{-}171)$$

[15] One important exception is the Benedict-Webb-Rubin equation of state [M. Benedict, G. B. Webb, and L. C. Rubin, "An Empirical Equation for Thermodynamic Properties of Light Hydrocarbons and Their Mixtures—Constants for Twelve Hydrocarbons", *Chem. Eng. Prog.*, **47**, (1951), 419]. This equation is widely used to calculate both vapor and liquid fugacities with the same set of constants. It is, however, only applicable for mixtures of light hydrocarbons.

or

$$a_j = \frac{\hat{f}_j^\alpha}{\hat{f}_j^0} \tag{8-172}$$

where $\hat{f}_j^0$ is the fugacity of j in the standard state.[16]

A number of standard states are in common use. The most prevalent one is the *pure material standard state* or *pure solvent standard state:* pure j at T and P of the mixture and in the same state of aggregation as that of the mixture. For this case,

$$\hat{f}_j^0 = f_j^\alpha(T, P, y_j = 1) \tag{8-173}$$

and

$$RT \ln a_j = \bar{G}_j^\alpha - G_j^\alpha = \overline{\Delta G_j} \tag{8-174}$$

If the solution were ideal, then

$$\overline{\Delta G_j} = \overline{\Delta G_j^{ID}} = RT \ln x_j \qquad \text{(ideal solution)} \tag{8-175}$$

or

$$a_j = x_j \qquad \text{(ideal solution)} \tag{8-176}$$

For most solutions, a_j/x_j is not unity; the difference of the ratio from unity is a measure of the nonideality. This ratio is defined as the *activity coefficient*, γ_j.[17]

$$\gamma_j \equiv \frac{a_j}{x_j} \tag{8-177}$$

or

$$\gamma_j = \frac{\hat{f}_j}{f_j x_j} \tag{8-178}$$

Note that Eq. (8-178) represents the deviation from the Lewis-Randall rule, as given in Eq. (8-149).

When the pure solvent standard state is employed, the activity coefficient can be related to the partial molal Gibbs free energy of mixing. Substituting Eq. (8-177) into Eq. (8-174) and using Eq. (8-105) for $\overline{\Delta G_j}$ and Eq. (8-97) for $\overline{\Delta G_j^{ID}}$, we obtain:

$$RT \ln (\gamma_j x_j) = \overline{\Delta G_j} = \overline{\Delta G_j^{ID}} + \overline{\Delta G_j^{EX}} = RT \ln x_j + \overline{\Delta G_j^{EX}} \tag{8-179}$$

or, simply,

$$\overline{\Delta G_j^{EX}} = RT \ln \gamma_j \tag{8-180}$$

[16] Eq. (8-172) is sometimes written as $a_j/a_j^0 = \hat{f}_j^\alpha/\hat{f}_j^0$, where $a_j^0 = 1$ in the standard state.

[17] Activity coefficients based on standard states other than pure solvent are discussed at the end of this section.

Multiplying Eq. (8-180) by x_j and summing over all x_j,

$$\Delta G^{EX} = RT\left(\sum_{j=1}^{n} x_j \ln \gamma_j\right) \tag{8-181}$$

Thus, we see that $RT \ln \gamma_j$ is nothing more than the excess partial molal Gibbs free energy of mixing and, therefore, derivatives of $\ln \gamma_j$ can be immediately evaluated from the derivatives of $\overline{\Delta G_j^{EX}}$. In particular,

$$\left(\frac{\partial \ln \gamma_j}{\partial T}\right)_{P,x} = \frac{1}{R}\left[\frac{\partial\left(\frac{\overline{\Delta G_j^{EX}}}{T}\right)}{\partial T}\right]_{P,x} = \frac{-\overline{\Delta H_j^{EX}}}{RT^2} = \frac{-\overline{\Delta H_j}}{RT^2} \tag{8-182}$$

$$\left(\frac{\partial \ln \gamma_j}{\partial P}\right)_{T,x} = \frac{1}{RT}\left[\frac{\partial(\overline{\Delta G_j^{EX}})}{\partial P}\right]_{T,x} = \frac{\overline{\Delta V_j^{EX}}}{RT} = \frac{\overline{\Delta V_j}}{RT} \tag{8-183}$$

$$\left(\frac{\partial \ln \gamma_j}{\partial x_i}\right)_{T,P,x[i,n]} = \frac{1}{RT}\left[\frac{\partial(\overline{\Delta G_j^{EX}})}{\partial x_i}\right]_{T,P,x[i,n]} = \int_{P^*}^{P}\left[\frac{\partial(\overline{\Delta V_j})}{\partial x_i}\right]_{T,P,x[i,n]} dP \tag{8-184}$$

By comparing Eqs. (8-182) and (8-183) to Eqs. (8-120c) and (8-119c), we see that, for condensed phases, the activity coefficient is a weak function of temperature and pressure in comparison to the fugacity (i.e., $\overline{\Delta H_j} \ll \bar{H}_j - H_j^*$; $\overline{\Delta V_j} \ll V_j$). Eq. (8-184) involves an integration from the ideal gas state to the state representing the condensed phase and, although correct, it is rarely used to determine the effect of composition on the activity coefficient.

A number of other useful relations for activity coefficients can be derived by utilizing the relationship between γ and $\Delta G^{EX}/RT$. The total derivative,

$$d\left(\frac{\Delta G^{EX}}{RT}\right) = \frac{1}{RT} d(\Delta G^{EX}) - \frac{\Delta G^{EX}}{RT^2} dT \tag{8-185}$$

can be evaluated by using Eq. (8-22) with B replaced by ΔG^{EX}:

$$\begin{aligned} d(\Delta G^{EX}) &= \left[\frac{\partial(\Delta G^{EX})}{\partial T}\right]_{P,x} dT + \left[\frac{\partial(\Delta G^{EX})}{\partial P}\right]_{T,x} dP + \sum_{i \neq k} (\overline{\Delta G_i^{EX}} - \overline{\Delta G_k^{EX}})\, dx_i \\ &= -\Delta S^{EX}\, dT + \Delta V\, dP + RT \sum_{i \neq k} (\ln \gamma_i - \ln \gamma_k)\, dx_i \end{aligned} \tag{8-186}$$

Substituting Eq. (8-186) into Eq. (8-185) yields:

$$\begin{aligned} d\left(\frac{\Delta G^{EX}}{RT}\right) &= \frac{1}{RT^2}(-\Delta G^{EX} - T\,\Delta S^{EX})\, dT + \frac{\Delta V}{RT} dP + \sum_{i \neq k} (\ln \gamma_i - \ln \gamma_k)\, dx_i \\ &= -\frac{\Delta H}{RT^2} dT + \frac{\Delta V}{RT} dP + \sum_{i \neq k} (\ln \gamma_i - \ln \gamma_k)\, dx_i \end{aligned} \tag{8-187}$$

Thus,

$$\left[\frac{\partial\left(\frac{\Delta G^{EX}}{RT}\right)}{\partial T}\right]_{P,x} = -\frac{\Delta H}{RT^2} \tag{8-188}$$

$$\left[\frac{\partial\left(\frac{\Delta G^{EX}}{RT}\right)}{\partial P}\right]_{T,x} = \frac{\Delta V}{RT} \tag{8-189}$$

$$\left[\frac{\partial\left(\frac{\Delta G^{EX}}{RT}\right)}{\partial x_i}\right]_{T,P,x[i,k]} = \ln \gamma_i - \ln \gamma_k \tag{8-190}$$

The generalized Duhem relation, Eq. (8-41), with B replaced by $\Delta G^{EX}/RT$ yields

$$\sum_{j=1}^{n} x_j\, d\left(\frac{\overline{\Delta G}_j^{EX}}{RT}\right) = \sum_{j=1}^{n} x_j\, d\ln \gamma_j = -\frac{\Delta H}{RT^2}\, dT + \frac{\Delta V}{RT}\, dP \tag{8-191}$$

or, in the form of Eq. (8-47):

$$\sum_{j=1}^{n} x_j \left(\frac{\partial \ln \gamma_j}{\partial x_i}\right)_{T,P,x[i,k]} = 0 \tag{8-192}$$

The methods described previously for obtaining $\bar{B}_j$ from B as a function of composition can also be used to obtain $\ln \gamma_j$ from ΔG^{EX}. For example, Eq. (8-34) becomes

$$\ln \gamma_i = \frac{\Delta G^{EX}}{RT} - \sum_{j \neq i} x_j \left[\frac{\partial\left(\frac{\Delta G^{EX}}{RT}\right)}{\partial x_j}\right]_{T,P,x[j,i]} \tag{8-193}$$

We will see in Chapter 9 that in most instances activity coefficients are found from experimental vapor-liquid or liquid-liquid equilibrium data. These are then expressed in analytical form as a function of composition (and occasionally, temperature) for interpolation or extrapolation purposes. Alternatively, activity coefficients may be estimated from some liquid model. For example, we have already seen that if the liquid mixture were to be modelled as an ideal solution, then the activity coefficient would be unity. Normally when selecting a model, however, the assumptions chosen can be traced back to some specification on ΔH or ΔS^{EX}. To illustrate such an approach we shall synthesize ΔG of a mixture from the enthalpies and entropies of mixing,

$$\Delta G = \Delta H - T\,\Delta S \tag{8-194}$$

then determine ΔG^{EX} from

$$\Delta G^{EX} = \Delta G - \Delta G^{ID} = \Delta H - T\,\Delta S^{EX} \tag{8-195}$$

and finally use Eq. (8-193), which, for a binary, becomes:

$$\ln \gamma_1 = \frac{\Delta G^{EX}}{RT} - x_2 \left[\frac{\partial\left(\frac{\Delta G^{EX}}{RT}\right)}{\partial x_2}\right]_{T,P} \tag{8-196}$$

For an ideal binary solution, the Gibbs free energy, entropy, and enthalpy of mixing are given by Eqs. (8-94) through (8-96); these functions are shown in Figure 8.8a, along with the activity and activity coefficient.

The simplest nonideal solution is the *regular* solution. As discussed in Section 8.5,

$$\Delta S^{EX} = 0 \tag{8-197}$$

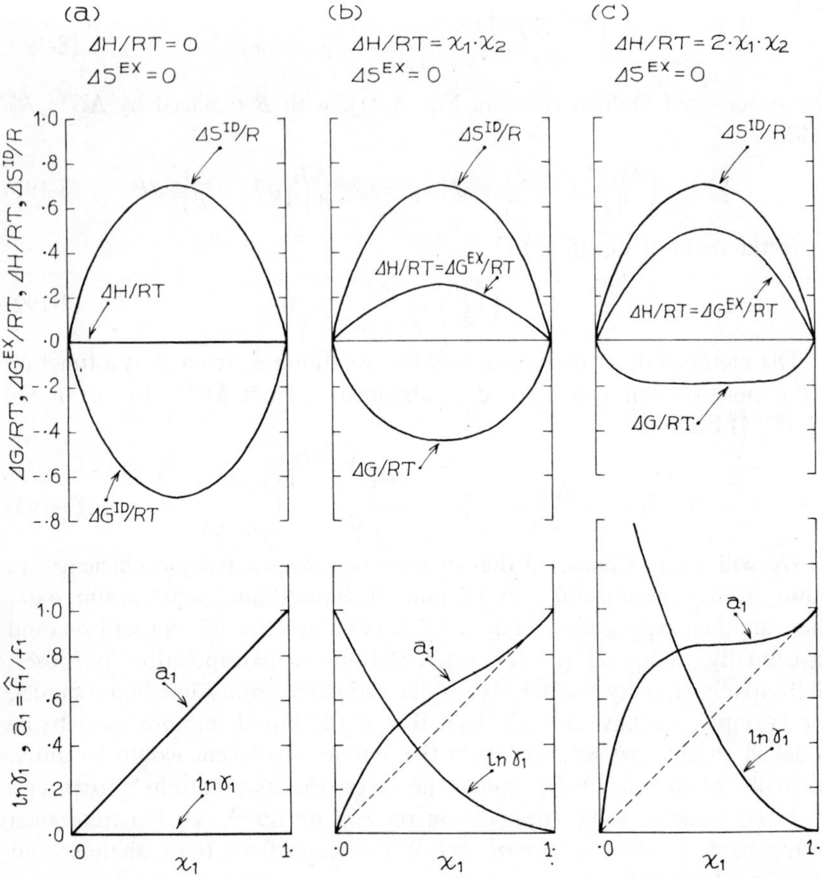

Figure 8.8 Thermodynamic functions of regular solutions.

and

$$\Delta G^{EX} = \Delta H \tag{8-198}$$

For many regular solutions, ΔH is symmetrical in x and a good correlation is often given by:[18]

$$\frac{\Delta H}{RT} = A x_1 x_2 \tag{8-199}$$

where A is a constant for a given binary mixture. Substituting Eqs. (8-197)

[18] J. H. Hildebrand and R. L. Scott, *The Solubility of Nonelectrolytes*, 3rd ed. (New York: Reinhold Publishing Corp., 1950).

and (8-199) into Eq. (8-195), we obtain

$$\frac{\Delta G^{EX}}{RT} = Ax_1x_2 \tag{8-200}$$ [19]

or

$$\frac{\Delta G}{RT} = Ax_1x_2 + x_1 \ln x_1 + x_2 \ln x_2 \tag{8-201}$$

Substituting Eq. (8-200) into Eq. (8-196), we obtain

$$\ln \gamma_1 = Ax_1x_2 - x_2(Ax_1 - Ax_2) = Ax_2^2 \tag{8-202}$$

or

$$a_1 = \gamma_1 x_1 = x_1 e^{Ax_2^2}$$

These functions are shown in Figure 8.8 for $A = 1$ in (b) and $A = 2$ in (c). Note the increasing nonideality as reflected in the activity and activity coefficient. These cases are examples of *positive deviations* from ideal solution behavior because $\gamma_1 > 1$ or $\ln \gamma_1 > 0$. Negative deviations refer to cases in which $\gamma_1 < 1$ or $\ln \gamma_1 < 0$.

In general, solutions that exhibit positive deviations tend to be partially immiscible over certain ranges of temperature. Case (c) of Figure 8.8 is, in fact, at the critical solution temperature, as can be shown by applying the binary mixture stability criteria, Eqs. (7-144) through (7-146). For $A = 2$ in

[19] For a regular solution, it can be shown that A must be proportional to $1/T$: from Eqs. (8-188) and (8-199),

$$\left[\frac{\partial\left(\frac{\Delta G^{EX}}{RT}\right)}{\partial T}\right]_{P,x} = -\frac{\Delta H}{RT^2} = -\frac{Ax_1x_2}{T} \tag{8-200a}$$

Differentiating Eq. (8-200),

$$\left[\frac{\partial\left(\frac{\Delta G^{EX}}{RT}\right)}{\partial T}\right]_{P,x} = x_1x_2\left(\frac{\partial A}{\partial T}\right)_{P,x} \tag{8-200b}$$

Equating Eqs. (8-200a) and (8-200b),

$$\left(\frac{\partial A}{\partial T}\right)_{P,x} = -\frac{A}{T}$$

or

$$\left(\frac{\partial \ln A}{\partial \ln T}\right) = -1$$

Thus,

$$\ln A = -\ln T + \text{constant}$$

or

$$AT = \text{constant} \tag{8-200c}$$

Note that Eq. (8-200*c*), when compared to Eq. (8-199), implies that ΔH is independent of temperature for a regular solution.

Eq. (8-202),

$$\mu_1 = RT \ln (\gamma_1 x_1) = RT(2x_2^2 + \ln x_1) \tag{8-203}$$

$$\left(\frac{\partial \mu_1}{\partial x_1}\right)_{T,P} = RT\left(-4x_2 + \frac{1}{x_1}\right) = 0 \text{ at } x_1 = 0.5 \tag{8-204}$$

$$\left(\frac{\partial^2 \mu_1}{\partial x_1^2}\right)_{T,P} = RT\left(4 - \frac{1}{x_1^2}\right) = 0 \text{ at } x_1 = 0.5 \tag{8-205}$$

$$\left(\frac{\partial^3 \mu_1}{\partial x_1^3}\right)_{T,P} = RT\left(\frac{2}{x_1^3}\right) > 0 \text{ for all } x_1 \tag{8-206}$$

Thus, $x_1 = 0.5$ is the critical solution concentration corresponding to $A = 2$. If the enthalpy of mixing were greater than $2x_1x_2(RT)$, the solution would exhibit an immiscible region centered about $x_1 = 0.5$.

Many mixtures are more complex than regular solutions. Enthalpies and entropies of mixing may have highly complex functionalities, as illustrated in Figure 8.9 for two mixtures of fairly dissimilar components, but for many systems there appear to be compensating factors in ΔH and ΔS^{EX} such that ΔG^{EX} is only slightly asymmetrical (see Figure 8.9). Thus, even for dissimilar components, ΔG^{EX} can often be correlated satisfactorily by a two- to five-term power series in mole fraction. For example, the four-constant Redlich-Kister expansion is:

$$\frac{\Delta G^{EX}}{RT(x_1x_2)} = A + B(x_1 - x_2) + C(x_1 - x_2)^2 + D(x_1 - x_2)^3 \tag{8-207}$$

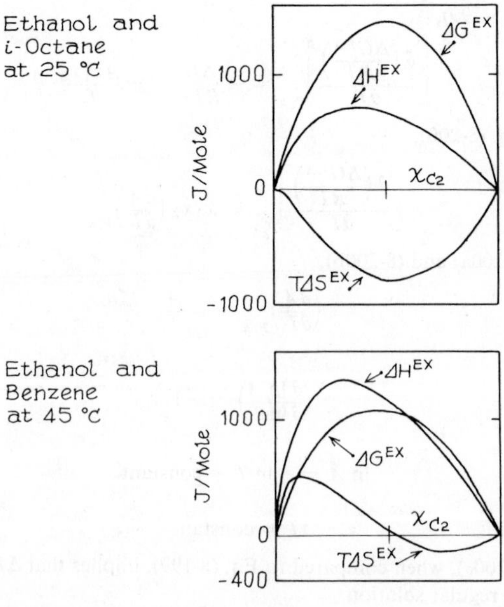

Figure 8.9 Excess functions for some binary mixtures.

where the factor (x_1x_2) is incorporated to ensure that ΔG^{EX} goes to zero at the extremes of pure 1 and pure 2. Substituting Eq. (8-207) into Eq. (8-196) and simplifying, we obtain

$$\ln \gamma_1 = x_2^2[A - B(4x_2 - 3) + C(2x_2 - 1)(6x_2 - 5) + D(2x_2 - 1)^2(8x_2 - 7)]$$

The applicability of expressing $\ln \gamma$ in a power series can be determined from a limited number of experimental data as follows. Given γ_1 and γ_2 for several different concentrations, ΔG^{EX} can be calculated using Eq. (8-181). These data may then be plotted as Figure 8.10a. By comparison to one of the curves given therein, it can be determined if one of the forms is more appropriate. Furthermore, if the one- or two-constant forms (i.e., two-suffix

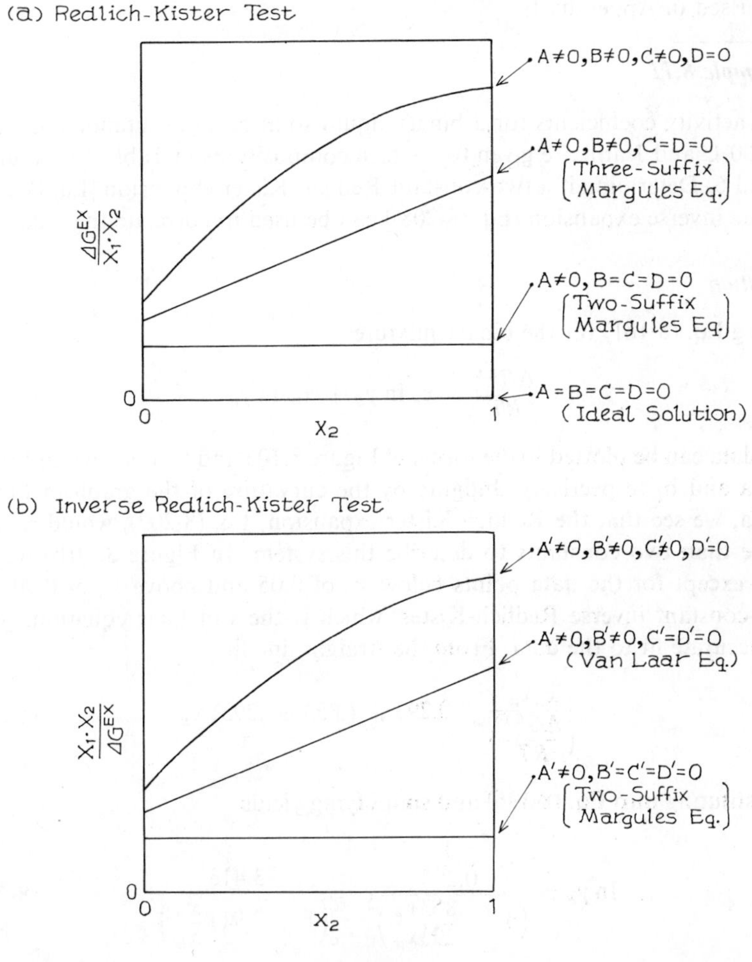

Figure 8.10

or three-suffix Margules equations) are appropriate, the values of the constants can be determined directly from the slope and intercept of the line best-fit through the data.

In a similar manner, another set of functions could be obtained by expanding ΔG^{EX} in an inverse power series, or

$$\frac{x_1 x_2}{\left(\frac{\Delta G^{EX}}{RT}\right)} = A' + B'(x_1 - x_2) + C'(x_1 - x_2)^2 + D'(x_1 - x_2)^3 \quad (8\text{-}208)$$

We shall refer to this expansion as the inverse Redlich-Kister equation. In this case, the van Laar equation corresponds to the two-constant expansion (see Figure 8.10b). Other correlating equations for activity coefficients are discussed in Appendix F.

Example 8.11

The activity coefficients for a binary liquid solution of *n*-butanol and water at 100°C and 1 atm are given for sixteen compositions in Table 9.1, columns 5 and 6. Determine if a two-constant Redlich-Kister expansion [Eq. (8-207)] or the inverse expansion [Eq. (8-208)] can be used to correlate these data.

Solution

Using Eq. (8-181) for the binary mixture,

$$\frac{\Delta G^{EX}}{RT} = x_B \ln \gamma_B + x_W \ln \gamma_W \quad (8\text{-}209)$$

the data can be plotted in the forms of Figure 8.10a and b, as shown in Figure 8.11a and b, respectively. Judging by the curvature of the graph in Figure 8.11a, we see that the Redlich-Kister expansion, Eq. (8-207), would require more than two constants to describe this system. In Figure 8.11(b) we see that except for the data points below x_B of 0.05 and above x_B of 0.90, the two-constant inverse Redlich-Kister, which is the van Laar equation, gives an accurate fit to the data. From the straight line fit,

$$\frac{x_B x_W}{\left(\frac{\Delta G^{EX}}{RT}\right)} = 0.293 + (.853 - .293)\, x_B \quad (8\text{-}210)$$

Substituting into Eq. (8-196) and simplifying yields

$$\ln \gamma_B = \frac{\frac{1}{0.293}}{\left(1 + \frac{.853 x_B}{.293 x_W}\right)^2} = \frac{3.413}{\left(1 + 2.911 \frac{x_B}{x_W}\right)^2} \quad (8\text{-}211)$$

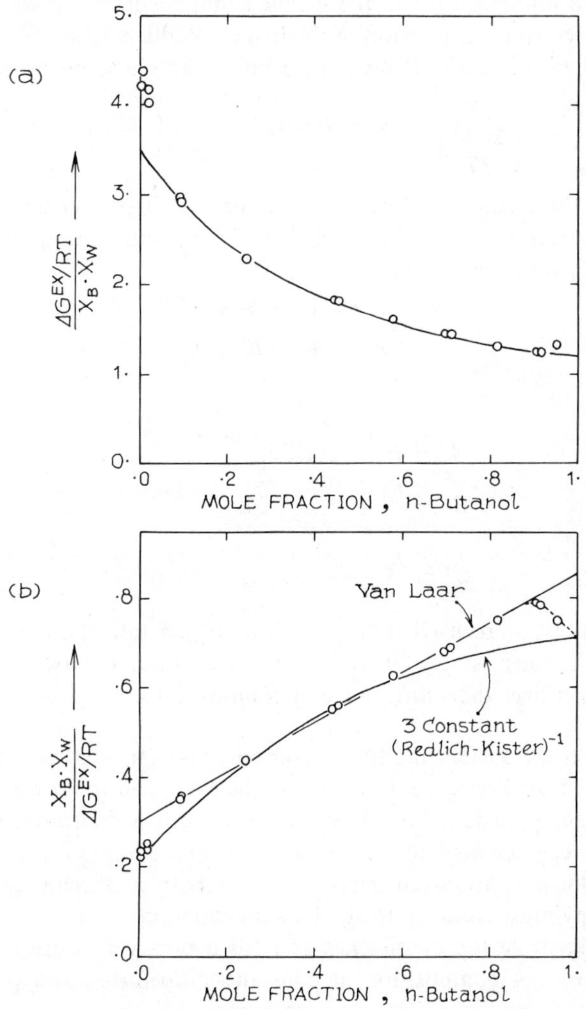

Figure 8.11 Redlich-Kister tests for *n*-butanol-water system at 100°C.

and

$$\ln \gamma_W = \frac{\dfrac{1}{0.853}}{\left(1 + \dfrac{.293}{.853}\dfrac{x_W}{x_B}\right)^2} = \frac{1.172}{\left(1 + 0.343\dfrac{x_W}{x_B}\right)^2} \tag{8-212}$$

The deviations at the two extremes may be real, or they may be anomalies because of the inaccuracies in analysis of the fairly low concentration of the lean component.

Although not called for in the problem statement, it is instructive to see how a three-constant expansion of the inverse Redlich-Kister form compares to the van Laar equation. If we rewrite Eq. (8-208) in terms of x_B,

$$I = \frac{x_B x_W}{\left(\frac{\Delta G^{EX}}{RT}\right)} = A' + B'(2x_B - 1) + C'(2x_B - 1)^2 \tag{8-213}$$

we can use three compositions to determine the three constants. Choosing the two intercepts, $I_0 = 0.225$ and $I_1 = 0.710$, and the equimolar value, $I_{1/2} = 0.575$, we have

$$I_0 = A' - B' + C$$
$$I_1 = A' + B' + C'$$
$$I_{1/2} = A'$$

or

$$A' = I_{1/2} = 0.575$$
$$B' = \frac{I_1 - I_0}{2} = 0.243$$
$$C' = \frac{I_1 + I_0}{2} - I_{1/2} = -0.107$$

These constants were used in Eq. (8-213) to generate the curve shown in Figure 8.11b. Surprisingly, it is less accurate than the two-constant van Laar equation over the entire range of compositions.

When activity coefficients for all components in a mixture are calculated or reported in the literature, the data can be examined for consistency. That is, the Duhem equation, Eq. (8-191) or (8-192), is a relationship involving all n activity coefficients and, hence, any one can be found if the other $n - 1$ activity coefficients are given. Hence, all n activity coefficients are not independent; they must conform to the Duhem equation.

Let us illustrate the consistency test for a binary mixture given γ_1 and γ_2 as functions of composition at constant temperature and pressure (see Figure 8.12). From Eq. (8-192) for the binary mixture,

$$x_1 \left(\frac{\partial \ln \gamma_1}{\partial x_1}\right)_{T,P} + x_2 \left(\frac{\partial \ln \gamma_2}{\partial x_1}\right)_{T,P} = 0 \tag{8-214}$$

or

$$x_1\, d\ln \gamma_1 + x_2\, d\ln \gamma_2 = 0 \qquad \text{(constant } T, P) \tag{8-215}$$

Eq. (8-215) is a differential test of the data; over any interval, dx_1, the shaded areas shown in Figure 8.12 must be equal.

A consistency test over all compositions can be made by integrating Eq. (8-215):

$$\int_{\ln \gamma_1|_{x_1=0}}^{\ln \gamma_1|_{x_1=1}} x_1\, d\ln \gamma_1 + \int_{\ln \gamma_2|_{x_2=1}}^{\ln \gamma_2|_{x_2=0}} x_2\, d\ln \gamma_2 = 0 \tag{8-216}$$

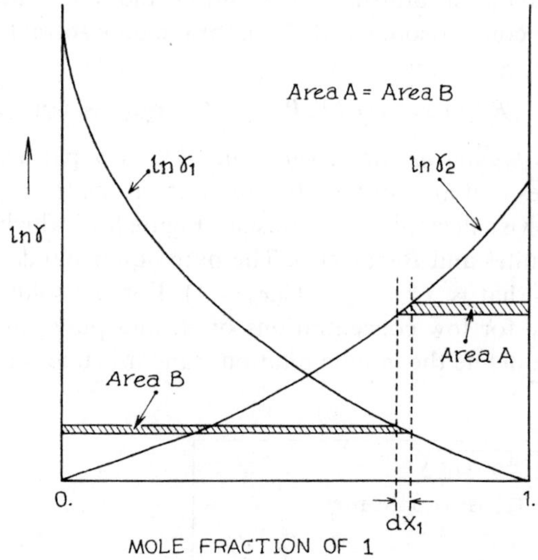

Figure 8.12 Differential test using the Duhem equation.

which is equivalent to:

$$\int_0^1 \ln \gamma_1 \, dx_1 + \int_1^0 \ln \gamma_2 \, dx_2 = 0$$

or, simply,

$$\int_0^1 \ln \frac{\gamma_1}{\gamma_2} \, dx_1 = 0 \tag{8-217}$$

If a set of data fails to satisfy the consistency test, it is always possible to smooth them to obtain a consistent set by first calculating ΔG^{EX} [i.e., using Eq. (8-181)], and then applying Eq. (8-196).

In the discussion above we have concentrated on the pure solvent activity coefficient, which is the form most commonly employed. As mentioned at the beginning of this section, there are other forms of activity coefficients that may be more convenient to use if one of the components does not exist as a pure liquid at the temperature and pressure in question.

In general, an activity coefficient can still be defined for any other standard state by the equation:

$$\hat{f}_i^{\alpha} = \hat{f}_i^0 \gamma_i^0 x_i \tag{8-218}$$

where $\hat{f}_i^0$ is the fugacity of i in the standard state. By analogy to Eq. (8-169),

$$RT \ln \gamma_i^0 + RT \ln x_i = \bar{G}_i^{\alpha}(T, P, x_1, \ldots, x_{n-1}) - \bar{G}_i^0(T, P^0, x_1^0, \ldots, x_{n-1}^0) \tag{8-219}$$

Note again the requirement that the standard-state temperature be equal to that of the system but the pressure, composition, and state of aggregation may be different. For comparison, Eq. (8-219), for the pure solvent standard state, is:

$$RT \ln \gamma_i + RT \ln x_i = \bar{G}_i^\alpha(T, P, x_1, \ldots, x_{n-1}) - G_i^\alpha(T, P) \qquad (8\text{-}220)$$

The *infinite dilution activity coefficient*, γ_i^{**}, is a particularly common choice for solutes that are not liquids when pure at the temperature in question. In order to visualize this state, consider Figure 8.13 which is drawn for a binary of A (solute) and B (solvent). The pure liquid standard state is used for the solvent, that is, $\hat{f}_B^0 = f_B = \hat{f}_B(x_B = 1)$. For the solute A, since data for $\hat{f}_A$ exist only for low concentrations of A, and pure liquid A does not exist at T, P, we define the infinite dilution standard state for A as the inter-

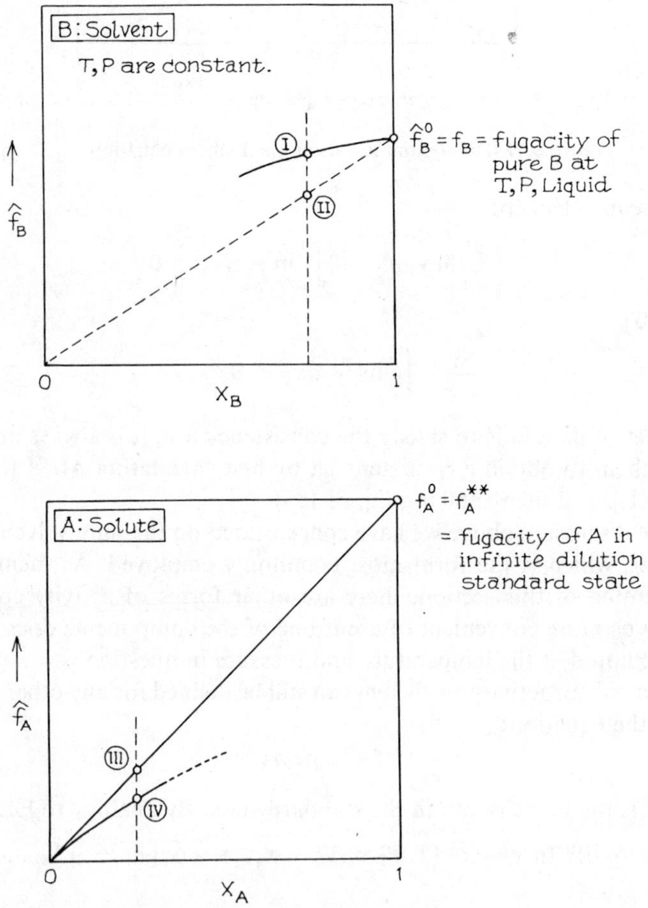

Figure 8.13 Standard states.

section of a tangent drawn to the $\hat{f}_A - x_A$ curve as $x_A \longrightarrow 0$ with the ordinate $x_A = 1$. Thus, to determine $\hat{f}_A^0 = \hat{f}_A^{**}$, data at a low concentration of A are all that are required.

In both cases one can express the activity coefficient in terms of the reference state fugacity:

$$\gamma_B = \frac{\hat{f}_B}{x_B f_B} \tag{8-221}$$

$$\gamma_A^{**} = \frac{\hat{f}_A}{x_A f_A^{**}} \tag{8-222}$$

A pictorial representation of the activity coefficient may also be noted on Figure 8.13 as the ratios:

$$\gamma_B = \frac{\hat{f}_B^{\mathrm{I}}}{\hat{f}_B^{\mathrm{II}}}; \quad \gamma_A^{**} = \frac{\hat{f}_A^{\mathrm{IV}}}{\hat{f}_A^{\mathrm{III}}} \tag{8-223}$$

In these two cases, the use of a standard state as a pure solvent is often called symmetrical normalization, but if the infinite dilution standard state is used for the solute, we say this component is unsymmetrically normalized. It is obvious that:

$$\left.\begin{array}{lll} \text{symmetrical:} & \gamma_i \longrightarrow 1, & x_i \longrightarrow 1 \\ \text{unsymmetrical:} & \gamma_i^{**} \longrightarrow 1, & x_i \longrightarrow 0 \end{array}\right\} \tag{8-224}$$

Other choices of standard states may be used. Many are discussed by Prausnitz.[20] The basic thermodynamic relations for all cases are similar but care must be taken to denote the reference state. For example, if the infinite dilution reference state were chosen for component j, Eq. (8-182) would become:

$$\left(\frac{\partial \ln \gamma_j^{**}}{\partial T}\right)_{P,x} = -\frac{\overline{\Delta H}_j}{RT^2} = -\frac{[\bar{H}_j - \bar{H}_j^{**}]}{RT^2}. \tag{8-225}$$

where $\bar{H}_j^{**}$ is the partial molal enthalpy of j in an infinitely dilute solution. For binary mixtures, this simple definition suffices; for multicomponent mixtures, a more explicit definition of the infinite-dilution standard state is necessary since as $x_j \longrightarrow 0$, $\bar{H}_j^{**}$ can vary depending on the relative amounts of the other constituents.

PROBLEMS

8.1. Comment on the ideality of a water-acetic acid solution at 15°C. The density of the solution is given below at this temperature.

[20] J. M. Prausnitz, *Molecular Thermodynamics of Fluid-Phase Equilibria* (Englewood Cliffs, N. J.: Prentice-Hall, Inc., 1969), Chap. 6.

Weight percent acetic acid	Density g/cm³
0	0.9991
10	1.0141
20	1.0283
30	1.0411
40	1.0522
50	1.0613
60	1.0684
70	1.0732
80	1.0747
90	1.0708
100	1.0545

8.2. The compressibility of nitrogen-helium gas mixtures is shown in Figure P8.2 at 3000 psia and 70°F. Estimate the value of $\bar{V}_{N_2}$, $\bar{V}_{He}$, $\overline{\Delta V}_{N_2}$, and $\overline{\Delta V}_{He}$ at a composition of 40 mole percent nitrogen.

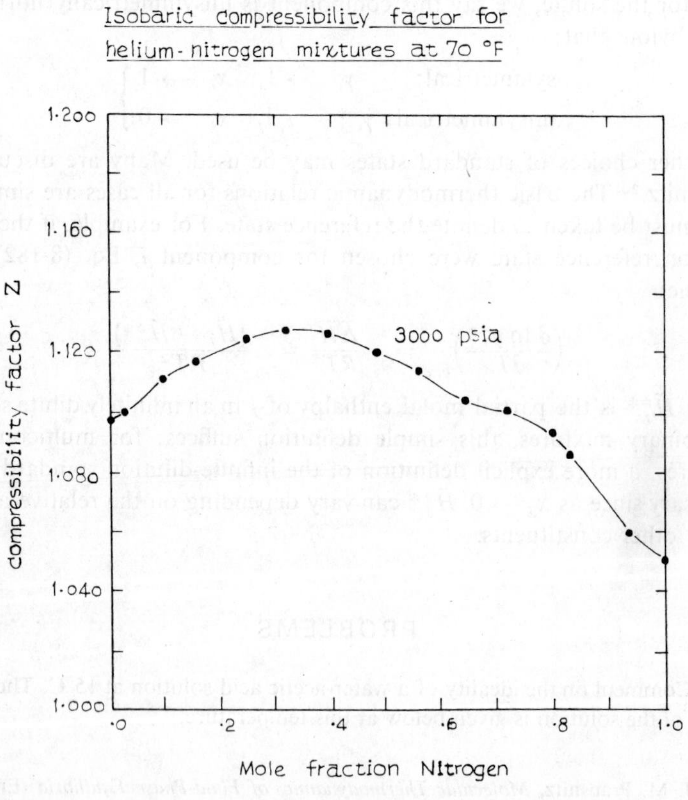

Figure P8.2

8.3. In a reported experiment a mixture of helium and ammonia was prepared as described below. With only this description and the given data, what is your best engineering estimate of the composition of the mixture removed from valve *C*?

As shown in Figure P8.3, there are separate supply manifolds for the helium and ammonia. The mixing tank is first evacuated to a very low pressure. Helium gas is then admitted very rapidly until the tank pressure is at 2 atm. The helium supply valve is then closed. Ten minutes later, the ammonia supply valve is opened to allow ammonia to flow rapidly into the tank. The valve is closed when the tank pressure reaches 3 atm. A day later, after diffusive mixing, the gas mixture is drawn off through valve *C*.

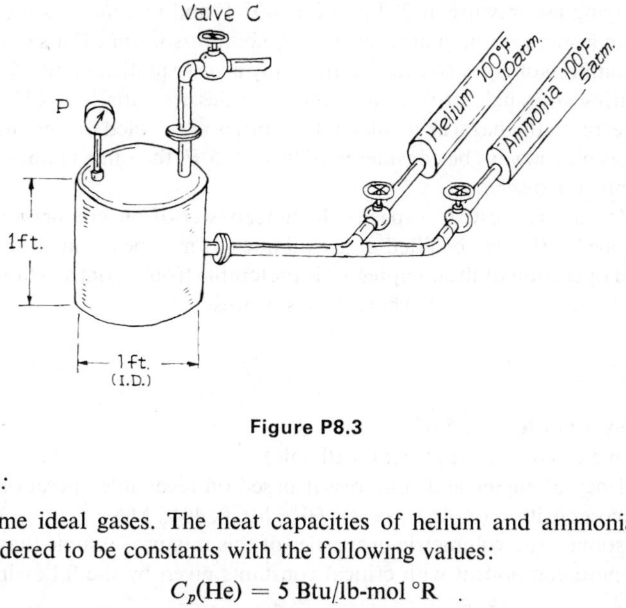

Figure P8.3

Data:

Assume ideal gases. The heat capacities of helium and ammonia may be considered to be constants with the following values:

$$C_p(\text{He}) = 5 \text{ Btu/lb-mol °R}$$
$$C_p(\text{NH}_3) = 8.5 \text{ Btu/lb-mol °R}$$

The tank dimensions are shown in Figure P8.3. Also, the tank wall thickness is 0.5 in.; it is originally at 100°F. C_p (aluminum) = 0.24 Btu/lb °R, ρ = 170 lb/ft³.

8.4. In Chapter 6 derivative compressibility factors were introduced and a correlation was given to relate these to the reduced temperature, pressure, and acentric factor in Appendix D. If, for a mixture, the mixture acentric factor and critical properties are assumed to be mole fraction averages, derive an expression relating $\bar{V}_j$ to Z_T, Z_p, Z, the critical properties, and acentric factor of the mixture.

Apply this relation to estimate the partial volume of methane in a 70.7 mole % methane, 29.3 mole % *n*-butane mixture at 192°F and 509 psia. Can you also suggest a way to determine the mixture specific volume?

Data:

Component	T_c, °R	P_c, psia	ω
CH_4	343.2	673	0.013
n-C_4H_{10}	765.5	550.7	0.201

8.5. In the design of a petrochemical plant, a mixture of methylcyclopentane and methylacetylene is to be compressed into storage bottles for shipment to customers. It is planned to take suction from a constant pressure header supplying the mixture at 200 psia and 405°F and to compress the gas to 400 psia in a small isothermal reciprocating compressor unit. For safety purposes the compressor is constructed so that only a very small amount of gas is taken in during any single stroke. The receiver bottles are initially at 200 psia, 405°F. To decrease the hazards involved, the bottles are cooled during filling so that the gas may always be considered to be at 405°F, the same temperature as the compressor discharge.

You are requested to specify the horsepower of the compressor motor so that one bottle may be filled in five minutes. It may be assumed that constant speed operation of the compressor is preferable from a control standpoint. No liquid phase appears anywhere in this process.

Data:

Receiver bottle size: 5 ft^3
Volume of connecting lines: (negligible)
Efficiency of engine and compressor based on reversible operation: 90%
Gas composition (mole percent): 60% MCP, 40% MA
To estimate the volumetric properties of this mixture, assume that it behaves as a pure component with critical constants given by the following rules:

T_c, V_c, Z_c are mole fraction averages of the pure components; $P_c = Z_c R T_c / V_c$

8.6. In a mixture of two gases, A and B, at 40°C and 70 atm,

$$\ln f_m = 4.25 - 0.235 x_A$$

with f_m in atm and x_A the mole fraction A.

(a) What are the fugacities of A and B in a mixture containing 20 mole % A?
(b) Derive an expression relating f_m to x_A for both an ideal solution and an ideal gas mixture.

8.7. Using the data in the plot of solution enthalpies of ethyl alcohol-benzene mixtures at 200 psia and 300°F, calculate and plot the partial molal enthalpies of both components as a function of composition (see Figure P8.7).

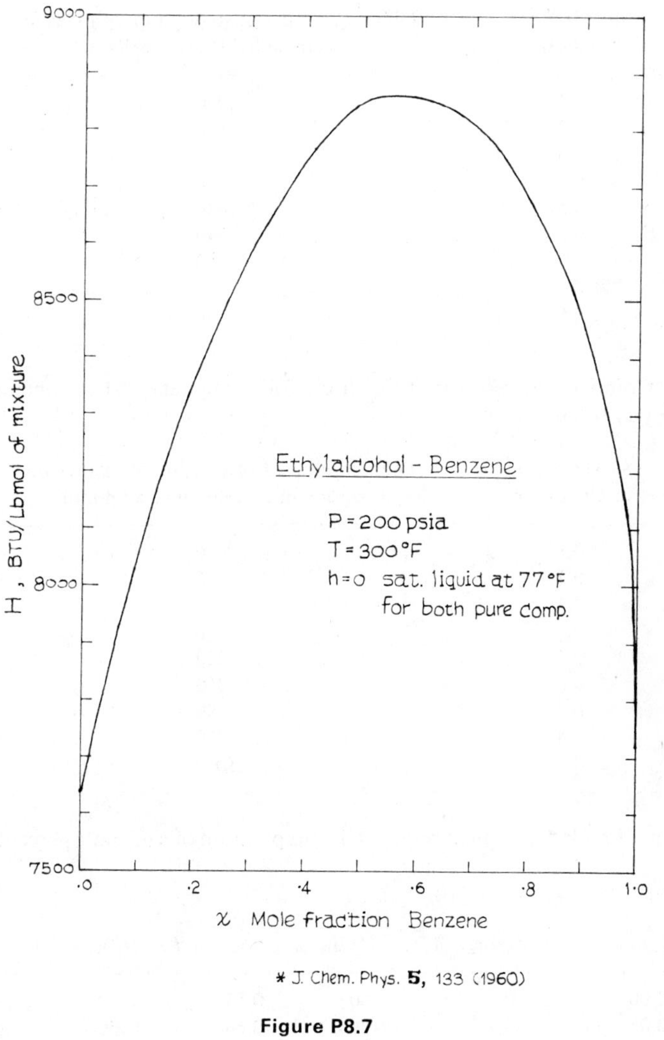

Figure P8.7

8.8. Given only the differential heat of solution of cyclohexane in methylethyl ketone at 18°C as determined by M. B. Donald and K. Ridgway, "The Binary System Cyclohexane-Methylethyl Ketone," *Chem. Eng. Sci.*, **5**, (1956), 188. What is:

(a) the integral heat of solution as a function of mole fraction cyclohexane.

(b) the differential heat of solution of methylethyl ketone in cyclohexane as a function of mole fraction cyclohexane.

Mole fraction cyclohexane	Differential heat of solution of cyclohexane in methylethyl ketone, cal/g-mol
0.1	990
0.2	850
0.3	700
0.4	570
0.5	440
0.6	340
0.7	210
0.8	115
0.9	60

(c) Compare the result in (b) with the following data. What conclusions do you reach?

Mole fraction cyclohexane	Differential heat of solution of methylethyl ketone in cyclohexane, cal/g-mol
0.1	20
0.2	45
0.3	80
0.4	130
0.5	220
0.6	350
0.7	490
0.8	720
0.9	1070

8.9. A small-scale experiment requires the preparation of a 40 mole percent sulfuric

Data:

$x_{H_2SO_4}$	$(\bar{H}-\bar{H}^0)_{H_2O}$	$(\bar{H}-\bar{H}^0)_{H_2SO_4}$	$x_{H_2SO_4}$	$(\bar{H}-\bar{H}^0)_{H_2O}$	$(\bar{H}-\bar{H}^0)_{H_2SO_4}$
0.00	0	0	0.55	−5730	19040
0.05	−43.7	4130	0.60	−6300	19530
0.10	−293.3	7730	0.65	−6690	19740
0.15	−580.0	9310	0.70	−7010	19910
0.20	−1000	11190	0.75	−7280	20027
0.25	−1450	12680	0.80	−7490	20098
0.30	−1910	13970	0.85	−7700	20136
0.35	−2470	15130	0.90	−7870	20153
0.40	−3060	16160	0.95	−8050	20172
0.45	−3880	17240	1.00	−8225	20200
0.50	−4850	18310			

acid solution from pure water and 80 mole percent sulphuric acid. The mixture is to be prepared by metering one fluid at a constant molal rate into a tank containing the second fluid. Cooling is to be provided to maintain the bath at a constant temperature. Which fluid should be charged to the reactor first in order to minimize the *rate* of heat release? What is the peak differential heat load per mole of fluid added?

What is the total heat of mixing per mole of final solution for both methods? At the temperature of operation, partial molal enthalpies are listed above. The $\bar{H}^0$ values refer to a state of infinite dilution (i.e., $x_{H_2O} \gg x_{H_2SO_4}$). The values of $\bar{H}$ are in cal/g-mol and x is mole fraction.

8.10. One is faced with the problem of diluting a 90 weight percent H_2SO_4 solution with water in the following manner. A tank contains 1000 lb of pure water at 77°F; it is equipped with a cooling device to remove the heat of mixing. This cooling device operates with a boiling refrigerant reflux condenser

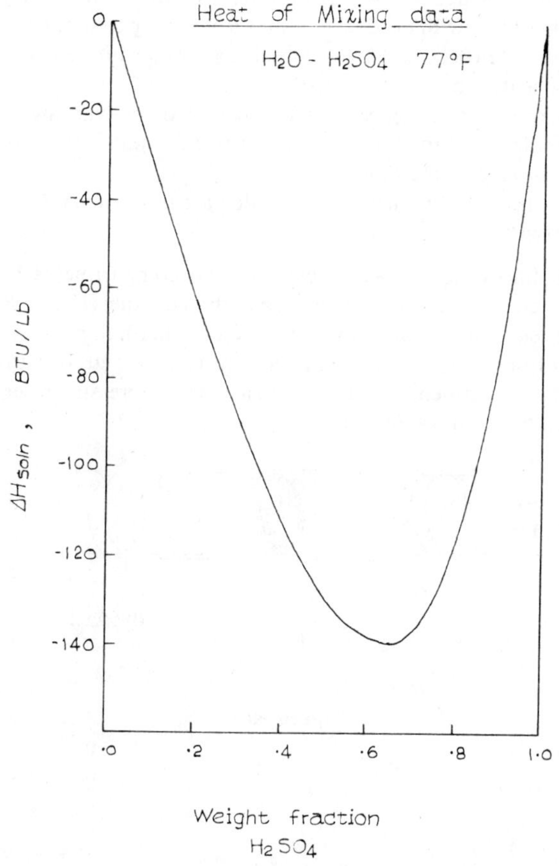

Figure P8.10

system to keep the temperature always at 77°F. Because of the peculiarities of the system, the rate of heat transfer, Btu/hr, must be constant. We wish to add 3000 lb of the acid solution (at a variable rate) in one hour. The acid is at 77°F. Heat of mixing data are shown in Figure P8.10.

(a) What is the total heat transferred from the system?

(b) Derive a differential equation to express the mass flow of 90 weight percent acid, lb/min, as a function of the concentrations of acid and water in the solution.

(c) What is the mass flow rate of 90 weight percent acid that you would specify when the overall tank liquid is 65 weight percent acid?

8.11. A constant volume bomb is available for use in experiments to study heat effects in the mixing of gases. The bomb is divided internally by a thin diaphragm and may be filled initially with pure gases in each section. The diaphragm is movable and different ratios of feed gases may be used.

A constant temperature bath at T_0 surrounds the bomb and is so equipped that any flow of heat to or from the bomb may be accurately determined. The bomb is filled with pure gases A and B at some pressure P and temperature T_0, the bomb closed, and the internal diaphragm broken so that the fluids mix. Any heat transferred is noted.

This experiment is repeated a number of times, varying only the initial ratio of moles of A to B. In no case is there a liquid phase present. Assume complete mixing of the fluids.

From the data obtained, can one calculate $\overline{\Delta U}_A$ and $\overline{\Delta U}_B$ as a function of composition? Explain.

8.12. A horizontal cylinder is separated into two compartments by a rigid, diathermal membrane permeable only to carbon dioxide (Figure P8.12). The left compartment is in contact—by means of a diathermal, movable piston—with a constant-pressure, constant-temperature reservoir at 100 atm and 373°K. The right compartment is connected to a constant-pressure reservoir of 50 atm by an adiabatic, movable piston.

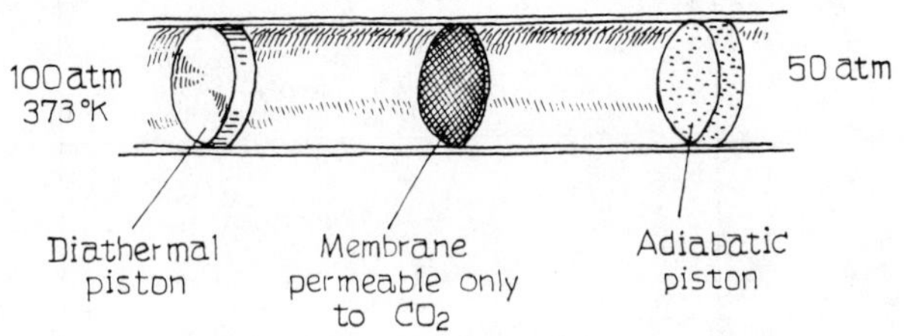

Figure P8.12

Both compartments are at equilibrium and contain mixtures of carbon dioxide and ethane. The mole fraction of carbon dioxide in the left compartment is 0.25. What is your best engineering estimate of the mole fraction of carbon dioxide in the right compartment? Use each of the following assumptions:

(a) Both gases are ideal and form ideal solutions.

(b) Gases form ideal solutions, but both pure gases and the mixture follow the Berthelot's equation of state:

$$Z = \frac{PV}{RT} = 1 + \left(\frac{9}{128}\right)\left(\frac{P_r}{T_r}\right)\left(1 - \frac{6}{T_r^2}\right)$$

(c) The pure gases are not ideal nor do the gases form an ideal solution.

8.13. It is often convenient to represent approximately the volumetric properties of a binary gas mixture with a virial equation of state as follows:

$$Z_{\text{mix}} = 1 + \frac{B_{\text{mix}} P}{RT}$$

Similar relations may be written for the pure components using B_A and B_B for components A and B. The second virial coefficient for the mixture is given as:

$$B_{\text{mix}} = y_A^2 B_A + y_A y_B B_{AB} + y_B^2 B_B$$

Here y is a mole fraction and B_{AB} the interaction second virial. Often a crude approximation is used to estimate these second virials in terms of reduced temperature, i.e.,

$$\frac{B_A}{V_{c_A}} = f_1\left(\frac{T}{T_{c_A}}\right)$$

$$\frac{B_B}{V_{c_B}} = f_1\left(\frac{T}{T_{c_B}}\right)$$

$$\frac{B_{AB}}{V_{c_{AB}}} = f_1\left(\frac{T}{T_{c_{AB}}}\right)$$

where f_1 is a universal function and

$$V_{c_{AB}} = \frac{1}{8}(V_{c_A}^{1/3} + V_{c_B}^{1/3})^3$$

$$T_{c_{AB}} = (T_{c_A} T_{c_B})^{1/2}$$

With these relations, derive an expression to determine the activity coefficients of A and B. How is this relationship changed if the gases form an ideal solution?

8.14. A recent text on thermodynamics states that for a single-phase mixture of k components,

$$\sum_k N_k \left(\frac{\partial P}{\partial N_k}\right)_{T,\underline{V}} = \frac{1}{\kappa_T}$$

where N_k = moles of component k present.

Derive this relation and state any assumptions regarding the properties of the mixture. What is your physical interpretation of the term

$$N_k\left(\frac{\partial P}{\partial N_k}\right)_{T,\underline{V},N_1,N_2,\ldots}$$

for an ideal gas mixture?

8.15. Show for any mixture containing $N_1, N_2, \ldots, N_n$ moles of each component

$$\left(\frac{\partial T}{\partial N_k}\right)_{P,\underline{V},N_i[k]} = \frac{1}{y_k N}\left(\frac{\partial T}{\partial \ln V}\right)_{N,P} - \sum_{j\neq k}\frac{y_j}{y_k}\left(\frac{\partial T}{\partial N_j}\right)_{P,\underline{V},N_i[j]}$$

What is the derivative for an ideal gas mixture?

8.16. Derive a general expression for the rate of temperature change per mole of component j added to a multicomponent mixture in which the total entropy and pressure of the mixture are held constant. Assume that the tank contents are well-mixed.

In a tank of steam at 30 psia and 400°F, the total entropy is 1.7937 Btu/°R using the Keenan and Keyes Steam Tables as a reference source (and, thus, the base value of $S = 0$, saturated liquid water at 32°F). Suppose a small amount of steam at 30 psia and 400°F were injected into the tank under conditions that the total entropy and pressure remained constant.

(a) Estimate the instantaneous rate of temperature change, °F/lb of steam added.
(b) What would this value be if, with a suitable choice of the base state of entropy, we kept the *total* entropy constant at 0.0 Btu/°R? Is this answer different from that of part (a)? How do you reconcile your answer?

8.17. (a) Derive an expression to compute the change in chemical potential of component k for a multicomponent, nonideal mixture as the moles of k are varied, keeping temperature, pressure, and the moles of all other components constant, i.e.,

$$\left(\frac{\partial \mu_k}{\partial N_k}\right)_{T,P,N_i[k]}$$

Express your results in terms of quantities that may be easily measured in laboratory experiments. Explain how each is determined.

(b) For a mixture of CO, H_2, and CH_4 containing 3 : 1 : 4 g-mol of each, determine

$$\left(\frac{\partial \mu_{CH_4}}{\partial N_{CH_4}}\right)_{T,P,N_{H_2},N_{CO}}$$

at 300°K, 1 atm. Assume ideal solutions.

8.18. Calculate the osmotic pressure across an ideal semipermeable membrane (permeable to water) for the system NaCl-H_2O: H_2O at 25°C and at pressures of 1 bar and 500 bars. Consider the concentration range from 0 to 25 weight percent NaCl.

Carry out the calculations for the following cases:

(a) No additional assumptions allowed.
(b) Assume that the NaCl-H_2O solution is ideal.

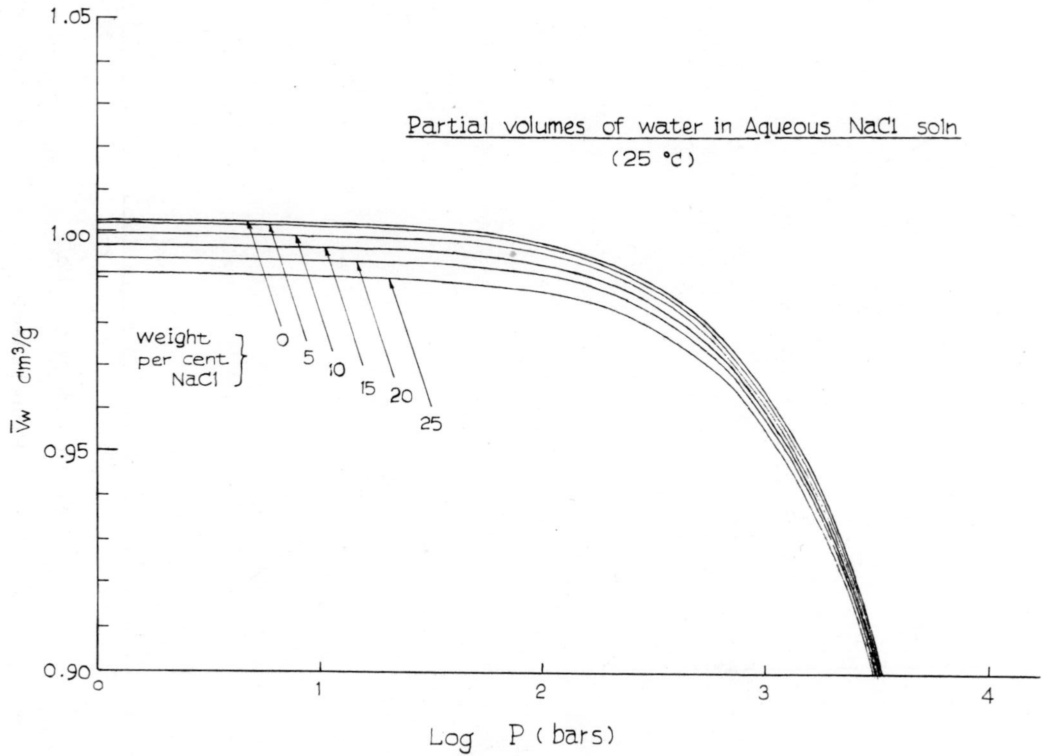
Partial volumes of water in Aqueous NaCl soln
(25 °C)
$\bar{V}_w$ cm³/g
1.05
1.00
0.95
0.90
0
1
2
3
4
Log P (bars)
weight per cent NaCl
0
5
10
15
20
25

Figure P8.18a

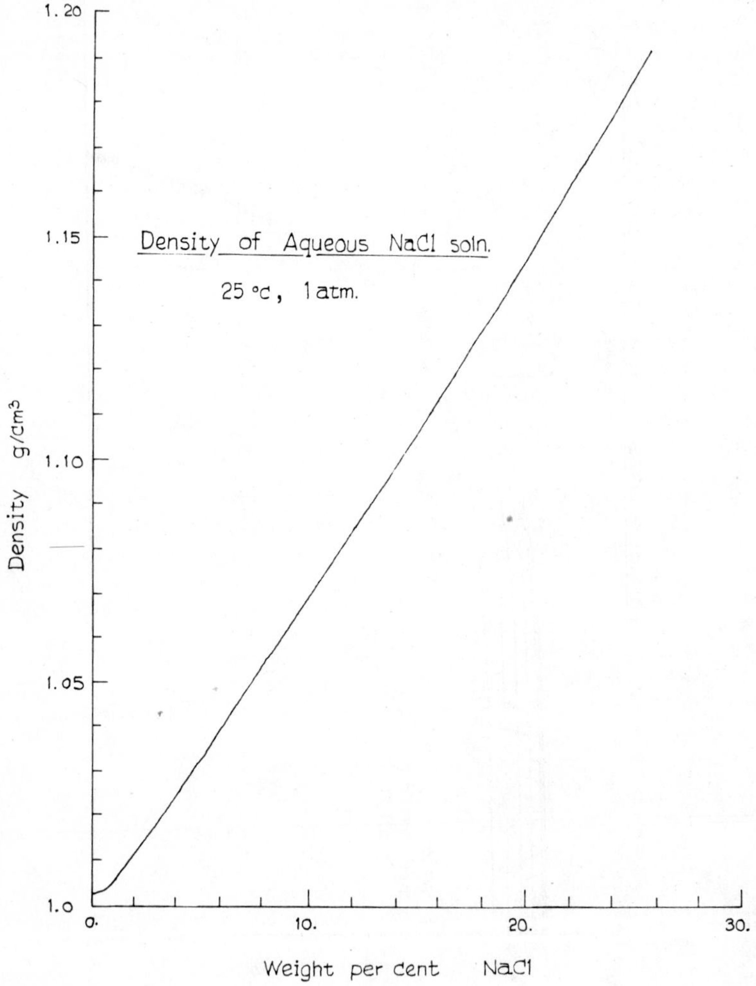

Figure P8.18b

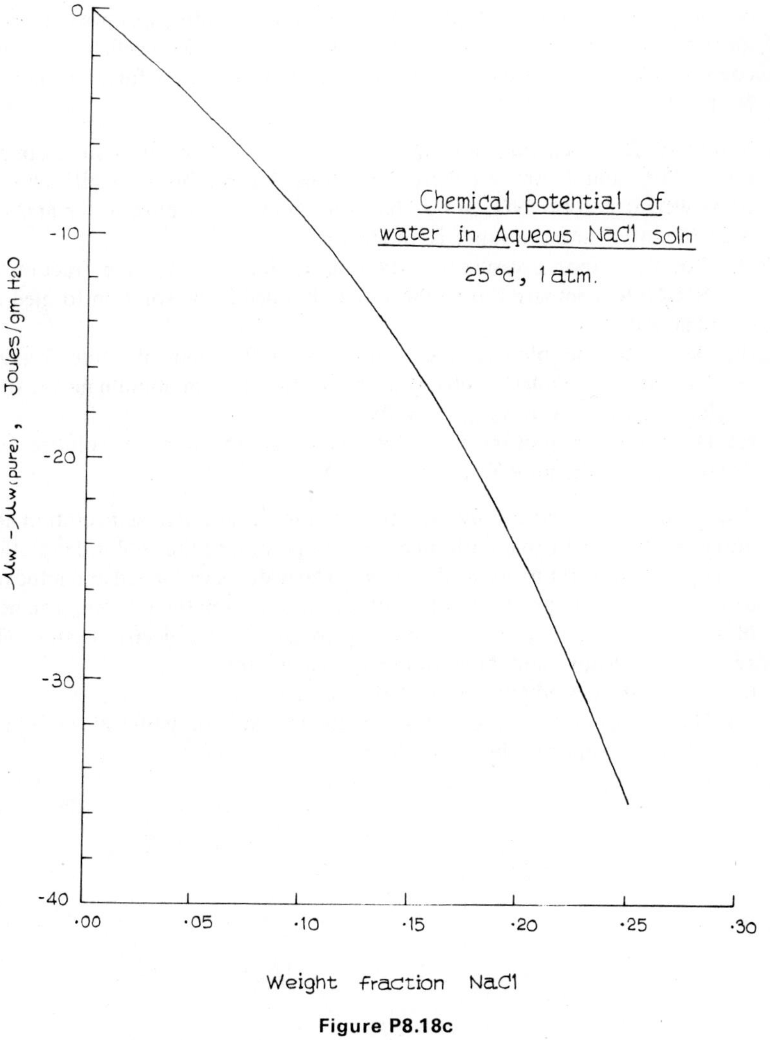

Figure P8.18c

(c) Assume that over the entire pressure and concentration range $\bar{V}_w = V_w$.
(d) Assume (b), (c), and that $V_w \neq f(P)$.
(e) Assume (b), (c), (d), and that $\ln x_w \cong -x_s$.
(f) Assume (b), (c), (d), (e), and that $\underline{V} = N_{\text{total}} V_w$.
Figures P8.18a–d provide useful data.

8.19. A "regular" solution is a term often applied to a liquid mixture in which the entropy of mixing for all components is equal to the entropy of mixing for an ideal solution. From this definition, express the activity coefficient of any component, j, at constant composition and pressure, as a function only of temperature.

8.20. Values of $(\mu - \mu_{\text{pure}})_{H_2O}$ and $(\mu - \mu_{\text{sat soln}})_{NH_4NO_3}$ are given in Example 8.8 at 25°C and 1 atm. At these conditions, the solubility of NH_4NO_3 is 67.63 weight percent NH_4NO_3. The vapor pressure of pure water at 25°C is 23.76 mmHg and the vapor is an ideal gas.
(a) Plot the vapor pressure of water as a function of mole fraction of NH_4NO_3; compare this to the value obtained if the solution formed an ideal solution.
(b) Determine and plot γ_w and $\gamma_{NH_4NO_3}$ as a function of mole fraction NH_4NO_3. Use unsymmetrical normalization for ammonium nitrate (i.e., let $\gamma_{NH_4NO_3} \longrightarrow 1$ as $x_{NH_4NO_3} \longrightarrow 0$).
(c) Determine and plot the excess free energy of mixing for the solution as a function of the mole fraction NH_4NO_3.

8.21. A constant volume crystallizer vessel containing a saturated salt solution at 1 atm is being fed with an unsaturated 5 weight percent sodium chloride solution at the same temperature as inside the pot. The water is vaporized and removed at 1 atm, and simultaneously solid salt is removed from the bottom. The heat of vaporization and crystallization is provided by an electric heater. The system is in steady state. How would you calculate:
(a) The entropy production rate of the universe.
(b) The rate of entropy production for the system, consisting of the liquid, solid, and vapor in the crystallizer.

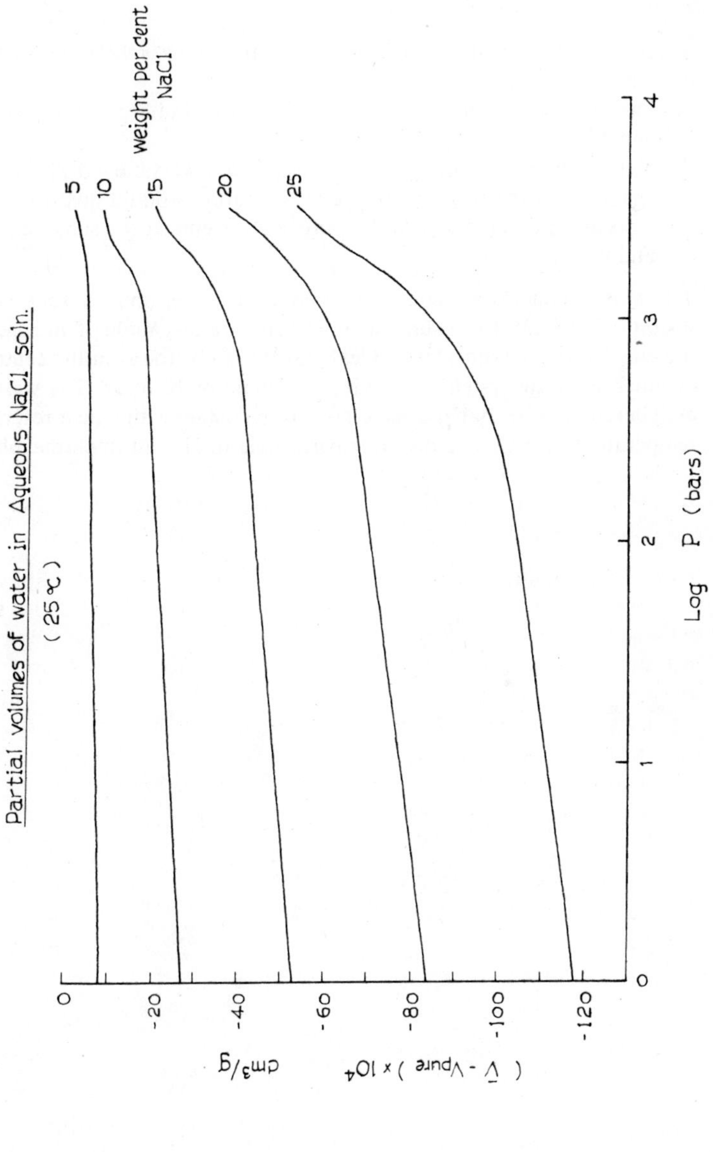

Partial volumes of water in Aqueous NaCl soln.
(25 °C)
Weight per cent NaCl
5
10
15
20
25
0
-20
-40
-60
-80
-100
-120
(V̄ - Vpure) x 10^4 dm^3/g
0
1
2
3
4
Log P (bars)

Figure P8.18d

(c) The "internal" entropy production rate.

8.22. The activity coefficient of stannic chloride in solution in normal heptane can be described to a good approximation by

$$T \ln \gamma = 820x^2$$

where T is in °K, x is the mole fraction heptane, and γ is the activity coefficient of stannic chloride.

(a) Obtain an expression for the activity coefficient of heptane at a constant T and P.

(b) Derive an expression giving the isothermal reversible work, the enthalpy change, and the entropy change in separating a liquid mixture of stannic chloride and heptane into the pure components at the same temperature and pressure.

8.23. Two simple systems are contained within a cylinder and are separated by a piston (Fig P8.23). Each subsystem is a mixture of $\frac{1}{2}$ mole of nitrogen and $\frac{1}{2}$ mole of hydrogen (consider as ideal gases). The piston is in the center of the cylinder, each subsystem occupying a volume of 10 liters. The walls of the cylinder are diathermal and the system is in contact with a heat reservoir at a temperature of 0°C. The piston is permeable to H_2 but impermeable to N_2.

How much work is required to push the piston to such a position that the volumes of the subsystems are 5 and 15 liters?

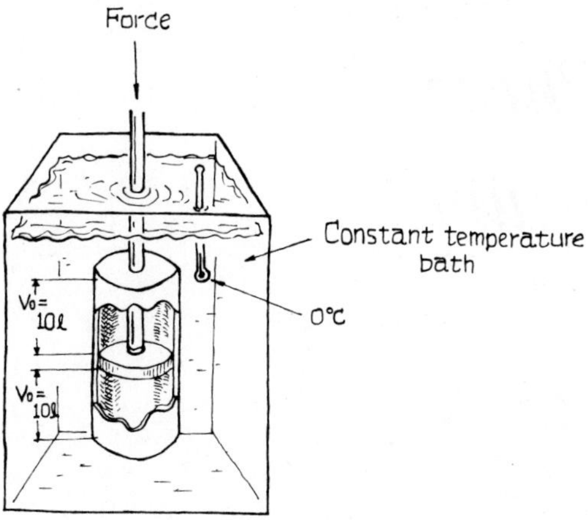

Figure P8.23

Phase Equilibrium

9

The criteria for coexistence of phases in equilibrium were developed in Section 7.3, and it was shown there that the temperatures, pressures, and component chemical potentials were equal in all phases. These criteria are valid even if chemical reactions occur in one or more phases. They may not, however, necessarily be valid if there are any constraints to the flow of mass or energy between phases. Thus, it is often stated that such criteria only apply in the general sense for *simple*, multiphase systems. Example 9.1, discussed later, illustrates a *nonsimple* system.

It is imperative to note that the equilibrium criteria stated above always form the starting point for further developments in phase equilibrium. This will become apparent in the treatment in this chapter.

9.1 The Phase Rule

In Section 2.7 it was stated in Postulate I that $(n + 2)$ *independently variable* properties characterize completely the stable equilibrium state of a simple system. We admitted that there might arise cases in which properties were *not* independently variable, although we have not yet met such cases (with the trivial exception of excluding one mole fraction from a set of n in a mixture). In fact, we found in Section 4.7 that *any* $(n + 2)$ properties were independently variable for a single-phase, nonreacting system.

We are now interested in extending our treatment to multiphase systems in order to determine the minimum set of variables to describe completely such systems both in extent and intensity. We will find even here that a general result may only be found for simple, composite systems. For such a system with π phases let us describe each phase separately and then relate all properties by invoking the criteria of phase equilibrium. If we were to apply Postulate I to each phase separately, we could choose any $(n + 2)$ properties of each phase, provided that each set includes no more than $(n + 1)$ intensive variables (see Section 5.2).

For each phase, a particularly convenient set of $(n + 2)$ properties is

$$T^\gamma, P^\gamma, x_1^\gamma, \ldots, x_{n-1}^\gamma, N^\gamma \tag{9-1}$$

where superscript γ is used as a dummy index to denote a phase.

For a composite system containing π phases we have a set such as Eq. (9-1) for each phase, or $\pi(n + 2)$ properties. To determine which of these $\pi(n + 2)$ properties are not independent, we must apply the criteria of phase equilibria, which for this system are

$$T^\alpha = T^\beta = \ldots = T^\gamma = \ldots = T^\pi \tag{9-2}$$

$$P^\alpha = P^\beta = \ldots = P^\gamma = \ldots = P^\pi \tag{9-3}$$

$$\mu_j^\alpha = \mu_j^\beta = \ldots = \mu_j^\gamma = \ldots = \mu_j^\pi \quad (j = 1, \ldots, n) \tag{9-4}$$

There are $(\pi - 1)$ equalities in each of Eqs. (9-2) and (9-3) and $n(\pi - 1)$ in Eq. (9-4) for a total of $(n + 2)(\pi - 1)$. If we transform this set of $(n + 2)$ $(\pi - 1)$ equalities into relations containing only the properties in the $\pi(n + 2)$ set of the kind given in Eq. (9-1), then the $(n + 2)(\pi - 1)$ equalities can be used as restraining equations to determine the relationships between the $\pi(n + 2)$ properties of the coexisting phases.

Let us carry out this transformation. We see that the set of variables in Eq. (9-1) is particularly convenient because we need transform only the chemical potentials. Specifically, we can expand these by relations of the kind that we have used in Chapter 8:

$$\mu_j^\gamma = g_j^\gamma(T^\gamma, P^\gamma, x_1^\gamma, \ldots, x_{n-1}^\gamma) \tag{9-5}$$

Thus, each of the equalities of Eq. (9-4) takes the form

$$g_j^\alpha(T^\alpha, P^\alpha, x_1^\alpha, \ldots, x_{n-1}^\alpha) = g_j^\beta(T^\beta, P^\beta, x_1^\beta, \ldots, x_{n-1}^\beta) \tag{9-6}$$

and we have $n(\pi - 1)$ such relations. The $(n + 2)(\pi - 1)$ restraining equations of Eqs. (9-2), (9-3), and (9-6) involve only $\pi(n + 1)$ *intensive* variables of the $\pi(n + 2)$ properties in Eq. (9-1). Clearly, then, the extensive variables included in the set of Eq. (9-1) (i.e., $N^\alpha, N^\beta, \ldots, N^\gamma, \ldots, N^\pi$) are not related by the criteria of phase equilibria and, hence, are independently variable.

Of the remaining $\pi(n + 1)$ intensive variables of the set given by Eq.

(9-1), we have $(n + 2)(\pi - 1)$ restraining equations so that the number of independently variable intensive properties is $\pi(n + 1) - (n + 2)(\pi - 1) = (n + 2 - \pi)$. Thus, we conclude that in order to describe completely the composite simple system, there are $(n + 2 - \pi)$ independently variable intensive properties and π independently variable extensive properties for the total of $(n + 2)$ variables required by Postulate I.

Although we have derived this result using specific sets of properties, the result is of general validity and not restricted to any particular set of intensive and extensive variables. For example, one could readily show that instead of the set of Eq. (9-1), we could have started with the set $T^\gamma, P^\gamma, \bar{S}_1^\gamma, \ldots, \bar{S}_{n-1}^\gamma, \underline{S}^\gamma$. In place of the expansion of Eq. (9-5), we would then have to use $\mu_j^\gamma = g_j(T^\gamma, P^\gamma, \bar{S}_1^\gamma, \ldots, \bar{S}_{n-1}^\gamma)$. Additional transformations are required if the chosen set contains intensive variables other than T^γ and P^γ, but the final result is unchanged.

The general conclusion, which is one of the most important results of thermodynamics of multiphase systems, can be stated as follows: *For a composite simple system containing π phases and n components in which chemical reactions do not occur, there are $n + 2 - \pi$ independently variable intensive properties and, therefore, at least π extensive properties must be included in the set of $n + 2$ properties necessary to describe completely the composite system.*

This result was first expressed by J. Willard Gibbs in 1875 and is commonly referred to as the *Gibbs phase rule*. In most texts the number of *independently variable intensive* properties is referred to as the *variance* or degrees of freedom, denoted by $\mathcal{f}$, and the phase rule is written as

$$\mathcal{f} = n + 2 - \pi \tag{9-7}$$

In a binary system,

$$\mathcal{f} = 4 - \pi \tag{9-8}$$

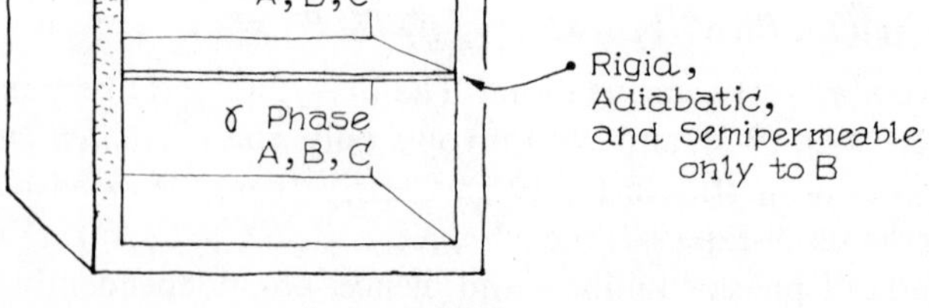

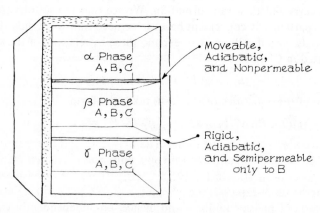

Figure 9.1

For a single phase, $\mathscr{f} = 3$ (i.e., we must specify three intensive properties such as T, P, x to describe completely all other intensive properties). We have dealt with such systems in Chapter 8.

Although the phase rule appears to be extremely simple, applying it to obtain particular relationships between dependent variables is often difficult. Thus, we shall treat the analytical procedures in detail in Sections 9.3 and 9.4 and present several applications in Section 9.5.

Before concluding the discussion of the phase rule, let us note that it does not apply to composite systems that are not simple systems. For non-simple systems, no general rule can be developed; each case must be analyzed separately with respect to the internal constraints that are present. Figure 9.1 and Example 9.1 illustrate this point.

Example 9.1

The isolated composite system shown in Figure 9.1 contains three phases with components A, B, and C in each phase. Determine the minimum number of properties necessary to describe completely the composite system and suggest how these properties should be chosen.

Solution

Because the composite system is not a simple system, Postulate I does not apply. To determine the number of independently variable properties, we proceed by choosing $n + 2$ or five properties for each phase and use the particular set of equilibrium criteria that applies to the given internal restraints. Let us choose the following $\pi(n + 2)$ or 15 properties:

$$T^\alpha, T^\beta, T^\gamma, P^\alpha, P^\beta, P^\gamma, x_A^\alpha, x_A^\beta, x_A^\gamma, x_B^\alpha, x_B^\beta, x_B^\gamma, N^\alpha, N^\beta, N^\gamma$$

The criteria of equilibrium can be shown to be (see Section 7.3):

$$T^\beta = T^\gamma$$

$$P^\alpha = P^\beta$$

$$\mu_B^\beta = \mu_B^\gamma$$

The last relation can be expanded:

$$\mu_B^\beta = g_B^\beta(T^\beta, P^\beta, x_A^\beta, x_B^\beta) = \mu_B^\gamma = g_B^\gamma(T^\gamma, P^\gamma, x_A^\gamma, x_B^\gamma)$$

The three restraining equations allow us to eliminate three variables from the set of fifteen. Nevertheless, we clearly cannot eliminate any arbitrary three. We can, however, eliminate one from each of the following three sets:

$$T^\beta, T^\gamma$$

$$P^\alpha, P^\beta$$

$$T^\beta, T^\gamma, P^\beta, P^\gamma, x_A^\beta, x_A^\gamma, x_B^\beta, x_B^\gamma$$

Thus, one satisfactory set of independently variable properties is T^α, P^α, P^γ,

x_A^α, x_A^β, x_A^γ, x_B^α, x_B^β, x_B^γ, N^α, N^β, N^γ. The required number of twelve may be compared to the value of five if the system were a simple one.

Numerous cases are encountered in phase equilibria in which we find it difficult to measure the low concentration of a component in one of the coexisting phases (e.g., for an aqueous solution of an inorganic salt in equilibrium with water vapor, the concentration of salt in the vapor phase is usually immeasurably small if not zero). In such cases in which there are no artificial walls separating the phases, we could either view the system as a simple one in which the concentration of insoluble component is very small, or we could treat it as if the phases were separated by a movable, diathermal barrier that is impermeable to the insoluble component. Either interpretation leads to the same practical result, provided that we are not interested in determining the concentration of the component in the phase in which its concentration is very small. In the first case (simple system), we would not use the restraining equation of equality of chemical potentials of the insoluble component because we are not interested in its concentration in one of the phases; in the second case (impermeable barrier), we do not have an equality of chemical potentials, and we assume that the concentration of the insoluble component is zero. Thus, from the pragmatic viewpoint, the two cases are equivalent.

9.2 Phase Diagrams

Later in this chapter we develop a number of useful equations that relate properties in different phases that are in equilibrium. These relationships can be represented graphically by property *phase diagrams*. The term phase diagram has many connotations, but we will almost exclusively limit ourselves to those diagrams that employ the variables temperature, pressure, and composition.

If we begin with the case of a pure component, composition is invariant and the pressure-temperature diagram is quite simple as illustrated in Figure 9.2. Each curve represents a boundary between single-phase domains. For states existing *on* any of these curves, the two bordering phases coexist. Thus, the curves are the P-T relationship that is predicted by the phase rule for monovariant systems. Figure 9.2 is typical of all pure substances, except that the slope of the liquid-solid equilibrium curve may be negative instead of positive, as drawn. As will become apparent in Section 9.3, a positive slope is consistent with a material that becomes more dense upon solidification, but a negative slope would mean that the solid phase is less dense (e.g., ice-water). Only a very few materials show this latter behavior.

Although not shown on Figure 9.2, there may be a number of different

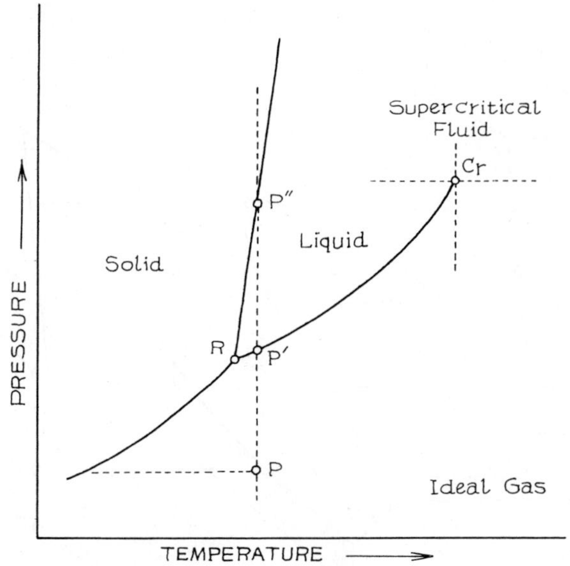

Figure 9.2 Single-component phase diagram.

solid phases each of which is the stable phase in some pressure-temperature domain.

The point shown as R is called the *triple point*, at which the three phases, gas, liquid, and solid, all coexist. From the phase rule, the variance at the triple point is zero for a pure substance; hence, it is a singular condition. The curve separating the gas and solid regions is the *sublimation* curve, the curve between liquid and gas is the *vaporization* curve, and the curve between liquid and solid is the *freezing* (or melting) curve.

Following the freezing or sublimation curves away from the triple point, we see that there is no definite point of termination (except, of course, $P \longrightarrow 0$, $T \longrightarrow 0$). However, for the vaporization curve, there is a temperature above which no liquid phase exists. This is shown as Cr in Figure 9.2 and is readily recognized as the critical point. The properties of the gas and liquid phases become identical at Cr. The naming of the domains above the critical point vary, but usually if $T < T_c$ and $P > P_c$, we say that the material is a supercritical fluid. For $P > P_c$ and $T > T_c$, we simply name the material a *fluid* and if $P < P_c$ and $T > T_c$, we are in the *gas* domain.

Since Figure 9.2 conveniently allows us to denote states of a system, we could show values of other properties at such states by adding an additional coordinate perpendicular to the plane of the paper. We illustrate this with the specific volume in Figure 9.3 in which an isometric view of the P-V-T space model is shown. To relate Figures 9.2 and 9.3, we also show the pressure-temperature phase diagram of the material in Figure 9.3. The P-T gas state of

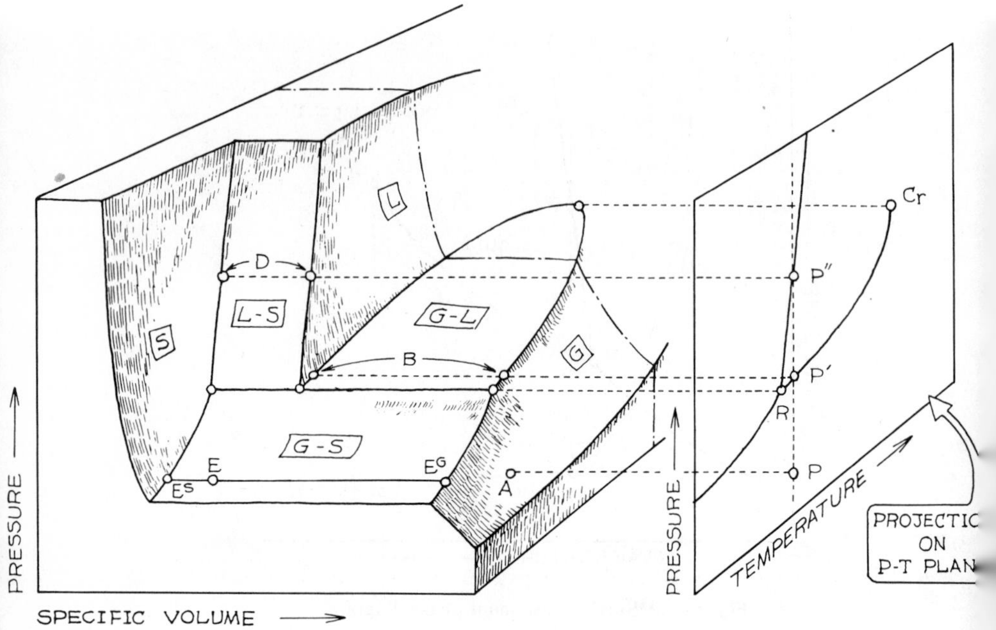

Figure 9.3 P-V-T diagram for a pure material.

Figure 9.2 is shown as point A on the isometric. The vapor-liquid state at P' is shown as points B and the solid-liquid state at P'' becomes points D. For the latter two states, two specific volumes are shown; each corresponds to a different equilibrium phase. Straight lines connecting points B or D project as points on the pressure-temperature plane. These lines are called *tie-lines* because they relate the properties of the phases that are in equilibrium.

All equilibrium states of the material must lie on the solid, liquid, or gas surfaces or on the curves of intersection between these surfaces. A state in the G-S domain, for example, point E, is in reality a mixture of gas and solid with properties of each phase as shown on the phase boundary curves at E^S and E^G. A state not on any surface shown is not an equilibrium state.

The dot-dashed curve shown in Figure 9.3 represents the trace of an intersection with a plane $T =$ constant. This trace is shown to illustrate the more customary way to show a three-dimensional diagram as that of Figure 9.3. Suppose one *projected* the space model onto a P-V plane and included a number of $T =$ constant traces. Then a figure such as 9.4 would result. (The solid-liquid volume change is greatly exaggerated for clarity; usually such volume changes are negligible compared to the volume change in going from a liquid to a gas.) Projections such as Figure 9.4 could be drawn using many different sets of intensive variables; some other kinds are illustrated in the figures included with the problems of Chapter 6. These two-dimensional

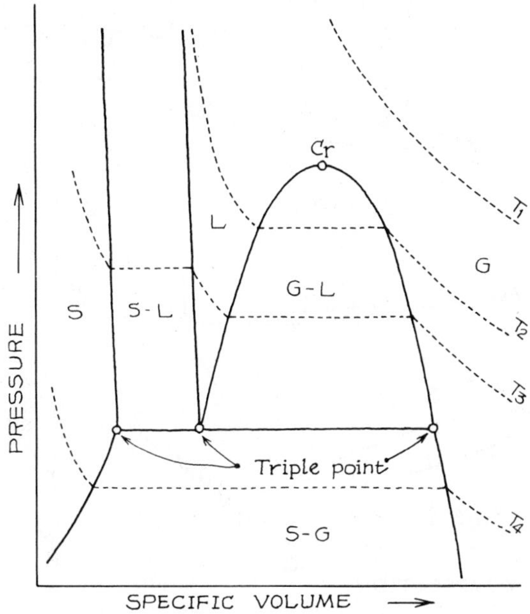

Figure 9.4 Isotherms projected on a pressure-volume plane.

projections of three-dimensional space models are very convenient to use in tracing property values during various processes.

To extend our pictorial study of phase diagrams to binary systems is considerably more difficult since, except on phase boundaries, three independent variables must be specified in order to delineate the state of the system. For the simplest case of a binary in which each component is completely soluble in all phases, we show the pressure-temperature-composition space model in Figure 9.5.[1] This model is really a set of surfaces that form a hollow figure.

To aid in visualizing this model, the pure component pressure-temperature plots are shown separately on the right and left sides of the model.

Any pressure-temperature-composition point occurs: (a) on a surface, (b) within the hollow figure, or (c) outside the hollow figure.[2] For the last case, the point is simply some state which is all gas, all liquid, or all solid, depending on the values chosen. If the point falls on a surface, it represents a state that is saturated and in equilibrium with at least one other phase.

[1] This diagram and many others shown in this chapter were based on those shown in the set of articles by C. E. Wales, *Chem. Eng.* (1963), May 27, p. 120; June 24, p. 111; July 22, p. 141; Aug. 19, p. 167; and Sept. 16, p. 187.

[2] There is a *G-L* surface, which is not shown in Figure 9.5 because it lies below the *L-G* surface. The hollow region between them terminates at the critical points curve, Cr_A-Cr_B.

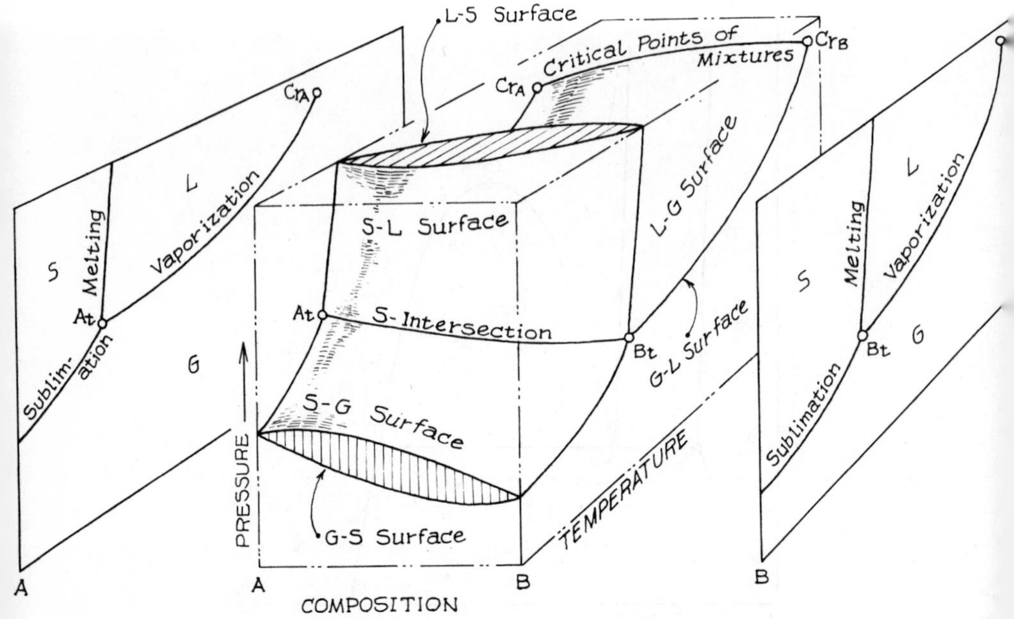

Figure 9.5 Space model of a binary system in which both components are miscible in all proportions.

Finally, points that fall within a hollow region represent a mixture of two (or more) phases in equilibrium, each of which would exist separately on a surface and which would be on a tie-line drawn through the point in question. Thus, single phases do not exist inside the hollow region.

To obtain a clearer picture of the shape of the space model, it is convenient to show isothermal and isobaric sections. This is readily accomplished, as suggested by Wales,[3] first by projecting all intersections onto a common pressure-temperature plane. Such projections are given in Figure 9.6a and 9.7a. Then at the various pressures shown in the former, and at the different temperatures shown in the latter, isobaric and isothermal sections are drawn in Figures 9.6b and 9.7b. That is, the set of curves separating two single-phase regions in Figures 9.6b and 9.7b are the loci of tie-lines bordering the hollow region of Figure 9.5. On each of these sections the pure component triple points (A_T and B_T), as well as the critical points (Cr_A and Cr_B), are shown for reference.

On the space model of Figure 9.5 there are six different surfaces and three curves of intersection. For the solid, there is the surface that represents solid states in equilibrium with gas (*S-G*) and with liquid (*S-L*). Similarly, there are (*L-G*), (*L-S*), (*G-L*), and (*G-S*) surfaces. Of this set of six, only the first two and

[3] C. E. Wales, *op. cit.*

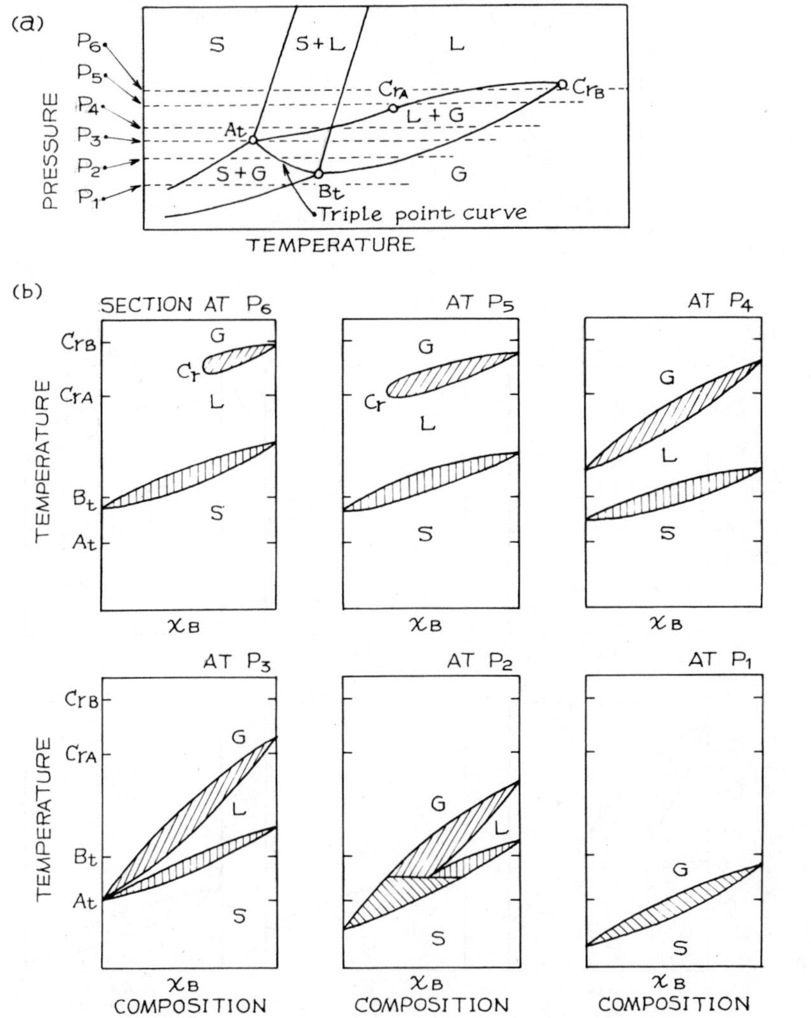

Figure 9.6 Isobaric cross-sections of the space model of Figure 9.5.

a portion of the third are visible in Figure 9.5. For the intersections there is one between (*S*-*G*) and (*S*-*L*), another between (*L*-*S*) and (*L*-*G*), and a third between (*G*-*S*) and (*G*-*L*). These are termed the *S*, *L*, and *G* intersections, respectively. They represent the loci of states for the triple points of mixtures; all begin and end at A_T and B_T. Only the *S* intersection is visible in the space model drawn.

All intersections project on a common triple point curve in the pressure-temperature plane as shown in Figures 9.6a and 9.7a. This must be so since, for three phases of a binary system to be in equilibrium, the variance is unity;

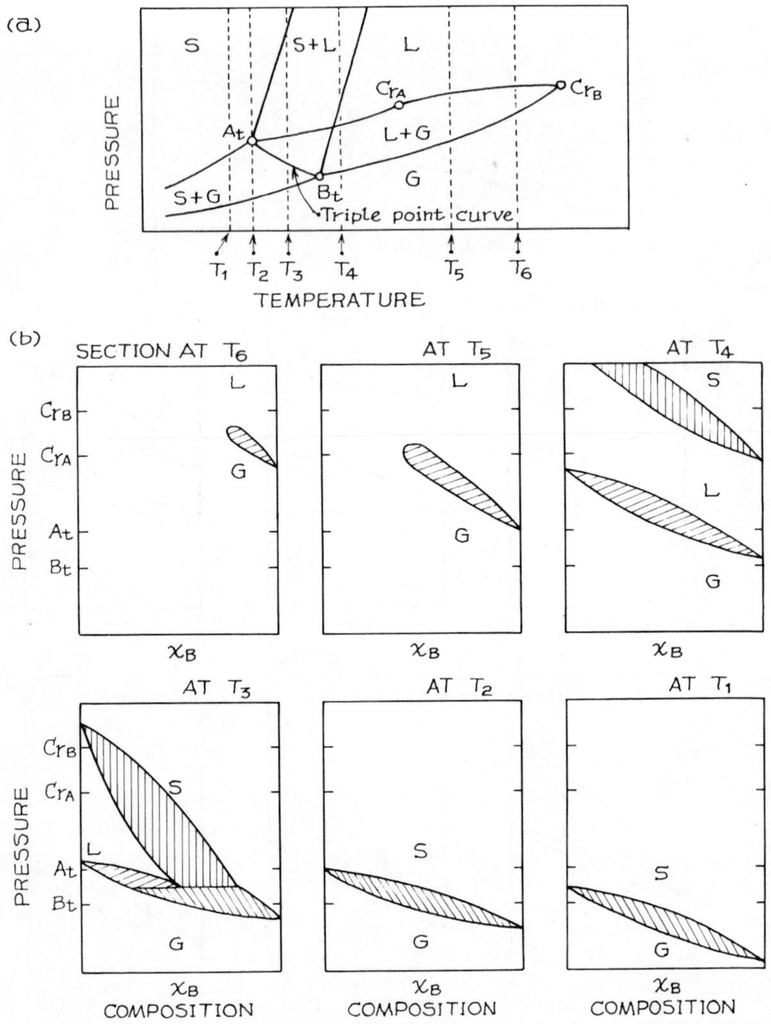

Figure 9.7 Isothermal cross-sections of the space model of Figure 9.5.

specification of *either* the pressure or temperature defines the state of the system completely.

The triple point region may be shown in another manner by employing Figures 9.6a or 9.6b along with projections of the pressure-composition and temperature-composition planes. In other words, view Figure 9.5 from the side, end, and top. These three views are given in Figure 9.8. At some chosen temperature, T, draw a vertical line through the top and side views. For this equilibrium state, the composition of the three phases may be immediately found as x^S, x^L, and x^G. The equilibrium pressure is read from the side view at

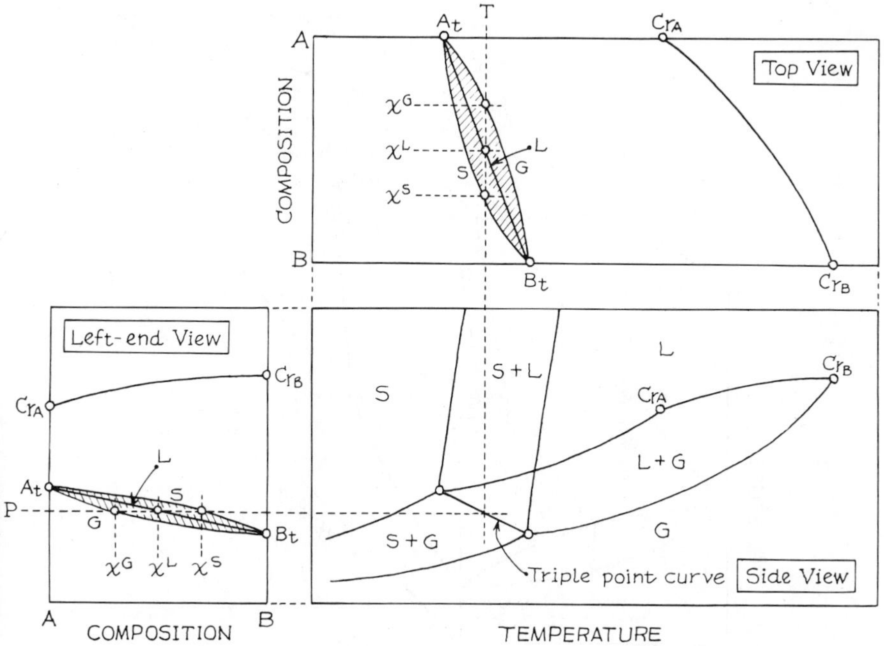

Figure 9.8 Triple point and critical point intersections for the space model of Figure 9.5.

the intersection of the isotherm with the triple point curve. The composition values may also be found in the left-end view at this equilibrium pressure.

Still another way to visualize the space model is to section it at constant composition as indicated in Figure 9.9. The hollow nature of the figure is clearly seen. In Figure 9.10 the pressure-temperature projection of the section has been drawn. Such constant composition sections are known as *isopleths*. No tie-lines can be drawn on isoplethal diagrams because two or more phases do not necessarily coexist at any one common composition.

If Figure 9.5 is reexamined, it will be noted that the hollow model is closed at the high-temperature gas-liquid end on a single space curve. Such a curve represents the loci of all mixture critical points and it connects Cr_A and Cr_B. The isoplethal section shown in Figure 9.10 is a particularly simple one and many variations exist. One of the more interesting is shown in Figure 9.11. Here the mixture critical point is marked at Cr_m, yet there are liquid and vapor states exceeding Cr_m both in pressure and temperature. Thus, an isothermal process may be visualized to occur from *a* to *b* (see Figure 9.11) in which the system is initially similar to a gas, then becomes a gas-liquid mixture, and finally ends as a gas. This process is called retrograde condensation. In an analogous manner, following the isobar *d* to *c*,

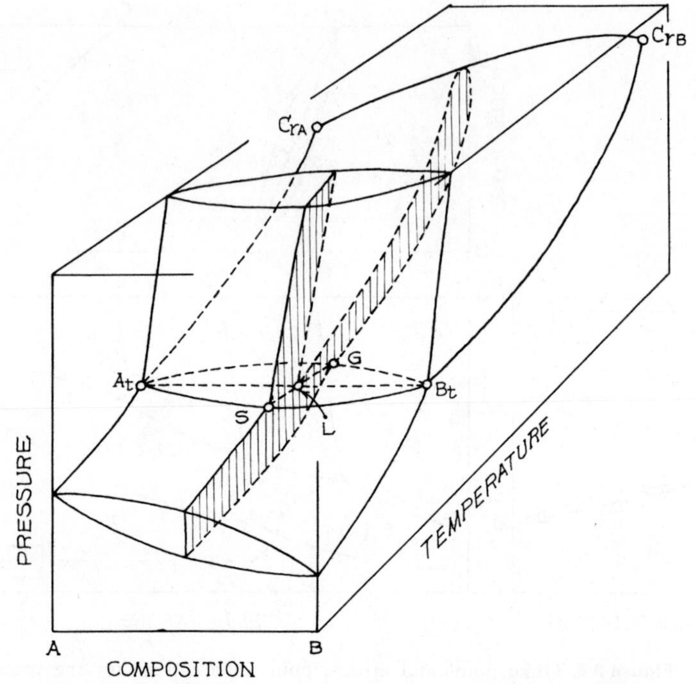

Figure 9.9 Constant composition section of the space model of figure 9.5.

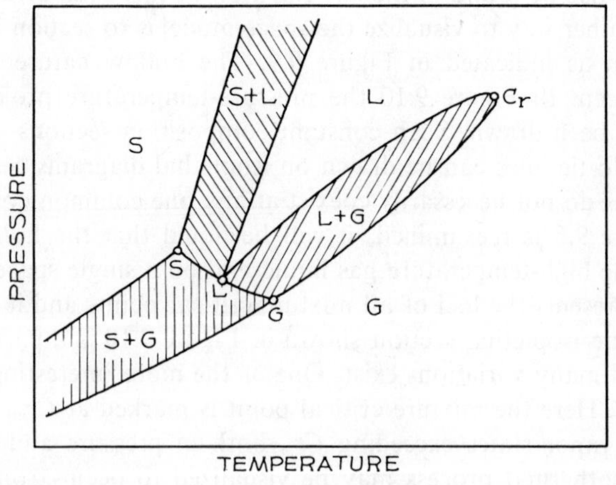

Figure 9.10 Pressure-temperature projection of the section shown in figure 9.9.

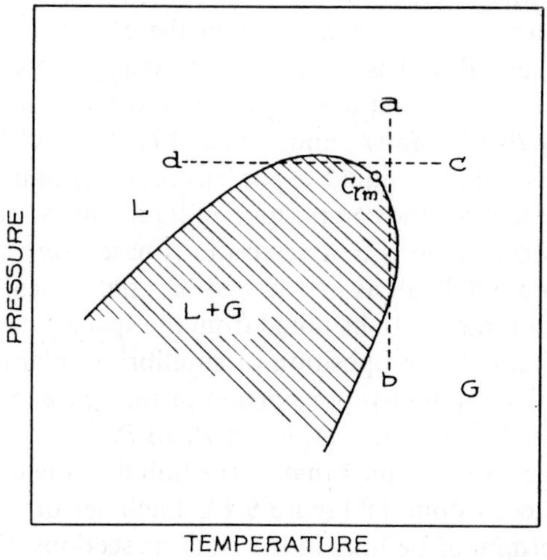

Figure 9.11 Isoplethal section for a mixture exhibiting retrograde phenomena.

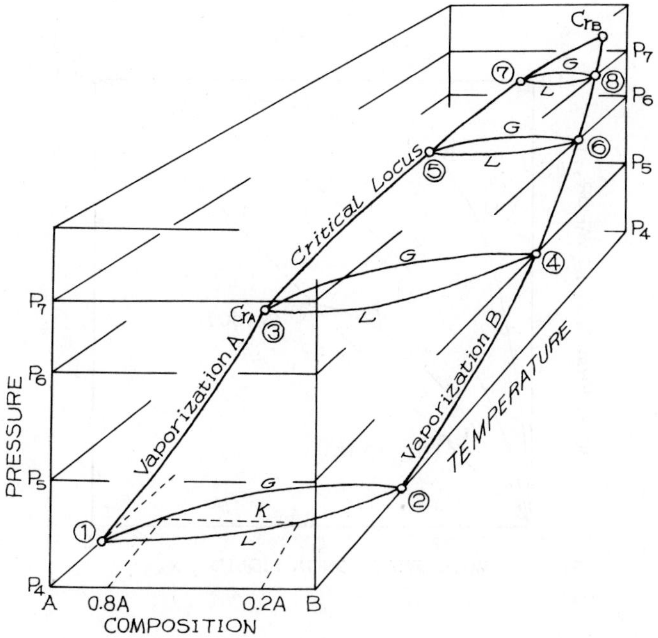

Figure 9.12 Isobaric sections of the liquid-gas region.

we would find retrograde vaporization in that we would have in succession, liquid, liquid-gas, and liquid.

For temperatures and pressures between the critical points of the pure components, isothermal and isobaric sections consist of closed loops that do not traverse the entire composition range. Several of these are shown in Figures 9.6b and 9.7b at P_5 and P_6 and at T_5 and T_6. Normally, as the pressure or temperature increases, the liquid-vapor lens becomes thinner and there is a smaller difference between the composition of liquid and vapor. The practical consequence of this change is that separations based on vapor-liquid equilibrium become more difficult as one approaches the critical region.

One last point of interest to be noted from the space model study involves relationships between the composition of equilibrium phases. For example, if we examine only the liquid-vapor portion of the space model as shown in Figure 9.12 and make isobaric sections at P_4 to P_8, we could then plot the composition of the vapor against that of the liquid at the different pressures shown. This has been done in Figure 9.13. Each set of compositions was taken from the termini of tie-lines in the isobaric sections. One tie-line (k) is shown at P_4 on Figure 9.12; this tie-line provides a single vapor-liquid equilibrium datum point as indicated in Figure 9.13. Above the critical pressure of A, the isobars on the $y - x$ plot are foreshortened.

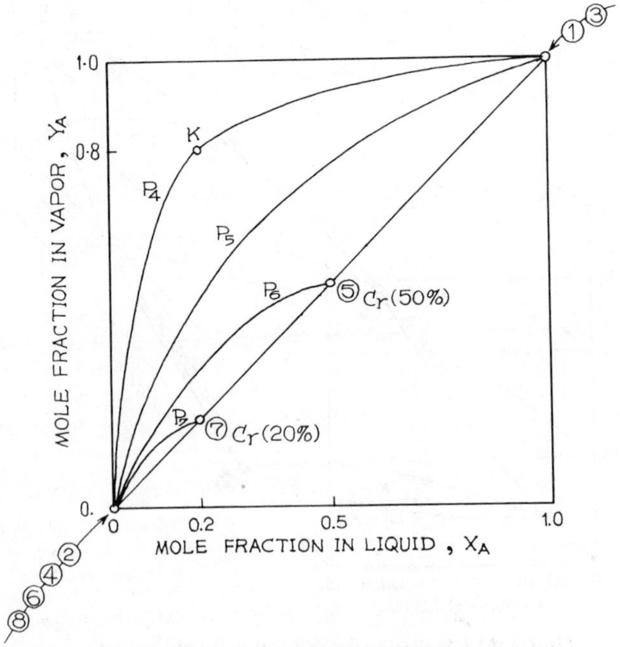

Figure 9.13 *y-x* or tie-line plot for figure 9.12.

Example 9.2

Given a system at P_2 in Figure 9.6b and as shown in Figure 9.14, trace and describe the equilibrium states as one cools a mixture originally at point 1.

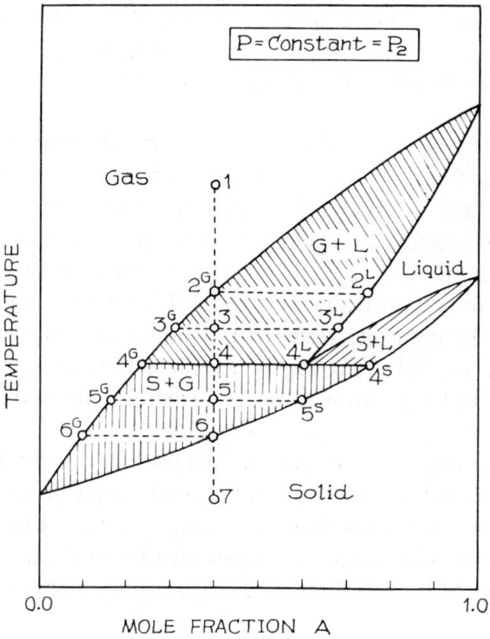

Figure 9.14 Isobaric section of figure 9.6(b) at P_2.

Solution

At point 1, the system is a gas. Cooling to point 2, the first liquid begins to form. This occurs at the *dew point* temperature. The first liquid appears with a composition appropriate to point 2^L. Cooling further until the system temperature is equal to 3, we now have a gas-liquid mixture with a vapor phase at 3^G and a liquid at 3^L. From a simple mass balance, it can be shown that the ratio: moles of gas to moles of liquid is in the same proportion as the lengths of the tie-lines $(\overline{3\text{-}3^L})/(\overline{3\text{-}3^G})$. At 4, we would find a gas phase at 4^G and a liquid at 4^L. At this temperature we have a triple point and we could have three phases in equilibrium (i.e., 4^G, 4^L, and 4^S); thus, the relative amounts of the three phases cannot be determined except within certain limits. To see this, assume that we had one mole of the mixture originally with a composition x_0. At 4, let there be V_4 moles of vapor, L_4 moles of liquid, and S_4 moles of solid.

Then,

$$V_4 + L_4 + S_4 = 1$$
$$[x(4^G)]V_4 + [x(4^L)]L_4 + [x(4^S)]S_4 = x_0$$

With only these two equations, the values of V_4, L_4, and S_4 cannot be found.

If we cool to point 5, we find a solid and a gas phase in equilibrium at 5^S and 5^G. The gas-to-solid mole ratio is $(\overline{5\text{-}5^S})/(\overline{5\text{-}5^G})$. Finally at 6, we are at a *bubble point* (though usually bubble points refer to liquid states with incipient vaporization). Below 6, the mixture is a solid.

In Figures 9.6 and 9.7 the dew and bubble points were shown to be monotonic functions of composition. There is no reason to expect this to always be the case. In fact, we frequently encounter systems in which the pressure, for an isothermal section, or the temperature, for an isobaric section, attains a maximum or minimum value. These systems are called azeotropic mixtures and the two kinds are illustrated in Figure 9.15 and 9.16. The azeotrope is that state in which the concentrations in the liquid and vapor phases become identical. Thus, for this state the relative volatility $[\alpha_{A\cdot B} \equiv (y_A/x_A)/(y_B/x_B)]$ becomes unity and separation processes such as simple distillation become impossible.

If we were to visualize the space model for an azeotropic system, the loci of azeotropes would be a curve that would result from making a seam down the gas-liquid hollow lens. Azeotropic compositions usually change with temperature or pressure variations so that one may often bypass the azeotropic limitation in distillation by operating columns at different pressures.

As shown in Figure 9.15 in the isothermal section at $T = T_z$, the pressure at the azeotrope is noted as P_z. The isobaric section is drawn for $P = P_z$ and again the azeotropic temperature is T_z. The azeotropic concentrations are drawn to be the same in both figures. We will discuss such phenomena further in Section 9.5.

The azeotropic mixture illustrates one of the limitations of the Gibbs phase rule. The variance of a two-phase binary is two. There are, however, cases in which we cannot necessarily describe completely the intensive properties of the phases in a mixture simply by specifying the temperature and pressure. For example, in Figure 9.15a, for the section at $T = T_z$, choose a pressure P' less than P_z. At T_z, P', there are *two* equally valid sets of coexisting liquid and vapor compositions: The Gibbs phase rule apparently breaks down. The problem is that we are expecting too much. As seen in the next section, the relation we obtain for $P = f(x)_T$ can be multivalued.

The apparent difficulty in interpreting the Gibbs phase rule is avoided if we reconsider the discussion in Section 9.1. We stated that there are *certain* sets of intensive properties which, when selected, will completely define the

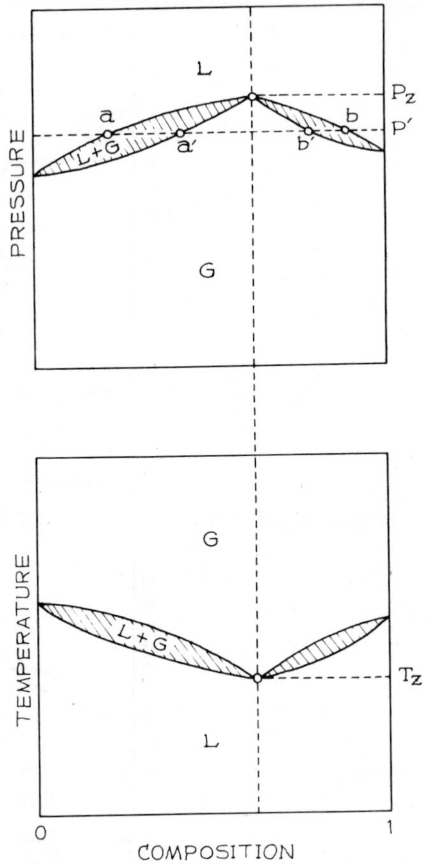

Figure 9.15 Isothermal and isobaric sections for a minimum boiling azeotropic system.

system. But this statement did not imply we could choose *any* set. Thus, for the case of an azeotropic system, we may not be able to use the pressure-temperature set. On the other hand, if we specify temperature and liquid composition, we always define a unique state for which pressure and vapor composition are determined.

It is also interesting to note that to specify the system completely (i.e., extensive as well as intensive properties), we could use pressure and temperature provided we also include N_A and N_B in the set of $(n + 2)$ properties required by Postulate I. In this case, the average mole fractions of both components are known and, hence, we can make an unambiguous choice regarding the branch of the azeotrope that is the correct one (e.g., in Figure 9.15a, a-a' or b-b').

Another variation in the behavior of binary systems that is often found is

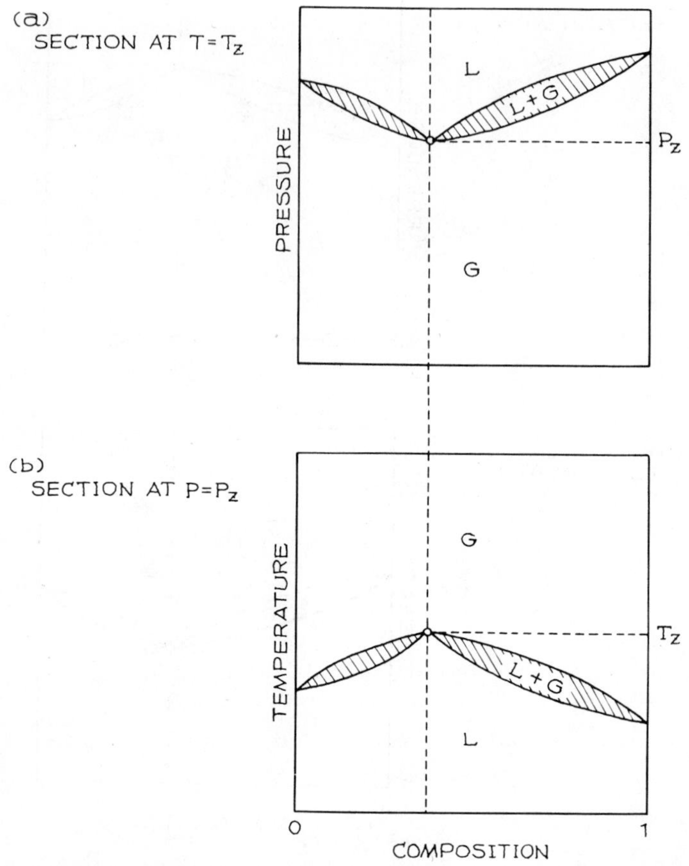

Figure 9.16 Isothermal and isobaric sections for a maximum boiling azeotropic system.

the *miscibility gap*. In some mixtures there exist regions in which a homogeneous phase separates into two distinct phases that are immiscible. This may occur in the liquid, solid, or, less frequently, in the gas phase. Each of the two phases so formed may themselves be mixtures, in which case the two components are said to be *partially miscible*. Alternatively, the phases may consist of pure components and in this case the two components are said to be *immiscible*.

Let us consider a liquid phase with a miscibility gap and in which the components are partially miscible. We make sections as shown in Figure 9.6b at P_4 and 9.7b at T_4, but as modified on Figure 9.17 to show a closed-loop gap in the liquid phase. Within the closed loop, two liquid phases exist; the composition of the two liquids would be those at the termini of tie-lines.

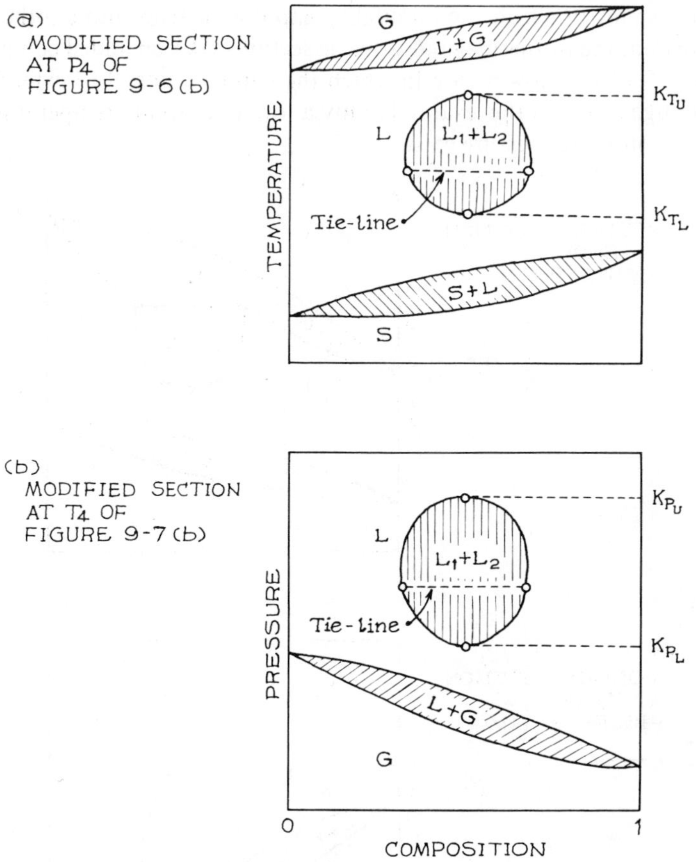

Figure 9.17 Isobaric and isothermal sections with a closed miscibility gap.

The temperatures and pressures labeled K_{TU}, K_{TL}, K_{PU}, and K_{PL} are called the upper and lower critical solution (or consolute) temperatures and pressures, respectively. Above an upper consolute limit or below a lower one, the two components are completely miscible and form only a single liquid phase.

Many systems only exhibit either an upper or lower consolute limit. These cases would be expected when the miscibility gap intersects a two-phase lens. For example, again using a liquid miscibility gap as an illustration, we would find only a lower consolute temperature when the miscibility gap intersects the liquid-gas lens and only an upper consolute temperature when the gap overextends the solid-gas lens. Carrying this reasoning further, when a miscibility gap is relatively large, we may not find either an upper or lower consolute limit.

For systems in which the miscibility gap does extend into another two-phase region, the isothermal and isobaric sections take on added complexity. In Figure 9.18 we show a case in which the gap of Figure 9.17 extends into the liquid-gas lens. In this example, only a lower consolute temperature and an upper consolute pressure exist.

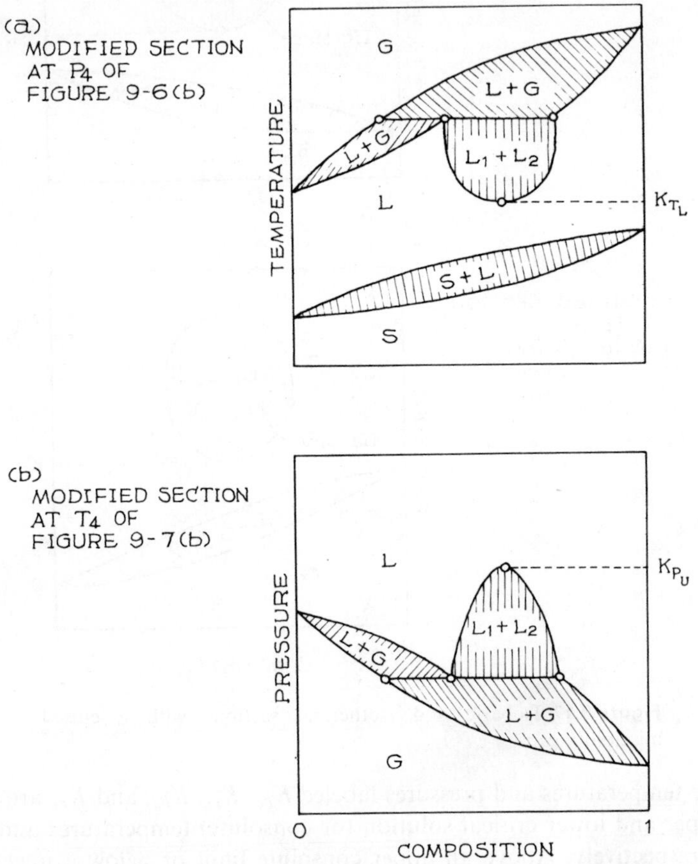

Figure 9.18 Isobaric and isothermal sections with a partial miscibility gap.

Should an azeotropic mixture also show a partial miscibility gap, the isothermal and isobaric sections might appear as in Figure 9.19. At the pseudo-azeotropic point the gas is now in equilibrium with two different liquid phases. Also, consider an isothermal compression of a gas at 1. Upon reaching 2, a liquid of composition 2′ forms. Increasing the pressure then results in the formation of additional liquid until 3 is reached. At this pressure we have a triple point with two liquid phases at 3′ and 3″ and a gas at 3‴. Further compression leads to the complete condensation of the vapor and there remains a

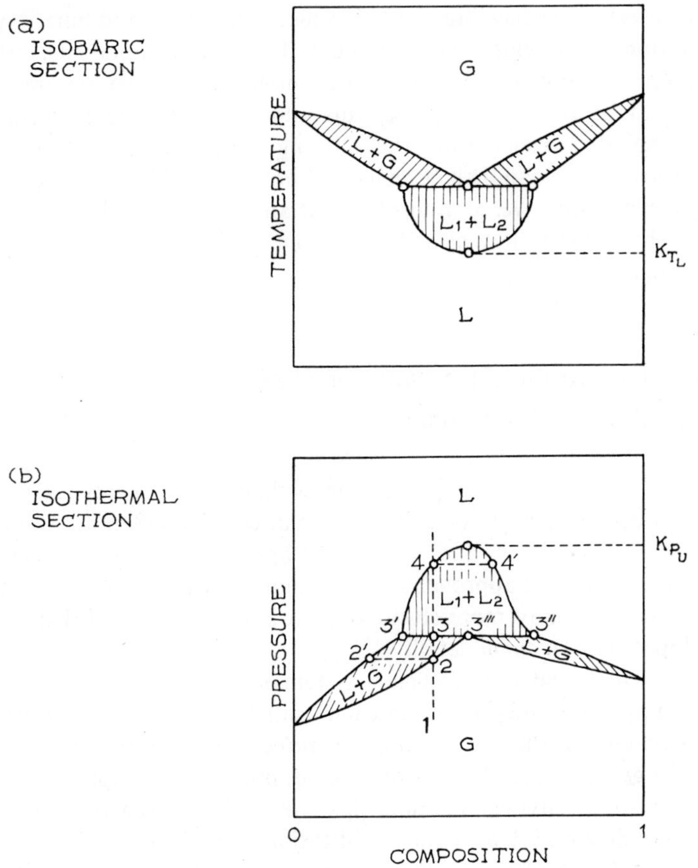

Figure 9.19 Isobaric and isothermal sections of an azeotropic system with a partial miscibility gap.

system containing only two liquid phases. Increasing the pressure further produces a change in the relative amounts and compositions of the two liquid phases until at 4 there exists only a single liquid.

For those systems with wide miscibility gaps, the concentrations of the lean and rich phases approach zero and unity, respectively. This occurs only rarely with liquid-phase gaps but is relatively common for solid-phase gaps.

We have introduced only a few simple examples in this interesting and often complex field of phase diagrams. For many other examples and a more detailed discussion, the interested reader should consult primary references.[4]

[4] C. E. Wales, *op. cit.*; J. E. Ricci, *The Phase Rule and Heterogeneous Equilibrium* (New York: D. Van Nostrand Co., Inc., 1951); L. O. Case, *Elements of the Phase Rule* (Ann Arbor, Mich.: Edwards Letter Shop, 1939); J. Zernike, *Chemical Phase Theory* (New York: Gregory Lounz, 1957); A. Findlay, A. N. Campbell, and N. O. Smith, *Phase Rule* (New York: Dover Publications, Inc., 1951).

In this section we have treated only single component and binary systems. For multiphase systems involving more than two components, graphical representation becomes unwieldy. For example, in a ternary system, composition could be plotted on a triangular diagram and the temperature or pressure included in such a way that a triangular prism is formed. Isothermal or isobaric sections can then be made. Although graphical representation of such systems is not covered in this text, the analytical procedures for describing their behavior follow directly from the methods described in Sections 9.3 and 9.5.

9.3 The Differential Approach for Phase Equilibrium Relationships

We shall develop two approaches for evaluating the analytical relationships between properties of coexisting phases: the differential approach, described in this section, and the integral approach, treated in Section 9.5. Although the integral approach is more commonly used, the differential approach is conceptually more elucidating. It is for this reason that the latter is developed here in some detail.

We shall first treat a simple binary system of components A and B with two coexisting phases α and β in which each component is present in each phase. The reasoning will then be extended to three coexisting phases of a binary and to the general case of multicomponent, multiphase simple systems.

To develop analytical relationships, we shall use the theoretical basis described in Section 9.1 with one modification: for the criteria of equilibrium between coexisting phases, instead of employing chemical potentials [as in Eq. (9-4)], we shall use the equivalent criteria of equality of the fugacity of each component in each phase.[5] Thus, the criteria of equilibrium for a binary mixture coexisting as α and β phases are:

$$T^{\alpha} = T^{\beta} \tag{9-9}$$

$$P^{\alpha} = P^{\beta} \tag{9-10}$$

$$\hat{f}_A^{\alpha} = \hat{f}_A^{\beta} \quad \text{or} \quad \ln \hat{f}_A^{\alpha} = \ln \hat{f}_A^{\beta} \tag{9-11}$$

and

$$\hat{f}_B^{\alpha} = \hat{f}_B^{\beta} \quad \text{or} \quad \ln \hat{f}_B^{\alpha} = \ln \hat{f}_B^{\beta} \tag{9-12}$$

To obtain differential equations involving P, T, and phase compositions,

[5] Note that while equality of component fugacities is completely equivalent to equality of chemical potentials, equality of component activities is not necessarily equivalent. The discrepancy exists because different reference states of *unit activity* are sometimes chosen for a component in different phases. For fugacities, the reference state is always chosen as an ideal gas at the system temperature.

we take the total derivatives of Eq. set (9-9) through (9-12) and expand all fugacities in terms of T, P, and the pertinent phase compositions. In view of Eqs. (9-9) and (9-10), no phase designation need be made for T and P.

For example, suppose we choose as the $(n + 1)$ variables in each phase, T, P, and x_A. Thus, for component A, we have

$$\begin{aligned} d\ln \hat{f}_A^\alpha &= \left(\frac{\partial \ln \hat{f}_A^\alpha}{\partial T}\right)_{P,x^\alpha} dT + \left(\frac{\partial \ln \hat{f}_A^\alpha}{\partial P}\right)_{T,x^\alpha} dP + \left(\frac{\partial \ln \hat{f}_A^\alpha}{\partial x_A^\alpha}\right)_{T,P} dx_A^\alpha \\ &= -\left(\frac{\bar{H}_A^\alpha - H_A^*}{RT^2}\right) dT + \left(\frac{\bar{V}_A^\alpha}{RT}\right) dP + \left(\frac{\partial \ln \hat{f}_A^\alpha}{\partial x_A^\alpha}\right)_{T,P} dx_A^\alpha \end{aligned} \tag{9-13}$$

$$\begin{aligned} d\ln \hat{f}_A^\beta &= \left(\frac{\partial \ln \hat{f}_A^\beta}{\partial T}\right)_{P,x^\beta} dT + \left(\frac{\partial \ln \hat{f}_A^\beta}{\partial P}\right)_{T,x^\beta} dP + \left(\frac{\partial \ln \hat{f}_A^\beta}{\partial x_A^\beta}\right)_{T,P} dx_A^\beta \\ &= -\left(\frac{\bar{H}_A^\beta - H_A^*}{RT^2}\right) dT + \left(\frac{\bar{V}_A^\beta}{RT}\right) dP + \left(\frac{\partial \ln \hat{f}_A^\beta}{\partial x_A^\beta}\right)_{T,P} dx_A^\beta \end{aligned} \tag{9-14}$$

where H_A^* refers to the enthalpy of A in an ideal gas state at the temperature of the system. Note that the expansions of Eqs. (9-13) and (9-14) are written separately for each phase. As discussed in Chapters 5 and 8, these expansions are always valid for each phase, regardless of whether or not the phase is in equilibrium with other phases.

The fugacity expansions for component B are more convenient if we use the same set of intensive variables. Thus, the equations corresponding to Eqs. (9-13) and (9-14) for component B are:

$$d\ln \hat{f}_B^\alpha = -\left(\frac{\bar{H}_B^\alpha - H_B^*}{RT^2}\right) dT + \left(\frac{\bar{V}_B^\alpha}{RT}\right) dP + \left(\frac{\partial \ln \hat{f}_B^\alpha}{\partial x_A^\alpha}\right)_{T,P} dx_A^\alpha \tag{9-15}$$

$$d\ln \hat{f}_B^\beta = -\left(\frac{\bar{H}_B^\beta - H_B^*}{RT^2}\right) dT + \left(\frac{\bar{V}_B^\beta}{RT}\right) dP + \left(\frac{\partial \ln \hat{f}_B^\beta}{\partial x_A^\beta}\right)_{T,P} dx_A^\beta \tag{9-16}$$

Equating differentials of $\ln \hat{f}_i$

$$-\left(\frac{\bar{H}_A^\alpha - \bar{H}_A^\beta}{RT^2}\right) dT + \left(\frac{\bar{V}_A^\alpha - \bar{V}_A^\beta}{RT}\right) dP + \left(\frac{\partial \ln \hat{f}_A^\alpha}{\partial x_A^\alpha}\right)_{T,P} dx_A^\alpha - \left(\frac{\partial \ln \hat{f}_A^\beta}{\partial x_A^\beta}\right)_{T,P} dx_A^\beta = 0 \tag{9-17}$$

and

$$-\left(\frac{\bar{H}_B^\alpha - \bar{H}_B^\beta}{RT^2}\right) dT + \left(\frac{\bar{V}_B^\alpha - \bar{V}_B^\beta}{RT}\right) dP + \left(\frac{\partial \ln \hat{f}_B^\alpha}{\partial x_A^\alpha}\right)_{T,P} dx_A^\alpha - \left(\frac{\partial \ln \hat{f}_B^\beta}{\partial x_A^\beta}\right)_{T,P} dx_A^\beta = 0 \tag{9-18}$$

where we must remember that $\bar{H}_i^\alpha$ and $\bar{V}_i^\alpha$ are evaluated at x_A^α, T, P whereas $\bar{H}_i^\beta$ and $\bar{V}_i^\beta$ are at x_A^β, T, P.

Eqs. (9-17) and (9-18) are two differential equations that must be satisfied simultaneously. We can solve by eliminating any one of the differentials in T, P, x_A^α, and x_A^β, thereby obtaining one differential equation in three variables. Integration of that equation yields the desired result.

Thus, the relationship $P = f(T, x_A^\beta)$ can be obtained by eliminating dx_A^α

simultaneously from Eqs. (9-17) and (9-18). This final result may be simplified by noting that the coefficients of the two dx_A^α terms and the two dx_A^β are related by the Duhem equation for component fugacities, i.e., for a binary system of phase π:

$$x_A^\pi \left(\frac{\partial \ln \hat{f}_A^\pi}{\partial x_A^\pi}\right)_{T,P} + x_B^\pi \left(\frac{\partial \ln \hat{f}_B^\pi}{\partial x_A^\pi}\right)_{T,P} = 0 \tag{9-19}$$

Thus, multiplying Eq. (9-17) by x_A^α and multiplying Eq. (9-18) by x_B^α, adding the two equations, and applying Eq. (9-19) to each phase, the result simplifies to:

$$-\left[\frac{x_A^\alpha(\bar{H}_A^\alpha - \bar{H}_A^\beta) + x_B^\alpha(\bar{H}_B^\alpha - \bar{H}_B^\beta)}{RT^2}\right] dT + \left[\frac{x_A^\alpha(\bar{V}_A^\alpha - \bar{V}_A^\beta) + x_B^\alpha(\bar{V}_B^\alpha - \bar{V}_B^\beta)}{RT}\right] dP$$
$$- \left[\left(x_A^\alpha - \frac{x_A^\beta x_B^\alpha}{x_B^\beta}\right)\left(\frac{\partial \ln \hat{f}_A^\beta}{\partial x_A^\beta}\right)_{T,P}\right] dx_A^\beta = 0 \tag{9-20}$$

Solving explicitly for dP,

$$dP = \frac{1}{T}\left[\frac{x_A^\alpha(\bar{H}_A^\alpha - \bar{H}_A^\beta) + x_B^\alpha(\bar{H}_B^\alpha - \bar{H}_B^\beta)}{x_A^\alpha(\bar{V}_A^\alpha - \bar{V}_A^\beta) + x_B^\alpha(\bar{V}_B^\alpha - \bar{V}_B^\beta)}\right] dT$$
$$+ \left[\frac{RT\left(x_A^\alpha - \frac{x_A^\beta x_B^\alpha}{x_B^\beta}\right)\left(\frac{\partial \ln \hat{f}_A^\beta}{\partial x_A^\beta}\right)_{T,P}}{x_A^\alpha(\bar{V}_A^\alpha - \bar{V}_A^\beta) + x_B^\alpha(\bar{V}_B^\alpha - \bar{V}_B^\beta)}\right] dx_A^\beta \tag{9-21}$$

Since P is a property, and since T and x_A^β form an independent set for P under conditions of two-phase equilibrium, Eq. (9-21) must be an exact differential equation. Therefore, it follows immediately that

$$\left(\frac{\partial P}{\partial T}\right)_{x_A^\beta, [\alpha-\beta]} = \frac{1}{T}\left[\frac{x_A^\alpha(\bar{H}_A^\alpha - \bar{H}_A^\beta) + x_B^\alpha(\bar{H}_B^\alpha - \bar{H}_B^\beta)}{x_A^\alpha(\bar{V}_A^\alpha - \bar{V}_A^\beta) + x_B^\alpha(\bar{V}_B^\alpha - \bar{V}_B^\beta)}\right] \tag{9-22}$$

and

$$\left(\frac{\partial P}{\partial x_A^\beta}\right)_{T, [\alpha-\beta]} = \frac{RT\left(x_A^\alpha - \frac{x_A^\beta x_B^\alpha}{x_B^\beta}\right)\left(\frac{\partial \ln \hat{f}_A^\beta}{\partial x_A^\beta}\right)_{T,P}}{x_A^\alpha(\bar{V}_A^\alpha - \bar{V}_A^\beta) + x_B^\alpha(\bar{V}_B^\alpha - \bar{V}_B^\beta)} \tag{9-23}$$

where the subscript $[\alpha - \beta]$ denotes conditions under which phases α and β coexist at equilibrium. Note that the partial derivatives in the left-hand side of Eqs. (9-22) and (9-23) do not imply that x_A^α is constant; x_A^α will in fact change as T is varied under conditions of $\alpha - \beta$ phase equilibrium and constant x_A^β. Similarly, x_A^α will change as x_A^β is varied with phase equilibrium and constant T. (Recall that for this case the variance is two, so that x_A^α can be expressed as a function of T and x_A^β.)

Since

$$\left(\frac{\partial T}{\partial x_A^\beta}\right)_{P, [\alpha-\beta]} \left(\frac{\partial P}{\partial T}\right)_{x_A^\beta, [\alpha-\beta]} \left(\frac{\partial x_A^\beta}{\partial P}\right)_{T, [\alpha-\beta]} = -1 \tag{9-24}$$

it follows that

$$\left(\frac{\partial T}{\partial x_A^\beta}\right)_{P,[\alpha-\beta]} = -RT^2\left[\frac{\left(x_A^\alpha - \frac{x_A^\beta x_B^\alpha}{x_B^\beta}\right)\left(\frac{\partial \ln \hat{f}_A^\beta}{\partial x_A^\beta}\right)_{T,P}}{x_A^\alpha(\bar{H}_A^\alpha - \bar{H}_A^\beta) + x_B^\alpha(\bar{H}_B^\alpha - \bar{H}_B^\beta)}\right] \tag{9-25}$$

Example 9.3

Prove that for a minimum boiling azeotrope at constant temperature the pressure maximizes at the azeotropic concentration.

Solution

An azeotrope occurs when $x_A^V = x_A^L$, at which point the term $(x_A^V - x_A^L x_B^V/x_B^L)$ in Eq. (9-25) vanishes. By inspection of Eq. (9-23), it is clear that $(\partial P/\partial x_A^L)_{T,[L-V]}$ also vanishes at the azeotropic concentration. To prove that a temperature minimum corresponds to a pressure maximum, all we need show is that $(\partial T/\partial x_A^L)_{P,[L-V]}$ and $(\partial P/\partial x_A^L)_{T,[L-V]}$ have opposite signs. The ratio of these two derivatives is $-[(\partial P/\partial T)_{x_A^L,[L-V]}]$ by Eq. (9-24). Eq. (9-22) shows that with $\alpha = V$ and $\beta = L$, the enthalpy and volume changes are always positive. Therefore, $(\partial P/\partial T)_{x_A^L,[L-V]}$ is always positive and, hence, $(\partial T/\partial x_A^L)_{P,[L-V]}$ and $(\partial P/\partial x_A^L)_{T,[L-V]}$ always have opposite signs.

In the procedure described above we evaluated the partial derivatives involving T, P, and x_A^β by eliminating the dx_A^α terms in solving Eqs. (9-17) and (9-18). Similarly, we could have worked with T, P, and x_A^α by eliminating dx_A^β [the results would be identical to Eqs. (9-22), (9-23), and (9-25) with x_A^α replacing x_A^β], or T, x_A^α, and x_A^β by eliminating dP, or P, x_A^α, and x_A^β by eliminating dT.

For example, to obtain the partial derivatives in T, x_A^α, and x_A^β, eliminating dP from Eqs. (9-17) and (9-18) yields:

$$\frac{1}{T}\left[\frac{(\bar{H}_A^\alpha - \bar{H}_A^\beta)}{(\bar{V}_A^\alpha - \bar{V}_A^\beta)} - \frac{(\bar{H}_B^\alpha - \bar{H}_B^\beta)}{(\bar{V}_B^\alpha - \bar{V}_B^\beta)}\right] dT - RT\left[\frac{\left(\frac{\partial \ln \hat{f}_A^\alpha}{\partial x_A^\alpha}\right)_{T,P}}{(\bar{V}_A^\alpha - \bar{V}_A^\beta)} - \frac{\left(\frac{\partial \ln \hat{f}_B^\alpha}{\partial x_A^\alpha}\right)_{T,P}}{(\bar{V}_B^\alpha - \bar{V}_B^\beta)}\right] dx_A^\alpha$$
$$+ RT\left[\frac{\left(\frac{\partial \ln \hat{f}_A^\beta}{\partial x_A^\beta}\right)_{T,P}}{(\bar{V}_A^\alpha - \bar{V}_A^\beta)} - \frac{\left(\frac{\partial \ln \hat{f}_B^\beta}{\partial x_A^\beta}\right)_{T,P}}{(\bar{V}_B^\alpha - \bar{V}_B^\beta)}\right] dx_A^\beta = 0 \tag{9-26}$$

Simplifying Eq. (9-26) by applying the Duhem equation for fugacity for each phase, we obtain

$$-\left[\frac{(\bar{H}_A^\alpha - \bar{H}_A^\beta)(\bar{V}_B^\alpha - \bar{V}_B^\beta) - (\bar{H}_B^\alpha - \bar{H}_B^\beta)(\bar{V}_A^\alpha - \bar{V}_A^\beta)}{RT^2}\right] dT$$
$$+\left[\frac{x_B^\alpha(\bar{V}_B^\alpha - \bar{V}_B^\beta) + x_A^\alpha(\bar{V}_A^\alpha - \bar{V}_A^\beta)}{x_B^\alpha}\right]\left(\frac{\partial \ln \hat{f}_A^\alpha}{\partial x_A^\alpha}\right)_{T,P} dx_A^\alpha \tag{9-27}$$
$$-\left[\frac{x_B^\beta(\bar{V}_B^\alpha - \bar{V}_B^\beta) + x_A^\beta(\bar{V}_A^\alpha - \bar{V}_A^\beta)}{x_B^\beta}\right]\left(\frac{\partial \ln \hat{f}_A^\beta}{\partial x_A^\beta}\right)_{T,P} dx_A^\beta = 0$$

From Eq. (9-27), it follows that

$$\left(\frac{\partial x_A^\beta}{\partial x_A^\alpha}\right)_{T,[\alpha-\beta]} = \left[\frac{x_B^\alpha(\bar{V}_B^\alpha - \bar{V}_B^\beta) + x_A^\alpha(\bar{V}_A^\alpha - \bar{V}_A^\beta)}{x_B^\beta(\bar{V}_B^\alpha - \bar{V}_B^\beta) + x_A^\beta(\bar{V}_A^\alpha - \bar{V}_A^\beta)}\right]\left(\frac{x_B^\beta}{x_B^\alpha}\right)\left[\frac{\left(\frac{\partial \ln \hat{f}_A^\alpha}{\partial x_A^\alpha}\right)_{T,P}}{\left(\frac{\partial \ln \hat{f}_A^\beta}{\partial x_A^\beta}\right)_{T,P}}\right] \tag{9-28}$$

$$\left(\frac{\partial T}{\partial x_A^\alpha}\right)_{x_A^\beta,[\alpha-\beta]} = \left[\frac{x_B^\alpha(\bar{V}_B^\alpha - \bar{V}_B^\beta) + x_A^\alpha(\bar{V}_A^\alpha - \bar{V}_A^\beta)}{(\bar{H}_A^\alpha - \bar{H}_A^\beta)(\bar{V}_B^\alpha - \bar{V}_B^\beta) - (\bar{H}_B^\alpha - \bar{H}_B^\beta)(\bar{V}_A^\alpha - \bar{V}_A^\beta)}\right] \left(\frac{RT^2}{x_B^\alpha}\right)\left(\frac{\partial \ln \hat{f}_A^\alpha}{\partial x_A^\alpha}\right)_{T,P} \tag{9-29}$$

and $(\partial T/\partial x_A^\beta)_{x_A^\alpha,[\alpha-\beta]}$ is given by Eq. (9-29) with x_A^α and x_A^β interchanged.

From the partial derivatives obtained from Eqs. (9-22), (9-23), (9-25), (9-28), and (9-29), we could construct a variety of isoplethal, isothermal, or isobaric cross sections. To evaluate an isopleth [e.g., $P = f(T)$ at constant x_A^β], we must integrate $(\partial P/\partial T)_{x_A^\beta,[\alpha-\beta]}$. To carry out the integration of Eq. (9-22), we must express the right-hand side as a function of P, T, and x_A^β, where x_A^β is held constant during integration. From physical property data of individual phases, we can express $\bar{H}_i^\alpha$, $\bar{V}_i^\alpha$ and $\bar{H}_i^\beta$, $\bar{V}_i^\beta$ as a function of P, T, x_A^α and P, T, x_A^β, respectively. However, x_A^α is not a constant during the integration since it varies with T. Therefore, in addition to physical property data, we must know x_A^α as a function of T and x_A^β under a $\alpha - \beta$ phase equilibrium. This relationship can be obtained by integration of Eq. (9-28), provided that we can express the right-hand side of the equation as a function of x_A^α, x_A^β, and T. But here we run into the same problem: the properties are functions of P also, and P is a floating variable in Eq. (9-28), and, therefore, P must be expressed as a function of x_A^α, x_A^β, and T. This functionality we originally sought by integration of Eq. (9-22). Thus, to obtain a rigorous solution to an isopleth requires simultaneous solution of Eqs. (9-22) and (9-28). Although the procedure is complicated for the rigorous case, in practice it can usually be simplified considerably by using judicious approximations. One example is given below, and others are treated in the next section.

Example 9.4

A binary system of components A and B coexists in liquid-vapor equilibrium at 175°C and at low pressure. The vapor phase can be considered an ideal mixture of ideal gases; the liquid-phase activity coefficients can be approximated by the van Laar equations:

$$\ln \gamma_A = \frac{A_{12}}{\left(1 + \frac{A_{12}}{A_{21}}\frac{x_A}{x_B}\right)^2} \tag{9-30}$$

and

$$\ln \gamma_B = \frac{A_{21}}{\left(1 + \frac{A_{21}}{A_{12}}\frac{x_B}{x_A}\right)^2} \tag{9-31}$$

where A_{12} and A_{21} are functions of temperature, and γ_A and γ_B can be considered independent of pressure.

(a). Determine the y-x relationship at a constant temperature of 175°C, and indicate how a $(P\text{-}x)_T$ diagram would be constructed.

(b). If isothermal P-x data were available instead and activity coefficients were not known, would it be possible to construct the y-x relationship? If so, indicate the procedure one would follow. *Note*: For a liquid phase obeying the van Laar equations, the following limiting law can be shown to pertain:

$$\lim_{x_A \to 0} \left(\frac{y_A}{x_A}\right) = \left(\frac{P_{vp_A}}{P_{vp_B}}\right) e^{A_{12}} \tag{9-32}$$

where P_{vp_A} and P_{vp_B} are the vapor pressures of pure A and B at the temperature of the system. For the system in question at 175°C, $P_{vp_A} = 5.65$ atm and $P_{vp_B} = 8.98$ atm.

Solution

(a). The y-x relationships at constant T can be determined from Eq. (9-28) if we let α = liquid and β = vapor. Thus,

$$\left(\frac{\partial y_A}{\partial x_A}\right)_{T,[\alpha-\beta]} = \left[\frac{x_A(\bar{V}_A^V - \bar{V}_A^L) + x_B(\bar{V}_B^V - \bar{V}_B^L)}{y_A(\bar{V}_A^V - \bar{V}_A^L) + y_B(\bar{V}_B^V - \bar{V}_B^L)}\right]\left(\frac{y_B}{x_B}\right)\left[\frac{\left(\frac{\partial \ln \hat{f}_A^L}{\partial x_A}\right)_{T,P}}{\left(\frac{\partial \ln \hat{f}_A^V}{\partial y_A}\right)_{T,P}}\right] \tag{9-33}$$

For an ideal vapor mixture of ideal gases, the following simplifications are applicable:

$$\bar{V}_i^V = V_i^V = \frac{RT}{P} \tag{9-34}$$

$$\left(\frac{\partial \ln \hat{f}_A^V}{\partial y_A}\right)_{T,P} = \frac{1}{y_A} \tag{9-35}$$

For conditions far removed from the critical point, there is an additional simplification:

$$\bar{V}_i^V = \frac{RT}{P} \gg \bar{V}_i^L \tag{9-36}$$

Expressing the liquid fugacity as a function of activity coefficient,

$$\left(\frac{\partial \ln \hat{f}_A^L}{\partial x_A}\right)_{T,P} = \frac{1}{x_A} + \left(\frac{\partial \ln \gamma_A}{\partial x_A}\right)_{T,P} \tag{9-37}$$

Substituting Eqs. (9-34) through (9-37) into Eq. (9-33) and simplifying, we obtain:

$$\left(\frac{\partial y_A}{\partial x_A}\right)_{T,[\alpha-\beta]} = \frac{y_A y_B}{x_A x_B}\left[1 + x_A\left(\frac{\partial \ln \gamma_A}{\partial x_A}\right)_{T,P}\right] \tag{9-38}$$

or

$$\int \frac{dy_A}{y_A(1-y_A)} = \int \frac{dx_A}{x_A(1-x_A)} + \int \frac{1}{(1-x_A)}\left(\frac{\partial \ln \gamma_A}{\partial x_A}\right)_{T,P} dx_A \quad (T \text{ constant}) \tag{9-39}$$

Eq. (9-39) can be integrated directly by evaluating the activity coefficient partial derivative from Eq. (9-30). It is, however, more convenient to rearrange terms prior to integration. Since

$$\frac{1}{1-x_A} = \frac{1}{x_B} = \frac{1-x_B+x_B}{x_B} = \frac{x_A}{x_B} + 1$$

$$\left(\frac{1}{1-x_A}\right)\left(\frac{\partial \ln \gamma_A}{\partial x_A}\right)_{T,P} = \frac{x_A}{x_B}\left(\frac{\partial \ln \gamma_A}{\partial x_A}\right)_{T,P} + \left(\frac{\partial \ln \gamma_A}{\partial x_A}\right)_{T,P} \tag{9-40}$$

From the Duhem expression for γ,

$$\frac{x_A}{x_B}\left(\frac{\partial \ln \gamma_A}{\partial x_A}\right)_{T,P} = -\left(\frac{\partial \ln \gamma_B}{\partial x_A}\right)_{T,P}$$

Thus,

$$\int \left(\frac{1}{1-x_A}\right)\left(\frac{\partial \ln \gamma_A}{\partial x_A}\right)_{T,P} dx_A = \int \left[\frac{\partial \ln\left(\frac{\gamma_A}{\gamma_B}\right)}{\partial x_A}\right]_{T,P} dx_A \tag{9-41}$$

Substituting Eq. (9-41) into Eq. (9-39), and using the indefinite integral method, we obtain:

$$\ln\left(\frac{y_A}{1-y_A}\right) = \ln\left(\frac{x_A}{1-x_A}\right) + \ln\left(\frac{\gamma_A}{\gamma_B}\right) + C \tag{9-42}$$

The constant of integration can be evaluated using the limiting condition given in the problem statement. That is, as $x_A \to 0$ and $y_A \to 0$, $\gamma_A \to e^{A_{12}}$, $\gamma_B \to 1$; thus from Eqs. (9-32) and (9-42),

$$\ln\left[\lim_{x_A \to 0}\left(\frac{y_A}{x_A}\right)\right] = \lim_{x_A \to 0}\left[\ln\left(\frac{y_A}{x_A}\right)\right] = \ln\left(\frac{e^{A_{12}}P_{vp_A}}{P_{vp_B}}\right) = A_{12} + C \tag{9-43}$$

and

$$C = \ln \frac{P_{vp_A}}{P_{vp_B}} \tag{9-44}$$

so

$$\left(\frac{y_A}{1-y_A}\right) = \left(\frac{x_A}{1-x_A}\right)\left(\frac{P_{vp_A}}{P_{vp_B}}\right)\left(\frac{\gamma_A}{\gamma_B}\right) \quad (T \text{ constant}) \tag{9-45}$$

Substituting Eqs. (9-30) and (9-31) into Eq. (9-45) and simplifying, we obtain the desired result:

$$\left(\frac{y_A}{1-y_A}\right) = g(x_A) = \left(\frac{P_{vp_A}}{P_{vp_B}}\right)\left(\frac{x_A}{x_B}\right) \exp\left[\frac{A_{12}A_{21}(A_{21}x_B^2 - A_{12}x_A^2)}{(A_{12}x_A + A_{21}x_B)^2}\right] \quad (T \text{ constant}) \tag{9-46}$$

or

$$y_A = \frac{g(x_A)}{1 + g(x_A)} \qquad (T \text{ constant}) \qquad (9\text{-}47)$$

To construct a P-x diagram at constant T, we must integrate Eq. (9-23). Substituting the simplifying assumptions given above into Eq. (9-23) (with $\beta =$ liquid), we have:

$$\left(\frac{\partial P}{\partial x_A}\right)_{T,[\alpha-\beta]} = \frac{RT\left(y_A - \frac{x_A y_B}{x_B}\right)\left[\frac{1}{x_B} + \left(\frac{\partial \ln \gamma_A}{\partial x_A}\right)_{T,P}\right]}{\frac{RT}{P}} \qquad (9\text{-}48)$$

or

$$\left(\frac{\partial \ln P}{\partial x_A}\right)_{T,[\alpha-\beta]} = \frac{y_A}{x_A}\left(1 - \frac{x_A y_A}{x_B y_B}\right)\left[\frac{\partial \ln\left(\frac{\gamma_A}{\gamma_B}\right)}{\partial x_A}\right]_{T,P} \qquad (9\text{-}49)$$

Thus,

$$\int d \ln P = \int \left\{\frac{1}{x_A}\left[\frac{1 - g(x_A)}{1 + g(x_A)}\right]\right\}\left[\frac{\partial \ln\left(\frac{\gamma_A}{\gamma_B}\right)}{\partial x_A}\right]_{T,P} dx_A \qquad (T \text{ constant}) \quad (9\text{-}50)$$

The integration is complex; the final result is much easier to obtain by the integral approach described in Section 9.5. The point to note is that simplifications resulting from the facts that one phase is a vapor (i.e., $\bar{V}^V \gg \bar{V}^L$) and that the vapor is ideal results in a procedure which, although complex, is manageable. Furthermore, we have seen that in the frequently occurring cases in which these assumptions are valid, knowledge of the condensed phase activity coefficient is all that is required to generate $(y\text{-}x)_T$ and $(P\text{-}x)_T$ diagrams.

(b). If we did not have an expression for the activity coefficient, and if we did have isothermal P-x data, we should be able to reverse the procedure above to construct the y-x relationship. Clearly, if we assumed that the liquid phase behaved as a van Laar liquid, all terms in the right-hand side of Eq. (9-50) could be expressed as functions of x_A and the van Laar constants, A_{12} and A_{21}. Thus, these constants could be evaluated by curve-fitting Eq. (9-50) to the experimental P-x data, and these constants could then be used in Eq. (9-46) to obtain the y-x relationship. Alternatively, we could use other applicable activity coefficient-liquid composition correlations and, in an analogous procedure, evaluate the coefficients in the expansion from the P-x data. Similarly, if the vapor phase were not ideal, an appropriate real-gas equation of state would have to be used for the vapor properties. Thus, it is possible to obtain y-x equilibrium information from the relatively simple measurement of equilibrium pressure without recourse to measurement of vapor phase composition.[6]

We have seen in the discussion of Eqs. (9-13) through (9-18) that in order to generate isothermal, isobaric, and isoplethal cross sections, we must have knowledge of the temperature-, pressure-, and concentration-dependency of the fugacities (or chemical potentials). If, however, we did not have knowledge of all property relations, we could develop some diagrams from knowledge of other cross sectional diagrams. The minimum number of such diagrams needed to specify all others is equivalent to determining the minimum number of independent partial derivatives involving temperature, pressure, and concentration.

For a binary system involving two phases, we require the minimum number of independent partials involving the variables, T, P, y_A, and x_A. In Section 6.3 we faced a similar problem for the four variables S, T, V, and P. There we found that all but four partials could be eliminated as independent by mathematical manipulation, and one of these four could be eliminated by thermodynamic reasoning [e.g., the Maxwell reciprocity relation was obtained from the fact that these variables satisfied the Fundamental Equation, $U = f(S, V)$]. Since the reciprocity condition does not apply here, it follows that there are four independent partials involving T, P, y_A, and x_A, but that no more than two can be chosen from any one set of the combinations of (T, P, x_A^α), (T, P, x_A^β), $(T, x_A^\alpha, x_A^\beta)$, and $(P, x_A^\alpha, x_A^\beta)$. Thus, any cross section could be obtained, for example, for a vapor-liquid binary system, from T-x, T-y, P-x, and P-y diagrams. Alternatively, in the general case in which simplifying assumptions are not applicable (see Example 9.3), we cannot obtain a y-x diagram from P-x data alone.

Let us now generalize the results obtained for the two-phase binary system to additional phases and components.

For a binary system involving the three phases α, β, and γ, the criteria of equilibrium become

$$T^\alpha = T^\beta = T^\gamma \tag{9-51}$$

$$P^\alpha = P^\beta = P^\gamma \tag{9-52}$$

$$\ln \hat{f}_A^\alpha = \ln \hat{f}_A^\beta = \ln \hat{f}_A^\gamma \tag{9-53}$$

$$\ln \hat{f}_B^\alpha = \ln \hat{f}_B^\beta = \ln \hat{f}_B^\gamma \tag{9-54}$$

These criteria are identical to solving simultaneously the two cases of

[6] For detailed procedures, see e.g., H. W. Preugla, Jr., and M. A. Pike, Jr. "Thermodynamics of Solutions—Numerical Method for Calculating Excess Free Energy and Activity Coefficients from Total Pressure Measurements," *J. Chem. Eng. Data.*, **6**, (1961) 400. or D. B. Myers, and R. L. Scott "Thermodynamic Functions for Nonelectrolyte Solutions," *Ind. Eng. Chem.*, **55**, 7, (1963), 43.

α-β and β-γ phase equilibria. For the α-β equilibrium case, the relationships developed previously [namely, Eqs. (9-17), (9-18), (9-21), and (9-27)] are still valid. For the β-γ equilibrium, we would obtain identical equations with α replaced by γ. We would then solve the two sets simultaneously.

For example, by eliminating x_A^α from Eqs. (9-17) and (9-18), we obtained Eq. (9-21), which we can write as

$$dP = \left(\frac{\partial P}{\partial T}\right)_{x_A^\beta,\,[\alpha-\beta]} dT + \left(\frac{\partial P}{\partial x_A^\beta}\right)_{T,\,[\alpha-\beta]} dx_A^\beta \tag{9-55}$$

where the partials are given by Eqs. (9-22) and (9-23). The analogous relation for the β-γ equilibrium in which we eliminate x_A^γ would then be

$$dP = \left(\frac{\partial P}{\partial T}\right)_{x_A^\beta,\,[\alpha-\gamma]} dT + \left(\frac{\partial P}{\partial x_A^\beta}\right)_{T,\,[\beta-\gamma]} dx_A^\beta \tag{9-56}$$

From Eqs. (9-55) and (9-56), we could solve simultaneously to obtain P as a function of T by eliminating dx_A^β, P as a function of x_A^β by eliminating T, or T as a function of x_A^β by eliminating P. For example, let us eliminate x_A^β. Thus,

$$\left(\frac{\partial x_A^\beta}{\partial P}\right)_{T,[\beta-\gamma]}\left[dP - \left(\frac{\partial P}{\partial T}\right)_{x_A^\beta,[\beta-\gamma]} dT\right] = \left(\frac{\partial x_A^\beta}{\partial P}\right)_{T,[\alpha-\beta]}\left[dP - \left(\frac{\partial P}{\partial T}\right)_{x_A^\beta,[\alpha-\beta]} dT\right] \tag{9-57}$$

or

$$dP = \frac{\left[\left(\frac{\partial P}{\partial T}\right)_{x_A^\beta,[\alpha-\beta]}\left(\frac{\partial x_A^\beta}{\partial P}\right)_{T,[\alpha-\beta]} - \left(\frac{\partial P}{\partial T}\right)_{x_A^\beta,[\beta-\gamma]}\left(\frac{\partial x_A^\beta}{\partial P}\right)_{T,[\beta-\gamma]}\right]}{\left[\left(\frac{\partial x_A^\beta}{\partial P}\right)_{T,[\alpha-\beta]} + \left(\frac{\partial x_A^\beta}{\partial P}\right)_{T,[\beta-\gamma]}\right]} dT$$

or

$$dP = -\left[\frac{\left(\frac{\partial T}{\partial x_A^\beta}\right)_{P,[\alpha-\beta]} - \left(\frac{\partial T}{\partial x_A^\beta}\right)_{P,[\beta-\gamma]}}{\left(\frac{\partial x_A^\beta}{\partial P}\right)_{T,[\alpha-\beta]} + \left(\frac{\partial x_A^\beta}{\partial P}\right)_{T,[\beta-\gamma]}}\right] dT \tag{9-58}$$

The bracket in Eq. (9-58) is clearly $(\partial P/\partial T)_{[\alpha-\beta-\gamma]}$, which can be written as $(\partial P/\partial T)$ for the monovariant system of three phases in equilibrium. Eq. (9-58) represents, of course, the locus of triple points shown in Figures 9.6 and 9.7. Note that for each set of T and P that satisfies Eq. (9-58), x_A^β will vary. We could have obtained this variation of x_A^β with T, for example, by eliminating dP from Eqs. (9-55) and (9-56). Other variables, such as x_A^α and x_A^γ, could be expressed as functions of T, P, or x_A^β in an analogous manner.

In the general case of n components distributed between π phases, we obtain, *for each component*, $\pi - 1$ equations of the form:

$$-\left(\frac{\bar{H}_i^\alpha - \bar{H}_i^\beta}{RT^2}\right) dT + \left(\frac{\bar{V}_i^\alpha - \bar{V}_i^\beta}{RT}\right) dP + \sum_{j\neq k}^{n}\left[\left(\frac{\partial \ln \hat{f}_i^\alpha}{\partial x_j^\alpha}\right)_{T,P,x_i^\alpha[x_j^\alpha,x_k^\alpha]} dx_j^\alpha - \left(\frac{\partial \ln \hat{f}_i^\beta}{\partial x_j^\beta}\right)_{T,P,x_i^\beta[x_j^\beta,x_k^\beta]} dx_j^\beta\right] = 0 \tag{9-59}$$

Thus, we have $n(\pi - 1)$ equations of this form relating the $[(n - 1)(\pi) + 2]$ variables involving T, P, and x. Solving these equations simultaneously, we can eliminate $n(\pi - 1) - 1$ variables, resulting in a differential equation involving $n + 3 - \pi$ variables. From this equation, the differential of any one variable is expressed as a function of the remaining $n + 2 - \pi$ variables.[7]

9.4 Pressure-Temperature Relations

It is well-known that in phase equilibria the system pressure is often a strong function of temperature. Also it turns out that the derivative dP/dT is related to the enthalpy (or entropy) and volume changes during a phase transformation. Since the latter properties are often of considerable interest to engineers, we shall illustrate a general approach by examining a few simple, but common systems, all of which involve a vapor phase in addition to one or more condensed phases.

The simplest case encountered is a pure liquid in equilibrium with its vapor. At equilibrium, we have $f^V = f^L$ in addition to temperature and pressure equalities. Thus,

$$d \ln f^V = d \ln f^L \tag{9-60}$$

Since the fugacity of a pure material is a function only of T and P, expanding,

$$\left(\frac{\partial \ln f^V}{\partial T}\right)_P dT + \left(\frac{\partial \ln f^V}{\partial P}\right)_T dP = \left(\frac{\partial \ln f^L}{\partial T}\right)_P dT + \left(\frac{\partial \ln f^L}{\partial P}\right)_T dP \tag{9-61}$$

Substituting Eqs. (6-40) and (6-41) into Eq. (9-61) and collecting terms,

$$\left(\frac{dP}{dT}\right)_{[L-V]} = \frac{(H^V - H^L)}{T(V^V - V^L)} = \frac{\Delta H^{\text{vap}}}{T\,\Delta V^{\text{vap}}} \tag{9-62}$$

Eq. (9-62) is commonly called the Clausius-Clapeyron equation. Expressing $\Delta V^{\text{vap}} = (RT/P)(Z^V - Z^L) = (RT/P)\,\Delta Z^{\text{vap}}$, then

$$\left[\frac{d \ln P}{d\left(\frac{1}{T}\right)}\right]_{[L-V]} = \frac{-\Delta H^{\text{vap}}}{R\,\Delta Z^{\text{vap}}} \tag{9-63}$$

The ratio $(\Delta H^{\text{vap}}/\Delta Z^{\text{vap}})$ is a weak but essentially linear function of temperature except near the critical point. Often, at low pressures, the assumption is made that over a nominal temperature range, the ratio is constant and,

[7] Note that the Duhem equations (one for each phase for a total of π) can be used to simplify the $n(\pi - 1)$ equations of the form of Eq. (9-59) but not to reduce the number of variables. That is, the Duhem equations reduce the number of coefficients of the dx_j terms and, therefore, reduce the amount of physical property data required in integrations to obtain the final result.

thus, Eq. (9-63) is readily integrated. When such an assumption is not valid or when high accuracy is desired, other integration techniques are available.[8]

Next, let us consider the case in which we have a nonvolatile solute, such as an inorganic salt, dissolved in a volatile solvent such as water. The vapor above the solution is then essentially pure water. Since the system is divariant, the pressure-temperature relation for such a system is not unique unless we place a further restriction on the system. Two kinds of restricted systems will be illustrated.

Let the salt concentration in the liquid be constant as the system temperature (or pressure) is varied. For the volatile component (denoted by subscript w),

$$d \ln f_w^V = d \ln \hat{f}_w^L \tag{9-64}$$

Expanding in terms of T and P for the pure vapor and in terms of T, P, and x_w for the liquid mixture, we obtain:

$$\begin{aligned} &\left(\frac{\partial \ln f_w^V}{\partial T}\right)_P dT + \left(\frac{\partial \ln f_w^V}{\partial P}\right)_T dP \\ &\quad = \left(\frac{\partial \ln \hat{f}_w^L}{\partial T}\right)_{P,x} dT + \left(\frac{\partial \ln \hat{f}_w^L}{\partial P}\right)_{T,x} dP + \left(\frac{\partial \ln \hat{f}_w^L}{\partial x_w}\right)_{T,P} dx_w \end{aligned} \tag{9-65}$$

Substituting for the partial derivatives of fugacity and noting that $dx_w = 0$ for this case of constant liquid composition, after simplification, one finds:

$$\left(\frac{\partial P}{\partial T}\right)_{x,[L-V]} = \frac{(H_w^V - \bar{H}_w^L)}{T(V_w^V - \bar{V}_w^L)} \tag{9-66}$$

The numerator represents the enthalpy change in vaporizing one mole of water from the solution at constant composition. The quantity H_w^V is not the enthalpy of saturated water vapor since, at a given T, the system pressure does not correspond to the equilibrium vapor pressure for pure water. The pressure correction, however, is ordinarily small and usually neglected. In fact, the numerator is often expanded by adding and subtracting H_w^L, the enthalpy of pure *liquid* water at the system temperature and pressure.[9] Then,

$$H_w^V - \bar{H}_w^L = (H_w^V - H_w^L) + (H_w^L - \bar{H}_w^L) \tag{9-67}$$

When pressure corrections to the enthalpy of pure water are neglected, the first term in Eq. (9-67) is the enthalpy of vaporization of pure water at the system temperature; the second term is then the partial molal enthalpy of mixing, $\overline{\Delta H_w^L}$.

[8] See, e.g., R. C. Reid and T. K. Sherwood, *The Properties of Gases and Liquids* (New York: McGraw-Hill Book Company, 1966) Sec. Ed., Chap. 4.

[9] Note that this state may not be a thermodynamically stable state if, at T, $P < P_{vp_w}$.

Then, if the additional assumptions that $V_w^V \gg \bar{V}_w^L$ and $V_w^V = RT/P$ are invoked,

$$\left[\frac{\partial \ln P}{\partial\left(\frac{1}{T}\right)}\right]_{x,[L-V]} = \frac{-(\Delta H_w^{\text{vap}} - \overline{\Delta H_w^L})}{R} \tag{9-68}$$

In theory, then, from P-T data at constant liquid composition, one could determine partial molal enthalpies of mixing. Both $\overline{\Delta H_w^L}$ and $\overline{\Delta H_s^L}$ can be found, the former from Eq. (9-68) and the latter from $\overline{\Delta H_w^L}$ and the Duhem equation. Nevertheless, neat as this approach appears, it is unfortunately not a very useful one. Even with very accurate $(P\text{-}T)_x$ data, differentiation leads to some error and since, usually $\overline{\Delta H_w^L} < \Delta H_w^{\text{vap}}$, it is difficult to extract accurate values of $\overline{\Delta H_w^L}$ from the data. Direct calorimetric determination of enthalpies of mixing is usually the preferred way to measure this property.

Instead of keeping the salt solution concentration constant, one could impose the restriction that the solution is at all times saturated with the non-volatile solute (i.e., there is always some undissolved salt present). In this case since $\pi = 3$ and $n = 2$, we have a univariant system.

Following the same general treatment, we have a liquid mixture, a pure water vapor phase, and a pure salt solid phase. Thus,

$$d \ln f_w^V = d \ln \hat{f}_w^L \tag{9-69}$$

and

$$d \ln f_s^S = d \ln \hat{f}_s^L \tag{9-70}$$

Expanding the fugacity in the pure phases in terms of T and P and the liquid phase in terms of T, P, and x_w, and substituting for the fugacity partials with respect to T and P, we obtain:

$$-\left(\frac{H_w^V - \bar{H}_w^L}{RT^2}\right) dT + \left(\frac{V_w^V - \bar{V}_w^L}{RT}\right) dP = \left(\frac{\partial \ln \hat{f}_w^L}{\partial x_w}\right)_{T,P} dx_w \tag{9-71}$$

and

$$-\left(\frac{H_s^S - \bar{H}_s^L}{RT^2}\right) dT + \left(\frac{V_s^S - \bar{V}_s^L}{RT}\right) dP = \left(\frac{\partial \ln \hat{f}_s^L}{\partial x_w}\right)_{T,P} dx_w \tag{9-72}$$

To determine the P-T relation, we must solve Eqs. (9-71) and (9-72) simultaneously for dx_w and then use the Duhem relation to eliminate one of the fugacity partials with respect to concentration. A simple procedure is to multiply Eqs. (9-71) and (9-72), respectively, by x_w and x_s. Upon adding the resultant equations, we see that terms in dx_w drop out by virtue of the Duhem equation. Thus,

$$\left(\frac{\partial P}{\partial T}\right)_{[S-L-V]} = \frac{x_w(H_w^V - \bar{H}_w^L) + x_s(H_s^S - \bar{H}_s^L)}{T[x_w(V_w^V - \bar{V}_w^L) + x_s(V_s^S - \bar{V}_s^L)]} \tag{9-73}$$

To compare this case to Eq. (9-68), assume that the predominant term in the denominator is $x_w V_w^V$ and that $V_w^V = RT/P$; then,

$$\left[\frac{\partial \ln P}{\partial\left(\frac{1}{T}\right)}\right]_{[S-L-V]} = -\frac{x_w(H_w^V - \bar{H}_w^L) + x_s(H_s^S - \bar{H}_s^L)}{Rx_w} \tag{9-74}$$

The term, $d \ln P/d(1/T)$, is again related to an enthalpy change (i.e., the change in evaporating x_w moles of water while simultaneously crystallizing x_s moles of salt). The division by x_w simply yields the result as the enthalpy change per mole of water evaporated.

We might carry Eq. (9-74) one step further. The enthalpy of mixing of water and salt is often of interest. This ΔH^{mix} is defined as

$$\Delta H^{\text{mix}} = x_w(\bar{H}_w^L - H_w^L) + x_s(\bar{H}_s^L - H_s^S) \tag{9-75}$$

where H_w^L is the enthalpy of the pure water at the system T and P. If one adds and subtracts $x_w H_w^L$ to the right-hand side of Eq. (9-74), then,

$$\left[\frac{d \ln P}{d\left(\frac{1}{T}\right)}\right]_{[S-L-V]} = \frac{\Delta H^{\text{mix}}}{Rx_w} - \frac{\Delta H_w^{\text{vap}}}{R} \tag{9-76}$$

Thus, from P-T data for saturated solutions and the heat of vaporization of the volatile component, one can determine, at least approximately, heats of mixing.

Now let us consider a binary liquid-vapor system in which both components exist in each phase. For this divariant system we shall use the additional restriction of constant liquid composition.

The pressure-temperature relation in this case follows directly from Eq. (9-22), if we denote liquid phase by β and vapor phase by α. Thus,

$$\left(\frac{\partial P}{\partial T}\right)_{x,[L-V]} = \frac{1}{T}\left[\frac{y_A(\bar{H}_A^V - \bar{H}_A^L) + y_B(\bar{H}_B^V - \bar{H}_B^L)}{y_A(\bar{V}_A^V - \bar{V}_A^L) + y_B(\bar{V}_B^V - \bar{V}_B^L)}\right] \tag{9-77}$$

Assuming that $\bar{V}_A^V \gg \bar{V}_A^L$, $\bar{V}_B^V \gg \bar{V}_B^L$, and noting that

$$y_A \bar{V}_A^V + y_B \bar{V}_B^V = V^V = \frac{RT}{P} \tag{9-78}$$

$$y_A \bar{H}_A^V + y_B \bar{H}_B^V = H^V \tag{9-79}$$

then

$$\left[\frac{\partial \ln P}{\partial\left(\frac{1}{T}\right)}\right]_{x,[L-V]} = -\frac{H^V - y_A \bar{H}_A^L - y_B \bar{H}_B^L}{R} \tag{9-80}$$

As expected, the slope of $\ln P$ vs. $1/T$ corresponds to an enthalpy change. This enthalpy change is, however, somewhat unusual. The term H^V represents the enthalpy of the saturated vapor mixture at y_A, y_B. The terms $\bar{H}_A^L$ and $\bar{H}_B^L$ represent the paritial molal enthalpy of the saturated liquid at x_A and x_B. To visualize the situation, consider Figure 9.20. The enthalpy of the saturated va-

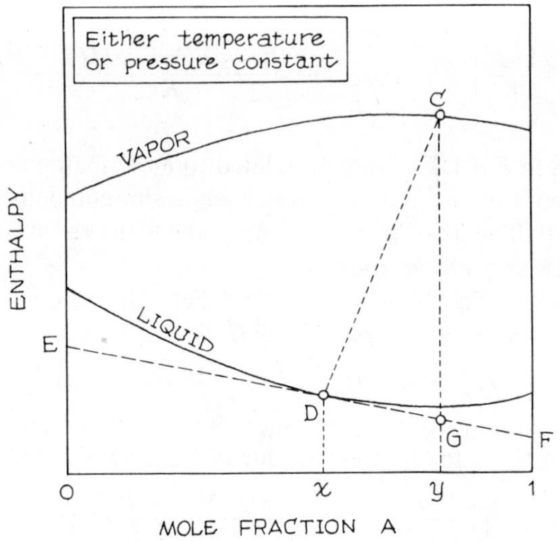

Figure 9.20 Enthalpy-concentration diagram.

por and liquid of a binary mixture of A and B is shown as a function of liquid and vapor mole fraction. Point C represents the enthalpy of a saturated vapor at y_A. This vapor is in equilibrium with a liquid of composition x_A with enthalpy H^L(point D). The diagram is drawn for either constant T or P. Consider a tangent to the lower curve at D. The tangent intersects the left-hand ordinate (pure B) at E and the right-hand ordinate (pure A) at F. Let us consider the significance of these points E and F.

If Figure 9-20 is drawn at constant temperature, then,

$$dH^L = \left(\frac{\partial H^L}{\partial P}\right)_{T,x_A} dP + \left(\frac{\partial H^L}{\partial x_A}\right)_{P,T} dx_A \tag{9-81}$$

Dividing Eq. (9-81) by dx_A and placing the restriction of saturated solution, then,

$$\left(\frac{\partial H^L}{\partial x_A}\right)_{T,[L-V]} = \left(\frac{\partial H^L}{\partial P}\right)_{T,x_A} \left(\frac{\partial P}{\partial x_A}\right)_{T,[L-V]} + \left(\frac{\partial H^L}{\partial x_A}\right)_{P,T} \tag{9-82}$$

but,

$$\bar{H}_A^L = H^L + x_B\left(\frac{\partial H^L}{\partial x_A}\right)_{T,P} \tag{9-83}$$

so by multiplying Eq. (9-82) by x_B and subtracting Eq. (9-83) from the result, we obtain:

$$\bar{H}_A^L = H^L + x_B\left(\frac{\partial H^L}{\partial x_A}\right)_{T,[L-V]} - x_B\left(\frac{\partial H^L}{\partial P}\right)_{T,x_A} \left(\frac{\partial P}{\partial x_A}\right)_{T,[L-V]} \tag{9-84}$$

A similar relation may be written for $\bar{H}_B^L$.

The $\bar{H}_A^L$ and $\bar{H}_B^L$ terms are those required for Eq. (9-80). The first two terms on the right-hand side of Eq. (9-84) yield point F on Figure 9.20. (If the equation were written for $\bar{H}_B^L$, the same terms would represent point E.) The last term on the right-hand side is a term that corrects for pressure changes in the system with composition; it is evaluated at point D. Ordinarily it is small and usually neglected. If so, then the numerator in Eq. (9-80) is clearly represented by the vertical distance $\overline{CG}$; the physical significance of this enthalpy term is then the change in enthalpy from a saturated vapor at C to a subcooled liquid at G with the same composition.

Similar reasoning yields for a constant pressure case

$$\bar{H}_A^L = H^L + x_B\left(\frac{\partial H^L}{\partial x_A}\right)_{P,[L-V]} - x_B C_p^L\left(\frac{\partial T}{\partial x_A}\right)_{P,[L-V]} \tag{9-85}$$

Since the last term on the right-hand side usually is not negligible, point G will have to be adjusted.

Many other cases requiring the interpretation of the P-T relations of liquid-vapor mixtures could be treated. The few discussed here point out the typical approach; all yielded, to a first approximation, an equation relating $\ln P$ to $1/T$ with the slope of the curve proportional to an enthalpy change between the phases.

9.5 The Integral Approach for Phase Equilibrium Relationships

The integral approach proceeds from the same starting point as the differential approach, namely, equating component fugacities in each coexisting phase. Instead of differentiating these relationships, however, we treat them directly in the integral form. In general, the procedure is to equate the fugacities of each component in each phase,

$$\hat{f}_i^\alpha = \hat{f}_i^\beta \tag{9-86}$$

and then to expand these fugacities as functions of temperature, pressure, and composition. First, for each phase, we expand in terms of activity coefficients; e.g., for phase α,

$$\hat{f}_i^\alpha = \hat{f}_i^0 \gamma_i^\alpha x_i^\alpha \tag{9-87}$$

$\hat{f}_i^0$ is the fugacity of component i in a reference state defined by T^0, P^0, and $x_1^0, \ldots, x_{n-1}^0$ and state of aggregation, which may be different from the α-state. It should be recalled from Chapter 8 that the specification of the reference state is necessary to define γ.

With the given reference state, each of the properties on the right-hand side of Eq. (9-87) can then be expanded as functions of the intensive variables. Thus,

$$\hat{f}_i^0 = g_{1i}(T^0, P^0, x_1^0, \ldots, x_{n-1}^0) \tag{9-88}$$

and

$$\gamma_i^\alpha = g_{2i}(T, P, x_1^\alpha, \ldots, x_{n-1}^\alpha, T^0, P^0, x_1^0, \ldots, x_{n-1}^0) \tag{9-89}$$

With these substitutions, Eq. (9-86) becomes a function in the $2n$ intensive variables $T, P, x_1^\alpha, \ldots, x_{n-1}^\alpha$, and $x_1^\beta, \ldots, x_{n-1}^\beta$. Furthermore, there are n such relations involving the α and β phases, one for each component. For a system of π phases, there are $(\pi - 1)n$ relations involving $\pi(n - 1) + 2$ intensive properties; these equations are the necessary set prescribed by the phase rule.

When the functions in Eqs. (9-88) and (9-89) are known, the integral approach yields directly a set of equations which, when solved simultaneously, provides relationships between intensive variables involving P, T, and x. In this case, the integral approach is clearly more direct than the differential approach.

In applying the integral approach, the choice of the reference state usually determines the ease in reaching a given objective. The activity coefficient has its greatest use when it can be related to parameters that describe only the mixing characteristics of the pure components. For example, it can be shown that in the general case,

$$\ln \gamma_i^\sigma = \ln\left(\frac{\hat{f}_i^\sigma}{\hat{f}_i^0}\right) + \frac{\overline{\Delta G}_i^{\sigma, EX}}{RT} \tag{9-90}$$

where $\overline{\Delta G}_i^{\sigma, EX}$ is the excess Gibbs free energy of mixing pure components at the same temperature and pressure and when these pure components are in the same state of aggregation as the mixture. This equation is valid for any choice of reference state at $T^0, P^0, x_1^0, \ldots, x_{n-1}^0$. Since the fugacity ratio in Eq. (9-90) is a function of S_i^0 when T^0 is not equal to T, the reference state temperature is invariably chosen as the system temperature. Usually (but not always) P^0 is chosen as the system pressure. In addition, whenever the pure component exists (or could possibly be imagined to exist) in the same state of aggregation as the mixture at T and P, this pure state is the most convenient reference state. In this case,

$$\hat{f}_i^0 = f_i^\sigma \tag{9-91}$$

and

$$\ln \gamma_i^\sigma = \frac{\overline{\Delta G}_i^{\sigma, EX}}{RT} \tag{9-92}$$

Thus, the activity coefficient can be expressed directly in terms of a model that describes the mixing process.

Let us now examine some common cases in which the reference state is taken as the pure component at the temperature, pressure, and state of aggregation of the mixture.

The fugacity of a component in a *vapor-phase* mixture can be expressed in the form of Eq. (9-87),

$$\hat{f}_i^V = f_i^V \gamma_i^V y_i \tag{9-93}$$

where, as noted above, f_i^V refers to the fugacity of pure i as a *vapor* at the system temperature and pressure. Eq. (9-93) is often used when the vapor mixture is an ideal solution (see Section 8.4) (i.e., where $\gamma_i = 1$). Nevertheless, it often proves to be troublesome if the stable state for i at T and P is not a vapor, for in this case, evaluating f_i as a vapor requires the introduction of the hypothetical state. This has led to a different approach in most treatments of vapor phase fugacity, i.e. , with the concept of the fugacity coefficient, ϕ_j,

$$\hat{f}_i^V = \phi_i P y_i \tag{9-94}$$

where ϕ_i is introduced by Eq. (8-151) and may be evaluated from a mixture equation of state. In Appendix E, for example, an expression for ϕ_i is shown for the Redlich-Kwong equation. With Eq. (9-94), no hypothetical states are introduced. For an ideal gas mixture, $\phi_i = 1$, and

$$\hat{f}_i^V = P y_i \tag{9-95}$$

The fugacity of components in liquid or solid mixtures is almost always expressed in terms of the activity coefficient form since to evaluate the fugacity coefficient for a component in a condensed phase would require an integration across the two phase region.[10] Thus, Eq. (9-87) becomes

$$\hat{f}_i^\sigma = f_i^\sigma \gamma_i x_i \tag{9-96}$$

and the reference state is pure i at the same T, P, and condensed state of aggregation as the mixture. To calculate f_i^σ, it is usually convenient to determine first the fugacity of pure i, in the condensed phase, at T under its equilibrium vapor pressure and then apply a correction term for the fact that $P_{vp_i} \neq P$.

$$\ln\left(\frac{f_{i,P}^\sigma}{f_{i,P_{vp}}^\sigma}\right) = \int_{P_{vp_i}}^{P} \left(\frac{\partial \ln f_i^\sigma}{\partial P}\right)_T dP = \frac{1}{RT}\int_{P_{vp_i}}^{P} V_i^\sigma \, dP \tag{9-97}$$

or

$$f_{i,P}^\sigma = (f_{i,P_{vp}}^\sigma) \exp\left[\frac{1}{RT}\int_{P_{vp_i}}^{P} V_i^\sigma \, dP\right] \tag{9-98}$$

If the condensed phase may be regarded as incompressible over the limits of integration, Eq. (9-98) simplifies to

$$f_{i,P}^\sigma = (f_{i,P_{vp}}^\sigma) \exp\left[\frac{V_i^\sigma (P - P_{vp_i})}{RT}\right] \tag{9-99}$$

The exponential term in Eqs. (9-98) and (9-99) is commonly called the *Poynting correction factor*. It can usually be neglected for pressures within an order

[10] Such integrations are made with some equations of state (e.g., the Benedict-Webb-Rubin for light hydrocarbons), but ordinarily the results yield inaccurate values for liquid fugacities.

of magnitude of P_{vp} and at temperatures not near the critical. [For example, if $(P - P_{vp}) = 10$ atm, $T = 400°K$, and $V_i^\sigma = 100$ cm³/g-mol, the correction factor is 1.031, or 3.1%]. The fugacity of pure i in the σ phase at T and P_{vp_i} is, of course, equal to the fugacity of pure i as a saturated vapor at T:

$$f_{i,P_{vp}}^\sigma = f_{i,P_{vp}}^V \qquad (9\text{-}100)$$

Under conditions in which the Poynting factor can be neglected and in which pure i vapor at T and P_{vp} is ideal,

$$\hat{f}_i^\sigma = P_{vp_i}^\sigma \gamma_i^\sigma x_i^\sigma \qquad \text{(vapor of pure } i \text{ is ideal at } P_{vp_i}\text{, Poynting factor negligible)} \qquad (9\text{-}101)$$

In addition, when the liquid mixture is ideal, γ_i^σ is unity and Eq. (9-101) simplifies to:

$$\hat{f}_i^\sigma = P_{vp_i}^\sigma x_i^\sigma \qquad (i\text{-vapor ideal at } P_{vp_i}\text{, Poynting factor negligible, ideal liquid mixture)} \qquad (9\text{-}102)$$

For equilibria involving a vapor and a condensed phase, three common cases are shown in Eqs. (9-103) through (9-105).

The Lewis and Randall rule,

$$f_i^V y_i = f_i^\sigma x_i^\sigma \qquad (9\text{-}103)$$

is valid when the vapor and condensed phases are ideal mixtures.

Raoult's law,

$$p_i = y_i P = P_{vp_i}^\sigma x_i^\sigma \qquad (9\text{-}104)$$

is valid when the vapor is an ideal mixture of ideal gases, the liquid is an ideal mixture, the saturated vapor of i at T and P_{vp_i} is ideal, and the Poynting correction is negligible.

A modification of Raoult's law in which the condensed phase is not ideal,

$$p_i = y_i P = P_{vp_i}^\sigma \gamma_i^\sigma x_i^\sigma \qquad (9\text{-}105)$$

is also commonly employed.

The following example attempts to illustrate many of the approximations and subtleties in applying the integral approach. Although it is quite lengthy, the reader is encouraged to follow the ideas closely because the procedure can be extremely helpful in understanding many phase equilibrium concepts.

Example 9.5

At atmospheric pressure, the binary system of n-butanol-water exhibits a minimum boiling azeotrope and partial miscibility in the liquid phase. The system has an upper consolute temperature at 124.8°C. Solubility limits as a function of temperature and x-y-T data at one atmosphere are tabulated below, along with pertinent data for the pure components.

(a). Using the x-y-T data, estimate the liquid phase activity coefficients of n-butanol and water as a function of composition at 1 atm and 100°C.

(b). The results of (a) were used in Example 8.11 to determine van Laar

DATA FOR EXAMPLE 9.5
VAPOR AND LIQUID MOLE FRACTIONS OF *n*-BUTANOL IN EQUILIBRIUM WITH WATER AT 1.009 ATM[11]

T, °C	y	x
100	0.0	0.0
95.8	0.150	0.008
95.4	0.161	0.009
92.8	0.237	0.019
92.8	0.240	0.020
92.7	0.246	0.098
92.7	0.246	0.099
92.7	0.246	0.247
93.0	0.250	0.454
93.0	0.247	0.450
96.3	0.334	0.697
94.0	0.276	0.583
96.6	0.340	0.709
106.4	0.598	0.903
100.8	0.444	0.819
106.8	0.612	0.908
110.9	0.747	0.950
117.5	1.000	1.000

DATA FOR EXAMPLE 9.5 (CONTINUED)
LIQUID-LIQUID SOLUBILITY LIMITS FOR THE *n*-BUTANOL-WATER SYSTEM
(x' AND x'' ARE LIQUID MOLE FRACTIONS OF *n*-BUTANOL IN THE TWO PHASES)

T, °C	x'	x''
80.0	0.0163	0.3927
90.0	0.0201	0.3595
100.0	0.0212	0.3243
105.0	0.0257	0.3040
110.0	0.0268	0.2795
115.0	0.0311	0.2496
120.0	0.0401	0.2116
123.0	0.0539	0.1760
124.8	0.1042	0.1042

expansions for γ. These are given in Eqs. (8-211) and (8-212). Plot y-x and T-x diagrams at 1 atm and compare these curves with the actual experimental data. Indicate the azeotropic composition and temperature as well as the solubility limits.

[11] T. E. Smith, and R. F. Bonner "Vapor-Liquid Equilibrium Still for Partially Miscible Liquids," *Ind. Eng. Chem.*, **41**, (1949) 2867.

(c). Make what you consider to be a reasonable assumption as a basis for modifying the van Laar expansions in order to predict the upper consolute temperature.

(d). Using the experimentally determined solubility limits, determine the van Laar coefficients at the various temperatures for which data are available. Describe how you would modify the van Laar expansion to cover a wider temperature range, and show how the liquid phase enthalpy of mixing could be calculated from your modified van Laar expansion.

DATA FOR EXAMPLE 9.5 (CONTINUED)
PURE COMPONENT DATA

	n-butanol	water
Enthalpy of vaporizaton at the normal boiling point, kcal/g-mol	10.47	9.729
Critical temperature, °K	563.0	647
Critical pressure, atm	43.6	218
Acentric factor	0.60	0.348

Solution

(a). Equating Eq. (9-94) to Eq. (9-96), for both components,

$$\phi_i P y_i = f_i^L \gamma_i x_i \tag{9-106}$$

The pressure is one atmosphere and y, x are given as a function of temperature in the table of data. The fugacity coefficients can be determined rigorously from Eq. (8-163), provided an equation of state for the mixture were available and applicable. For example, if the simple Redlich-Kwong equation of state were chosen, then ϕ_i is given by Eq. (E-19) of Appendix E. This particular equation is probably a poor choice for such a polar gas mixture and more specialized methods should be employed. O'Connell and Prausnitz[12] suggest a truncated virial equation and have outlined a procedure to determine the fugacity coefficient for components in polar gas mixtures. If their method is used, ϕ_{H_2O} and ϕ_{n-C_4OH} are, for all practical purposes, unity. Thus, by using these values, we are treating the gas phase as an ideal gas mixture. Although probably not a bad assumption at such a low pressure, one is cautioned that the ideal gas assumption may be very poor for systems at higher pressures.[13]

[12] J. P. O'Connell, and J. M. Prausnitz "Empirical Correlation of Second Virial Coefficients for Vapor-Liquid Equilibrium Calculations," *Ind. Eng. Chem. Proc. Des. Dev.* 6, (1967) 245.

[13] An excellent discussion of ways to calculate fugacity coefficients is given in J. M. Prausnitz, *Molecular Thermodynamics of Fluid Phase Equilibria* (Englewood Cliffs, N. J.: Prentice-Hall, Inc., 1969), Chap. 5.

Next, we must determine f_i^L, the fugacity of each pure component as a liquid at 1 atm and 100°C. We could use Eq. (9-98). The Poynting correction in this case is negligible and we may then write:

$$f^{\sigma}_{i,P_{vp}} = f^{L}_{i,P_{vp}} = f^{V}_{i,P_{vp}} = \left(\frac{f}{P}\right)_{i,P_{vp}} P_{vp_i} \tag{9-107}$$

The ratio (f/P) at 100°C and at the pure component vapor pressures is found from Appendix D, Eq. (D-18), Table D.3. The results are shown below:

	n-butanol	water
$T_r = T/T_c = 373.2/T_c$	0.663	0.577
$P_r = P/P_c = 1/P_c$	0.023	0.0046
$\log[(f/P)]^{(0)}$	~ 0	~ 0
$\log[(f/P)]^{(1)}$	~ 0	~ 0
$\log(f/P)$	~ 0	~ 0
(f/P)	~ 1.0	~ 1.0

As in the fugacity coefficient estimation, the low system pressure leads us to the conclusion that pure *n*-butanol or water vapor at 1 atm and 100°C is essentially an ideal gas, thus,

$$f^{L}_{i,P_{vp}} = f^{V}_{i,P_{vp}} = P_{vp_i} \tag{9-108}$$

With these assumptions, our problem has simplified to Eq. (9-105) and γ_i may be determined after the vapor pressures of both components are determined at 100°C. Such vapor pressures are found from data sources, from the normal boiling point and the Clausius-Clapeyron equation, Eq. (9-63), or by generalized methods outlined in Appendix D.

One final step remains, however. If we use the given *x*-*y*-*T* data, then we will have determined activity coefficients for a range of temperatures rather than at 100°C. To correct the activity coefficients to 100°C, use is made of Eq. (9-109).

$$\left(\frac{\partial \ln \gamma_i}{\partial T}\right)_{P,x} = -\frac{\overline{\Delta H_i}}{RT^2} \tag{9-109}$$

Over the small temperature range (92-117°C), $\overline{\Delta H_i}$ for both components can probably be assumed constant, i.e., integrating,

$$\ln\left(\frac{\gamma_{i,T_2}}{\gamma_{i,T_1}}\right) = \frac{\overline{\Delta H_i}}{R}\left(\frac{1}{T_2} - \frac{1}{T_1}\right) \qquad (P, x \text{ constant}) \tag{9-110}$$

We ordinarily lack data for component partial molal enthalpies of mixing and, thus, Eqs. (9-109) and (9-110) are normally of little value. They are usually small, however, and, over *small* ranges of ΔT, the effect of temperature on γ is normally neglected.[14] We could however, make an approximate correction, assuming that the solution exhibited regular behavior. In this case (see footnote 19 in Chapter 8), the product ($T \ln \gamma_i$), at constant pressure and composition, is constant. In Table 9.1 we show values of γ_B (*n*-butanol)

TABLE 9.1
ACTIVITY COEFFICIENTS FOR THE SYSTEM *n*-BUTANOL AND WATER AT ONE ATMOSPHERE

x_B	T (°C)	γ_B (at T)	γ_W (at T)	γ_B (at 100°C)	γ_W (at 100°C)
.008	95.8	41.83	1.004	40.11	1.004
.009	95.4	40.53	1.006	38.73	1.006
.019	92.8	31.22	1.014	29.23	1.014
.020	92.8	30.10	1.013	28.18	1.013
.098	92.7	6.32	1.096	6.10	1.094
.099	92.7	6.26	1.097	6.04	1.095
.247	92.7	2.51	1.313	2.46	1.306
.450	93.0	1.366	1.775	1.36	1.756
.454	93.0	1.369	1.780	1.36	1.761
.583	94.0	1.133	2.17	1.131	2.14
.697	96.3	1.047	2.52	1.046	2.50
.709	96.6	1.035	2.58	1.035	2.56
.819	100.8	0.997	3.01	0.997	3.02
.903	106.4	0.991	3.35	0.991	3.42
.908	106.8	0.992	3.36	0.992	3.43
.950	110.9	0.999	3.51	0.999	3.64

and γ_W (water) both as calculated at the system temperature and as corrected to 100°C assuming regular solutions.

These activity coefficients may be expressed in the van Laar form as:

$$\ln \gamma_B = \frac{\left(\dfrac{1274}{T}\right)}{\left[1 + 2.911\left(\dfrac{x_B}{x_W}\right)\right]^2}$$

$$\ln \gamma_W = \frac{\left(\dfrac{438}{T}\right)}{\left[1 + 0.343\left(\dfrac{x_W}{x_B}\right)\right]^2} \tag{9-111}$$

where T is in °K. If the system temperature is used, γ values in columns 3

[14] If we were merely interested in an analytical expression for γ to be used only under conditions of 1 atm and L-V equilibrium, it would not be necessary to include the temperature correction term since all we would do to calculate y-x, T-x, and T-y is reverse the procedure for calculating γ. Here, however, we are interested in applying the γ-expansions to other conditions.

and 4 of Table 9.1 are correlated; if T is set equal to 373.2°K, then the γ values of columns 5 and 6 are found.

(b). With Eqs. (9-105), (9-110), and (9-111) and vapor pressures of both n-butanol and water as a function of temperature,[15] one can determine γ values for various liquid compositions and temperatures. One way to perform the calculation is to solve Eq. (9-105) for both y_B and y_W, add to eliminate y_B and y_W, then substitute Eqs. (9-110) and (9-111) for γ_B and γ_W. The result is:

$$P = x_B P_{vp,B}(T)\gamma_B(T, x_B) + x_W P_{vp,W}(T)\gamma_W(T, x_B) \qquad (9\text{-}114)$$

For $P = 1$ atm, Eq. (9-114) must be solved by a trial-and-error procedure to relate T and x_B. With T and x_B values, Eq. (9-105) is used to obtain y_B and y_W.

The results are given as the smooth curves in Figures 9.21 and 9.22. The experimental points are also shown. The maximum and minimum in

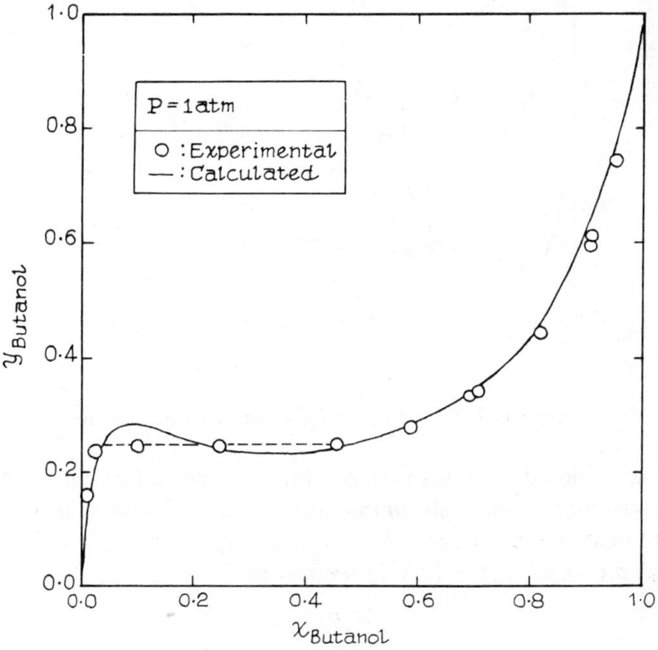

Figure 9.21 x-y diagram for n-butanol-water.

[15] Approximate expressions for vapor pressures (in atmospheres) are:

$$P_{vp,B} = \exp\left[-5{,}280\left(\frac{1}{T} - \frac{1}{390.7}\right)\right] \qquad (9\text{-}112)$$

$$P_{vp,W} = \exp\left[-4{,}890\left(\frac{1}{T} - \frac{1}{373.2}\right)\right] \qquad (9\text{-}113)$$

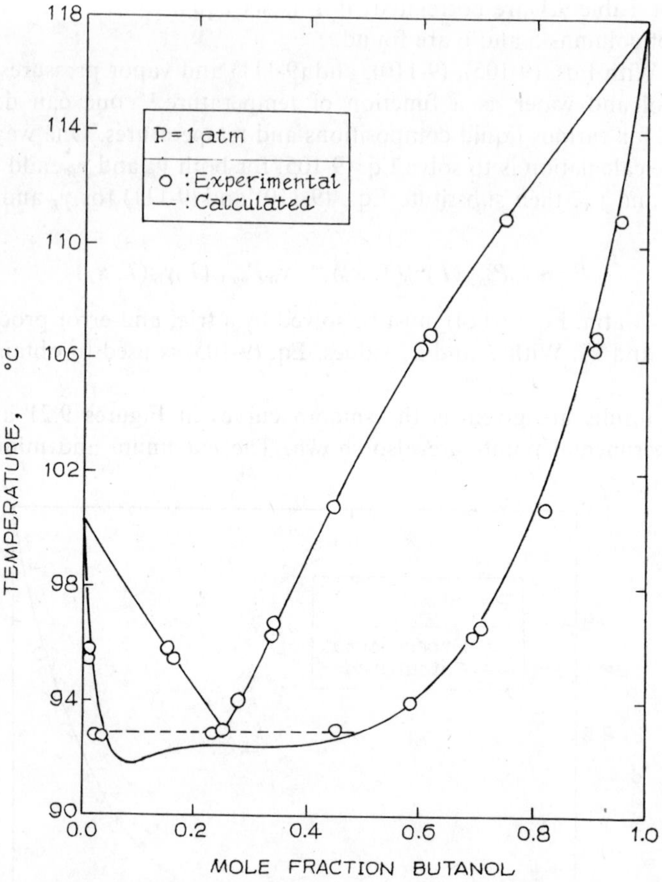

Figure 9.22 *T-x-y* diagram for *n*-butanol-water system.

the *y-x* curve should alert us to the possibility of phase instability. If we wish to check stability, a relatively simple procedure at this point in our calculations is to inspect the variation of the activity, $\gamma_i x_i$, with x_i. In terms of activity, for phase stability, Eq. (7-182) reduces to

$$\frac{d \ln a_B}{dx_B} > 0 \qquad (9\text{-}115)$$

and

$$\frac{d \ln a_W}{dx_B} < 0 \qquad (9\text{-}116)$$

Calculated values of a_B and a_W are shown in Figure 9.23. By inspection, the limits of stability are x_B less than 0.09 and greater than 0.34. To determine the concentrations of the two liquid phases, we must invoke the criteria of equilibrium, which in terms of activities are:

$$a'_B = a''_B \tag{9-117}$$

and

$$a'_W = a''_W \tag{9-118}$$

The conditions satisfying these relations can be found by trial-and-error inspection of Figure 9.23. An alternative procedure is to plot the concentrations of x'_B and x''_B which satisfy each of Eqs. (9-117) and (9-118), as shown in Figure 9.24. The concentrations of the coexisting liquid phases are found to be 0.0397 and 0.505 at the intersection of the two curves.

Given the concentrations of the two liquid phases, we can complete the construction of the y-x and T-x, T-y curves in Figures 9.21 and 9.22 by draw-

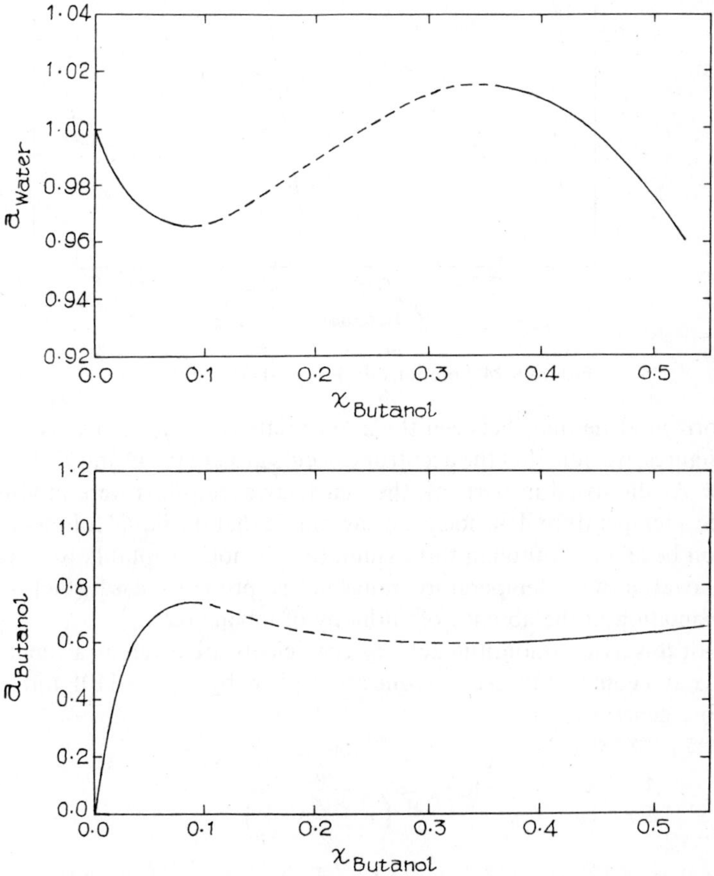

Figure 9.23 Activities in the *n*-butanol-water system at one atmosphere.

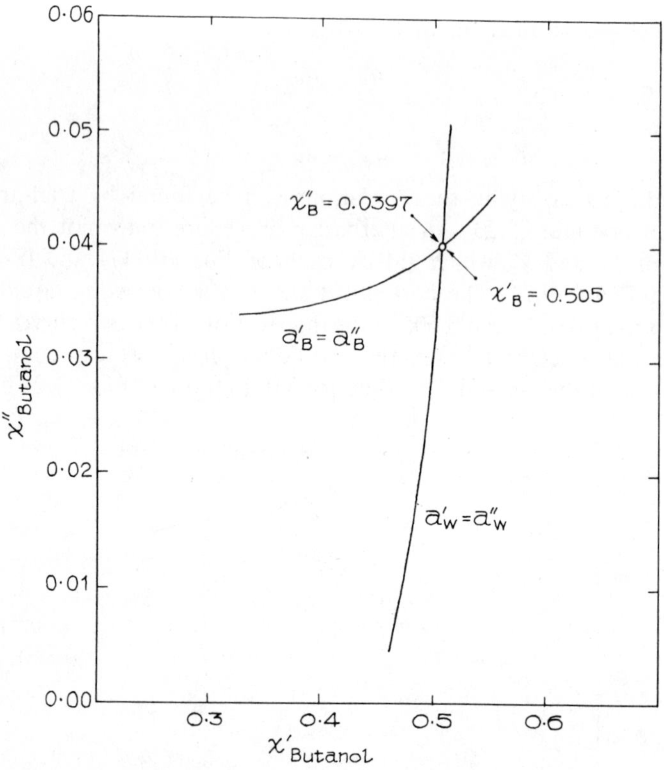

Figure 9.24 Solutions to Eqs. (9-117) and (9-118).

ing horizontal tie-lines between the appropriate concentrations. Thus, from these figures, we note that the azeotrope occurs at x_B of 0.249 and T of 92.9°C.

(c). As discussed in part (a), the van Laar expansions were modified to include a temperature-dependence by assuming that the liquid follows regular solution behavior. Although this assumption is not acceptable for extrapolation over a wide temperature range, it is probably a satisfactory first approximation in the absence of enthalpy of mixing data.

With this assumption, the activity coefficients are given as a function of T and x at a constant pressure of one atmosphere by Eqs. (9-110) and (9-111) or, more generally,

$$\ln \gamma_B = \frac{\dfrac{A_{12}}{T}}{\left(1 + \dfrac{A_{12}}{A_{21}}\dfrac{x_A}{x_B}\right)^2} \tag{9-119}$$

$$\ln \gamma_W = \frac{\dfrac{A_{21}}{T}}{\left(1 + \dfrac{A_{21}}{A_{12}}\dfrac{x_B}{x_A}\right)^2} \tag{9-120}$$

For equilibrium to exist between two liquid phases,

$$x'_B\gamma'_B(T, P, x'_B) = x''_B\gamma''_B(T, P, x''_B) \tag{9-121}$$

$$x'_W\gamma'_W(T, P, x'_B) = x''_W\gamma''_B(T, P, x''_B) \tag{9-122}$$

where, to distinguish the two liquid phases, the superscripts prime and double prime are used.

Eqs. (9-121) and (9-122) are to be used to determine the upper consolute temperature. If Eqs. (9-119) and (9-120) are to be substituted for activity coefficients, a pressure correction may have to be made since these γ-correlations were obtained at one atmosphere and the pressure at the consolute temperature may be quite different.

If such a pressure-correction to γ is necessary, as shown in Chapter 8,

$$\ln\left[\frac{\gamma_i(T, P_2, x_i)}{\gamma_i(T, P_1, x_i)}\right] = \frac{1}{RT}\int_{P_1}^{P_2} \overline{\Delta V_i^L}\, dP \tag{9-123}$$

In our case, P_1 would be one atmosphere and P_2 the pressure of the system at the consolute temperature. P_2 is, of course, not known.

Eqs. (9-121) and (9-122), as modified by Eq. (9-123), are two equations in three unknowns. The minimum pressure at which the two liquid phases can coexist is that corresponding to a bubble point at T. Thus, the third equation necessary to specify T, P, and x is related to the liquid-vapor equilibrium and would have a form similar to Eq. (9-114), but with the activity coefficient terms modified for T- and P-corrections.

Fortunately, the pressure correction to the activity coefficient is normally very small, except for temperatures near the critical. Thus, in this case, Eqs. (9-121) and (9-122) reduce to

$$x'_B\gamma'_B(T, x'_B) = x''_B\gamma''_B(T, x''_B) \tag{9-124}$$

and

$$x'_W\gamma'_W(T, x'_B) = x''_W\gamma''_W(T, x''_B) \tag{9-125}$$

or

$$\ln\frac{x'_B}{x''_B} = \frac{A_{12}}{T}\left[\frac{1}{\left(1+\dfrac{A_{12}}{A_{21}}\dfrac{x''_B}{x''_W}\right)^2} - \frac{1}{\left(1+\dfrac{A_{12}}{A_{21}}\dfrac{x'_B}{x'_W}\right)^2}\right] \tag{9-126}$$

and

$$\ln\frac{x'_W}{x''_W} = \frac{A_{21}}{T}\left[\frac{1}{\left(1+\dfrac{A_{21}}{A_{12}}\dfrac{x''_W}{x''_B}\right)^2} - \frac{1}{\left(1+\dfrac{A_{21}}{A_{12}}\dfrac{x'_W}{x'_B}\right)^2}\right] \tag{9-127}$$

Since these equations result in a trivial solution at a consolute temperature, we must find the consolute temperature by indirect means. One method is to solve for sets of x', x'', and T, and then plot x' vs. x'' and extrapolate to the point where $x' = x''$, as described below.

To obtain equilibrium sets of x', x'', and T from Eqs. (9-126) and (9-127), a trial-and-error procedure is required. For example, we can choose a value of x'_B and then vary x''_B and solve the two equations for T until we find a value of x''_B for which the two calculated temperatures agree. Again, this is a relatively simple process if a computer is available, but it is a very tedious task if done by hand calculation.

The results are shown by the smooth curves in Figure 9.25. The upper consolute solution temperature is predicted to be 177.6°C with a corresponding liquid concentration of 18.4 mole percent butanol.

Also shown in Figure 9.25 are the experimentally determined temperatures and concentrations. There is a sizable discrepancy between the experimental and predicted values. At 92.9°C [the bubble point in the two-phase region at 1 atm, as found in part (b)], the experimental concentrations are 2.0 and 35 mole percent n-butanol compared to 3.97 and 50.5 mole percent predicted from the vapor-liquid equilibrium.

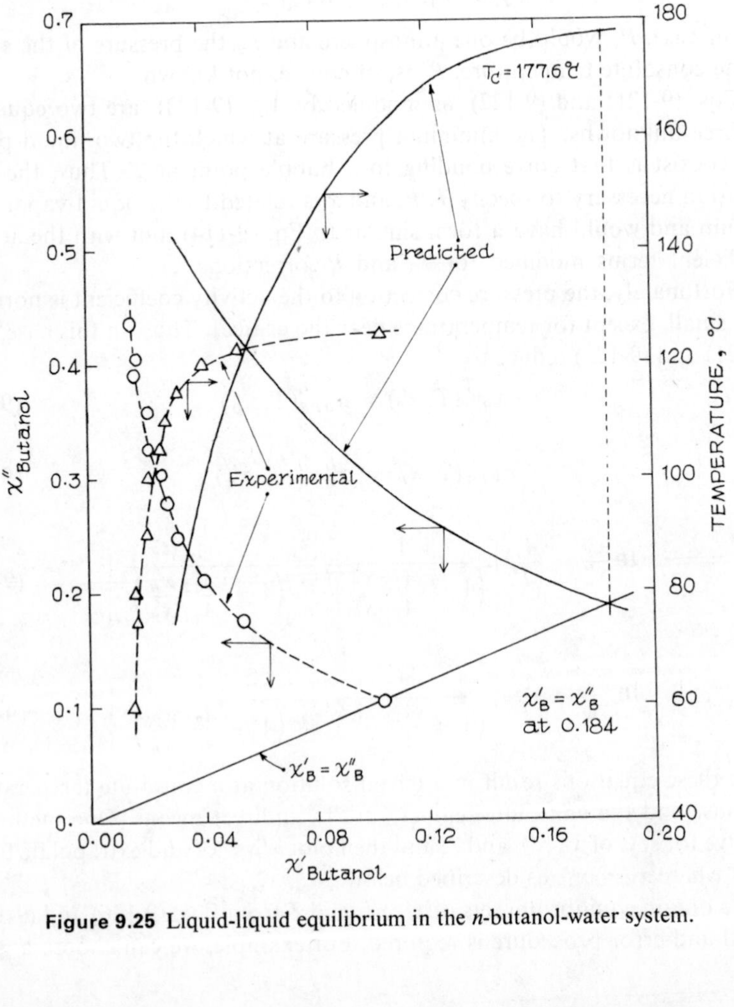

Figure 9.25 Liquid-liquid equilibrium in the n-butanol-water system.

It is doubtful if a higher-order expansion of the activity coefficient would reduce the discrepancy. As seen in Figures 9.21 and 9.22, the van Laar expansion gives a good prediction of the vapor-liquid equilibrium data. It is more probable that the temperature-correction applied to the van Laar expansions is not accurate.

(d). Given the data relating x'_B, x''_B, and T, we can reverse the procedure of part (c) to solve for the van Laar constants. Since we would like to determine the effect of temperature on γ, let us replace the terms A_{12}/T and A_{21}/T by A_B and A_W, respectively, and then determine the variation of the latter parameters with T. Making these substitutions in Eqs. (9-126) and (9-127), we must solve these expressions simultaneously for A_B and A_W as functions of x' and x''.

Let us first solve for the ratio of coefficients, $A = A_B/A_W$. Dividing Eq. (9-126) by Eq. (9-127), we obtain

$$\frac{\ln\left(\frac{x'_B}{x''_B}\right)}{\ln\left(\frac{x'_W}{x''_W}\right)} = A\left[\frac{\dfrac{1}{\left(1 + A\dfrac{x''_B}{x''_W}\right)^2} - \dfrac{1}{\left(1 + A\dfrac{x'_B}{x'_W}\right)^2}}{\dfrac{1}{\left(1 + \dfrac{1}{A}\dfrac{x''_W}{x''_B}\right)^2} - \dfrac{1}{\left(1 + \dfrac{1}{A}\dfrac{x'_W}{x'_B}\right)^2}}\right] \tag{9-128}$$

which can be readily simplified to

$$\frac{\ln\left(\frac{x'_B}{x''_B}\right)}{\ln\left(\frac{x'_W}{x''_W}\right)} = \frac{2 + A\left(\dfrac{x'_B}{x'_W} + \dfrac{x''_B}{x''_W}\right)}{\dfrac{x'_B}{x'_W} + \dfrac{x''_B}{x''_W} + 2A\left(\dfrac{x''_B}{x''_W}\right)\left(\dfrac{x'_B}{x'_W}\right)}$$

or

$$A = \frac{A_B}{A_W} = \frac{\left(\dfrac{x'_B}{x'_W} + \dfrac{x''_B}{x''_W}\right)\dfrac{\ln\left(\frac{x'_B}{x''_B}\right)}{\ln\left(\frac{x'_W}{x''_W}\right)} - 2}{\dfrac{x'_B}{x'_W} + \dfrac{x''_B}{x''_W} - 2\left(\dfrac{x''_B}{x''_W}\right)\left(\dfrac{x'_B}{x'_W}\right)\dfrac{\ln\left(\frac{x'_B}{x''_B}\right)}{\ln\left(\frac{x'_W}{x''_W}\right)}} \tag{9-129}$$

Having solved Eq. (9-129) for A, we can find A_B by rewriting Eq. (9-126) in the form:

$$A_B = \frac{\ln\left(\frac{x'_B}{x''_B}\right)}{\left(1 + A\dfrac{x''_B}{x''_W}\right)^2 - \left(1 + A\dfrac{x'_B}{x'_W}\right)^2} \tag{9-130}$$

From these equations, values of A_B and A_W were determined from the liquid-liquid equilibrium data. The results are shown in Figure 9.26, in which

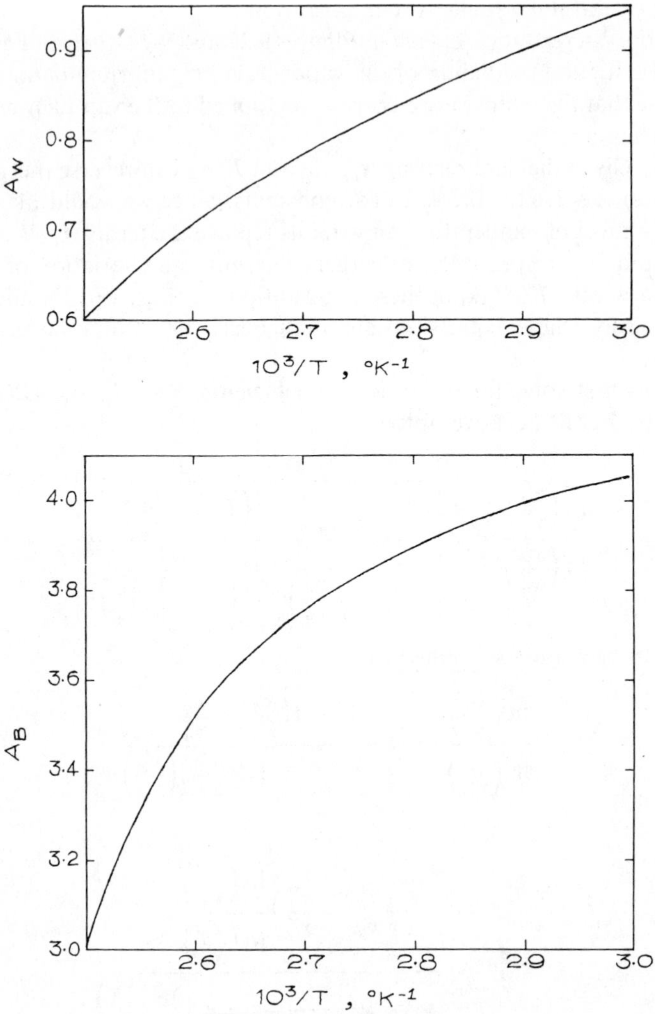

Figure 9.26 Effect of temperature on A_B and A_W.

the values were plotted as a function of $1/T$. If the partial molal enthalpies of solution were independent of temperature, such plots would be linear; if, in addition, the solutions were regular, the slopes would intersect the origin. It can be seen from Figure 9.26 that the solutions do not exhibit regular behavior and that the partial molal enthalpies of solution are not constant (although they do not vary by much more than a factor of two).

If we wished to modify the van Laar expansion to cover the temperature range from 80 to 125°C, we could have fitted power series expansions in T for A_B and A_W. Since the variations of γ_B and γ_W with T are related to partial molal enthalpies of mixing, $\overline{\Delta H_i}$, we should be able to calculate these quantities from van Laar expressions that contain the correct temperature functionality of A_B and A_W. Thus,

$$\overline{\Delta H_B} = -RT\left(\frac{\partial \ln \gamma_B}{\partial T}\right)_{P,x} \tag{9-131}$$

and

$$\begin{aligned}\left(\frac{\partial \ln \gamma_B}{\partial T}\right)_{P,x} &= \frac{\dfrac{dA_B}{dT}}{\left(1+\dfrac{A_B}{A_W}\dfrac{x_B}{x_W}\right)^2} - \frac{2A_B\left(\dfrac{x_B}{x_W}\right)\left[\dfrac{1}{A_W}\dfrac{dA_B}{dT} - \dfrac{A_B}{(A_W)^2}\dfrac{dA_W}{dT}\right]}{\left(1+\dfrac{A_B}{A_W}\dfrac{x_B}{x_W}\right)^3} \\ &= \frac{\left(\dfrac{dA_B}{dT}\right)}{\left(1+\dfrac{A_B}{A_W}\dfrac{x_B}{x_W}\right)^2}\left\{1 - \frac{2\dfrac{x_B}{x_W}\left[1-\left(\dfrac{A_B}{A_W}\right)\left(\dfrac{dA_W}{dA_B}\right)\right]}{1+\dfrac{A_B}{A_W}\dfrac{x_B}{x_W}}\right\}\end{aligned} \tag{9-132}$$

where the primes have been left off the mole fractions since either phase may be used.

$$\overline{\Delta H_B} = \frac{R\left[\dfrac{dA_B}{d\left(\dfrac{1}{T}\right)}\right]}{\left(1+\dfrac{A_B}{A_W}\dfrac{x_B}{x_W}\right)^2}\left\{1 - \frac{2\left(\dfrac{x_B}{x_W}\right)\left[1-\left(\dfrac{A_B}{A_W}\right)\left(\dfrac{dA_W}{dA_B}\right)\right]}{1+\dfrac{A_B}{A_W}\dfrac{x_B}{x_W}}\right\} \tag{9-133}$$

and

$$\overline{\Delta H_W} = \frac{R\left[\dfrac{dA_W}{d\left(\dfrac{1}{T}\right)}\right]}{\left(1+\dfrac{A_W}{A_B}\dfrac{x_W}{x_B}\right)^2}\left\{1 - \frac{2\left(\dfrac{x_W}{x_B}\right)\left[1-\left(\dfrac{A_W}{A_B}\right)\left(\dfrac{dA_B}{dA_W}\right)\right]}{1+\dfrac{A_W}{A_B}\dfrac{x_W}{x_B}}\right\} \tag{9-134}$$

Enthalpies of solution,

$$\Delta H_m = x_B\,\overline{\Delta H_B} + x_W\,\overline{\Delta H_W} \tag{9-135}$$

were calculated from Eqs. (9-133) and (9-134) by graphically differentiating the curves in Figure 9.26. The results for 90°C and 115°C are shown in Figure 9.27 (the dashed lines are the hypothetical lines in the two-phase region). Although some readers may find the shape of these curves unusual (and, hence, suspect), curves of this type wherein ΔH_m exhibits an inversion are not uncommon for organic-water solutions in which the organic has a functional group capable of hydrogen-bonding with water. In fact, the curves in Figure 9.27 are very similar to those determined experimentally for aqueous solutions of ethanol, propanol, and dioxane.[16]

[16] Although a detailed discussion of Figure 9.27 is beyond the scope of this book, the interested reader is referred to J. S. Rowlinson, *Liquids and Liquid Mixtures*, 2nd ed. (New York: Plenum Press), 1969.

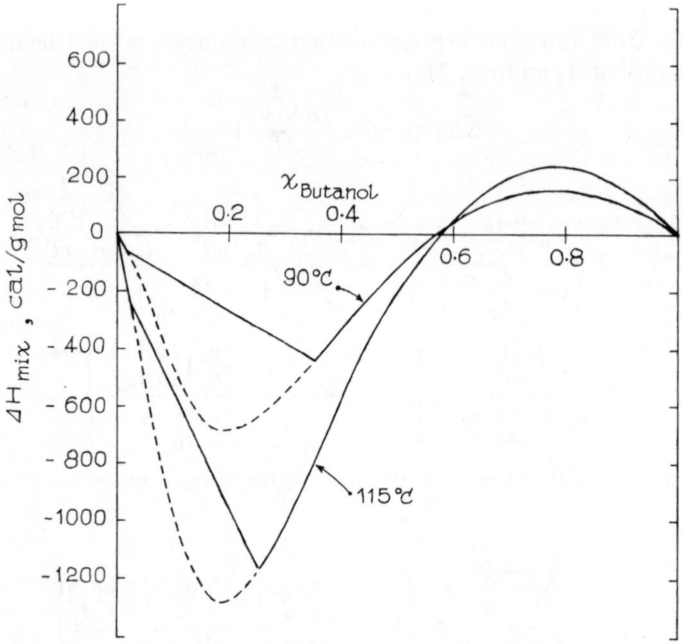

Figure 9.27 Heat of mixing for the *n*-butanol-water system.

Returning to our general discussion of the integral approach to phase equilibria, we have heretofore concentrated on cases for which the reference states could be chosen conveniently as pure materials at the temperature, pressure, and state of aggregation of the mixture. In this case, we have seen that the activity coefficients are directly related to the Gibbs free energy of mixing, Eq. (9-92).

There are many important areas of application in which a pure component of a mixture does not normally exist in the same state of aggregation as the mixture at the temperature and pressure of the mixture. For example, in a liquid mixture near the triple-point region, one of the components may be stable as a solid when pure at T and P. Similarly, at high temperatures, the mixture temperature may be above the critical temperature of one or more of the pure components. In such cases, recourse is usually made to one of two procedures: either we can use a hypothetical reference state of a pure component in the same state of aggregation of the mixture, or we can choose a reference state of aggregation different from that of the mixture. The latter case has already been treated in Section 8.7 in which different standard states for activity and activity coefficients were discussed; only the former case involving hypothetical standard states will be described here.

In the first case, we assign a value to the fugacity of a hypothetical pure component which we presume it would have attained if it had been able to exist at the temperature and pressure and in the same state of aggregation as the mixture. Thus, we would continue to express $\hat{f}_i^\sigma$ by Eq. (9-96), but we must treat f_i^σ as the fugacity of i in a hypothetical state. In the calculation of f_i^σ we will require pure component specific volumes in the hypothetical state if Poynting corrections are made. We will also require vapor pressures since relations such as Eq. (9-107) are normally employed. Let us refer to such hypothetical properties as V_i^{L+} and $P_{vp_i}^{L+}$.

When dealing with a hypothetical liquid state at conditions below the triple point of a pure component (i.e., when the pure component is stable as a solid at T and P), the hypothetical values $P_{vp_i}^{L+}$ and V_i^{L+} can usually be evaluated by extrapolation of the vapor-liquid equilibrium below the triple point, as shown in Figure 9.21. If a Poynting correction is necessary (it usually is not because V_i^{L+} can be evaluated from any known liquid volume at T, i.e., from liquid-solid equilibrium data,) it is corrected to P using the liquid coefficient of isothermal expansion. The latter correction is also not usually required because most liquids are quite incompressible below the triple point.

The problem of employing a hypothetical liquid standard state for a mixture component that is above the critical temperature is not as straightforward. We could again extrapolate the vapor pressure above the critical point to obtain $P_{vp_i}^{L+}$ (at $T > T_{c_i}$) (see Figure 9.28); the Poynting correction to the vapor pressure may not be negligible, however, in the high-pressure region, and a reasonable estimate of V_i^{L+} is required over the range from P to $P_{vp_i}^{L+}$. Since above the critical point there is no distinction between a liquid and a vapor, it is reasonable to assume that the vapor volume (i.e., the true fluid volume) at T could be used for the Poynting correction. This procedure is analogous to assuming that $f_i^L = f_i^V$ for the reference state fugacities, a relation which is consistent with the view that a critical point exists because the Fundamental Equations of liquid and vapor phases are congruent above the critical point.

Although the concept of a hypothetical reference state above the critical point is not very satisfying to many readers, we should keep in mind that the reference state concept is only a convenience for defining an activity coefficient. Furthermore, the activity coefficient has practical use only when we are dealing with systems that have relatively small perturbations from an ideal mixing law. Since thermodynamic investigation of high-pressure phase equilibria has only received significant attention in the last decade, more sophisticated approaches will undoubtedly evolve as mixing phenomena under such conditions are better understood and as models of the fluid phases at high pressure are refined.[17]

[17] For a more detailed discussion of high-pressure phase equilibria, see J. M. Prausnitz, *ibid.*, Chap. 10.

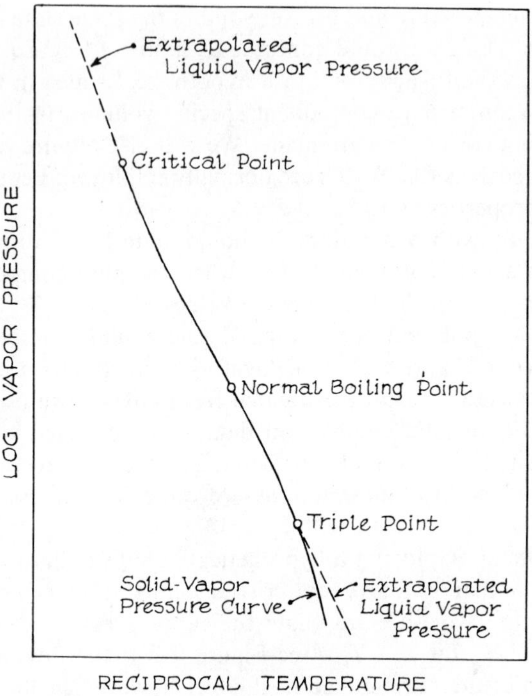

Figure 9.28 Generalized vapor pressure curve.

PROBLEMS

9.1. During an experiment an aqueous solution of magnesium chloride is cooled to $-10°C$. Water is then evaporated isothermally until a solid phase precipitates. An analysis indicates the solid phase to be pure water ice and the solution to contain 13.0 g $MgCl_2$ per 100 g of water.

What is the vapor pressure of water over the solution?

Data at $-10°C$:

Vapor pressure of liquid water = 2.149 mm Hg.
Vapor pressure of water ice = 1.950 mm Hg.
Latent heat of sublimation of water ice = 1,221 Btu/lb.
Volume change in sublimation = 32,100 ft³/lb.

9.2. A rigid tank, 1 ft³ in volume, is evacuated and placed in a constant temperature bath at $-20°F$. 10^{-4} lb of pure water ice is introduced.

Plot $\Delta \underline{U}$, $T\Delta \underline{S}$, $\Delta \underline{A}$ and the total heat flow into the tank as a function of the fraction of ice evaporated. What fraction has evaporated at equilibrium?

Data at −20°F:

ΔS of sublimation of water ice = 2.776 Btu/lb°R.
Vapor pressure of water ice = 0.0062 psia.

9.3. If a small diameter wire is passed over a block of ice and weights are attached to each end of the wire, the wire apparently cuts through the ice but leaves no trace of the path (Figure P9.3). This phenomenon is often called *regelation.*

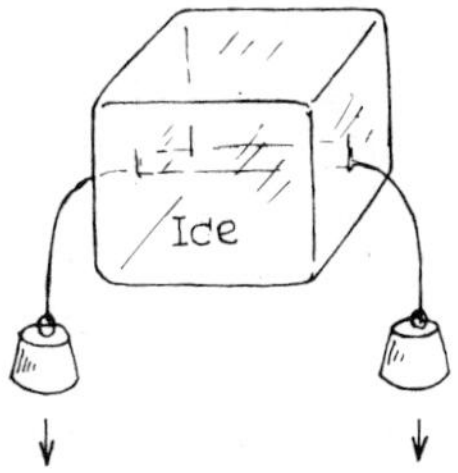

Figure P9.3

Many physics texts explain the phenomenon by the fact that the high pressure exerted by the wire lowers the ice freezing point and thus the wire "melts through." Similar reasoning is invoked to explain the occurrence of a water lubrication film beneath ice skates.

What is your opinion of this explanation?

9.4. For a pure component, a first-order transition is a change in which the first and all higher derivatives of μ have a discontinuity at the point of change. A second-order transition is one in which only the second and higher derivatives of μ with respect to P and T have a discontinuity. Solid methane, for instance, shows, at a particular temperature, a discontinuous C_p because of the onset of free rotation of the molecules.

In other words, in the first-order transition, the chemical potential is continuous at the point of change, but it has a discontinuous first derivative. In a second-order transition, it is the specific entropy and the specific volume which are continuous at the point of change and yet have discontinuous first derivatives.

Derive the analogues to the Clapeyron equation for a second-order phase transition.

9.5. A calorimeter has recently been constructed for NASA to determine the thermal conductivities of various superinsulations. In essence, this calorimeter is a thick-walled copper sphere with a radius of 1.66 ft; the superinsulation is placed on the outside of the sphere. The insulation is 0.5 in. thick and consists of many layers of 0.00025-in. aluminized Mylar separated by thin spacers made of glass-fiber paper, silk netting, or other such materials. The tests are carried out in high-vacuum environmental chambers. The outer layer of insulation is maintained at 530°R. The calorimeter is filled with liquid hydrogen.

Heat transfer through the insulation is manifested by the boiling of the liquid hydrogen, and the mass flow of vapor is measured as a function of time. The vapor vent line may be considered adiabatic (actually, it has a separate liquid hydrogen guard to prevent axial conduction) and is of sufficient cross sectional area so that the pressure in the calorimeter is equal to the prevailing atmospheric pressure.

In any given test, the calorimeter is filled with liquid hydrogen and the liquid level held about constant until the calorimeter heat leak reaches a steady-state value. At this time, the approximate head of liquid hydrogen is measured and the test started. During a test the flow rate of vented hydrogen vapor is measured at various times and the barometric pressure recorded. The assumption that the copper shell and both hydrogen phases are at the same temperature is believed to be quite good; also, the liquid and vapor phases are always in equilibrium with each other at the prevailing atmospheric pressure existing at the exit end of the vent line.

Some typical test data are shown below. After an appropriate thermodynamic analysis, determine the effective thermal conductivity of the insulation.

TEST XX-3
MASS OF LIQUID HYDROGEN IN TANK AT START OF TEST = 84.7 LB

Time hrs	Mass flow rate of vented hydrogen vapor, lb/hr (±0.001)	Barometric pressure mm Hg
0	0.090	762
2	0.091	758
4	0.093	755
6	0.098	742*
8	0.096	743
10	0.099	740
12	0.095	750
14	0.087	759
16	0.088	762
18	0.090	763
20	0.091	759

* *Remarks:* A severe thunderstorm and squall occurred between the hours of about 6 and 10; hence, the drop in barometric pressure.

SATURATION PROPERTIES OF HYDROGEN

T, °K	P, mm Hg	Enthalpy, Btu/lb Liquid	Enthalpy, Btu/lb Vapor
20.0	699.2	86.0	282.3
20.4	760	88.1	283.4
21	904.4	90.4	284.4

9.6. It is a fact that occasionally during rapid loading of liquid oxygen and hydrogen into missile tanks, one or the other of the tanks has imploded with catastrophic results (see Figure P9.6). Such tanks are constructed to withstand an internal pressure somewhat above atmospheric but will collapse if the external pressure significantly exceeds the internal pressure.

During the initial part of the loading cycle, the cryogen may be all gas, a mixture of gas and liquid, or all liquid. The cryogen is pumped through a side port and has intimate contact with the gas already in the tank.

(a) Demonstrate that a pressure decrease cannot occur if only gas flows into the tank.

(b) Derive a general relation to relate the fractional change in tank pressure and temperature with the fractional increase in the mass of gas in the tank for the case in which you think the pressure drops the maximum amount. The feed is all liquid, saturated at the pressure existing in the transfer line, and when this liquid enters, it is immediately vaporized by contact with gas present in the tank. Clearly state any assumptions made.

(c) If the feed were a mixture of liquid and gas, saturated at the transfer line pressure, derive a relation to determine the critical quality of the feed. The critical quality is defined as that fraction of vapor above which no pressure drop is possible irrespective of any rate process.

(d) Determine the fractional change in pressure for both hydrogen and oxygen for step (b) when the mass of gas in the tank has doubled. The liquid in the transfer line is saturated at 37 psia and the initial tank gas pressure and temperature are 1 atm, 500°R, respectively. Assume that the initial gas in the oxygen tank is oxygen and that the initial gas in the hydrogen tank is hydrogen.

(e) For the same transfer line pressure and initial tank conditions as in (d), determine the critical quality for a saturated mixture of gas and liquid cryogen using the relation derived in (c). Do the calculation for both hydrogen and oxygen.

(f) As an engineer, what recommendations could you make to minimize implosion hazards during loading?

Data:

Tank volume: 10,000 ft^3.

	Vapor pressure, psia	Saturation temperature, °R	Heat of vaporization Btu/lb
O_2	37.0	180	86
H_2	37.0	43.2	197

C_p (vapor) = 5.8 (H_2); 7.0 (O_2), Btu/lb-mol°R
Assume ideal gases.

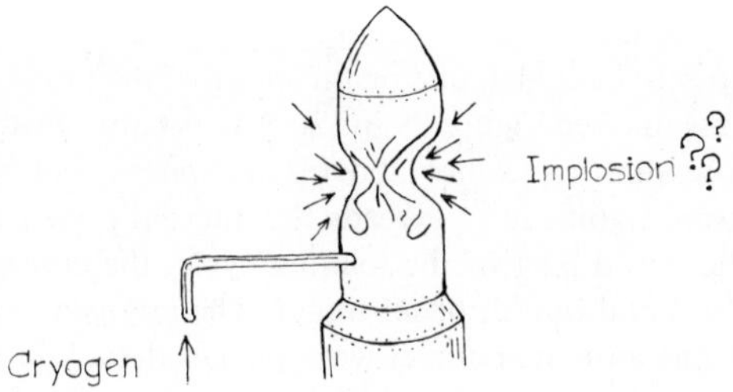

Figure P9.6

9.7. At 100 psia, the following data for *n*-pentane are available:
Saturated liquid volume = 0.03045 ft³/lb.
Vapor volume at 460°F = 1.2975 ft³/lb.
Saturation temperature = 224.5°F.
Enthalpy of vaporization = 122.4 Btu/lb.
Pressure effect on enthalpy:

T, °F	$\left(\frac{\partial H}{\partial P}\right)_T$, Btu/lb-psi
460	0.0592
430	0.0618
400	0.0647
370	0.0688
340	0.0749
310	0'0836
280	0.0957
250	0.1126

What is the slope of the vapor pressure curve, dP/dT, at 100 psia?

9.8. Bottles of carbon dioxide are often used as fire extinguishers (Figure P9.8). You are requested to analyze the blowdown process and determine the state of the gas both in the bottle and upon leaving the bottle as a function of time.

Assume for purposes of discussion that the bottle is a right circular cylinder about 8-in. in diameter and between 4.5 and 5 ft tall. The volume is about 1.7 ft³ and the inside area 10 ft². The steel wall is 0.25-in. thick; the heat capacity of steel is about 0.1 Btu/lb°R and the density 529 lb/ft³.

These bottles are stored at ambient temperature (around 40°F) and are charged initially to 500 psia. For emergency use the valve is opened quickly, and it is found experimentally that the bottle pressure decays by a factor of 2 every 2.0 min.

(a) Suppose that the bottle were adiabatic (i.e., no heat transfer between the walls and gas). What is the presssure in the cylinder when solid carbon dioxide *first* can form (1) in the gas flowing anywhere in the valve and (2) in the gas within the cylinder? (Neglect any kinetic energy effects.)

(b) Do not neglect heat transfer between the cylinder walls and gas, but assume that such heat transfer occurs by a natural convection process with $h \sim 6$ Btu/hr-ft²-°F. Plot the cylinder gas and bottle wall temperatures and exit gas temperature as a function of time; also answer the same two questions formulated in (a).

Data for carbon dioxide are given with Problem 6.12.

Figure P9.8

9.9. Some consideration has been given to evaluating activity coefficients in a binary liquid mixture from the isothermal integral expression:

$$RT \ln \gamma_i = \int_0^P \overline{\Delta V_i}\, dP$$

Discuss how this integral might be evaluated across the two-phase region (i.e., from the dew point to the bubble point). What experimental data would be required in the integration?

9.10. A saturated solution of naphthalene in chlorobenzene is prepared at 20°C and 1 atm. The liquid phase is decanted and the liquid mole fraction of naphthalene is determined to be 25.6%.

The solid-free solution is placed in a piston-cylinder and compressed isothermally to 500 atm. Estimate the mole fraction of naphthalene in the liquid under these conditions.

Data:

For solutions of naphthalene in chlorobenzene, the molal liquid volume, cm³/g-mol, is independent of pressure and is given as:

$V = 102 + 26x$, x = mole fraction naphthalene.

Ideal liquid solution properties may be assumed.

	Naphthalene	Chlorobenzene
Melting point, °C	80.05	−45.2
Molal volume, cm³/g-mol	115 (solid)	102 (liquid)
Latent heat of fusion, cal/g-mol	4,440	—

9.11. A research engineer, who is working on the properties of some high-boiling systems that are to be vacuum distilled, is having a bitter argument with one of the design engineers. The research engineer claims that his isothermal data indicate that the partial pressures of the two components in the vapor over the binary solutions are proportional to the volume fractions of the components in the liquids computed as though Amagat's law (no volume change on mixing) applied to the liquid phase.

Actually, he has not determined whether a volume change does or does not occur upon mixing but he, nevertheless, writes for these systems:

$$p_1 = P_{vp_1}\left[\frac{V_1x_1}{(V_1x_1 + V_2x_2)}\right]$$

$$p_2 = P_{vp_2}\left[\frac{V_2x_2}{(V_1x_1 + V_2x_2)}\right]$$

p is the partial pressure.
P_{vp} is the vapor pressure.
x is the mole fraction.
V is the pure molal liquid volume.

The design engineer says that these equations cannot be correct and he adamantly refuses to look at the data given to him as substantiation.

Is he right in taking this stand? Discuss and indicate your reasoning.

9.12. The following data have been reported for the vapor-liquid equilibrium between ethyl alcohol and n-hexane at 1 atm.

T°C	Mole fraction alcohol in liquid	Mole fraction alcohol in vapor
76.0	0.990	0.905
73.2	0.980	0.807
67.4	0.940	0.635
65.9	0.920	0.580
61.8	0.848	0.468
59.4	0.755	0.395
58.7	0.667	0.370
58.35	0.548	0.360
58.1	0.412	0.350
58.0	0.330	0.340
58.25	0.275	0.330
58.45	0.235	0.325
59.15	0.102	0.290
60.2	0.045	0.255
63.5	0.010	0.160
66.7	0.006	0.065

What does thermodynamics tell us about the validity of these data? What can one conclude about the heat of solution and the entropy of solution? The vapor pressures of ethyl alcohol and n-hexane are:

T°C	P_{vp} (alcohol)
50.0	220.00 mm Hg
52.0	242.50
56.0	291.85
58.0	319.95
60.0	350.30
62.0	383.10
64.0	418.35
66.0	456.45
68.0	497.25
70.0	541.20
72.0	588.35
74.0	638.95
76.0	693.10
78.4	760.00

T°C	P_{vp} (n-hexane)
−53.9	1 mm Hg
−34.5	5
−25.0	10
−14.1	20
−2.3	40
5.4	60
15.8	100
31.6	200
49.6	400
68.7	760
93.0	2 atm
131.7	5
166.6	10

9.13. The elements cadmium and bismuth are miscible in all proportions in the liquid state but practically immiscible in the solid state. Assuming that the liquid solutions are ideal, estimate the eutectic temperature T_e and compositions x_e for the system. Compare your results with the experimental values of $T_e = 417$°K and x_e (Cd) $= 0.57$.

	Cd	Bi
ΔH_{fusion}, cal/g-mol	2600	1530
T_f, °K	594	544

9.14. In a binary solution of two components, the eutectic point is the lowest freezing point of the mixture. It is usually less than the freezing point of either pure component.

You are asked to estimate the eutectic point (i.e., composition and temperature) for a liquid-air mixture (O_2 and N_2). Assume that the liquid phase forms an ideal solution and that all solid phases are pure components (i.e., no mixed crystals form). Any gas phase is ideal. Some data that may be of use are given below. Neglect any pressure effects.

	Nitrogen	Oxygen
Freezing point, °R	114	98
$\Delta H_{\text{vaporization}}$, Btu/lb-mol	2580	3220
$\Delta H_{\text{sublimation}}$, Btu/lb-mol	2890	3412
ΔH_{fusion}, Btu/lb-mol	310	192

Assume that ΔH values do not vary with temperature.

9.15. The partial pressure of water over a sodium nitrate solution is given below. The concentration of the salt is constant at 80 g/100 g H_2O. What is the differential heat of solution of water, $\bar{H}_w^L - H_w^L$, at 10°C?

T, °C	Partial pressure of water, mm Hg
125	1294
100	568
75	218
50	70.2
25	18.2
10	7.2 (saturated solution)

9.16. The partial pressures of water over sodium nitrate solutions are shown below at various temperatures and compositions.

Concentration,	Partial pressure of water, mm Hg		
g $NaNO_3$/100 g H_2O	0°C	25°C	50°C
0	4.58	23.76	92.54
10	4.42	22.93	89.2
20	4.28	22.14	86.1
30	4.15	21.39	83.1
40	4.04	20.69	79.6
50	3.93	20.04	77.5
60	3.83	19.42	74.9
70	3.73	18.83	72.56
73	3.70*		
80		18.29	70.25
90		17.77	68.1
92		17.67*	
100			66.1
110			64.2

* Indicates a saturated solution.

In addition, the following data are available for saturated solutions of $NaNO_3$.

T, °C	Partial pressure of water, mm Hg	Concentration of $NaNO_3$ g/100 g H_2O
0	3.7	73.0
20	13.2	88.2
25	17.67	92.0
40	39.1	104.8
60	98.6	124.0
80	216.1	148.0
100	422.0	176.0
120	748.0	210.6

Determine the heat of crystallization at 25°C. Also determine separately $(H^S - \bar{H}^L)_{NaNO_3}$ at 25°C, saturated solution. Note

$$\Delta H_{cry} = x_{H_2O}(H^L - \bar{H}^L)_{H_2O} + x_{NaNO_3}(H^S - \bar{H}^L)_{NaNO_3}$$

9.17. The *International Critical Tables*, Vol. III, p. 313, lists the boiling points for mixtures of acetaldehyde and ethyl alcohol at various pressures. The data are summarized below for mixtures containing 80 mole percent ethyl alcohol in the liquid. From these data determine the molal heat of vaporization of ethyl alcohol at 47.5°C, from a liquid mixture containing 80 percent ethyl alcohol [i.e., what is $(\bar{H}^V - \bar{H}^L)$ alcohol].

$\bar{H}^V$ = partial molal enthalpy of saturated vapor at 47.5°C.
$\bar{H}^L$ = partial molal enthalpy of saturated liquid at 47.5°C.

Data:

Mole Fraction Ethyl Alcohol in Liquid = 0.80

T°C	P (mm Hg)	Mole fraction ethyl alcohol in vapor
58.1	699	.318
47.5	398	.385
22.7	77	.330

Note: The vapor phase may be considered as an ideal gas, but the liquid phase is a nonideal solution.

9.18. A liquid mixture of acetic acid and water has the following properties:

At 50°C, and with the liquid composition at 50 weight percent acetic acid, the weight fraction acetic acid in the vapor = 0.412. Assume that the vapor phase is an ideal gas mixture.

At 50°C, the vapor pressures are: H_2O = 87 mm Hg, acetic acid = 55 mmHg. Also, at 50°C, $\Delta H^{vap}_{H_2O}$ = 10,000 cal/g-mol, $\Delta H^{vap}_{acetic\ acid}$ = 9,500 cal/g-mol.

Liquid composition = 50 weight percent of each

T°C	Total pressure over solution mm Hg
20	15.7
25	21.4
30	28.8
35	38.3
40	50.2
45	65.0
50	85.0
55	107
60	138
65	172
70	216
75	269
80	331
85	407
90	602
100	725

What is the physical significance of the value of the tangent slope when referred to the curve of ln P vs. $1/T$ from the data in (a) above? What can you infer about enthalpies of mixing in the liquid phase from these data?

9.19. Some vapor-liquid equilibrium data for the system acetone-water are shown below:

$P = 760$ mm Hg

	Mole fraction acetone		Differential enthalpy of mixing for the saturated liquid, cal/g-mol	
T(°C)	Liquid	Vapor	$\overline{\Delta H_A}$	$\overline{\Delta H_w}$
100	0	0	1265.0	0
84.75	0.02	0.4451	—	—
75.13	0.05	0.6340	887.5	10.3
68.19	0.10	0.7384	563.0	36.8
65.02	0.15	0.7813	—	—
63.39	0.20	0.8047	118.2	115.0
61.45	0.30	0.8295	−134.8	195.3
60.39	0.40	0.8426	−236.5	248.0
59.91	0.50	0.8518	−246.0	254.0
59.55	0.60	0.8634	−202.0	200.0
58.79	0.70	0.8791	−135.3	20.7
58.07	0.80	0.9017	−68.5	−131.0
57.07	0.90	0.9371	−18.8	−415.0
54.14	1.00	1.0000	0	−782.0

In addition to these data, the vapor pressures of the pure components are shown below:

T, °C	P_{vp}, acetone mm Hg	P_{vp}, water mm Hg
15	145	12.8
20	182.8	17.5
30	282.7	31.8
40	422.5	55.3
50	612.9	92.5
60	865.7	149.4
70	1193.7	233.7
80	1610.9	355.1
90	2132.4	525.8
100	2773.1	760.0

(a) Assuming the vapor phase to be ideal, determine the entropy change in mixing pure liquid water and liquid acetone, at one atmosphere, to form a 40 mole percent solution. The pure components and final solution are at 60.39°C.

(b) Calculate the total entropy change for a process in which 0.6 mole of liquid water and 0.4 mole of liquid acetone, both at 60.39°C, are separately vaporized, expanded reversibly and isothermally to a pressure so that they may be added to a vapor mixture at one atmosphere (with 84.26 mole percent acetone) reversibly across semipermeable membranes. One mole of this vapor is then condensed to a liquid with a composition of 40 mole percent acetone at 60.39°C. Compare your result with the relation determined in (a).

(c) Comment on the consistency of the data.

9.20. The following scheme has been proposed to measure enthalpies of mixing in binary liquid mixtures.

For one of the components, for example B, the partial pressure is measured as a function of temperature for several liquid compositions. For each liquid composition, the logarithm of this measured pressure of B is plotted against the logarithm of the vapor pressure of pure B. Curves are drawn through the plotted points and the slope at any point is called m. The differential enthalpy of mixing of B, $\overline{\Delta H}_B^L$ is calculated as $\Delta H_B^{\text{vap}}(1 - m)$ for many different liquid compositions. (Note that m may also be a function of temperature.) The integral enthalpy of mixing per mole of A in the solution is then determined by the isothermal integration of:

$$\frac{\Delta H}{x_A} = \text{integral enthalpy of solution per mole of } A = \int_0^{x_B/x_A} \overline{\Delta H}_B \, d\left(\frac{x_B}{x_A}\right)$$

Assuming ideal gases, prove that this scheme is thermodynamically correct.

9.21. In C. G. Houser, and J. H. Weber "Liquid Phase Enthalpy Values for the Methane-Ethane System," *J. Chem. Eng. Data*, **6**, (1961), 510, the authors state the following relations:

$$\left(\frac{\partial P}{\partial T}\right)_x = \left(\frac{\Delta H^{\text{vap}}}{T\,\Delta V^{\text{vap}}}\right) \tag{1}$$

$$\Delta V^{\text{vap}} = V_d - V_b - (y - x)\left(\frac{\partial V_L}{\partial x}\right)_{T,P} \tag{2}$$

$$\Delta H^{\text{vap}} = H_d - H_b - (y - x)\left(\frac{\partial H_L}{\partial x}\right)_{T,P} \tag{3}$$

where "ΔH^{vap} represents the differential enthalpy of vaporization, i.e., the heat required to vaporize a mole of mixture of composition y_1 from a large quantity of liquid of composition x_1. The composition of the liquid remains unchanged in the process and the relationship between y_1 and x_1 is the equilibrium condition."

Subscripts d and b refer to dew point and bubble point conditions.

(a) Derive these relations for a binary mixture.

(b) In Figure 5 of the article noted above, saturated liquid enthalpy at constant pressure is plotted as a function of composition; the isotherms are also drawn. Determine the partial molal enthalpy of methane and ethane in the saturated liquid at $x_{CH_4} = 0.2$ at 200 psia, and verify your results with the values given by the authors in their Table 1.

9.22. Use your own creative imagination to devise a two-constant arithmetic expression that you think will properly express ΔG^{EX} as a function of composition for a binary liquid mixture.

Remember that $\Delta \underline{G}^{EX}$ is proportional to the total number of moles and should go to zero as $x_A, x_B = 0$.

Using this proposed equation, determine algebraic (constant temperature, constant pressure) expressions for the activity coefficients as a function of composition and estimate the x-y curve for ethanol and water at one atmosphere from the azeotropic point,

T°C	Vapor pressure, mm Hg Ethanol	Water
76	693	301
78	750	327
80	812	355
82	877	384
84	950	416
86	1026	450
88	1102	487
90	1187	525
92	1280	566
94	1373	610
96	1473	657
98	1581	707
100	1693	760

$$x_{\text{alcohol}} = y_{\text{alcohol}} = 0.8943 \text{ at } 78.17°\text{C}$$

The vapor pressures of the pure components as a function of temperature are shown above:

You may wish to test your calculated x-y results with the following isobaric liquid-vapor equilibrium data obtained from the Sc.D. thesis of J. S. Carey, M.I.T., 1929.

T°C	x Mole fraction ethanol, liquid	y Mole fraction ethanol, vapor
95.7	0.0190	0.1700
90.0	0.0600	0.3560
86.4	0.1000	0.4400
84.3	0.1600	0.5040
83.3	0.2000	0.5285
82.3	0.2600	0.5570
81.8	0.3000	0.5725
81.2	0.3600	0.5965
80.7	0.4000	0.6125
80.2	0.4600	0.6365
79.8	0.5000	0.6520
79.4	0.5600	0.6775
79.13	0.6000	0.6965
—	0.6600	0.7290
78.6	0.7000	0.7525
—	0.7600	0.7905
78.3	0.8000	0.8175
—	0.8600	0.8640
78.17	0.8943	0.8943

9.23. The values of ΔG^{EX} and ΔH^{EX} for a particular binary liquid mixture are shown in Figure P9.23 as a function of the mole fraction of component A. One can describe these curves by: $\Delta G^{EX} = \eta\, \Delta H^{EX}$. It was also found for T, P constant that: $\overline{\Delta G}_A^{EX} = \eta_A \overline{\Delta H}_A^{EX}$ and $\overline{\Delta G}_B^{EX} = \eta_B \overline{\Delta H}_B^{EX}$. The constants η, η_A, and η_B are not functions of temperature.

(a) Derive a relationship to show how the activity coefficient varies with temperature for component A if the pressure and composition were fixed.

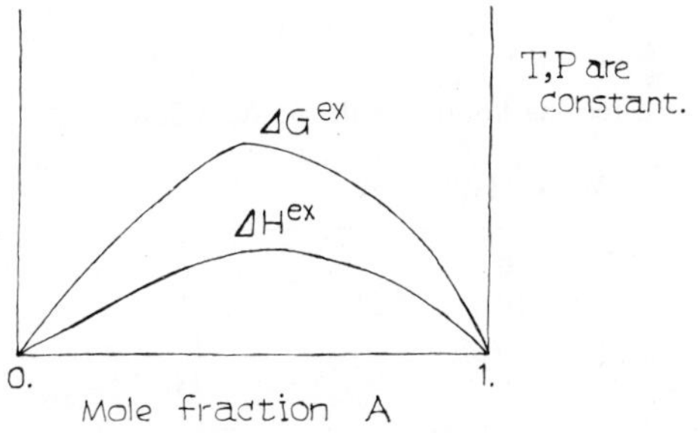

Figure P9.23

(b) Using the expression derived in (a), keeping the pressure and composition constant, and supposing that $\gamma_A = 1.2$ at 300°K, what is γ_A at 400°K if $\eta = 2$?

(c) Discuss the cases $\eta_A > 1$, $\eta_A = 1$, and $0 < \eta_A < 1$ from the standpoint of $\overline{\Delta S}_A^{EX}$ and indicate what type of liquid solution one might expect in each of these cases. Assume $\gamma_A > 1$.

9.24. It is claimed that a low-temperature liquid mixture may be made by mixing ice and pure sulfuric acid in a Dewar flask. If the ice and acid were originally at 0°C, what would be the final temperature and liquid composition when one pound of acid is poured over four pounds of ice?

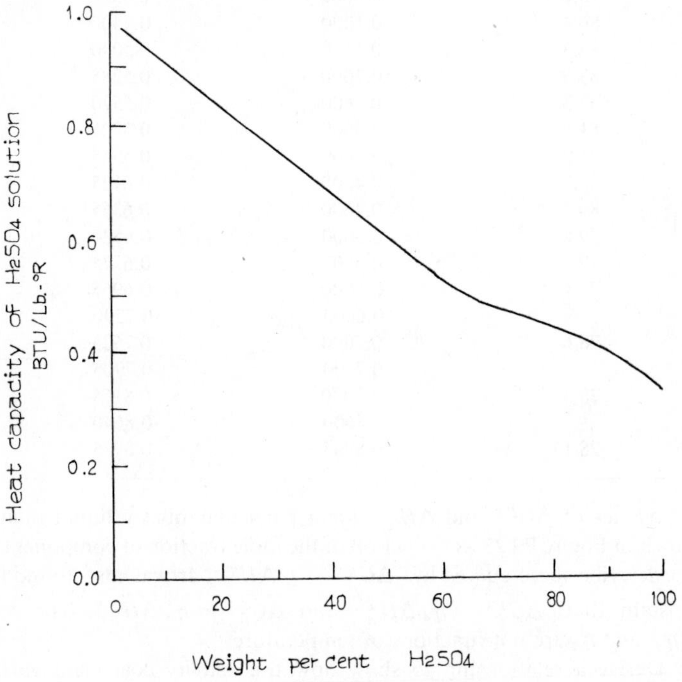

Figure P9.24a

Data:

Freezing Point of Sulfuric Acid Solutions

Wt % H_2SO_4	T, °C
0	0
10	−5
20	−13
30	−35
40	−62
50	−35

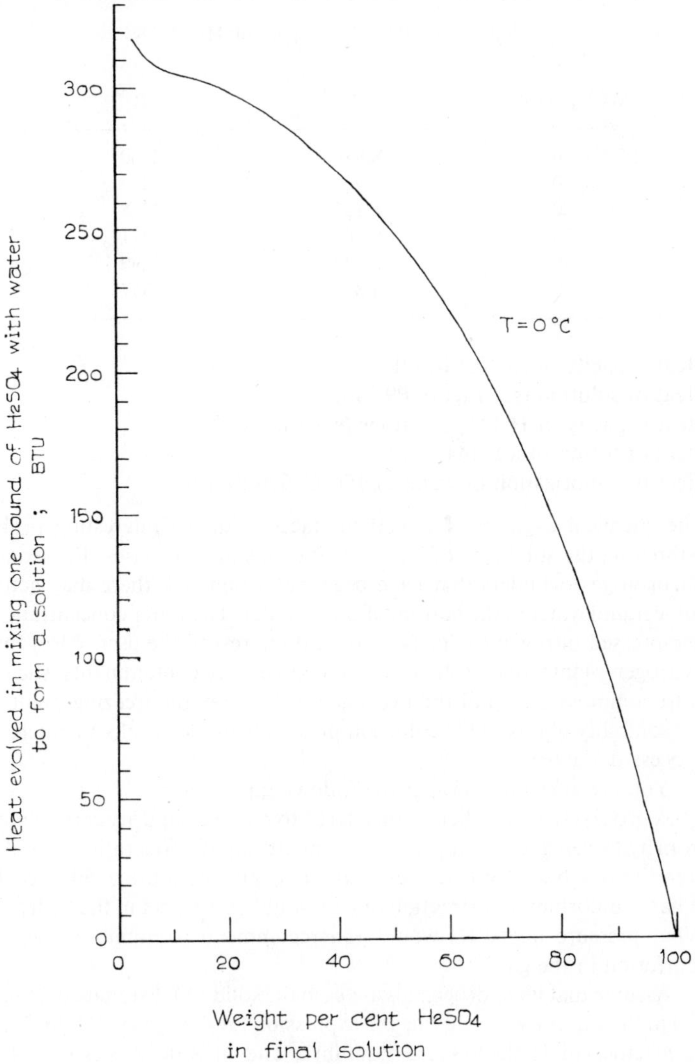

Figure P9.24b

Vapor Pressure of Ice

P, mm Hg	*T*, °C
4.58	0
1.95	−10
0.776	−20
0.286	−30
0.0966	−40

Partial Pressure of Water over Sulfuric Acid Solutions

$$\log_{10} p = A - \frac{B}{T}, \qquad (p, \text{mm Hg}; T, {}^\circ\text{K})$$

Wt % H_2SO_4	A	B
0	8.946	2260
10	8.925	2259
20	8.925	2268
30	8.864	2271
40	8.844	2299
50	8.832	2357

Heat capacity of ice, 0.5 Btu/lb°F.
Heat of solution (see Figure P9.24b).
Heat capacity of H_2SO_4 solutions (see Figure P9.24a).
Heat of fusion of ice, 144 Btu/lb.
Heat of vaporization of water at 100°C, 970 Btu/lb.

9.25. The chemical engineer is sometimes faced with the interesting problem of estimating the solubility of liquids and solids in inert gases. For example, in nitrogen gas cylinders that have been water-pumped, there may actually be some liquid water in the bottom of the cylinder. The water concentration in the compressed nitrogen gas phase is of real interest to the user. Also, for liquid hydrogen plants, one of the ways to remove trace contaminants, such as CO, is to compress and cool the hydrogen well below the freezing point of CO; the solubility of this CO in cold, compressed hydrogen is of vital interest to the process designer.

You are asked to consider the following problem:

A process scheme is being considered to remove impurity carbon monoxide from hydrogen gas by compressing the mixture and refrigerating it until the CO deposits as a solid. A pressure of 20 atm and a temperature of 50°K is planned. What equilibrium mole fraction of CO would you expect in the hydrogen gas? What pressure (at 50°K) would *you* recommend to minimize the CO concentration in the gas?

Assume that no hydrogen dissolves in the solid CO. Estimate the fugacity of CO in the vapor by simplifying the expression for $\hat{f}_j$ as given in J. M. Prausnitz, "Fugacities in High-Pressure Equilibria and in Rate Processes," *AIChE J.* **5**, (1959), 3.

$$\ln\left(\frac{\hat{f}_j}{y_j P}\right) = -\ln Z + \frac{2}{V}\sum_k y_k B_{jk} + \frac{3}{2V^2}\sum_i\sum_k y_i y_k C_{ijk}$$

This equation was derived by using the virial equation of state to express the volumetric properties of the mixture. The terms are:

y = mole fraction.
Z = mixture compressibility factor.
V = mixture volume.
B_{jk} = second virial coefficient for the interaction between components j and k.
C_{ijk} = third virial coefficient for an ijk interaction.

In the simplification of this equation for this problem, assume that Z (mixture) = Z (pure hydrogen) and $y_{H_2} \gg y_{CO}$. The vapor in equilibrium with *pure* solid CO at these low temperatures is ideal.

To obtain the virial coefficients B and C, the article referenced above suggests one method. Alternatively, Chapter 7 of the *Properties of Gases and Liquids*, 2nd ed., by R. C. Reid and T. K. Sherwood (McGraw-Hill Book Company, 1966) discusses such estimations.

At 50°K, the vapor pressure of CO is 1.1×10^{-3} atm and the solid has a density of about 1.03 g/cm³. Compressibility data for pure hydrogen are given with Problem 6.1.

9.26. Jolley and Hildebrand studied the solubility and entropy of solution of gases in nonpolar solvents. They have plotted the equilibrium solubility of various

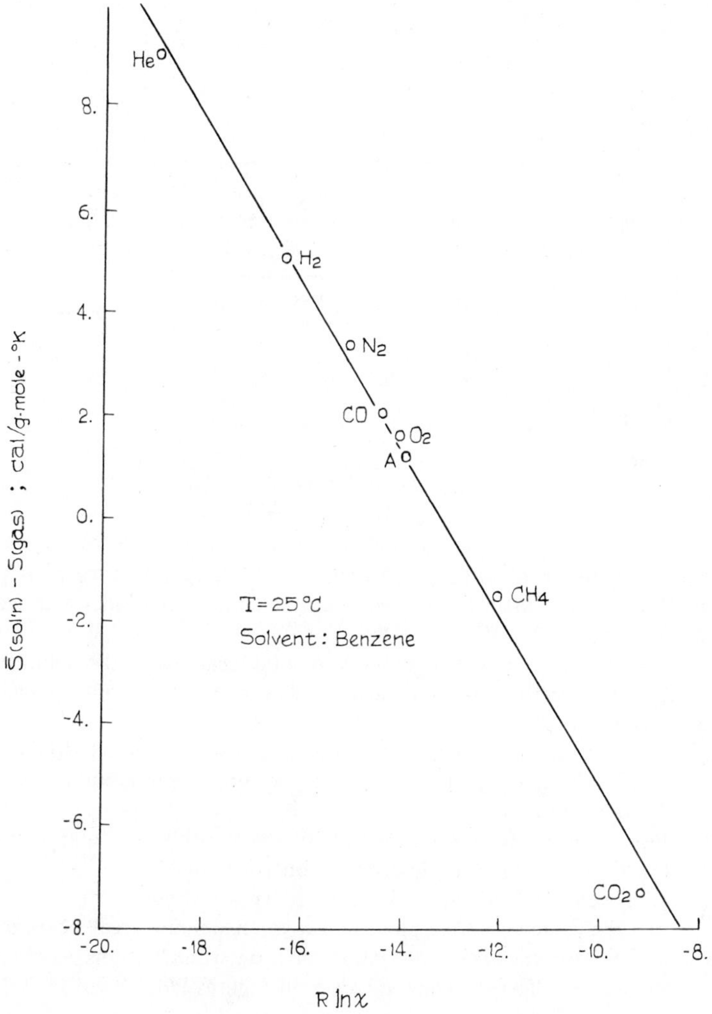

Figure P9.26

inert gases in benzene at 25°C when there is a partial pressure of 1 atm of the inert gas over the solution. x is the equilibrium solubility (mole fraction) of the gas in benzene, $\bar{S}$ is the partial molal entropy of the gas in solution, and S (gas) is the entropy of the pure gas at 25°C and 1 atm. (See Figure P9.26.)

(a) Which gases become more soluble with an increase in temperature at temperatures greater than 25°C? Prove your reasoning and state clearly any simplifying assumptions you make in the proof.

(b) Estimate ΔS (vaporization) of pure benzene at 25°C and 1 atm.

9.27. A well-insulated tank is vented to the atmosphere. Initially, it contains a hot

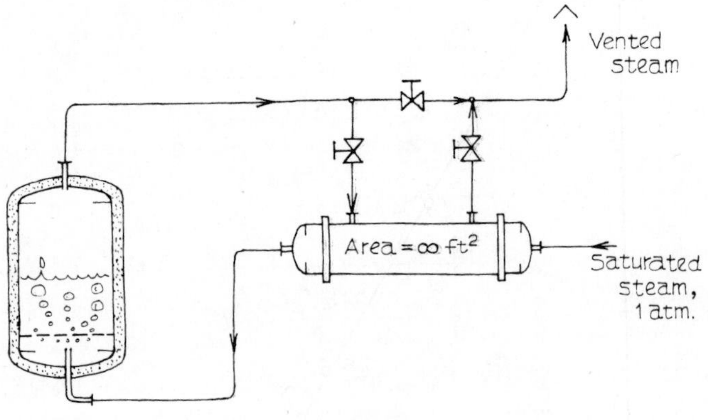

Figure P9.27a

aqueous solution of sodium hydroxide at 194°F, with a concentration of 45 weight per cent caustic. Connected to the tank is a steam header with saturated steam at 1 atm, 212°F.

Assume that the vented steam is in equilibirium with the solution at all times and neglect any sensible heat effects for the vessel walls or inert gases over the solution.

(a) If the steam line were opened and steam allowed to bubble slowly into the tank, plot the resulting solution temperature as a function of sodium hydroxide concentration.

(b) Plot a curve showing the pounds of steam condensed and vented as a function of the sodium hydroxide concentration.

(c) If the vent line from the tank were arranged (as shown in Figure P9.27a) so that the vented gas were passed through an "infinite-area" heat exchanger counter-current to the entering steam, how would the answers in part (a) and (b) be affected? Data are given in Figures P9.27b and P9.27c.

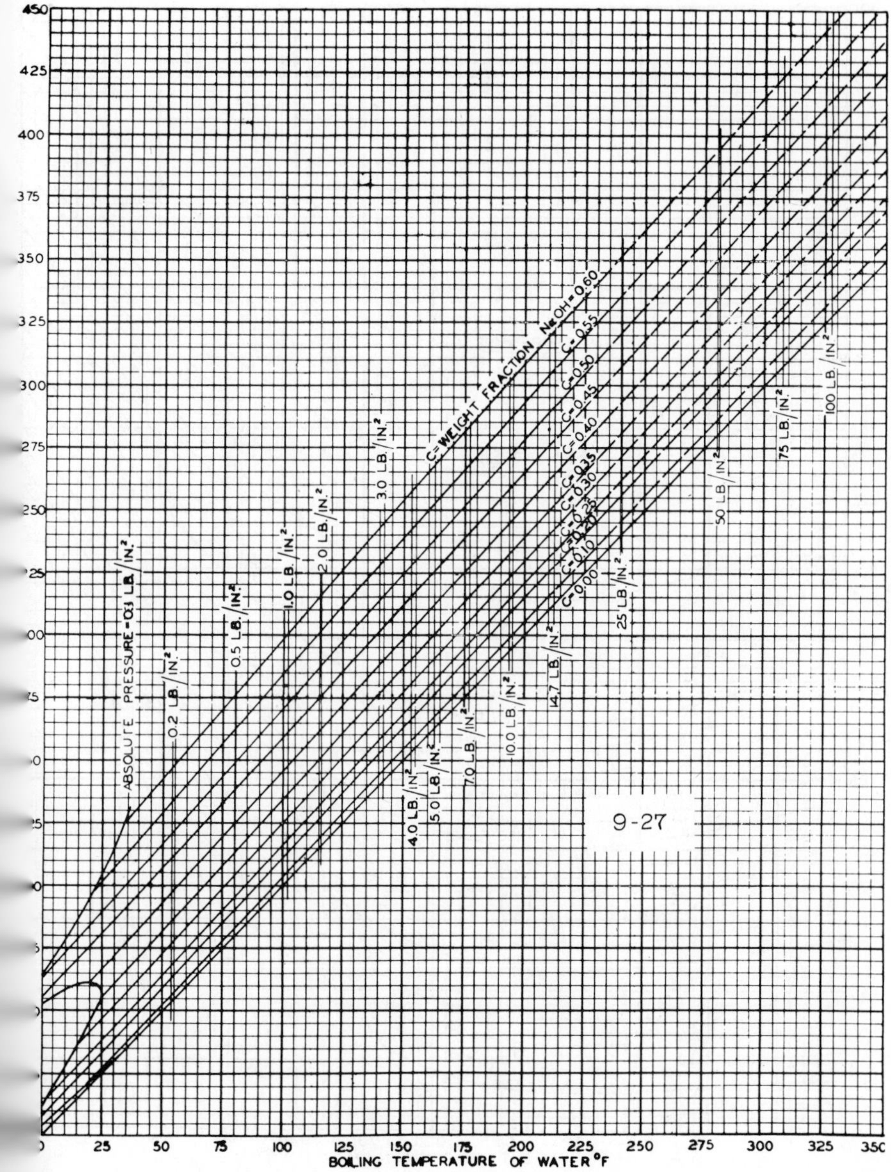

Figure P9.27b

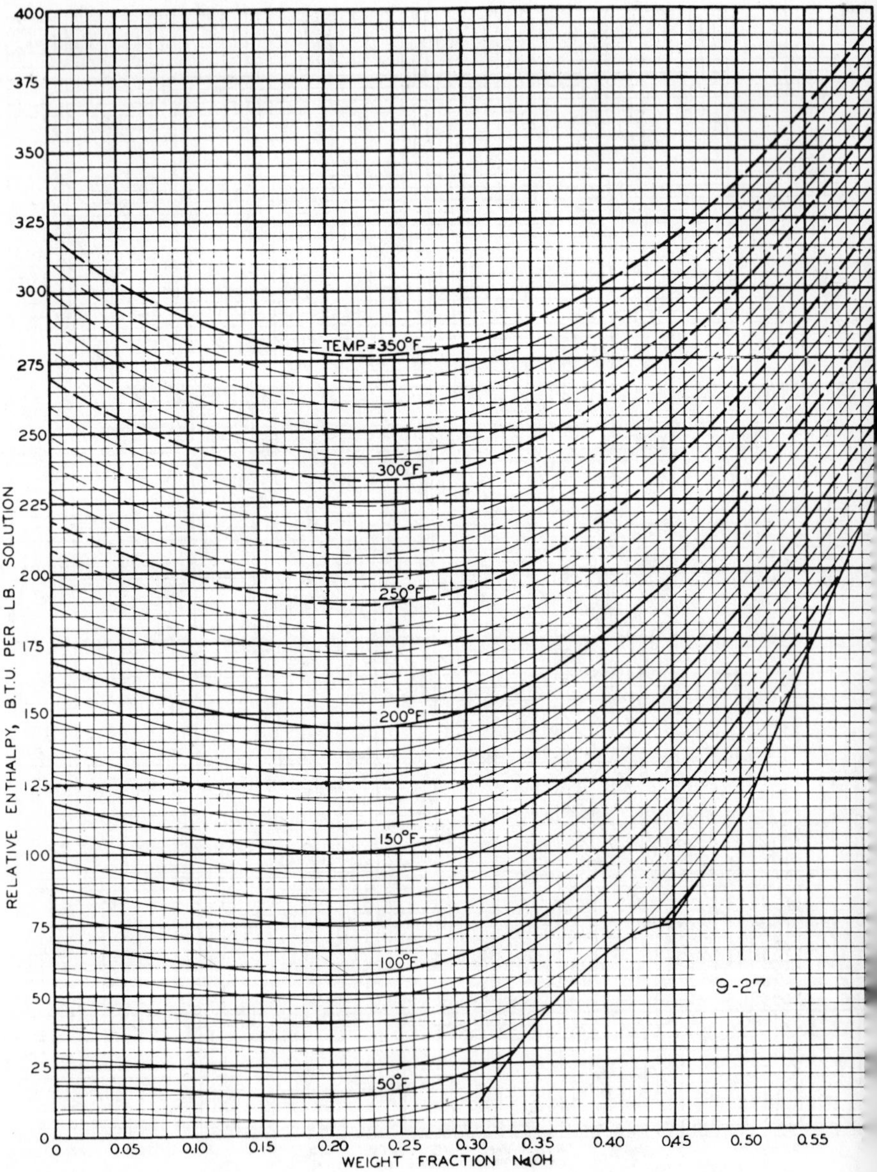

Figure P9.27c

9.28. An aqueous solution of 15 weight percent NH_4NO_3 is to be concentrated to 30 weight percent NH_4NO_3 by reverse osmosis. The feed solution is pressurized to 1500 atm and passed through a cell containing a membrane permeable to water but not to NH_4NO_3. Based on results of laboratory tests, a commercial process design has been formulated. The proposed process is illustrated in Figure P9.28 along with the proposed operating conditions. There is a significant pressure drop in the high-pressure chamber since a high fluid velocity is necessary to minimize the mass transfer resistance in the bulk fluid phase.

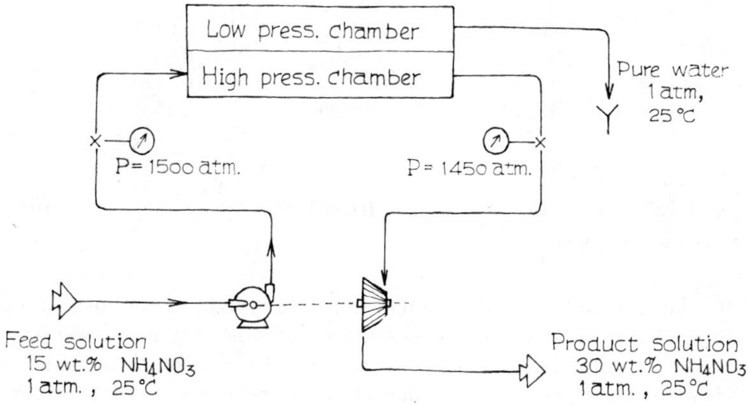

Figure P9.28

(a) Describe in detail a *reversible* process that will accomplish the same *overall* separation as that described (i.e., will accept the same feed solution and produce the same product solutions).

You are free to use any kind of any finite number of pumps, membranes, heat exchangers, or any other device dictated by your creativity. Present a flow sheet summarizing your process; indicate the operating conditions (i.e., pressure and composition) at each point in the process.

(b) For your reversible process, calculate the work requirements for each compressor and/or turbine.

(c) The net work requirement for the proposed reverse osmosis process is 86 J/g of feed solution. What is the overall efficiency of this process?

Data:

At 1 atm and 25°C the chemical potentials of both species have been determined as a function of composition. These data are shown in the figures given with Example 8.8. For the chemical potential, ammonium nitrate is referred to a saturated solution, while the reference for H_2O is pure water.

At 25°C and for pressures of 0.1 to 2,000 atm, the following assumptions may be made:

V (pure water) $= 1\ cm^3/g$.

$\bar{V}_W$ (independent of NH_4NO_3 conc.) = 1 cm³/g.
V (pure crystalline NH_4NO_3) = 0.58 cm³/g.
$\bar{V}_{NH_4NO_3}$ (15 weight percent NH_4NO_3) = 0.622 cm³/g.
$\bar{V}_{NH_4NO_3}$ (30 weight percent NH_4NO_3) = 0.632 cm³/g.
Molecular weight of NH_4NO_3 = 80.04.

Weight fraction NH_4NO_3	$T = 25°C, P = 1$ atm $p/P_{vp\,(water)}$	$p/P_{vp\,(NH_4NO_3)}$
0	1.0	0
0.10	0.9601	0.0483
0.15	0.9405	0.0750
0.20	0.9215	0.1120
0.25	0.8990	0.1565
0.30	0.8746	0.2115

At 25°C and 1 atm, $P_{vp\,(water)} = 0.0312$ atm, $P_{vp\,(NH_4NO_3)(pure)} = 10^{-5}$ atm. Assume ideal gases.

9.29. It is late fall and you suddenly remember that you have not put antifreeze in the radiator of your car. All service stations nearby are closed and it is predicted that the temperature will drop to 10°F tonight. In desperation you visit the chemical engineering laboratories to borrow some suitable chemical to mix with water to make a noncorrosive antifreeze. There is foramide ($HCONH_2$), urea (NH_2CONH_2), methanol, ethanol, glycerol, sucrose, and other such chemicals. Which one to use and how much? Then you dimly remember that in physical chemistry someone told you that in dilute solutions there was the same freezing point depression per mole of solute for all such materials. Make any and all assumptions necessary to estimate rapidly the value of this constant ($\Delta T/N$). ΔT is the freezing point depression in °C when N gram moles of solute are dissolved in 1,000 grams of water. $\Delta H_{f_{H_2O}}$ = 144 cal/g.

9.30. Comment on the general rule often quoted in texts:

"If the solution of a solute in a solvent is endothermic, then the solubility of the solute increases with an increase in temperature; the converse is also true."

Prepare a case in detail to show if the above stated rule is true; if there are any cases where it is false, describe clearly the conditions whereunder the violation occurs. Your analysis should include, but not be limited to, the special case of an ideal liquid solution with neglible pressure effects.

Finally, illustrate your analysis with the case of lithium chloride or sodium hydroxide dissolving into water. For these systems, there are data in the International Critical Tables and the book, *Selected Values of Chemical and Thermodynamic Properties, Circular 500*. Of course, if desired, other literature sources may be used.

9.31. WATER LIMITED

Office of the President

HOW MUCH CLEAN WATER DID YOU PASS TODAY?

Dr. Adolph H. Haw
Director of Corporate Research

My dear A. H. Haw:

I had a most interesting chat this morning with a young lad from the States. He is president of a small, but enterprising firm which promotes new inventions. He is eager to have us join with him to finance the development of a novel concept in water desalination that his firm, MITY, Inc., has proposed. I have my doubts about the feasibility of his invention, but I did not want to pull the plug on the chap until you had an opportunity to look at his idea.

Summarizing from my notes, he feels that our reverse osmosis systems will soon become obsolete. As you well know, the systems we now manufacture to yield water from a 3.5% sodium chloride brine require us to compress the brine to at least 29 atmospheres (the osmotic pressure) to raise the fugacity of water the necessary 2.15% so that it will equal that of fresh water at 77°F.

Now this chap, Mr. Rocky Jones, tells me that if one increases the brine temperature from only 77 to 77.65°F, we can still raise the fugacity of the water the necessary 2.15%. Thus, he proposes we work with him to construct a device to accomplish reverse osmosis *without any compression* but simply by heating the brine and maintaining the fresh water side of the membrane a few degrees cooler.

The questions you should consider are:

1. Is the process as outlined possible?
2. If so, please evaluate its potential and suggest, if possible, the type of membrane that might be used, i.e., would the cellulose acetate membranes now used in pressure-driven reverse osmosis be suitable?
3. Would such a small temperature difference ($\sim$ 0.65°F) be sufficient or would we need much larger temperature differences between the brine and fresh water sides of the membrane?

Mr. Jones will be in to see me again tomorrow, so your attention to this matter should receive priority.

The

Arnold A. Top

Chemical Equilibria 10

The theoretical framework for equilibria in chemically reacting systems has been established in earlier chapters. With the exception of the Third Law of thermodynamics, which is an empirical generalization that simplifies the determination of physical property data, there are no new fundamental concepts introduced in this chapter. Indeed, the developments to follow are nothing more than applications of principles, which we have already established, to systems in which molecular rearrangements can occur.

The fundamental basis for chemical-reaction equilibrium is contained in Postulates I and II (see Section 2.7). Postulate I, which applies to a closed, simple system, states that for *given internal restraints*, there exist stable equilibrium states that can be *characterized completely* by two independently variable properties plus the moles charged initially. Postulate II states that the isolated system will tend to a unique stable equilibrium state, consistent with the internal restraints. Thus, if N_{CO_2} moles of CO_2 and N_{H_2} moles of H_2 were charged initially to a vessel of total volume $\underline{V}$ and total energy $\underline{U}$ and the vessel were then isolated, the system would tend to a stable equilibrium state which is consistent with the internal restraints. Since the final equilibrium state is completely characterized by N_{CO_2}, N_{H_2}, $\underline{V}$ and $\underline{U}$, we should be able to calculate all other properties of the final state from the information given.

It is well-known that CO_2 and H_2 can react in a variety of ways to form a number of different products. For example, the Sabatier reaction,

$$CO_2 + 4\,H_2 = CH_4 + 2\,H_2O$$

proceeds under some conditions, and the Fisher-Tropsch synthesis of methanol,

$$CO_2 + 3\,H_2 = CH_3OH + H_2O$$

can be made to occur under other conditions. Although we shall see that thermodynamics can be used to determine the equilibrium extent of any hypothesized reaction, it is of little help in deciding whether or not the hypothesized reaction will in fact occur. *Indeed, thermodynamics can be applied to determine the final equilibrium conditions only after one specifies which reactions occur within the time span of interest.* If a chemical reaction is too slow to proceed under the conditions of interest, then the barrier to reaction is treated as an internal restraint. Thus, to specify a system as required by Postulate I, the internal restraints should indicate which reactions have appreciable rates under the conditions of interest. These internal restraints must be supplied by the engineer and scientist from consideration of experimental kinetic data or from experience with similar systems. These considerations are outside the realm of classical thermodynamics.

10.1 Stoichiometry and Independent Sets of Reactions

This section is a brief review of the elements of stoichiometry.

In the absence of nuclear splitting, elements must be conserved in chemical reactions. This is achieved when reactions are stoichiometrically balanced. In such cases, of course, the total system mass is also constant.

Although most readers are familiar with the common chemical notation for a reaction, e.g.,

$$CO(g) + 3\,H_2(g) = CH_4(g) + H_2O(l) \tag{10-1}$$

we shall find it convenient to use the algebraic notation:

$$0 = \nu_1 \mathcal{C}_1(p_1) + \nu_2 \mathcal{C}_2(p_2) + \nu_3 \mathcal{C}_3(p_3) + \nu_4 \mathcal{C}_4(p_4) \tag{10-2}$$

or, simply,

$$0 = \sum_{i=1}^{k} \nu_i \mathcal{C}_i(p_i) \tag{10-3}$$

where ν_i is the *stoichiometric coefficient* or *stoichiometric number* of species i (by convention, positive for products and negative for reactants), $\mathcal{C}_i$ is the chemical formula and p_i is the physical state of i.

The stoichiometric coefficients in Eq. (10-3) must satisfy elemental material balances. For example, if we are considering a chemical reaction between $CO(g)$, $H_2(g)$, $H_2O(l)$, and $CH_3OH(l)$, Eq. (10-3) would be:

$$0 = \nu_1 CO(g) + \nu_2 H_2(g) + \nu_3 H_2O(l) + \nu_4 CH_3OH(l) \tag{10-4}$$

The element balances yield three equations in four unknowns:

$$\text{C-balance:} \quad +\nu_1 \qquad\qquad + \nu_4 = 0 \tag{10-5}$$

$$\text{O-balance:} \quad +\nu_1 \quad + \nu_3 + \nu_4 = 0 \tag{10-6}$$

$$\text{H-balance:} \quad 2\nu_2 + 2\nu_3 + 4\nu_4 = 0 \tag{10-7}$$

These equations may be solved for any three coefficients in terms of the fourth one. Choosing ν_1 as the independent variable, we find:

$$\nu_2 = 2\nu_1; \qquad \nu_3 = 0; \qquad \nu_4 = -\nu_1 \tag{10-8}$$

Thus, H_2O does not enter into this reaction; it is treated as an inert. The reaction is:

$$0 = \nu_1[CO(g) + 2\,H_2(g) - CH_3OH(l)] \tag{10-9}$$

or

$$CH_3OH(l) = CO(g) + 2\,H_2(g) \tag{10-10}$$

In general, we can solve for the stoichiometric coefficients *uniquely* provided that the number of species involved is no more than one greater than the number of elements. For example, consider a chemical reaction between CO, H_2, H_2O, CH_4, and CH_3OH. In this case, there are five compounds but only three elements. The element balances would then yield:

$$\begin{aligned} 0 = {} & \nu_4[-CO(g) - 3\,H_2(g) + H_2O(l) + CH_4(g)] \\ & + \nu_5[-CO(g) - 2\,H_2(g) + CH_3OH(l)] \end{aligned} \tag{10-11}$$

so that any linear combination of the two reactions given by Eqs. (10-1) and (10-10) would satisfy the element balances. Unless we have reason to believe that these two reactions are coupled,[1] we should treat them as independent reactions instead of adding them to obtain one reaction. That is, the reaction

$$2\,CO(g) + 5\,H_2(g) = H_2O(l) + CH_4(g) + CH_3OH(l) \tag{10-12}$$

is more restrictive because it implies that the moles of methane and methanol are in a 1:1 ratio in a product obtained from CO and H_2. Note, however, that Eq. (10-12) plus either Eq. (10-1) or Eq. (10-10) comprise a satisfactory set of two reactions to describe the transformations between the five species.

If the number of species involved is equal to or less than the number of elements, then two or more of the element balances are redundant. Such is the case for the methanol decomposition reaction, Eq. (10-10). By inspection, one can see that the C- and O-balances are redundant [see Eqs. (10-5) and

[1] There may be cases in which two reactions of the form A + B = C and A + B = D are coupled (e.g., by catalytic or enzymatic necessity) such that C and D always form in a 1:1 ratio. In this case, there is only one independent reaction of the form 2 A + 2 B = C + D. Since this reaction is more restrictive in defining the final composition, however, and since such a restriction is not imposed by stoichiometry, we must have evidence beyond thermodynamic reasoning to impose such a restriction.

(10-6) with $\nu_3 = 0$]. Such cases are typical of formation or decomposition reactions of the form A + B = C, where A, B, and C are compounds. Isomerization reactions of the form A = B also fall into this group.

There is a simple method to determine the maximum number of *independent reactions* necessary to describe a chemical system. The procedure is based on the fact that *formation reactions* in which compounds are synthesized from their elements are always independent. For example, the formation reactions for the compounds involved in Eq. (10-10) are:

$$C(s) + \frac{1}{2}O_2(g) = CO(g) \tag{10-13}$$

$$C(s) + \frac{1}{2}O_2(g) + 2\,H_2(g) = CH_3OH(l) \tag{10-14}$$

where the stoichiometric multipliers have been determined by inspection. (Note that formation reactions always involve only one species that is not an element.) For reasons that will become apparent later, the physical state of elements in formation reactions are usually taken to be the most stable physical state under the conditions of interest.

If all of the elements are present in the system, then clearly the number of independent reactions is equal to the number of compounds. Thus, for a system containing C, H_2, O_2, CO, and CH_3OH, there are only two independent reactions, as given by Eqs. (10-13) and (10-14) or any other linear combination of these reactions.

If one or more elements is missing in the final system, then the number of independent reactions is less than the number of compounds present. To determine the independent reactions, we start from the formation reactions and then eliminate those elements that are not present. For example, if a system contained only H_2, CO, and CH_3OH, the C and O_2 would have to be eliminated from Eqs. (10-13) and (10-14). In this case, subtraction of one from the other yields Eq. (10-10), which would be the only independent reaction. Note that if this system also contained elemental carbon, there would still be only one reaction, Eq. (10-10), because the elimination of O_2 also eliminated C. Thus, the carbon would be considered as an inert.

Example 10.1

Determine independent sets of reactions for each of the following systems:

(a) C, O_2, H_2, CO, H_2O, CH_4, CH_3OH

(b) O_2, H_2, CO, H_2O, CH_4, CH_3OH

(c) H_2, CO, H_2O, CH_4, CH_3OH

(d) CO, H_2O, CH_4, CH_3OH

Solution

In all cases there are four compounds for which we have the formation reactions:

$$C(s) + \frac{1}{2}O_2(g) = CO(g)$$
$$\frac{1}{2}O_2(g) + H_2(g) = H_2O(l)$$
$$C(s) + 2\,H_2(g) = CH_4(g)$$
$$C(s) + \frac{1}{2}O_2(g) + 2\,H_2(g) = CH_3OH(l)$$

We can use the Gauss reduction technique[2] to transform this set into another set of four reactions, each containing one less element. For example, C can be eliminated from the third and fourth reactions by subtracting from them the first reaction:

$$C(s) + \frac{1}{2}O_2(g) = CO(g)$$
$$\frac{1}{2}O_2(g) + H_2(g) = H_2O(l)$$
$$-\frac{1}{2}O_2(g) + 2\,H_2(g) = CH_4(g) - CO(g)$$
$$2\,H_2(g) = CH_3OH(l) - CO(g)$$

Repeating the procedure to eliminate successively O_2 and H_2, we obtain:

$$C(s) + \frac{1}{2}O_2(g) = CO(g)$$
$$\frac{1}{2}O_2(g) + H_2(g) = H_2O(l)$$
$$3\,H_2(g) = CH_4(g) - CO(g) + H_2O(l)$$
$$0 = 3\,CH_3OH(l) - CO(g) - 2\,CH_4(g) - 2\,H_2O(l)$$

Thus, for case (d), we have only one reaction:

$$CO(g) + 2\,CH_4(g) + 2\,H_2O(l) = 3\,CH_3OH(l)$$

If H_2 is also present, as in case (c), we have a second reaction:

$$CO(g) + 3\,H_2(g) = CH_4(g) + H_2O(l)$$

If O_2 is also present, as in case (b), we have a third reaction:

$$\frac{1}{2}O_2(g) + H_2(g) = H_2O(l)$$

[2] See, e.g., F. B. Hildebrand, *Methods of Applied Mathematics* (Englewood Cliffs, N.J.: Prentice-Hall, Inc., 1965), p. 4.

and if elemental carbon is also present, as in case (a), we have a fourth reaction:

$$C(s) + \frac{1}{2}O_2(g) = CO(g)$$

Example 10.2

In the catalytic oxidation of ammonia, the following reactions have been suggested:

$$4\,NH_3 + 5\,O_2 = 4\,NO + 6\,H_2O \quad \text{(a)}$$

$$4\,NH_3 + 3\,O_2 = 2\,N_2 + 6\,H_2O \quad \text{(b)}$$

$$4\,NH_3 + 6\,NO = 5\,N_2 + 6\,H_2O \quad \text{(c)}$$

$$2\,NO + O_2 = 2\,NO_2 \quad \text{(d)}$$

$$2\,NO = N_2 + O_2 \quad \text{(e)}$$

$$N_2 + 2\,O_2 = 2\,NO_2 \quad \text{(f)}$$

From these, determine a set of independent reactions to describe the system.

Solution

There are four compounds (NH_3, NO, NO_2, H_2O) containing three elements (N_2, H_2, O_2), one of which is not present (H_2). Thus, H_2 must be eliminated from the two formation reactions that involve hydrogen:

$$N_2 + 3\,H_2 = 2\,NH_3$$

$$2\,H_2 + O_2 = 2\,H_2O$$

or

$$2\,N_2 + 6\,H_2O = 4\,NH_3 + 3\,O_2$$

which is the reverse of reaction (b). Since reactions (e) and (f) are formation reactions for the other two compounds present, the set of (b), (e), and (f) are independent and any other set can be obtained as a linear combination of these three.

10.2 Criteria of Chemical-Reaction Equilibrium

In Section 7.3 the criteria of equilibrium were determined for a closed, simple system containing π phases in which a single chemical reaction occurred. As shown, if the entire system were maintained at constant $\underline{S}$ and $\underline{V}$, then minimization of $\underline{U}$ required the equality of T, P, and μ_i in each phase and, in addition,

$$\sum \nu_i \mu_i = 0 \tag{10-15}$$

where the summation includes only those species which take part in the chemical reaction. Since μ_i is the same in all phases, there is no need to superscript this property.

Equation (10-15) is not restricted to systems at constant $\underline{S}$, $\underline{V}$, and M. It can also be obtained by $\underline{H}$-minimization at constant $\underline{S}$, P, M, or, for example, at constant T, P, M, $\underline{G}$ is a minimum and $\delta \underline{G}$ must vanish:

$$\delta \underline{G} = \sum_{s=1}^{\pi} \sum_{i=1}^{n} \mu_i^{(s)} \, \delta N_i^{(s)} = 0 \tag{10-16}$$

Following the notation of Section 7.3, we see that the *extent of reaction*, ξ, is defined as:

$$\delta \xi \equiv \frac{\delta N_i}{\nu_i} \tag{10-17}$$

or

$$\delta N_i = \sum_{s=1}^{\pi} \delta N_i^{(s)} = \nu_i \, \delta \xi \tag{10-18}$$

Adopting the convention that $\nu_i = 0$ for all species that do not take part in the reaction, conservation of mass requires:

$$\sum_{s=1}^{\pi} \delta N_i^{(s)} - \nu_i \, \delta \xi = 0 \qquad (i = 1, 2, \ldots, n) \tag{10-19}$$

Multiplying Eq. (10-19) by $\mu_i^{(1)}$, summing over all i, and subtracting from Eq. (10-16) yields:

$$\delta \underline{G} = \sum_{s=2}^{\pi} \sum_{i=1}^{n} (\mu_i^{(s)} - \mu_i^{(1)}) \, \delta N_i^{(s)} + \left(\sum_{i=1}^{n} \nu_i \mu_i^{(1)} \right) \delta \xi = 0 \tag{10-20}$$

or

$$\mu_i^{(1)} = \mu_i^{(2)} = \ldots \mu_i^{(\pi)} \qquad (i = 1, \ldots, n) \tag{10-21}$$

and

$$\sum_{i=1}^{n} \nu_i \mu_i = 0 \tag{10-22}$$

If r *independent* reactions could occur, then Eq. (10-19) becomes:

$$\sum_{s=1}^{\pi} \delta N_i^{(s)} - \nu_i^{(1)} \, \delta \xi^{(1)} - \ldots - \nu_i^{(r)} \, \delta \xi^{(r)} = 0 \tag{10-23}$$

which, after multiplying by $\mu_i^{(1)}$, summing over i, and substituting in Eq. (10-16) yields:

$$\delta \underline{G} = \sum_{s=2}^{\pi} \sum_{i=1}^{n} (\mu_i^{(s)} - \mu_i^{(1)}) \, \delta N_i^{(s)} + \sum_{m=1}^{r} \left[\sum_{i=1}^{n} \nu_i^{(m)} \mu_i^{(1)} \right] \delta \xi^{(m)} = 0 \tag{10-24}$$

Since the r reactions are independent, each $\delta \xi^{(m)}$ can be varied independently and, thus, the coefficient of each $\delta \xi^{(m)}$ must vanish. Hence, we obtain r independent criteria of the form of Eq. (10-22), one for each reaction.

For a system in which r independent reactions occur, the equilibrium composition can be determined if the r variables, $\xi^{(m)}$, are known. Integrating Eq. (10-23) from the initial condition [for which $\xi^{(m)}$ is normally defined as zero] to the equilibrium condition,

$$N_i = N_{i_0} + \sum_{m=1}^{r} \nu_i^{(m)} \xi^{(m)} \tag{10-25}$$

Defining

$$\nu^{(m)} = \sum_{i=1}^{n} \nu_i^{(m)}$$

the final total mole number is:

$$N = N_0 + \sum_{i=1}^{n} \sum_{m=1}^{r} \nu_i^{(m)} \xi^{(m)} = N_0 + \sum_{m=1}^{r} \nu^{(m)} \xi^{(m)} \tag{10-26}$$

For a single-phase system, the equilibrium concentrations are then:

$$x_i = \frac{N_i}{N} = \frac{N_{i_0} + \sum_{m=1}^{r} \nu_i^{(m)} \xi^{(m)}}{N_0 + \sum_{m=1}^{r} \nu^{(m)} \xi^{(m)}} \tag{10-27}$$

For each reaction there is a *limiting reactant* and a *limiting product* which define, respectively, the maximum and minimum allowable values of ξ. Consider species j as a reactant; if essentially all N_{j_0} initially present reacts to form products, then $\xi = -N_{j_0}/\nu_j$, which is a positive number because ν_j is negative. The maximum value of ξ then corresponds to the smallest positive value of $(-N_{j_0}/\nu_j)$, where j varies over all reactants. Similarly, if j were a product, and if essentially all j were converted to reactants, then $\xi = -N_{j_0}/\nu_j$ would be negative. Thus, the limiting product is the species for which $(-N_{j_0}/\nu_j)$ attains the smallest negative value. The limits on ξ are then:

$$-\left(\frac{N_{j_0}}{\nu_j}\right)_{\text{products}} \leqslant \xi \leqslant -\left(\frac{N_{j_0}}{\nu_j}\right)_{\text{reactants}} \tag{10-28}$$

10.3 Equilibrium Constants

The criterion of chemical-reaction equilibrium, Eq. (10-22), can be viewed as an equation relating intensive variables. Since the chemical potential can be expressed as a function of $T, P, x_1, \ldots, x_{n-1}$, Eq. (10-22) can be written as:

$$\sum_{i=1}^{n} \nu_i \mu_i(T, P, x_1, \ldots, x_{n-1}) = 0 \tag{10-29}$$

Given the temperature, pressure, and moles charged, all mole fractions can be expressed as a function of the extent of reaction, as given by Eq. (10-27). In this case, Eq. (10-29) is one equation in the one unknown, ξ. To express the chemical equilibrium criterion in terms of fugacities, from Eq. (8-117c),

$$\mu_i = RT \ln \hat{f}_i + \lambda_i(T) \tag{10-30}$$

Before substituting Eq. (10-30) into Eq. (10-29), we need to eliminate $\lambda_i(T)$. To accomplish this, we can rewrite Eq. (10-30) for a particular state:

$$\mu_i^0 = RT \ln \hat{f}_i^0 + \lambda_i(T) \tag{10-31}$$

The state chosen is called a standard state and its properties are noted by a superscript zero. The pressure, composition, and state of aggregation are arbitrary; the temperature, however, must be chosen as that of the system under consideration. Then, subtracting Eq. (10-31) from Eq. (10-30) and substituting into Eq. (10-29):

$$\prod_{i=1}^{n}\left(\frac{\hat{f}_i}{\hat{f}_i^0}\right)^{\nu_i} = \exp\left(-\frac{\Delta G^0}{RT}\right) \equiv K_a \tag{10-32}$$

where:

$$\Delta G^0 = \sum_{i=1}^{n} \nu_i \mu_i^0 = \Delta H^0 - T\,\Delta S^0 \tag{10-33}$$

and K_a is the chemical equilibrium constant. It is clear that the right-hand side of Eq. (10-32) depends only on the properties of the reactants and products in their standard states. The left-hand side involves the fugacities of all components in the mixture at the pressure and equilibrium composition of the system.

In most cases, it is expedient to choose the standard state as *pure i* in the most stable physical state at the system temperature. Thus, $\hat{f}_i^0 \longrightarrow f_i^0$. Further, if pure i is normally a gas at T and 1 atm, then usually one would modify the standard-state pressure to be that value where $f_i^0 = 1$ atm. This choice simplifies Eq. (10-32). If pure i at T and 1 atm were an ideal gas, then the standard-state pressure would also be 1 atm. Should i normally be a liquid or solid at the system temperature and around atmospheric pressure, then the usual standard state is defined as the pure condensed phase at T under the existing vapor pressure.

To convert Eq. (10-32) to a function of concentrations, the mixture fugacity, $\hat{f}_i$, is expanded as a function of concentration by using Eq. (8-151) or Eq. (8-178) if i is present in a vapor phase or in a condensed phase, respectively. If a component is present in two or more phases, either fugacity expansion may be used since $\hat{f}_i$ is equal in all phases at equilibrium.

Should all species taking part in the reaction be present as vapors, substitution of Eq. (8-151) into Eq. (10-32) yields:

$$K_a = \left(\prod_{i=1}^{n} \phi_i^{\nu_i}\right)\left(\prod_{i=1}^{n} y_i^{\nu_i}\right)P^{\nu} = K_\phi K_y P^{\nu} \tag{10-34}$$

where $\nu = \sum_{i=1}^{n} \nu_i$, and the standard state fugacities, f_i^0, have been taken as unit fugacity. If the vapor mixture forms an ideal solution, then (see Section 8.7):

$$\phi_i = \frac{f_i}{P} \qquad \text{(ideal solution)} \tag{10-35}$$

where f_i is the fugacity of pure vapor i at the temperature and pressure of the mixture. Substituting Eq. (10-35) into Eq. (10-34) yields:

$$K_a = K_{f/P} K_y P^{\nu} \qquad \text{(ideal vapor solution)} \tag{10-36}$$

For liquid or solid mixtures, Eq. (8-178) can be used for $\hat{f}_i$:

$$\hat{f}_i = f_i \gamma_i x_i \tag{10-37}$$

Thus,

$$\frac{\hat{f}_i}{f_i^0} = \left(\frac{f_i}{f_i^0}\right)\gamma_i x_i \tag{10-38}$$

If the standard-state pressure is taken as P_{vp_i} or 1 atm and the system pressure as P, then:

$$RT \ln\left(\frac{f_i}{f_i^0}\right) = \int_{(P_{vp_i} \text{ or } 1 \text{ atm})}^{P} V_i \, dP \tag{10-39}$$

The integral of Eq. (10-39) is usually quite small; if it were neglected, then substitution of Eq. (10-38) into Eq. (10-32) yields:

$$K_a = \left(\prod_{i=1}^{n} \gamma_i^{\nu_i}\right)\left(\prod_{i=1}^{n} x_i^{\nu_i}\right) = K_\gamma K_x \qquad \text{(condensed phase, pressure correction neglected)} \tag{10-40}$$

The relations presented above have shown how equilibrium concentrations may be related to the equilibrium constant. That is, if one could locate or calculate an equilibrium constant, then equilibrium conversions may be determined.

We recall that K_a was dependent upon the standard state that was chosen. With few exceptions this standard state is selected as the pure material, at unit fugacity if the component usually exists as a gas at the temperature in question and at a pressure around atmospheric. For components that are normally in a condensed phase at this temperature, the standard state may be chosen in a number of ways. A unit-fugacity state could still be selected, although in such cases it would be a hypothetical state. More commonly we would select the pure condensed phase at T as the standard state (i.e., the material under its own vapor pressure). In other cases, we might even find it convenient to define the standard state as a solution in which the concentrations of the compounds are clearly delineated. Any of these states is perfectly satisfactory from a thermodynamic point of view. The selection, nevertheless, is made in a more pragmatic sense (i.e., it is determined by the available data). For a limited number of reactions, K_a has been expressed as a temperature function as will be shown later in Figure 10.1. When such data are unavailable, K_a can be calculated from ΔG^0 by using Eq. (10-32) or from ΔH^0 and ΔS^0 by using Eqs. (10-32) and (10-33). Since it is impractical to list such functions for every reaction, tables are available for a large number of compounds showing the free energy and enthalpy of formation of the species from the elements. In these tables the function ΔG^0 becomes, for each species, ΔG_f^0 and, likewise, ΔH^0 becomes ΔH_f^0. To obtain ΔG^0 and ΔH^0, at the temperature of interest,

$$\Delta G^0 = \sum_{i=1}^{n} \nu_i \, \Delta G_{f_i}^0 \tag{10-41}$$

and,

$$\Delta H^0 = \sum_{i=1}^{n} \nu_i \, \Delta H^0_{f_i} \tag{10-42}$$

where, again, $\nu_j = 0$ for those species that do not participate in the reaction. For elements, by convention,[3]

$$\Delta H^0_f = \Delta G^0_f = 0 \text{ at all temperatures} \tag{10-43}$$

Several tabulations[4] form an excellent reference source for values of ΔH^0_f and ΔG^0_f over a wide range of temperatures. These same compilations also show values of C^0_p, the heat capacity in the ideal gas standard state.

If ΔG^0 and ΔH^0 are available at one temperature, ΔG^0 can be found at any other temperature as follows:

$$\frac{d\left(\frac{\Delta G^0}{T}\right)}{dT} = \frac{1}{T}\frac{d\,\Delta G^0}{dT} - \frac{\Delta G^0}{T^2} = -\frac{\Delta S^0}{T} - \frac{\Delta G^0}{T^2} = -\frac{\Delta H^0}{T^2} \tag{10-44}$$

Eq. (10-44) may be integrated if ΔH^0 is known at one temperature T_1 since:

$$\Delta H^0_T = \Delta H^0_{T_1} + \int_{T_1}^{T} \Delta C^0_p \, dT \tag{10-45}$$

where:

$$\Delta C^0_p = \sum_{i=1}^{n} \nu_i C^0_{p_i} \tag{10-46}$$

Before completing this discussion, it is appropriate to describe briefly the *Third Law of thermodynamics*. Experimental evidence indicates that the entropy change in a chemical reaction becomes negligible as the absolute temperature approaches zero, that is,

$$\Delta S^0_0 = 0 \tag{10-47}$$

With the additional stipulation that the entropy state of elements is zero at $T = 0$, then it follows that $\Delta S^0_{0_f} = 0$ for all materials; or, as more often stated, $S^0_{0_i} = 0$. With this base, it is possible to refer to an *absolute* entropy that can be calculated by integrating with *actual* heat capacity data through the solid phase (at very low temperatures), the liquid phase, and into the vapor phase. Entropy changes that result from phase transitions are also included. The determination of such *absolute* entropies obviously requires considerable data[5] and will not be considered further in this text.

[3] For elements such as oxygen, nitrogen, etc., there is no problem with this convention. For elements that are solids at the temperature of interest, a clear statement of the crystal form is also necessary. For example, the standard state for carbon is based on graphite. Should other forms of carbon be present in a system, the ΔG^0_f and ΔH^0_f for such forms are not zero.

[4] D. R. Stull, E. F. Westrum, Jr., and G. C. Sinke, *The Chemical Thermodynamics of Organic Compounds* (New York: John Wiley & Sons, Inc., 1969). D. R. Stull and H. Prophet, *JANEF Thermochemical Tables*, 2nd ed., NSRDS-NBS 37, June, 1971.

[5] For example, since as $T \to 0$, $C_p(s) \to 0$, special caution must be used to integrate at low temperatures. All phase transformations (first and higher order) also must be included.

In many instances the equilibrium constant of a reaction is known at one temperature and an extrapolation to other temperatures is necessary. From Eqs. (10-32) and (10-44),

$$\frac{d \ln K_a}{dT} = -\frac{1}{R}\frac{d\left(\frac{\Delta G^0}{T}\right)}{dT} = \frac{\Delta H^0}{RT^2} \qquad (10\text{-}48)$$

Often ΔH^0 does not vary appreciably with temperature and, therefore, $\ln K_a$ is nearly linear in $1/T$. A number of reaction equilibrium constants have been plotted in this fashion on Figure 10.1. From the slope one can find a temperature-average value of ΔH^0, i.e.,

$$-\Delta H^0 = \frac{R \ln\left(\frac{K_{T_2}}{K_{T_1}}\right)}{\left(\frac{1}{T_2} - \frac{1}{T_1}\right)} \qquad (10\text{-}49)$$

Positive slopes on Figure 10.1 then correspond to reaction with $\Delta H^0 < 0$ (i.e., exothermic reactions).

In summary, to determine the equilibrium constant of a reaction:

1. If the standard states of all components are chosen as pure materials in an ideal gas-unit fugacity state,
   (a) Use Figure 10.1 if applicable. log plots
   (b) Determine the enthalpy and free energy of formation for all components (see footnote 4) and use Eqs. (10-41) and (10-42).
   (c) If other sources are utilized that allow the free energy and enthalpy of formation to be determined at only a single temperature, then Eq. (10-48) may be integrated to find ΔG^0 and $\ln K_a$ at other temperatures with ΔH^0 either assumed constant or expressed as a function of temperature as in Eq. (10-45).
   For such standard states, $\hat{f}_i^0$ in Eq. (10-32) is then set at one atmosphere and, of course, $\hat{f}_i$ must then also be expressed in atmospheres.
2. If one or more of the components is chosen with a standard state that differs from the pure component, ideal gas-unit fugacity state, then $\hat{f}_i^0$ in Eq. (10-32) must be the true fugacity of i in this chosen state and, to obtain ΔG^0, the chemical potential of i in the same state must be used.
   (a) Some tabulations list the free energy of formation of species in states that are different from ideal gas-unit fugacity states, and these may be used.
   (b) If the free energy of formation of the species is known in *any* reference state, it may be converted to the desired state by devising a reversible process and calculating the change in Gibbs free energy. For example, suppose that ΔG_f^0 were available for an ideal gas-unit

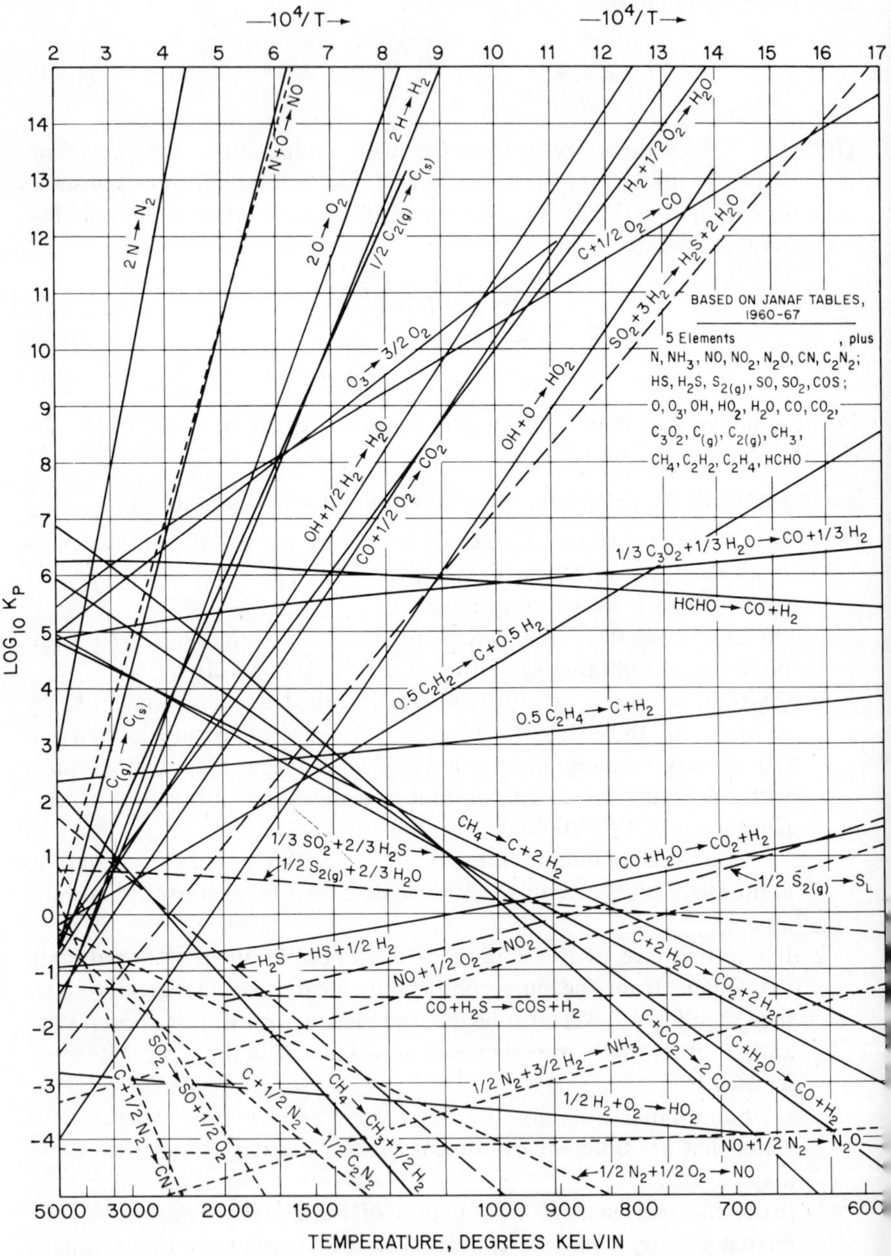

—10⁴/T→
—10⁴/T→
LOG10 KP
TEMPERATURE, DEGREES KELVIN
BASED ON JANAF TABLES, 1960-67
5 Elements, plus N, NH3, NO, NO2, N2O, CN, C2N2; HS, H2S, S2(g), SO, SO2, COS; O, O3, OH, HO2, H2O, CO, CO2, C3O2, C(g), C2(g), CH3, CH4, C2H2, C2H4, HCHO
2N → N2
N+O → NO
2H → H2
2O → O2
1/2 C2(g) → C(s)
H2+1/2 O2 → H2O
C+1/2 O2 → CO
SO2+3H2 → H2S+2H2O
O3 → 3/2 O2
OH+O → HO2
OH+1/2 H2 → H2O
CO+1/2 O2 → CO2
1/3 C3O2+1/3 H2O → CO+1/3 H2
HCHO → CO+H2
0.5 C2H2 → C+0.5 H2
0.5 C2H4 → C+H2
C(g) → C(s)
CH4 → C+2H2
1/3 SO2+2/3 H2S → 1/2 S2(g)+2/3 H2O
CO+H2O → CO2+H2
1/2 S2(g) → SL
H2S → HS+1/2 H2
NO+1/2 O2 → NO2
C+2H2O → CO2+2H2
CO+H2S → COS+H2
SO2 → SO+1/2 O2
C+1/2 N2 → CN
C+1/2 N2 → 1/2 C2N2
CH4 → CH3+1/2 H2
1/2 N2+3/2 H2 → NH3
C+CO2 → 2CO
C+H2O → CO+H2
1/2 H2+O2 → HO2
NO+1/2 N2 → N2O
1/2 N2+1/2 O2 → NO

Figure 10.1

fugacity state but the desired standard state for this material was a liquid, pure at a pressure P which is greater than P_{vp} at the same temperature. We could correct ΔG_f^0 as follows:

$$\Delta G_f^0(\text{liquid}, P) = \Delta G_f^0 \text{ (ideal gas-unit fugacity)}$$
$$+ \int_{1\text{ atm}}^{P_{vp}} V^V\, dP + \int_{P_{vp}}^{P} V^L\, dP$$

The first integral represents the change in free energy in an isothermal variation from one atmosphere to the vapor pressure (also in atmospheres). There is no free energy change in condensation at P_{vp}. The third term reflects the free energy change of the liquid as the pressure changes from P_{vp} to the system pressure.

3. If no values of the free energy of formation can be located, they may be approximated by group-contribution methods.[6]

In most of the cases discussed above, the standard state was chosen at a fixed pressure; thus, neither K_a nor ΔG^0 were functions of pressure. If, however, the standard state for any of the reactants or products were to be chosen at the system pressure, P, then both K_a and ΔG^0 become functions of this pressure. From Eq. (10-33),

$$\left(\frac{\partial\, \Delta G^0}{\partial P}\right)_T = \sum_{i=1}^{n} \nu_i \left(\frac{\partial \mu_i^0}{\partial P}\right)_T = -RT\left(\frac{\partial \ln K_a}{\partial P}\right)_T \tag{10-50}$$

If μ_i^0 is not a function of P, this derivative vanishes. If μ_i^0 is a function of P, then $(\partial \mu_i^0/\partial P)_T = V_i^0$, the molal volume of i in the chosen standard state. It is usually more convenient to use a ΔG^0 (or K_a) that is independent of pressure. In this regard, referring to the liquid-phase example shown in 2(b) above, we see that if the standard state were chosen as the pure liquid at its vapor pressure, then ΔG_f^0 would be pressure-independent.

Example 10.3

A closed system at constant temperature and pressure is in equilibrium when it attains a minimum Gibbs free energy. Consider a system initially charged with 1 g-mol of pure I_2, which is maintained at 800°C and 1 atm, in which the following dissociation reaction occurs:

$$I_2(g) = 2\, I(g)$$
$$\Delta H^0 = 37{,}400 \text{ cal/g-mol}$$
$$\Delta S^0 = 26.9 \text{ cal/g-mol °K}$$

The standard states are pure vapors at unit fugacity.

Determine the equilibrium composition by first calculating the Gibbs free energy of the mixture as a function of the moles of I_2 dissociated (denoted

[6] R. C. Reid and T. K. Sherwood, *Properties of Gases and Liquids*, 2nd ed. (New York: McGraw-Hill Book Company, 1966), Chap. 5.

by x) and then determining the minimum in $\underline{G}$ with respect to x. Assume that the gas mixture is ideal.

Solution

Let us form a mixture containing $(1 - x)$ g-mol of I_2 and $2x$ g-atom of I by a two-step process. First select x g-mol of I_2 in its standard state and assume that it decomposes to $2x$ g-atom of I in its standard state. Then let us mix the remaining $(1 - x)$ g-mol I_2 with the newly formed $2x$ g-atom of I. The changes in enthalpy and entropy are:

$$\Delta \underline{H}_1 = x\,\Delta H^0$$
$$\Delta \underline{S}_1 = x\,\Delta S^0$$
$$\Delta \underline{H}_2 = \Delta \underline{H}_{mix} = 0 \text{ (ideal gas mixture)}$$
$$\Delta \underline{S}_2 = \Delta \underline{S}_{mix} = \Delta \underline{S}^{ID} = -R[(1 - x)\ln y_{I_2} + (2x)\ln y_I]$$
$$= -R\left[(1 - x)\ln\left(\frac{1 - x}{1 + x}\right) + (2x)\ln\left(\frac{2x}{1 + x}\right)\right]$$

Since

$$\Delta \underline{G} = \Delta \underline{H} - T\,\Delta \underline{S}$$
$$\Delta \underline{G} = x\,\Delta H^0 - Tx\,\Delta S^0 + RT\left[(1 - x)\ln\left(\frac{1 - x}{1 + x}\right) + (2x)\ln\left(\frac{2x}{1 + x}\right)\right]$$

Values for the enthalpy and entropy changes are shown in Table 10.1 and they are plotted in Figures 10.2 and 10.3.

The free energy $\Delta \underline{G}$ attains a minimum value when x is about 0.052 (i.e., when some 5.2% of the original iodine has decomposed). Returning to the $\Delta \underline{G}$ expression derived above, to find the minimum value of $\Delta \underline{G}$ as x is varied,

TABLE 10.1

FREE ENERGY AS A FUNCTION OF I_2 DISSOCIATED

x	$\Delta \underline{H} = x\,\Delta H^0$	$xT\,\Delta S^0$	$T\,\Delta S_{mix}$	$T\,\Delta \underline{S} = xT\,\Delta S^0 + T\,\Delta S_{mix}$	$\Delta \underline{G} = \Delta \underline{H} - T\,\Delta \underline{S}$
0	0	0	0	0	0
0.02	748	555	360	915	−167
0.04	1,496	1,110	599	1,709	−213
0.05	1,870	1,388	705	2,093	−223
0.06	2,244	1,666	798	2,464	−220
0.08	2,992	2,221	964	3,185	−193
0.1	3,740	2,775	1,089	3,865	−125
0.3	11,220	8,328	1,911	10,239	+981
0.5	18,700	13,880	2,033	15,913	+2,787
0.7	26,180	19,432	1,687	21,119	+5,061
0.9	33,660	24,934	833	25,817	+7,843
1.0	37,400	27,760	0	27,760	+9,640

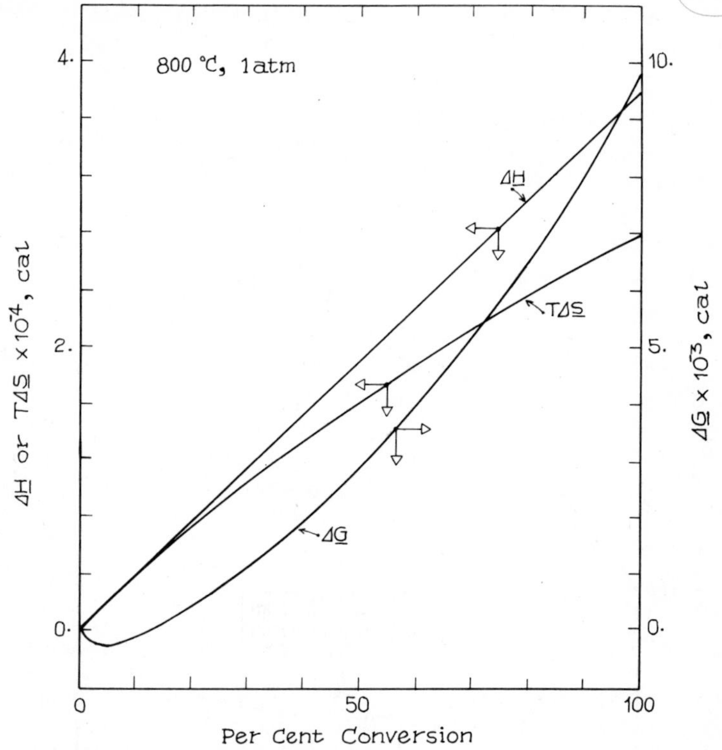

Figure 10.2 $\Delta \underline{H}$, $T\,\Delta \underline{S}$, and $\Delta \underline{G}$ for the reaction $I_2(g) \rightleftharpoons 2\,I(g)$.

we can also differentiate with respect to x and set the result equal to zero, i.e.,

$$0 = (\Delta H^0 - T\,\Delta S^0) + RT \ln \left(\frac{4x^2}{1 - x^2}\right) = \Delta G^0 + RT \ln \left(\frac{y_I^2}{y_{I_2}}\right)$$

This is, of course, the same result that we would obtain from Eqs. (10-32) and (10-34) using unit-fugacity standard states and assuming ideal gases and one atmosphere pressure (i.e., $K_\phi = 1$, $P^\nu = 1$), with $\Delta G^0 = \Delta H^0 - T\,\Delta S^0 = 37{,}400 - (1073)(26.9) = 9640$ cal/g-mol I_2.

The important point of this simple example is to note that the entropy change in mixing was the term that led to the minimization of $\Delta \underline{G}$. Without it, in this case, the equilibrium mixture would have contained only pure iodine as I_2.

With only a slight increase in complexity, gas-phase nonidealities may be included. For such a case:

$$\Delta \underline{H} = x\,\Delta H^0 + \Delta \underline{H}^{EX}$$
$$\Delta \underline{S} = x\,\Delta S^0 + \Delta \underline{S}^{ID} + \Delta \underline{S}^{EX}$$

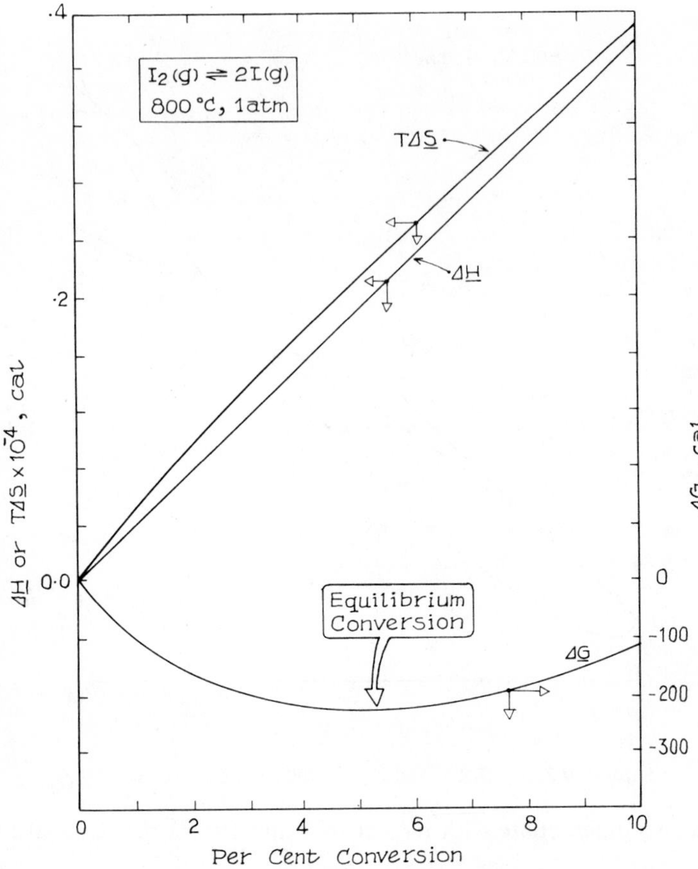

Figure 10.3 Expanded view of figure 10.2.

where,

$$\Delta \underline{H}^{EX} = 2x\bar{H}_{\mathrm{I}} + (1 - x)\bar{H}_{\mathrm{I}_2} - 2xH_{\mathrm{I}} - (1 - x)H_{\mathrm{I}_2}$$

$$\Delta \underline{S}^{EX} = 2x\bar{S}_{\mathrm{I}} + (1 - x)\bar{S}_{\mathrm{I}_2} - 2xS_{\mathrm{I}} - (1 - x)S_{\mathrm{I}_2}$$

With $\Delta \underline{S}^{ID}$ as given previously, using the $\Delta \underline{G}$ expression derived for ideal gases but including $\Delta \underline{H}^{EX} - T\,\Delta \underline{S}^{EX}$ where:

$$\Delta \underline{H}^{EX} - T\,\Delta \underline{S}^{EX} = \Delta \underline{G}^{EX} = 2x(\bar{G}_{\mathrm{I}} - \bar{G}_{\mathrm{I}}^{ID}) + (1 - x)(\bar{G}_{\mathrm{I}_2} - \bar{G}_{\mathrm{I}_2}^{ID})$$

$$= 2xRT \ln \gamma_{\mathrm{I}} + (1 - x)RT \ln \gamma_{\mathrm{I}_2}$$

then,

$$\frac{d\,\Delta \underline{G}}{dx} = 0 = \Delta G^0 + RT \ln \frac{y_{\mathrm{I}}^2}{y_{\mathrm{I}_2}} + RT \ln \frac{\gamma_{\mathrm{I}}^2}{\gamma_{\mathrm{I}_2}}$$

or

$$\Delta G^0 = -RT \ln \left[\frac{(y_I \gamma_I)^2}{(y_{I_2} \gamma_{I_2})}\right] = -RT \ln K_a$$

since $\gamma_i y_i = \hat{f}_i$. To proceed further in this case requires data for the variation of γ_I and γ_{I_2} (or $\Delta \underline{H}^{EX}$, $\Delta \underline{S}^{EX}$) with composition.

Although ΔH^0 is commonly referred to as the enthalpy of reaction, it is strictly speaking the enthalpy change between products and reactants in their respective standard states. ΔH^0 is *not* the enthalpy change when the reactants and products are present in an equilibrium mixture. For an equilibrium mixture,

$$\Delta H_r \equiv \sum_{i=1}^{n} v_i \bar{H}_i \tag{10-51}$$

where $\bar{H}_i$ is the partial molal enthalpy of component i at the system temperature, pressure, and in the equilibrium mixture. The relation between ΔH^0 and ΔH_r is shown schematically in Table 10.2. For ideal gas mixtures, there

TABLE 10.2
ΔH^0 AND ΔH_r FOR A REACTION
$aA + bB = cC + dD$

1. Start with a moles of A and b moles of B in their standard states, pure, at P^0.
2. Expand or compress each pure reactant isothermally to the system pressure P.

$$\Delta H_1 = a(H_{A,P} - H_A^0) + b(H_{B,P} - H_B^0)$$

3. Mix these pure reactants isothermally with a large system of A, B, C, D, in chemical equilibrium at P.

$$\Delta H_2 = a(\bar{H}_{A,P} - H_{A,P}) + b(\bar{H}_{B,P} - H_{B,P})$$

4. Let a moles of A react isothermally with b moles of B to form c moles of C and d moles of D in the equilibrium mixture at P.

$$\Delta H_3 = \Delta H_r$$

5. Remove c moles of C and d moles of D from the equilibrium mixture to pure components at P.

$$\Delta H_4 = c(H_{C,P} - \bar{H}_{C,P}) + d(H_{D,P} - \bar{H}_{D,P})$$

6. Expand or compress each pure product isothermally to the standard state pressure, P^0.

$$\Delta H_5 = c(H_C^0 - H_{C,P}) + d(H_D^0 - H_{D,P})$$

Then, since

$$\begin{aligned} \Delta H^0 &= \Delta H_1 + \Delta H_2 + \Delta H_3 + \Delta H_4 + \Delta H_5 \\ &= \Delta H_r - \sum_{i=1}^{n} v_i(\bar{H}_{i,P} - H_{i,P}) - \sum_{i=1}^{n} v_i(H_{i,P} - H_i^0) \\ &= \Delta H_r - \sum_{i=1}^{n} v_i \, \overline{\Delta H}_{i,P} - \sum_{i=1}^{n} v_i(H_{i,P} - H_i^0) \end{aligned}$$

are no heats of mixing or pressure-effects on enthalpy so that in this special case, $\Delta H^0 = \Delta H_r$.

Example 10.4

A student working in the laboratory has placed a small quantity of dry calcium carbonate in an evacuated one-liter bomb. She wishes to heat this bomb to 1000°C and, at this temperature, attain a CO_2 pressure of one atmosphere. Neglecting solid volumes and assuming that CO_2 at 1000°C and one atmosphere behaves as an ideal gas, determine how many grams of original $CaCO_3$ should be added. At 898°C,

$$\Delta G^0 = 0, \quad \Delta H^0 = 42{,}000 \text{ cal for the reaction:}$$

$$CaCO_3(s) = CaO(s) + CO_2(g)$$

Solution

No standard states were given. Usually in such a case it is implied that the standard states for gases are unit fugacity and for solids, pure components at their vapor pressure. (However, it is always advisable to refer to the literature to verify the standard states in question.) Thus, the activities of all solids are unity and:

$$K_a = K_p = p_{CO_2}$$

At 898°C, $\Delta G^0 = 0$; thus, $K_a = \exp(-\Delta G^0/RT) = 1$ atm. Since the reaction is endothermic, an increase in temperature will increase K_a and at 1000°C, with chemical equilibrium, the pressure of CO_2 would exceed one atmosphere. Since this higher pressure is not allowed, all the $CaCO_3$ must have been decomposed and only CaO remains as the solid phase.
Thus:

$$N^V_{CO_2} = \frac{P\underline{V}}{RT} = \frac{(1)(1000)}{(82.07)(1273)} = 9.6 \times 10^{-3} \text{ g-mol } CO_2$$

$$= \text{g-mol initial } CaCO_3 = 0.96 \text{ g initial } CaCO_3$$

10.4 The Phase Rule for Chemically Reacting Systems

We have seen in Section 10.2 that the criteria of phase equilibrium are valid even when chemical reactions occur within the system. In addition, we find one more restraining equation of the form of Eq. (10-22) or Eq. (10-32) for each independent chemical reaction.

Since the criteria of phase equilibrium were used previously to develop the phase rule, Eq. (9-7), we need only modify the rule to account for chemical equilibrium. That is, the variance should be reduced by one for each independent chemical reaction.

$$\mathscr{f} = n + 2 - \pi - r \tag{10-52}$$

where r is the number of independent chemical reactions.

As discussed in Sections 9.1 and 9.2 (see the discussion following Figures 9.14 and 9.15), the phase rule *does not* imply that any set of $\mathscr{f}$ intensive variables can be used to describe completely all other intensive variables. It does imply that *there are certain sets of $\mathscr{f}$ intensive variables that can be used to describe completely all other intensive variables.* Consider, for example, a two-phase system containing five species in which one isomerization reaction occurs between species 1 and 2. For this system, $\mathscr{f} = 5 + 2 - 2 - 1 = 4$. If one of the phases present were an ideal vapor phase, then the chemical equilibrium expression,

$$K_a = K_y = \frac{y_1}{y_2} \tag{10-53}$$

can be interpreted as a function of y_1, y_2, and T. Thus, only two of these three intensive variables are independent and, hence, a set of four intensive variables necessary to satisfy the phase rule must contain at least two variables other than y_1, y_2, and T.[7]

Since Postulate I is valid for systems in which chemical reactions occur, $n + 2$ independently variable properties are still required to specify the intensity and *extent* of the system. Thus, it follows that a minimum of $\pi + r$ extensive variables are necessary in addition to the maximum of $n + 2 - \pi - r$ intensive variables. As illustrated in the following problem, there are special cases in which some intensive variables are set by specifying certain extensive variables. In such cases, the dependent intensive variables cannot be used in the set of $n + 2$ that are necessary to describe completely the system.

Example 10.5

For each of the following cases, determine a set of $n + 2$ independent variables which include $n + 2 - \pi - r$ intensive variables. Consider also the special case in which only the reactants are initially charged and these are fed in the ratio of their stoichiometric coefficients. Assume that the vapor phase is ideal in each case.

[7] If the phase containing the reactant species were not ideal, then Eq. (10-53) would contain a K_ϕ term which, in theory, is a function of all y_i and, hence, the restriction on y_1, y_2, and T would appear to be removed. If, however, the nonideality were not large, we might find that astronomical pressures are required in order to satisfy a set of y_1, y_2, and T which was far removed from that necessary to satisfy Eq. (10-53).

(a) $A(g) + B(g) = C(s)$ (A and B not soluble in C)

(b) $A(g) + B(g) = C(g) + D(g)$

Solution

Case (a): Since $n = 3$, $\pi = 2$, and $r = 1$,

$$\mathcal{f} = n + 2 - \pi - r = 3 + 2 - 2 - 1 = 2$$

Possible intensive variables are T, P, y_A (species C is present as a pure solid). The equilibrium relationship is:

$$K_a = \frac{\dfrac{\hat{f}^S_C}{f^0_C}}{\left(\dfrac{\hat{f}^V_A}{f^0_A}\right)\left(\dfrac{\hat{f}^V_B}{f^0_B}\right)} = f(T) \tag{10-54}$$

Choose f^0_C as the fugacity of pure solid C at T and the vapor pressure of C(s) at T and neglect any effect of pressure in the fugacity of the pure solid. Then, $\hat{f}^S_C/f^0_C = 1$. Choosing the standard states of A and B as unit fugacity, $\hat{f}^V_A/f^0_A = y_A P$ and $\hat{f}^V_B/f^0_B = y_B P$. Thus, Eq. (10-54) becomes:

$$K_a = (y_A y_B)^{-1} P^{-2} \tag{10-55}$$

or, simply,

$$g(y_A, T, P) = 0 \tag{10-56}$$

Thus, any two of the three intensive variables can be used in conjunction with Eq. (10-56) to determine the third intensive variable.

To determine the extent of the system, we must specify $\pi + r = 3$ extensive variables. Let us choose these as N_{A_0}, N_{B_0}, and N_{C_0}. Defining ξ as the extent of reaction, Eq. (10-55) can be written as:

$$y_A y_B = \left(\frac{N_{A_0} - \xi}{N_{A_0} + N_{B_0} - 2\xi}\right)\left(\frac{N_{B_0} - \xi}{N_{A_0} + N_{B_0} - 2\xi}\right) = K_a^{-1} P^{-2} \tag{10-57}$$

If T and P were the two intensive variables specified, Eq. (10-57) could be used to determine ξ and, thence, y_A and y_B. Thus, T, P, N_{A_0}, N_{B_0}, and N_{C_0} would be an independent set of $n + 2$ variables.

For the special case in which A and B are fed in the stoichiometric ratio of 1:1, then $N_{A_0} = N_{B_0}$ and Eq. (10-57) reduces to:

$$\frac{1}{4} = K_a^{-1} P^{-2} \tag{10-58}$$

Clearly, both T and P could not be considered independent. Only one can be selected as independent since Eq. (10-58) can then be used to calculate the other. Since $\nu_A = \nu_B$, y_A and y_B are fixed at 0.5 by the stoichiometric feed condition. Thus, the mole fraction is an intensive variable which is set by the stoichiometric feed condition. Only one other intensive variable can be set independently for this special case, and that intensive variable must be either

T or P. A set of $n + 2$ variables would then have to include three extensive variables in addition to T or P (e.g., $T, y_A, N_{B_0}, N_{C_0}, \underline{V}$).

Case (b): Since $n = 4, \pi = 1, r = 1$,

$$\mathscr{f} = n + 2 - \pi - r = 4 + 2 - 1 - 1 = 4$$

One set of intensive variables is T, P, y_A, y_B, y_C. Choosing the standard-state fugacities as unity, the equilibrium relationship becomes:

$$K_a = \frac{y_C y_D}{y_A y_B} \tag{10-59}$$

or, simply,

$$g(T, y_A, y_B, y_C) = 0 \tag{10-60}$$

Thus, any three variables from the set T, y_A, y_B, y_C are independent and three of these, in addition to P, can be used to specify the intensity of the system. Let us assume that we are given T, P, y_A, y_B. To define the extent of the system, we require $\pi + r = 2$ extensive variables. Let us choose these as N_{A_0} and N_{B_0}.

Given $T, P, y_A, y_B, N_{A_0}, N_{B_0}$, we could first calculate K_a from T and then solve for y_C using Eq. (10-59) in the form:

$$\frac{y_C(1 - y_A - y_B - y_C)}{y_A y_B} = K_a \tag{10-61}$$

Knowing y_C, we can solve next for y_D by difference. The total moles initially fed, N_0, and the extent of reaction at equilibrium, ξ, can be determined by solving simultaneously the two equations for A and B:

$$y_A = \frac{N_{A_0} - \xi}{N_0} \tag{10-62}$$

$$y_B = \frac{N_{B_0} - \xi}{N_0} \tag{10-63}$$

The initial moles of C and D can then be found from the following equations:

$$y_C = \frac{N_{C_0} + \xi}{N_0} \tag{10-64}$$

or

$$N_{C_0} = y_C N_0 - \xi \tag{10-65}$$

and

$$N_{D_0} = N_0 - N_{A_0} - N_{B_0} - N_{C_0} \tag{10-66}$$

For the special case in which A and B are fed in the stoichiometric ratio of 1:1, then $N_{A_0} = N_{B_0}$, and $N_{C_0} = N_{D_0} = 0$, and Eq. (10-61) reduces to:

$$\frac{(0.5 - y_A)^2}{y_A^2} = K_a \tag{10-67}$$

Thus, y_A and T cannot be considered as independent. We are free to choose y_A and P or T and P, but not y_A, T, and P. A set of $n + 2$ variables would then have to include four extensive variables in addition to y_A, P or T, P. Such a set could be N_{A_0}, N_{B_0}, N_{C_0}, N_{D_0}, y_A, P.

In Example 10.5, the special cases of stoichiometric feeds are sometimes cited as examples of cases that do not obey the phase rule, as stated in Eq. (10-52). If we recall that the phase rule states that there are *certain sets* of $\mathscr{f}$ intensive variables that describe the intensity of the system, then the special cases do not disobey the rule. For example, in case (a) of Example 10.5, when we specify stoichiometric feed conditions, then y_A is set and must be included as one of the two degrees of freedom allowed by the phase rule. Similarly, in case (b), the two variables y_B and y_C are set by specifying y_A, P, and the stoichiometric feed condition and, hence, y_B and y_C must be included as two of the four degrees of freedom allowed by the phase rule.

It should be noted that for a *single-phase* system, specifying the moles charged plus T and P is always sufficient to determine $\mathscr{f}$ intensive variables. In this case, we need only determine the r values of $\xi^{(m)}$ (one for each independent reaction) to calculate the $(n - 1)$ equilibrium compositions. Given T and P, there is one equilibrium relationship for each reaction, or r equations in the r unknown values of ξ.

A final word of caution is in order. When the phase rule is applied to a given system, it is assumed that we know the number of species and phases present and the number of reactions that occur. We can, however, always charge materials to a "black box" and control the temperature and pressure; nature will then decide for us which reactions occur and which species and phases are present at equilibrium. If we try to specify T and P and also n, π, and r, we run the risk of overspecifying the system. In such cases, one or more phases and species will disappear by one or more reactions going to completion.

The system discussed in Example 10.4 illustrates this point. If we charged $CaCO_3(s)$, $CaO(s)$, and $CO_2(g)$ to a vessel and set T and P, it is unlikely that the three phases will coexist in chemical equilibrium. In this case, $\mathscr{f} = n + 2 - \pi - r = 3 + 2 - 3 - 1 = 1$. Thus, we can control the temperature or the pressure, but not both if we are to have three phases present at equilibrium. If we set the temperature at 898°C, then the equilibrium pressure of CO_2 is 1 atm. If we set the pressure higher than 1 atm, all the CO_2 will disappear by reaction with CaO; if we set the pressure lower than 1 atm, all the $CaCO_3$ will decompose to CO_2 and CaO. In either case, $\mathscr{f} = 2 + 2 - 2 - 0 = 2$, which is in agreement with the fact that we have set T and P independently.[8]

[8] Note that while reactions in a heterogeneous system can go to completion, reactions in a homogeneous phase can never be complete. If ξ were equal to $-(N_{i_0}/\nu_i)$ where i is the limiting reactant or product [see Eq. (10-28)], K_a would be ∞ or 0, respectively, and ΔG^0 would be $+\infty$ or $-\infty$, respectively. Since ΔG^0 is finite, the reaction can never go to completion for a single-phase system.

Example 10.6

Barium sulfide is made by reducing barium sulfate-rich ore with coke in a rotary kiln. It has been proposed that the reduction be carried out by contacting the ore with CO in a fluidized bed. At the design temperature, the equilibrium constant for the reduction reaction,

$$BaSO_4(s) + 4\ CO(g) = BaS(s) + 4\ CO_2(g) \tag{1}$$

was determined experimentally by equilibrating a mixture of CO and CO_2 with a mixture of BaS and $BaSO_4$ solids. The applicability of the experimental results to the industrial process has been questioned for the following reason: the barium sulfate ore contains appreciable amounts of Fe_2O_3 and Fe_3O_4. Since the reaction

$$3\ Fe_2O_3(s) + CO(g) = 2\ Fe_3O_4(s) + CO_2(g) \tag{2}$$

is known to proceed at the temperature under consideration, it is suggested that the gas phase in equilibrium with $BaSO_4$—BaS will be different from that in equilibrium with the ore ($BaSO_4$—BaS—Fe_2O_3—Fe_3O_4).

(a) How many intensive variables can be set independently for an equilibrium system containing $BaSO_4(s)$, $BaS(s)$, $Fe_2O_3(s)$, $Fe_3O_4(s)$, $CO(g)$, $CO_2(g)$? Assume that no solid solutions are formed.

(b) If the following initial system was maintained at a fixed temperature and one-atmosphere pressure, what would be the final equilibrium composition?

1 mole ore (70-mol % $BaSO_4$, 5 mol % Fe_2O_3, 25 mol % Fe_3O_4)
0.5 mole CO (no CO_2 initially present)

Assume that at this temperature $K_{a_1} = 10^8$, $K_{a_2} = 20$, and the vapor phase is an ideal gas mixture.

Solution

(a) There are six species, five phases, and two independent reactions. Thus $\mathscr{f} = 1$. In this particular case, however, the two equilibria cannot be satisfied simultaneously because:

$$\frac{y_{CO_2}}{y_{CO}} = (K_{a_1})^{1/4} = K_{a_2}$$

Since $(K_{a_1})^{1/4}$ is 100 and K_{a_2} is 20, at equilibrium, one of the solids must disappear, leaving five species, four phases, and only one reaction. Thus, $\mathscr{f} = 2$.

(b) Since the temperature and pressure are set, we must determine which solid phase disappears. The initial (y_{CO_2}/y_{CO}) ratio is zero; as Fe_2O_3 and $BaSO_4$ are reduced, (y_{CO_2}/y_{CO}) increases. When this ratio reaches 20, reaction (2) is satisfied, but reaction (1) is not. Thus, the ratio will lie between 20 and 100 while Fe_3O_4 is oxidized to Fe_2O_3 and $BaSO_4$ is reduced to BaS. The cycle will continue until Fe_3O_4 or $BaSO_4$ is depleted. The first to vanish will be the limiting reactant for the overall reaction

$$BaSO_4(s) + 8\, Fe_3O_4(s) = BaS(s) + 12\, Fe_2O_3(s)$$

Since N_{i_0}/ν_i is 0.7 for $BaSO_4$ and 0.031 for Fe_3O_4, the limiting reactant is Fe_3O_4 and it will disappear. Thus, at equilibrium, we will have 0.05 + (.25) (3)/(2) = 0.425 mole of Fe_2O_3. From K_{a_1},

$$\frac{y_{CO_2}}{y_{CO}} = \frac{1 - y_{CO}}{y_{CO}} = 100$$

or $y_{CO} = 0.01$, $y_{CO_2} = 0.99$. Thus $N_{CO} = (.01)(.5) = .005$ and $N_{CO_2} = .495$. Originally, by an oxygen balance,

$$4\, N_{BaSO_4} + 3\, N_{Fe_2O_3} + 4\, N_{Fe_3O_4} + N_{CO}$$
$$= (4)(.7) + (3)(.05) + (4)(.25) + 0.5 = 4.45$$

At equilibrium,

$$4.45 = 4\, N_{BaSO_4} + 3\, N_{Fe_2O_3} + N_{CO} + 2\, N_{CO_2}$$
$$= 4\, N_{BaSO_4} + (3)(0.425) + 0.005 + (2)(0.495)$$

or,

$$N_{BaSO_4} = 0.545 \text{ mole and } N_{BaS} = 0.7 - 0.545 = 0.155 \text{ mole}$$

Since we have found that reactions (1) and (2) cannot be satisfied simultaneously, we might question if they are truly independent. If we apply the criteria of Section 10.2 to the system of $BaSO_4(s)$, $BaS(s)$, $Fe_2O_3(s)$, $Fe_3O_4(s)$, $CO(g)$, and $CO_2(g)$, we would indeed find that there are two independent reactions. Applying the phase rule to this system, we find that it is univariant. Thus, if we specify the pressure, we can determine the temperature for which $(K_{a_1})^{1/4} = K_{a_2}$. Alternatively, if we specify only the temperature, then we should be able to find a pressure and gas phase composition such that

$$K_{a_1} = \frac{a_{BaS}(a_{CO_2})^4}{a_{BaSO_4}(a_{CO})^4}$$

and

$$K_{a_2} = \frac{(a_{Fe_3O_4})^2 a_{CO_2}}{(a_{Fe_2O_3})^3 a_{CO}}$$

Since activities of the solids are weak functions of pressure, it may be necessary to go to extremely high or low pressures before these expressions are simultaneously satisfied.

10.5 Equilibria with Simultaneous Reactions

In many cases of practical interest, two or more chemical reactions may occur simultaneously. Consider the general case in which r reactions equilibrate simultaneously. At constant temperature and pressure, there are r equilibrium expressions of the form of Eq. (10-32) involving r unknowns (i.e., there is one ξ for each independent reaction). Since the equilibrium expressions are generally complex polynomials in ξ [see, e.g., Eq. (10-57)], simultaneous solution of the r equatioins is a tedious task. Two approaches to the solution of simultaneous equilibria have been used: the *series reactor method* and the *G-minimization method.* Each is amenable to machine computation; for three or more reactions, computer techniques are highly desirable. We shall describe each of these methods briefly, but we shall not pursue in any detail their computer applications.

The series reactor method[9] is conceptually easier to grasp. It is based on the fact that the Gibbs free energy of a system at constant T and P will always be reduced when any one reaction is allowed to proceed toward equilibrium. To illustrate, let us consider a quaternary system containing species A, B, C, D in which two reactions occur:

$$\nu_{\mathrm{A}}^{(1)}\mathrm{A} + \nu_{\mathrm{B}}^{(1)}\mathrm{B} = 0 \tag{10-68}$$

$$\nu_{\mathrm{B}}^{(2)}\mathrm{B} + \nu_{\mathrm{C}}^{(2)}\mathrm{C} + \nu_{\mathrm{D}}^{(2)}\mathrm{D} = 0 \tag{10-69}$$

The total Gibbs free energy of the system is

$$\underline{G} = \sum_{i=1}^{n} \mu_i N_i \tag{10-70}$$

If the first reaction is allowed to proceed to equilibrium at constant T and P, while C and D are treated as inert species, then during the approach to equilibrium,

$$d\underline{G} = \sum_{i=1}^{n} \mu_i \, dN_i + \sum_{i=1}^{n} N_i \, d\mu_i \tag{10-71}$$

$$= \mu_{\mathrm{A}} \, dN_{\mathrm{A}} + \mu_{\mathrm{B}} \, dN_{\mathrm{B}} + N_{\mathrm{A}} \, d\mu_{\mathrm{A}} + N_{\mathrm{B}} \, d\mu_{\mathrm{B}} + N_{\mathrm{C}} \, d\mu_{\mathrm{C}} + N_{\mathrm{D}} \, d\mu_{\mathrm{D}} \tag{10-72}$$

Although each μ_i varies during the process because composition is changing, the Gibbs-Duhem relation, Eq. (5-69), requires that:

$$\sum_{i=1}^{n} N_i \, d\mu_i = 0 \qquad (\text{constant } T, P) \tag{10-73}$$

and, thus, the last four terms in Eq. (10-72) vanish. Since

$$dN_{\mathrm{A}} = \nu_{\mathrm{A}}^{(1)} \, d\xi^{(1)} \tag{10-74}$$

and

$$dN_{\mathrm{B}} = \nu_{\mathrm{B}}^{(1)} \, d\xi^{(1)} \tag{10-75}$$

it follows that Eq. (10-72) reduces to:

$$d\underline{G} = (\nu_{\mathrm{A}}^{(1)} \mu_{\mathrm{A}} + \nu_{\mathrm{B}}^{(1)} \mu_{\mathrm{B}}) \, d\xi^{(1)} \tag{10-76}$$

or

$$d\underline{G} = \sum_{i=1}^{n} \nu_i^{(1)} \mu_i \, d\xi^{(1)} \tag{10-77}$$

[9] H. P. Meissner, C. L. Kusik, and W. H. Dalzell, "Equilibrium Composition with Multiple Reactions." *Ind. Eng. Chem. Fund.*, **8**, (1969), 659.

Integrating Eq (10-77) from the initial conditions to the final condition for which the first reaction is in equilibrium,

$$\underline{G} - \underline{G}_0 = \int_0^{\xi_e^{(1)}} \left(\sum_{i=1}^{n} \nu_i^{(1)} \mu_i \right) d\xi^{(1)} \tag{10-78}$$

Using Eqs. (10-32) and (10-33) in Eq. (10-29), we obtain:

$$\sum_{i=1}^{n} \nu_i \mu_i = \Delta G^0 + RT \ln \prod \left(\frac{\hat{f}_i}{f_i^0} \right)^{\nu_i} \tag{10-79}$$

$$= RT \ln \left(\frac{Q_a}{K_a} \right) \tag{10-80}$$

where:

$$Q_a \equiv \prod \left(\frac{\hat{f}_i}{f_i^0} \right)^{\nu_i} \tag{10-81}$$

is the "apparent equilibrium constant." That is, it is the value of the equilibrium expression using the nonequilibrium concentrations.

If Q_a is greater than K_a, or $Q_a/K_a > 1$, then reaction proceeds to the left and $\xi < 0$. If $Q_a/K_a < 1$, then reaction proceeds to the right and $\xi > 1$. Thus, the change in Gibbs free energy, $\underline{G} - \underline{G}_0$ is always negative and, hence, the total $\underline{G}$ decreases when any one reaction is allowed to proceed to equilibrium. It follows that, in an iterative process, if each reaction is allowed to proceed to equilibrium *sequentially*, $\underline{G}$ will always converge on a minimum value that corresponds to the equilibrium condition for the simultaneous reactions.

The series reactor method is an iterative procedure in which we envision r reactors in series, one for each independent reaction. In any given reactor, all species that do not take part in that reaction are treated as inert species. In reactor (1), the first reaction is allowed to equilibrate, in reactor (2), the second, and so forth until the last reaction is allowed to equilibrate in reactor (r). The procedure is repeated until the extent of reaction in every reactor is below some predetermined minimum value.

As pointed out by Meissner, *et al.*, the rate of convergence is highly dependent on the set of independent reactions chosen. By linear combinations of the independent reactions, it is usually possible to find a set that converges within from 3 to 10 iterations.

In choosing an independent set of reactions, convergence will be retarded if any one reaction is reagent-starved. That is, if A $\longrightarrow$ B with $K_y < 1$ and B $\longrightarrow$ C with $K_y > 1$, then the second reaction will be starved of B by the first reaction and many iterations will be required to build up the concentration of C. Thus, a better procedure would be to form C directly from A and then form B from C (i.e., A $\longrightarrow$ C followed by C $\longrightarrow$ B).

If the dominant products are known, they should always be formed first and then the secondary products should be formed in subsequent reactions from the dominant products and/or any of the feed components present in excess. As illustrated in Example 10.7, secondary products should be formed one at a time.

If the dominant products are not known, the following procedure, adapted from Meissner, for choosing the sequence of reactions is helpful.

1. To the extent possible, form species by sequential reactions with K_y no smaller than 0.1 and preferably greater than unity. (Sequential reactions are defined as a sequence in which the products of a reaction are consumed in the following reaction.)
2. When one of the species in the feed is in stoichiometric excess with reference to the first reaction, one of the later reactions should be of the mixed type in which this excess reactant combines with a product of an earlier reaction. This earlier reaction will desirably have a K_y value greater than unity.
3. If it is not possible to form a species by using a sequential reaction with K_y greater than unity, these species may be formed by parallel reactions having K_y less than unity. (Parallel reactions are reactions starting with the same reactants.) Insert these reactions in the series at those points where the reactants for the parallel reactions are not yet reduced to starvation levels.

The series reactor technique is illustrated in the following example.

Example 10.7

Most processes currently under study for gasifying coal to methane involve three steps: partial oxidation with O_2 and steam to form a CO- and H_2-rich stream, followed by the water-gas shift reaction to increase the H_2/CO ratio, and then the methanation reaction to form CH_4 from CO and H_2. The three-step process is inefficient because the endothermic first step is carried out at high temperature, and the exothermic third step is conducted at a much lower temperature.

Our Process Development Department has suggested an alternate scheme. By a combination of catalysts, they think that they can react coal with steam to form methane in a single reactor. They propose the following kinetic scheme, where the coal "molecule" is idealized as anthracene:

$$C_{14}H_{10} + 8H_2 = C_6H_6 + 2C_4H_{10} \tag{a}$$

$$C_6H_6 + 6H_2O = 9H_2 + 6CO \tag{b}$$

$$C_4H_{10} + 4H_2O = 9H_2 + 4CO \tag{c}$$

$$CO + 3H_2 = CH_4 + H_2O \tag{d}$$

$$CO + H_2O = CO_2 + H_2 \tag{e}$$

In a preliminary analysis determine the equilibrium yield of methane. Assume that the properties of coal can be approximated by those of anthracene. The temperature and pressure of kinetic interest are 1000°K and 50 atm. Assume that a 20% excess of steam is fed with anthracene, the excess based on the overall reaction,

$$4C_{14}H_{10} + 46H_2O = 23CO_2 + 33CH_4 \qquad \text{(f)}$$

For the purpose of this example assume that all species are present as ideal gases. The Gibbs free energies of formation based on pure vapors at one atm are as follows.

	$\Delta G^0_{f\,1000}$
$C_{14}H_{10}$	65.64
C_6H_6	57.39
C_4H_{10}	57.30
H_2O	−47.19
H_2	0.00
CO	−47.78
CO_2	−94.76
CH_4	+1.40

Solution

The five reactions listed above are clearly independent because each one contains a chemical species not present in another reaction. That is, comparing (d) to (e), the former contains CH_4 while the latter does not and, hence, these two are independent. Similarly, (c) contains C_4H_{10}, (b) C_6H_6, and (a) $C_{14}H_{10}$. Thus, the five are independent, although they may not be a convenient set for use in the series reactor technique. On a basis of 1 g-mol $C_{14}H_{10}$ fed, with a 20% excess of H_2O, using reaction (f), the g-mol H_2O are (46)(1.2)/4 = 13.8 g. Since these are the only two species fed, the five reactions given in the problem statement cannot be used with the series reactor method because there is no path by which these two species can react. This situation can be remedied by replacing (a) by a combination of (a), (b), and (c).

$$C_{14}H_{10} + 14H_2O = 14CO + 19H_2 \qquad \text{(g)}$$

If (g) is used as the first reaction, then the inverse of (b) and (c) can be used as a parallel set to form C_6H_6 and C_4H_{10}, respectively, from the CO and H_2 products of the first reaction. Similarly, (d) would be used to form CH_4 by a parallel reaction and CO_2 would be formed in the mixed reaction (e).

Combining Eqs. (10-32) and (10-36) with $K_{f/P} = 1$,

$$K_y = K_a P^{-\nu} = \exp\left\{\frac{-\Delta G^0}{RT}\right\} P^{-\nu} \qquad (10\text{-}82)$$

Values for K_y for the five reactions (g), (−b), (−c), (d), (e) at 1000°K and 50 atm are given in Table 10.3, Set A. Note that reaction (g) is slightly unfavorable; based on one mole of H_2O, which is the limiting reactant for (g), log K_y is −14.49/14 or K_y is slightly less than 0.1. Thus, we might antici-

TABLE 10.3
K_y AT 1000°K, 50 ATM

	$\log_{10} K_y$
Set A	
(g) $C_{14}H_{10} + 14H_2O = 14CO + 19H_2$	−14.490
(−b) $6CO + 9H_2 = C_6H_6 + 6H_2O$	0.322
(−c) $4CO + 9H_2 = C_4H_{10} + 4H_2O$	0.586
(d) $CO + 3H_2 = CH_4 + H_2O$	2.970
(e) $CO + H_2O = CO_2 + H_2$	−0.042
Set B	
(f) $4C_{14}H_{10} + 46H_2O = 23CO_2 + 33CH_4$	39.058
(b) $3CO_2 + 13CH_4 = 4C_4H_{10} + 6H_2O$	−36.138
(i) $9CO_2 + 15CH_4 = 4C_6H_6 + 18H_2O$	−42.872
(j) $3CO_2 + CH_4 = 4CO + 2H_2O$	−2.836
(k) $CH_4 + 2H_2O = CO_2 + 4H_2$	−3.011
Set C	
(a) $C_{14}H_{10} + 8H_2 = C_6H_6 + 2C_4H_{10}$	−12.994
(b) $C_6H_6 + 6H_2O = 6CO + 9H_2$	−0.322
(c) $C_4H_{10} + 4H_2O = 4CO + 9H_2$	−0.586
(d) $CO + 3H_2 = CH_4 + H_2O$	2.970
(e) $CO + H_2O = CO_2 + H_2$	−0.042

pate that one of the subsequent parallel reactions, (−b), (−c), or (d) may be somewhat starved, resulting in slow convergence.

Since we have some indication that CO_2 and CH_4 may be the dominant products (see the problem statement above), we might consider reaction (f) first and then form C_4H_{10}, C_6H_6, CO, and H_2, respectively, in four parallel reactions using CO_2 and CH_4 as reactants. These reactions are shown in Table 10.3 as Set B. Note that this sequence should converge rapidly because the first reaction has a large K_y based on one mole of $C_{14}H_{10}$, which is the limiting reactant, and all subsequent reactions have small values of K_y.

Using Set B, we see that the series reactor method gave the results shown in Table 10.4. Note that on the second iteration through the five reactors, the extent of conversion in each of the five reactors is low, the maximum percentage mole change of the major products being around 10% (for CO). The values obtained after two iterations are within 1% of the equilibrium values.

The rapid conversion of the series reactor technique, as obtained for Set B, is highly dependent on the choice of reactions. To illustrate, the method was used for Sets A, B, and C of Table 10.3. For Set C, the feed conditions were modified by converting half of the initial $C_{14}H_{10}$ to CO and H_2 using reaction (g). Thus, the feed for this case was 0.5 mole $C_{14}H_{10}$, 6.8 mole H_2O, 9.5 mole H_2 and 7.0 mole CO.

TABLE 10.4

RESULTS OF THE SERIES REACTOR TECHNIQUE FOR SET B OF EXAMPLE 10.7

		First iteration					Second iteration					Final values (seventh iteration)
Species	Moles Fed	R1	R2	R3	R4	R5	R1	R2	R3	R4	R5	
$C_{14}H_{10}$	1.0	8.2E-5	8.2E-5	8.2E-5	8.2E-5	8.2E-5	2.1E-6	2.1E-6	2.1E-6	2.1E-6	2.1E-6	2.1E-6
C_6H_6	0.0	0.0	0.0	1.5E-8	1.5E-8	1.5E-8	1.5E-8	1.5E-8	6.5E-9	6.5E-9	6.5E-9	6.5E-9
C_4H_{10}	0.0	0.0	1.7E-8	1.7E-8	1.7E-8	1.7E-8	1.7E-8	2.7E-9	2.7E-9	2.7E-9	2.7E-9	2.7E-9
H_2O	13.8	2.307	2.307	2.307	3.379	2.683	2.683	2.683	2.683	2.817	2.793	2.798
H_2	0.0	0.0	0.0	0.0	0.0	1.391	1.391	1.391	1.391	1.391	1.439	1.441
CO	0.0	0.0	0.0	0.0	2.144	2.144	2.144	2.144	2.144	2.412	2.412	2.438
CO_2	0.0	5.747	5.747	5.747	4.139	4.487	4.487	4.487	4.487	4.286	4.298	4.281
CH_4	0.0	8.246	8.246	8.246	7.710	7.362	7.362	7.362	7.362	7.295	7.283	7.279
N_{total}	14.8	16.300	16.300	16.300	17.372	18.068	18.068	18.068	18.068	18.202	18.226	18.240
ξ	—	0.250	4.3E-9	3.8E-9	0.536	0.348	2.0E-5	−3.6E-9	−2.1E-9	0.067	0.012	0.000

Note: R1 = reactor 1, R2 = reactor 2, etc.

The rate of convergence for each of these three cases can be seen by following the rate of decrease of $\underline{G}$, as shown in Figure 10.4. For Set A, the rate of convergence is slow, as expected, because the first step [reaction (g)] results in very little conversion of $C_{14}H_{10}$ per pass. Seven iterations are required to reduce the concentration of $C_{14}H_{10}$ to 10% of the initial value. For Set C, the rate of convergence is even slower than that for Set A because the first step [reaction (a)] is again unfavorable. In fact, the rate of convergence is so slow that, after the second iteration, we might have been misled into thinking we had reached equilibrium if we had not first seen the results of Sets A or B. To determine if the asymptote in Set C is an equilibrium condition, we could try to approach this asymptote from the product side. That is, we could convert all of the $C_{14}H_{10}$ to CO_2 and CH_4 using reaction (f) such that the feed would be (23/4) moles CO_2, (33/4) moles CH_4, and (13.8-46/4) moles H_2O, and then rerun the Set C calculations. When this case is rerun, it converges to the true equilibrium condition as rapidly as Set B.

The second method for solving simultaneous chemical equilibria, the $\underline{G}$-minimization technique, does not require knowledge of an independent set of chemical reactions. The procedure described in Section 10.1 for evaluating the independent reactions from elemental balances is incorporated implicitly in the $\underline{G}$-minimization method.

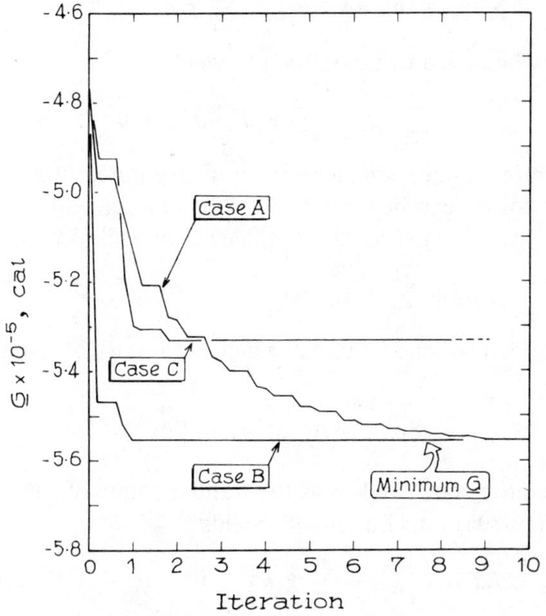

Figure 10.4 Change in $\underline{G}$ for various cases.

For a system at constant temperature and pressure, the criterion of equilibrium is

$$\delta \underline{G} = \sum_{i=1}^{n} \mu_i \, \delta N_i = 0 \tag{10-83}$$

where the variations in N_i are not independent; they must conform to the element balances, which are treated as restraining equations. Let A_k be the *total* g-atom of the *kth* element present in the system, as determined by the initial composition of the system. Let a_{ik} be the number of atoms of the *kth* element present in a molecule of chemical species *i*. The element balance can then be written as

$$\sum_{i=1}^{n} N_i a_{ik} - A_k = 0 \qquad (k = 1, \ldots, m) \tag{10-84}$$

where the constituents are made up of *m* elements.

The restraining equations for the variations in N_i can be solved simultaneously with Eq. (10-83) using the Lagrangian method of undetermined multipliers. Differentiating Eq. (10-84), we obtain:

$$\sum_{i=1}^{n} a_{ik} \, \delta N_i = 0 \qquad (k = 1, \ldots, m) \tag{10-85}$$

Multiplying Eq. (10-85) by χ_k, the Lagrangian multiplier, and summing over all elements,

$$\sum_{k=1}^{m} \chi_k \sum_{i=1}^{n} a_{ik} \, \delta N_i = \sum_{i=1}^{n} \left(\sum_{k=1}^{m} \chi_k a_{ik} \right) \delta N_i = 0 \tag{10-86}$$

which can now be added to Eq. (10-83) to yield:

$$\sum_{i=1}^{n} \left(\mu_i + \sum_{k=1}^{m} \chi_k a_{ik} \right) \delta N_i = 0 \tag{10-87}$$

Since the restraining equations have been incorporated in Eq. (10-87), each of the δN_i variations can be considered to be independently variable. Thus, for Eq. (10-87) to be satisfied, the coefficients of each δN_i must vanish, or

$$\mu_i + \sum_{k=1}^{m} \chi_k a_{ik} = 0 \qquad (i = 1, \ldots, n) \tag{10-88}$$

The chemical potential can be transformed to fugacity by using Eqs. (10-30) and (10-31):

$$\mu_i = \Delta G^0_{f_i} + RT \ln \left(\frac{\hat{f}_i}{f^0_i} \right) \tag{10-89a}$$

where the reference state is taken as the standard state of component *i*. Substituting Eq. (10-89a) into Eq. (10-88) yields:

$$\Delta G^0_{f_i} + RT \ln \left(\frac{\hat{f}_i}{f^0_i} \right) + \sum_{k=1}^{m} \chi_k a_{ik} = 0 \qquad (i = 1, \ldots, n) \tag{10-89b}$$

which can then be transformed to a composition-dependent equation by expanding $\hat{f}_i$. For example, if species i were a vapor, then $\hat{f}_i = y_i\phi_i P$. Taking $f_i^0 = 1$, Eq. (10-89b) becomes:

$$\Delta G^0_{f_i} + RT \ln (y_i\phi_i P) + \sum_{k=1}^{m} \chi_k a_{ik} = 0 \qquad (i = 1, \ldots, n) \qquad (10\text{-}90)$$

There are n equations of the form of Eq. (10-90); the set contains $n + m - 1$ unknowns [i.e., $(n - 1)$ values of y_i and m values χ_k]. In addition to Eq. (10-90), we have m equations of the form of Eq. (10-84) or, in terms of mole fractions,

$$\sum_{i=1}^{m} y_i a_{ik} = \frac{A_k}{N} \qquad (k = 1, \ldots, m) \qquad (10\text{-}91)$$

These equations contain an additional unknown, N, making a total of $n + m$ unknowns and $n + m$ equations. These equations are solved simultaneously for χ_k and y_i, yielding the equilibrium composition. The procedure is tedious for systems containing more than three species; iterative computer methods have been used for more complex systems.[10]

10.6 Le Chatelier's Principle in Chemical Equilibria

If a system in thermal, mechanical, and chemical equilibrium were perturbed from the equilibrium state, changes would usually occur within the system to reestablish a new equilibrium state. For example, if the system temperature were suddenly increased at constant pressure and extent of reaction, the system would no longer be in chemical equilibrium. If the system were then isolated from the environment, reaction would occur in such a way as to decrease the temperature (i.e., the system would adjust itself to reduce the effect of the initial perturbation). This simple example can easily be proved from Eq. (10-48), irrespective of whether the reaction is exothermic or endothermic. Almost all equilibrium systems behave in this manner for all kinds of perturbations and the general rule that defines such behavior is called the Le Chatelier principle. One of the simplest statements of this principle is given by Maass and Steacie[11]: "If an attempt is made to change the pressure, temperature, or concentration of a system in equilibrium, then the equilibrium will shift, if possible, in such a manner as to diminish the magnitude of the alteration in the factor that was varied."

A system that follows Le Chatelier's principle is said to *moderate* with respect to the perturbation. It is interesting to examine the behavior of a chemically reacting system in the light of this principle. The simple method

[10] See e.g., F. J. Zeleznik and Sanford Gorden, "Calculation of Complex Chemical Equilibria," *Ind. Eng. Chem.*, **60** (6), (1968), 27 or Y. H. Ma and C. W. Shipman, "On the Computation of Complex Equilibria," *AIChE J.*, **18**, (1972), 299.

[11] Otto Maass and E. W. R. Steacie, *Introduction to the Principles of Physical Chemistry* (New York: John Wiley & Sons, Inc., 1939).

proposed by Katz[12] will first be used. For those interested, other treatments are available in Prigogine and Defay and Callen.[13]

To develop the Katz technique, the mathematics are considerably simplified if only ideal gas mixtures are considered and standard states are limited to those corresponding to a pure ideal gas at one atmosphere. A brief discussion is presented later to indicate the extension to other cases.

Consider the general reaction:

$$\text{aA}(g) + \text{bB}(g) = \text{cC}(g) + \text{dD}(g)$$

Define the following ratios:

$$Q_p \equiv \frac{p_{\text{C}}^{\text{c}} p_{\text{D}}^{\text{d}}}{p_{\text{A}}^{\text{a}} p_{\text{B}}^{\text{b}}} \tag{10-92}$$

$$Q_N \equiv \frac{N_{\text{C}}^{\text{c}} N_{\text{D}}^{\text{d}}}{N_{\text{A}}^{\text{a}} N_{\text{B}}^{\text{b}}} = \frac{Q_p N^{\nu}}{P^{\nu}} \tag{10-93}$$

$$Q_y \equiv \frac{y_{\text{C}}^{\text{c}} y_{\text{D}}^{\text{d}}}{y_{\text{A}}^{\text{a}} y_{\text{B}}^{\text{b}}} = \frac{Q_N}{N^{\nu}} = \frac{Q_p}{P^{\nu}} \tag{10-94}$$

where $\nu = \text{c} + \text{d} - \text{a} - \text{b}$. At equilibrium, $K_p = Q_p$, $K_y = Q_y$, etc., but the Q definitions are applicable whether or not equilibrium has been established.

If the reaction proceeds to the right, Q_N increases. We can easily prove this by taking the logarithm of Eq. (10-93) and differentiating with respect to the extent of reaction, ξ. The result is:

$$\frac{dQ_N}{d\xi} = Q_N\left(\frac{\text{c}^2}{N_{\text{C}}} + \frac{\text{d}^2}{N_{\text{D}}} + \frac{\text{a}^2}{N_{\text{A}}} + \frac{\text{b}^2}{N_{\text{B}}}\right) \tag{10-95}$$

Since $dQ_N/d\xi$ is clearly positive, then Q_N increases as the reaction proceeds to form more C and D. One may also show that Q_y increases with ξ by a similar technique.

The key to the entire treatment is to consider the system in chemical equilibrium; if a perturbation should occur, the equilibrium constants are unaffected (except for temperature changes) but the values of Q may change. The subsequent increase or decrease in Q_N or Q_y indicates the direction of the shift in the reaction. Several examples are treated below. All cases are isothermal.

Addition of an inert gas at constant pressure

$$K_p = \text{constant} = \frac{Q_N P^{\nu}}{N^{\nu}}$$

As N increases, P is a constant, and Q_N will then increase if $\nu > 0$ and decrease

[12] Lewis Katz, "A Systematic Way to Avoid Le Chatelier's Principle in Chemical Reactions," *J. Chem. Ed.*, **38**, 1961, 375.

[13] I. Prigogine and R. Defay, *Chemical Thermodynamics* trans., D. H. Everett (New York: Longmans, Green & Co., Inc., 1954), Chap. 17. H. B. Callen, *Thermodynamics* (New York: John Wiley & Sons, Inc., 1960), Chap. 8.

if $\nu < 0$. Thus, reaction occurs to form more C and D if $\nu > 0$ and less C and D if $\nu < 0$.

Addition of an inert gas at constant volume

$$K_p = \text{constant} = \frac{Q_N P^\nu}{N^\nu} = Q_N\left(\frac{RT}{\underline{V}}\right)^\nu$$

Since T and $\underline{V}$ are constant, Q_N does not change and there is no change in the moles of reactants and products present at equilibrium.

Variation of system pressure

$$K_p = \text{constant} = Q_y P^\nu$$

For an increase in P, the reaction shifts to the right if $\nu < 0$ and to the left if $\nu > 0$. Similar, but opposite, conclusions hold if P is decreased.

Addition of a reactant or product at constant volume

$$K_p = \text{constant} = Q_N\left(\frac{RT}{\underline{V}}\right)^\nu$$

If C or D were added, Q_N increases initially, but since all other terms are constant, reaction must take place to the left to decrease Q_N to its original value. Likewise, reaction occurs to the right if A or B were to be added.

Addition of a reactant or product at constant pressure

This case requires a little more consideration. Since $Q_p = Q_y P^\nu$, and P is constant, a change in Q_p indicates a corresponding change in Q_y.

If the addition of one of the components leads to an increase in the ratio denoted by Q_p, then reaction must occur in such a way to lower Q_p to its original value since, at equilibrium, $K_p = Q_p$. Now

$$Q_p = \frac{P^\nu Q_N}{N^\nu}$$

If component j is added, with all the moles of other components constant:

$$\left(\frac{\partial Q_p}{\partial N_j}\right)_{T,P,N_i[j]} = \frac{P^\nu}{N^{2\nu}}\left[N^\nu\left(\frac{\partial Q_N}{\partial N_j}\right)_{T,P,N_i[j]} - (Q_N)\left(\frac{\partial N^\nu}{\partial N_j}\right)_{T,P,N_i[j]}\right]$$

But

$$Q_N = \prod_{j=1}^{k} N_j^{\nu_j}; \quad \left(\frac{\partial Q_N}{\partial N_j}\right)_{T,P,N_i[j]} = \frac{\nu_j Q_N}{N_j}; \quad \text{and} \quad \left(\frac{\partial N^\nu}{\partial N_j}\right)_{T,P,N_i[j]} = \nu N^{(\nu-1)}$$

Substituting,

$$\left(\frac{\partial \ln Q_p}{\partial N_j}\right)_{T,P,N_i[j]} = \left(\frac{1}{N_j}\right)(\nu_j - y_j\nu) \tag{10-96}$$

From Eq. (10-94), Q_p increases with the addition of N_j providing $(\nu_j - y_j\nu) > 0$. In this case, *reaction* then occurs to decrease Q_p to its original value. For example, consider the reaction,

$$2NH_3 = N_2 + 3H_2 \qquad \nu = 2$$

For ammonia, the term, $[-2 - y_{NH_3}(2)]$, is certainly never greater than zero and Q_p decreases with addition of NH_3. Reaction then occurs to increase Q_p (i.e., the amounts of N_2 and H_2 will increase because of ammonia decomposition). For hydrogen addition, it is easily shown that reaction will occur to form more ammonia.

For nitrogen, $\nu_{N_2} - y_{N_2}\nu = 1 - 2y_{N_2}$. This term is greater than zero if $y_{N_2} < 0.5$, but less than zero if $y_{N_2} > 0.5$. For the case, $y_{N_2} < 0.5$, Q_p increases with the addition of nitrogen, so reaction will occur to reduce Q_p (i.e., N_2 and H_2 react to form more NH_3). If, however, $y_{N_2} > 0.5$, Q_p decreases with nitrogen addition and additional NH_3 must decompose to reestablish Q_p at its original value. Such behavior as noted in the last case is often referred to as a case in which the system does not "moderate" with respect to the variation. It is interesting to note that, although the ammonia system may not moderate with respect to the addition of nitrogen (i.e., react in such a way that the effect of the nitrogen addition is decreased), the final concentration of nitrogen, expressed, say, as mole/liter, is still less than the original nitrogen concentration before addition. Thus, to this extent, the system moderates even with respect to nitrogen.[14]

Another example involves the methanol synthesis reaction:

$$CO(g) + 2H_2(g) = CH_3OH(g)$$

We can define the following mole numbers:

	Initial moles	Moles in an equilibrium state
CO	A	$A - x$
H_2	B	$B - 2x$
CH_3OH	0	x
		$A + B - 2x$

$$K_p = \frac{p_{CH_3OH}}{p_{CO}p_{H_2}^2} = \frac{y_{CH_3OH}}{P^2 y_{CO} y_{H_2}^2}$$
$$= \frac{(x)(A + B - 2x)^2}{(A - x)(B - 2x)^2 P^2}$$

At about 500°C, $K_p \sim 10^{-6}$; unless the pressure is high, the amount of methanol formed is small. To illustrate the effect of adding H_2 or CO, assume

[14] This interesting observation was pointed out by Professor Norman Smith of Fordham University.

a low pressure of 1 atm. The equilibrium constant expression may then be approximated as:

$$\frac{x}{K_p} \sim \frac{AB^2}{(A + B)^2}$$

Consider two cases; in the first, start with one mole of CO and allow chemical equilibrium to be attained with various amounts of hydrogen. As shown in Figure 10.5, the amount of CH_3OH present at equilibrium continually increases with an increase in hydrogen. This is what is predicted since:

$$\nu_{H_2} - y_{H_2}\nu = -2 - (-2)y_{H_2} < 0$$

for all values of y_{H_2} up to unity. From Eq. (10-96), Q_p decreases and reaction occurs to raise Q_p to its original value. For the second case, start with one mole of H_2 and add different quantities of CO. The value of x/K_p increases as CO is first added, but then decreases if more than one mole is added. Again we note that

$$\nu_{CO} - y_{CO}\nu = -1 - (-2)y_{CO} = 2y_{CO} - 1$$

If $y_{CO} < 0.5$, from Eq. (10-96) Q_p decreases with CO addition and subsequent reaction occurs to form more CH_3OH and bring Q_p to its original level.

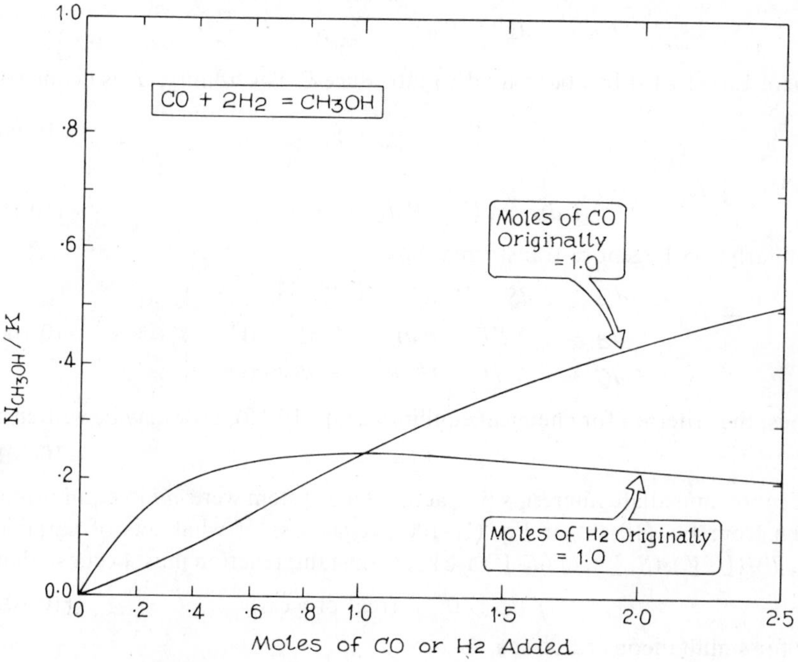

Figure 10.5 Variation of methanol produced with the addition of CO or H_2.

If $y_{CO} > 0.5$, the opposite effect occurs, and the addition of CO leads to a decrease in the yield of CH_3OH. The maximum yield of CH_3OH occurs when one mole of CO has been added to one mole of H_2.

In the more general sense, moderation may be examined from the point of view of starting with a system in equilibrium. The system is perturbed by changing one of the variables. We are interested in following the behavior of the system back to an equilibrium state (which may be different from the original state). If, in the change from the nonequilibrium state to the equilibrium state, there is a corresponding change in the perturbing variable opposite to that which originally caused the upset, then we speak of the system as moderating with respect to that variable. We have already seen, for example, in the ammonia-hydrogen-nitrogen equilibria, if the system were perturbed at constant temperature and pressure by adding hydrogen, the system will react such that more ammonia is formed. This change reduces the amount of hydrogen and the system moderates. Many other examples could be cited. Instead of considering many in detail, however, it is interesting to develop the moderation criteria, Eq. (10-96), in another way and, in so doing, introduce the concept of the *affinity*, Y.

From Section 7.3 for a closed, single-phase, single-reaction system, equilibrium is defined as:

$$T\,d\underline{S} = d\underline{U} + P\,d\underline{V} - \sum_{j=1}^{n} \mu_j \nu_j\, d\xi \leq 0 \tag{10-97}$$

where Eq. (10-18) has been used to introduce ξ. The affinity, Y, is defined as

$$Y \equiv -\sum_{j=1}^{n} \mu_j \nu_j \tag{10-98}$$

Thus,

$$T\,d\underline{S} = d\underline{U} + P\,d\underline{V} + Y\,d\xi \leq 0 \tag{10-99}$$

Similarly, by Legendre transformations,

$$\left.\begin{aligned} d\underline{H} &= T\,d\underline{S} - P\,d\underline{V} - Y\,d\xi \geq 0 \\ d\underline{A} &= -\underline{S}\,dT - P\,d\underline{V} - Y\,d\xi \geq 0 \\ d\underline{G} &= -\underline{S}\,dT + \underline{V}\,dP - Y\,d\xi \geq 0 \end{aligned}\right\} \tag{10-100}$$

Thus, the criterion for chemical equilibria, Eq. (10-15), now may be written as

$$Y = 0 \tag{10-101}$$

Of more immediate interest is the fact that if a system were *not* in equilibrium, then from Eq. (10-48) or Eq. (10-100), regardless of what sets of variables $(P, T), (\underline{U}, \underline{V}), (\underline{S}, \underline{V})$, or $(T, \underline{V})$ are kept constant, reaction must occur so that:

$$Y\,d\xi > 0 \qquad \text{(real process)} \tag{10-102}$$

or for simultaneous reactions,

$$\sum_{m=1}^{r} Y^{(m)}\, d\xi^{(m)} > 0 \tag{10-103}$$

Presumably, some reaction could take place so that viewed singly, Eq. (10-102) is violated, but for the entire system Eq. (10-103) still holds. Considering only a single reaction a generalized reaction rate R may be defined as:

$$R = \frac{d\xi}{dt} \tag{10-104}$$

so that for any actual chemical reaction,

$$YR > 0 \tag{10-105}$$

Returning to the original problem, we start with a system in chemical equilibria and perturb it by addition of a small amount of one of the reactive components. We are interested in the behavior of the perturbed system as it approaches the new equilibrium state. We hold the pressure and temperature constant. Let i be the added component. The change in affinity is:

$$Y \sim Y_0 + \left(\frac{\partial Y}{\partial N_i}\right)_{T,P,N_j[i]} \delta N_i \tag{10-106}$$

for small values of δN_i. Eq. (10-101) requires that $Y_0 = 0$ because the derivative is evaluated at the original equilibrium state. Then from Eqs. (10-105) and (10-106):

$$\left[\left(\frac{\partial Y}{\partial N_i}\right)_{T,P,N_j[i]} \delta N_i\right]\left(\frac{1}{\nu_i}\frac{dN_i}{dt}\right) > 0 \tag{10-107}$$

where δN_i is the amount of i added (positive in this case) and (dN_i/dt) is the response of the system to the addition. If $(dN_i/dt) < 0$, then the system moderates with respect to an addition of N_i. For (dN_i/dt) to be negative, however, and still satisfy the inequality (10-107),

$$\frac{1}{\nu_i}\left(\frac{\partial Y}{\partial N_i}\right)_{T,P,N_j[i]} < 0 \qquad \text{(for moderation)} \tag{10-108}$$

To explore the consequences of Eq. (10-108), from Eq. (10-98):

$$\begin{aligned}\frac{1}{\nu_i}\left(\frac{\partial Y}{\partial N_i}\right)_{T,P,N_j[i]} &= -\frac{1}{\nu_i}\sum_{j=1}^{n}\nu_j\left(\frac{\partial \mu_j}{\partial N_i}\right)_{T,P,N_j[i]} \\ &= -\frac{1}{\nu_i}\sum_{j\neq i}\nu_j\left(\frac{\partial \mu_j}{\partial N_i}\right)_{T,P,N_j[i]} - \left(\frac{\partial \mu_i}{\partial N_i}\right)_{T,P,N_j[i]}\end{aligned}$$

But

$$\left(\frac{\partial \mu_i}{\partial N_i}\right)_{T,P,N_j[i]} = -\frac{1}{N_i}\sum_{j\neq i} N_j\left(\frac{\partial \mu_j}{\partial N_i}\right)_{T,P,N_j[i]}$$

so

$$\frac{1}{\nu_i}\left(\frac{\partial Y}{\partial N_i}\right)_{T,P,N_j[i]} = -\frac{1}{\nu_i}\sum_{j=1}^{n} N_j\left(\frac{\nu_j}{N_j} - \frac{\nu_i}{N_i}\right)\left(\frac{\partial \mu_j}{\partial N_i}\right)_{T,P,N_j[i]} < 0$$

$$\text{(for moderation)} \tag{10-109}$$

The right-hand side of Eq. (10-109) must be < 0 for the system to moderate with additional N_i. Note that this expression may be greatly simplified for the case of ideal gases. Here,

$$\left(\frac{\partial \mu_j}{\partial N_i}\right)_{T,P,N_j[i]} = RT\left(\frac{\partial \ln y_j}{\partial N_i}\right)_{T,P,N_j[i]} \tag{10-110}$$

Eq. (10-109) then simplifies to:

$$\frac{1}{\nu_i}\left(\frac{\partial Y}{\partial N_i}\right)_{T,P,N_j[i]} = \frac{RT}{N_i}\left[\left(\frac{y_i \nu}{\nu_i}\right) - 1\right] \tag{10-111}$$

Thus, for moderation:

$$\frac{y_i \nu}{\nu_i} - 1 < 0$$

or

$$\frac{y_i \nu - \nu_i}{\nu_i} < 0 \qquad \text{(to moderate)} \tag{10-112}$$

Eq. (10-112) is a more compact form than Eq. (10-96) which requires subsequent steps to determine changes in Q_p.

This technique, in which the affinity concept is used to study the effects of system upsets is very powerful and not limited to ideal gases, although the final results may be of such a form as to hinder immediate physical visualization of the consequences [viz. Eq. (10-109)].

Some of the concepts discussed in this chapter are illustrated by the examples shown below.

Example 10.8

Apply Eq. (10-112) to the decomposition reaction:

$$NH_2COONH_4(s) = 2NH_3(g) + CO_2(g)$$

Solution

The amount of solid present is unimportant. Thus

$$\nu_{CO_2} = 1, \quad \nu_{NH_3} = 2, \qquad \nu = 3$$

For CO_2 addition, to have the reaction proceed in order to form solid,

$$\frac{(3)(y_{CO_2}) - 1}{1} < 0; \qquad y_{CO_2} < \frac{2}{3}$$

Similarly, to form more solid by ammonia addition,

$$\frac{(3)(y_{NH_3}) - 2}{2} < 0; \qquad y_{NH_3} < \frac{2}{3}$$

Example 10.9

A gas mixture containing 1 g-mol HI, 0.1 g-mol I_2, and 0.01 g-mol H_2 is pumped into a reactor that is equipped with a movable piston to adjust the volume. The temperature of the reactor and contents is rapidly increased to 127°C and maintained at this level by means of a constant temperature bath. A suitable catalyst is present to insure that chemical equilibrium is maintained.

$$2\,\text{HI} = \text{H}_2 + \text{I}_2$$

The gases form an ideal gas mixture. At 400°K,

$$\Delta H^0 = 2700 \text{ cal/g-mol}$$

and the absolute entropies of the three components in the standard state (unit fugacity) are shown below:

	H_2	I_2	HI
S^0_{400} cal/g-mol °K	33.23	64.89	51.48

These three components are also reported to have heat capacities in the standard state as shown:

$$\text{H}_2\text{: } C_p^0 = 6.52 + 0.78 \times 10^{-3}T + 0.12 \times 10^5 T^{-2}$$

$$\text{I}_2\text{: } C_p^0 = 8.94 + 0.14 \times 10^{-3}T - 0.17 \times 10^{-5} T^{-2}$$

$$\text{HI: } C_p^0 = 6.70 + 0.46 \times 10^{-3}T + 1.04 \times 10^6 T^2$$

where T is in °K. The vapor pressure of liquid I_2 at 127°C is 144 mm Hg.

(a) As the reactor and contents are heated to 127°C, the pressure is maintained constant at one atmosphere. Will the HI that is present decompose to form more H_2 and I_2?

(b) After chemical equilibrium has been established at 127°C, the mixture is compressed isothermally to 3 atm. Will any reaction occur?

(c) After equilibrium has been established at 3 atm, 127°C, the piston is locked in place and helium gas is introduced to increase the total pressure to 4 atm. Will any further reaction occur? How many moles of helium were introduced?

(d) Evaluate the equilibrium constant at 327°C.

Solution

(a) The actual free energy change is indicative of the direction of a chemical reaction. If $\Delta G < 0$, the reaction may proceed as written; if $\Delta G = 0$, the constituents are in chemical equilibrium; if $\Delta G > 0$, reaction can proceed only in the reverse direction. The value of ΔG is calculated from:

$$\Delta G = \Delta G^0 + RT \ln \prod_{j=1}^{n} a_j^{\nu_j}$$

For the case of interest, with standard states of unit fugacity and ideal gases, $a_j = \hat{f}_j/1 = p_j$. Then:

$$\Delta G = \Delta G^0 + RT \ln \left(\frac{p_{\text{H}_2} p_{\text{I}_2}}{p_{\text{HI}}^2} \right)$$

$$\Delta G^0 = \Delta H^0 - T\,\Delta S^0$$

$$\Delta G^0_{400} = 2700 - 400[(33.23) + (64.89) - (2)(51.38)]$$

$$= 4560 \text{ cal/g-mol H}_2$$

Since

$$p_{H_2} = \frac{0.01}{1.11}\ \text{atm}, \qquad p_{I_2} = \frac{0.1}{1.11}\ \text{atm}, \qquad \text{and} \quad p_{HI} = \frac{1.0}{1.11}\ \text{atm},$$

$$\Delta G = 4560 + (1.987)(400)\ln\left[\frac{(0.01)(0.1)}{(1.0)^2}\right] = -930\ \text{cal/g-mol } H_2$$

Thus, the reaction proceeds via the decomposition of HI to form more H_2 and I_2.

(b) Ordinarily, for ideal gases, with $\nu = 0$, there is no effect of pressure upon the equilibrium composition. In this case, however, one must verify that no liquid iodine condenses. Initially, $p_{I_2} = 0.1/1.11 = 0.099$ atm. This value will increase as noted in (a) because of HI decomposition at the higher temperature. The vapor pressure of I_2 at 127°C is 144 mm Hg = 0.19 atm. During compression of the mixture to 3 atm, liquid I_2 is certain to condense. The partial pressure of I_2 then remains at 0.19 atm after condensation begins; thus, more HI must decompose to satisfy the equilibrium value of K.

(c) Since the standard states (unit fugacity) are not a function of the system pressure, the equilibrium constant remains constant during the helium addition. Also, in this case, pressure variations do not affect the partial pressures since,

$$K_a = K_p = \frac{p_{H_2} p_{I_2}}{p_{HI}^2}$$

From (b), $p_{I_2} = 0.19$ atm. Let $x = p_{H_2}$, then $p_{HI} = 3 - 0.19 - x = (2.81 - x)$ atm.

$$K = \exp\left(\frac{-\Delta G^0}{RT}\right) = \exp\left[-\frac{(4560)}{(1.987)(400)}\right]$$

$$= 3.20 \times 10^{-3} = \frac{p_{H_2} p_{I_2}}{p_{HI}^2}$$

$$= \frac{(x)(0.19)}{(2.81 - x)^2}$$

$$x = 0.125\ \text{atm}$$

$$P_{HI} = 2.69\ \text{atm}$$

From a mass balance on H_2,

$$(0.5)(N_{HI})_0 + (N_{H_2})_0 = (0.5)(1.0) + 0.01 = 0.51 = (0.5)(N_{HI}) + N_{H_2}$$

Also,

$$\frac{N_{HI}}{N_{H_2}} = \frac{p_{HI}}{p_{H_2}} = \frac{2.69}{0.125}$$

Thus,

$$0.51 = \left[1 + (0.5)\left(\frac{2.69}{0.125}\right)\right] N_{H_2}$$

$$N_{H_2} = 0.0434\ \text{g-mol}$$

$$N_{HI} = 0.934\ \text{g-mol}$$

$$N_{I_2}\,(\text{gas}) = \left(\frac{0.19}{0.125}\right)(0.0434) = 0.0660 \text{ g-mol}$$

$$N_{I_2}\,(\text{liquid}) = 0.1 + \frac{1.0 - 0.934}{2} - 0.0660 = 0.0670 \text{ g-mol}$$

$$N_{He} = \left(\frac{1}{0.125}\right)(0.0434) = 0.347 \text{ g-mol}$$

(d) The variation in the equilibrium constant with temperature is given as:

$$\left(\frac{\partial \ln K}{\partial T}\right)_P = \frac{\Delta H^0}{RT^2}$$

$$\Delta H^0 = \Delta H^0_{400} + \int_{400}^{T} \Delta C_p^0\, dT$$

$$\Delta C_p^0 = C^0_{p_{H_2}} + C^0_{p_{I_2}} - 2C^0_{p_{HI}}$$

Substituting for the heat capacity values as given,

$$\Delta H^0 = \Delta H^0_{400} + (2.06)(T - 400) + \left(\frac{0}{2}\right)(T - 400)^2 + \left(\frac{2.08}{3}\right)(10^{-6})(T - 400)^3 - \frac{(5)(10^3)}{T}$$

Integrating

$$\ln\left(\frac{K_{600}}{K_{400}}\right) = \int_{400}^{600} \left(\frac{\Delta H^0}{RT^2}\right) dT$$

$$\ln K_{600} = -4.54$$

$$K_{600} = 1.06 \times 10^{-2}$$

This value of K is over three times larger than the value at 127°C; equilibrium at 600°K would result in further decomposition of HI to H_2 and I_2.

PROBLEMS

10.1. Assume that the equations shown below represent chemical reactions. Compute the value of ΔG^0_{298} for each. The standard states are as follows: I_2 gas, unit fugacity; I_2 liquid, pure liquid, 1 atm; I_2 solid, pure solid, 1 atm. The densities of liquid and solid iodine may be taken to be 1.03 g/cm^3 and the gas iodine to be an ideal gas.

(a) I_2, solid $\rightleftharpoons$ I_2, gas

(b) I_2, solid $\rightleftharpoons$ I_2, liquid

(c) I_2, liquid $\rightleftharpoons$ I_2, gas

Some vapor pressure data for iodine are given below.

T, °C	P_{vp}, mm Hg	T, °C	P_{vp}, mm Hg
−50	0.000037	60	4.31
−40	0.00019	70	8.22
−30	0.0008	80	15.1
−20	0.0030	90	26.8
−10	0.0099	100	45.5
0	0.0299	114.15*	90.1
10	0.0808	120	111
20	0.202	130	157
30	0.471	150	294
40	1.03	160	394
50	2.16	184.35**	760

* Melting point.
** Normal boiling point.

10.2. An inventor claims to be able to produce diamonds from β-graphite at room temperature by a process involving the application of 540,000 psia pressure. In view of the data shown below, are his claims to be taken seriously?

Specific gravity, β-graphite = 2.26.
Specific gravity, diamond = 3.51.

$$C_{\beta\text{-graphite}} \rightleftharpoons C_{\text{diamond}}, \qquad \Delta G^0_{298} = 686 \text{ cal/g-atom}$$

Both solids are incompressible and no solid solutions are formed.

10.3. One g-mol of A is mixed with 2 g-mol of B and the reaction,

$$A(g) + B(g) \rightleftharpoons C(g), \qquad \Delta G^0_{298} = 100 \text{ cal}$$

allowed to proceed to equilibrium at 25°C and 20 atm. Calculate the equilibrium mole fractions of all components.

Assume that the gases form an ideal solution and that each pure gas follows the equation of state,

$$P(V - b) = RT \text{ where } b \text{ values are:}$$

$$b_A = 45 \text{ cm}^3/\text{g-mol}$$

$$b_B = 45 \text{ cm}^3/\text{g-mol}$$

$$b_C = 60 \text{ cm}^3/\text{g-mol}$$

10.4. (a) For the general reaction at some temperature and pressure,

$$aA + bB = cC + dD$$

suppose one began initially with pure A and B and desired to vary the *initial* mole ratio (A/B) to maximize the concentration of C and D in the equilibrium mixture. Show for the case of an ideal gas mixture that the desired initial ratio $(N_{A_0}/N_{B_0}) = a/b$.

(b) What is the maximum conversion of N_2 to ammonia obtainable in a Haber ammonia unit operating at 300 atm and 500°C when stoichiometric quantities of hydrogen and nitrogen are used?

Use as a basis 1 mole of nitrogen and assume that an ideal solution forms.

$$\frac{1}{2}N_2(g) + \frac{3}{2}H_2(g) = NH_3(g)$$

$$\Delta G^0\left(\frac{\text{cal}}{\text{g-mol}}NH_3\right) = 12.31T\log T - 12.30T + \frac{2.526}{10^4}T^2 - \frac{8.45}{10^7}T^3 - 9170$$

T is in °K

(c) Ammonia at 50 psia and 100°C is passed through an adiabatic reversible turbine to an exit pressure of 1 atm. Assuming complete reversibility and that chemical equilibrium exists at all times, what is the maximum work that can be obtained?

NH_3 PROPERTIES*

P(psia)	T(°F)	S(Btu/lb °R)	H(Btu/lb)
50	212	1.4814	725.04
14.7	47.75	1.4814	642.10

* Reference: Nat. Bur. Standards Circular 142.

10.5. For a chemical reacting mixture of ideal gases, an equilibrium constant K_c is often used. The standard states for all components are unit concentration (i.e., that pressure where $P = RT$).

What is $(\partial \ln K_c/\partial T)_V$?

10.6. One gram mole of ethylene and one gram mole of benzene are fed to a constant volume batch reactor and heated to 600°K. On the addition of a Friedel-Crafts catalyst an equilibrium mixture of ethylbenzene, benzene, and ethylene is formed.

$$C_6H_6(g) + C_2H_4(g) \rightleftharpoons C_6H_5C_2H_5(g)$$

The pressure in the reactor, before the addition of the catalyst (i.e., before any reaction has occurred) is 2 atm. Calculate the total heat removed by the cooling fluid in a heat exchanger used to maintain the reactor temperature constant at 600°K as the reaction proceeds to equilibrium.

Data:

$C_6H_6(1)$: $\Delta G^0_{f_{298}} = 27{,}100$ cal/g-mol, $\Delta H^0_{f_{298}} = 12{,}470$ cal/g-mol
$C_2H_4(g)$: $\Delta G^0_{f_{298}} = 16{,}290$ cal/g-mol, $\Delta H^0_{f_{298}} = 12{,}500$ cal/g-mol
$C_6H_5C_2H_5(g)$: $\Delta G^0_{f_{298}} = 31{,}230$ cal/g-mol, $\Delta H^0_{f_{298}} = 7{,}120$ cal/g-mol

Physical Properties:

	Boiling point (°C)	ΔH_{vap} at 1 atm, cal/g-mol	Average C_p, cal/g-mol °C
C_6H_6	80.1	7,350	27
C_2H_4	−103.71	3,230	13
$C_6H_5C_2H_5$	136.19	8,600	40

10.7. A very efficient plant finds that it has two gas streams, the properties of which are tabulated below. The management would like to extract the maximum work possible from the streams before discarding the products in a collection vessel at 300°K and 1 atm. Rocky Jones, boy genius, has suggested a black box, which he claims will serve the purpose. (See Figure P10.7)

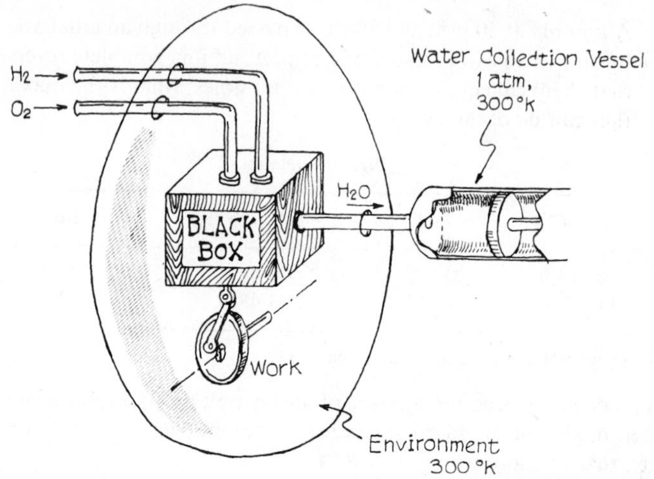

Figure P10.7

What is the maximum power obtainable from Rocky's box?

	Pure oxygen	Pure hydrogen
Temperature, °K	300	600
Pressure, atm	2	3
Flow rate, g-mol/sec	1	2
C_p, cal/g-mol °K	7	7

Data:

$\Delta G_f^0(H_2O) = -54{,}640$ cal/g-mol; $\Delta H_f^0(H_2O) = -57{,}780$ cal/g-mol at 300°K with products and reactants in the standard states of pure gases, unit fugacity. The vapor pressure of liquid water at 300°K is 0.035 atm.

10.8. Hydrogen exists in two modifications that depend on the nuclear spin. These forms are called ortho- and para-hydrogen. The properties of hydrogen can be divided into two groups: the first group depends on the ortho-para composition. These properties include the specific heat, velocity of sound, entropy, enthalpy, and thermal conductivity. The second group of properties is essentially independent of the ortho-para composition. Such properties include vapor pressure, *P-V-T* properties, and latent heats of vaporization.

The equilibrium composition of a hydrogen mixture depends on the temperature. The table below shows the variation:

Temperature, °K	Fraction para-hydrogen
<10	1.0
10	0.999999
20	0.99821
30	0.97021
40	0.88727
50	0.77054
60	0.65569
70	0.55991
80	0.48537
90	0.42882
100	0.38620
120	0.32959
150	0.28603
200	0.25974
250	0.25264
300	0.25072
>300	0.25

In addition, heat capacities are shown below for the para form, the so-called normal form (constant composition of 25% para, 75% ortho), and for the equilibrium form (i.e., the equilibrium mole fraction).

HEAT CAPACITY, C_p/R

T, °K	Normal (25% para)	Para	Equilibrium mixture
10	2.500	2.500	2.500
20	2.500	2.500	2.630
30	2.500	2.500	3.433
40	2.501	2.502	4.317
50	2.505	2.519	4.567
60	2.519	2.574	4.359
70	2.547	2.681	4.033
80	2.590	2.841	3.746
90	2.647	3.038	3.533
100	2.714	3.248	3.386
110	2.785	3.450	3.289
120	2.857	3.625	3.231
130	2.927	3.765	3.201
140	2.993	3.865	3.190
150	3.053	3.929	3.194
160	3.109	3.961	3.207
170	3.159	3.967	3.227

T, °K	Normal (25% para)	Para	Equilibrium mixture
180	3.204	3.955	3.251
190	3.244	3.929	3.276
200	3.280	3.896	3.301
210	3.312	3.859	3.326
220	3.341	3.820	3.350
230	3.366	3.782	3.372
240	3.388	3.747	3.392
250	3.407	3.715	3.409
260	3.423	3.685	3.425
270	3.438	3.659	3.439
280	3.450	3.636	3.451
290	3.461	3.616	3.461
300	3.470	3.600	3.470

(a) What is the change in entropy and enthalpy between normal hydrogen at 300°K and para-hydrogen at 20°K? Assume during the change that the pressure is constant and the gas mixture is ideal.

(b) What is the change in enthalpy and entropy when converting one g-mol of ortho-hydrogen to para-hydrogen at 300°K? At 20°K?

10.9. A constant volume bomb is charged with 2 g-mol of pure ethylene at 24°C and 60 atm. It is then heated to 140°C and, during the heating, the reaction

$$C_2H_4(g) \rightleftharpoons 2\ C(\text{graphite}) + 2\ H_2(g)$$

proceeds to equilibrium.

How much energy, as heat, is transferred?

Use Appendices D and E for estimating any volumetric properties and assume ideal solutions. Also, the free energy of formation of ethylene may be expressed as:

$$\Delta G_f^0 = 11{,}550 + 15.45T(°\text{K}),\ \text{cal/g-mol}$$

10.10. A reaction occurs between A and B to form C as follows:

$$A(l) + 3\ B(g) \rightleftharpoons C(l)$$

where the gas standard state denotes unit fugacity, and the liquid standard states are pure liquid at the vapor pressure corresponding to the reaction temperature.

(a) Set up an expression for the chemical equilibrium constant in terms of the compositions, fugacities of pure components, and activity coefficients. Assume that there are two phases (liquid and gas) at equilibrium and both phases are ideal solutions. Clearly define all quantities and describe briefly how they may be determined from theory or experiment.

(b) If the chemical reaction equilibrium constant were written so that all standard states were gas at unit fugacity, how would this constant be related to the K calculated in (a)?

(c) Which of the values of K calculated above would vary most greatly if the system pressure were doubled? Why?

(d) From the phase rule, how many independent variants are there for this system in chemical and phase equilibrium?

10.11. Gilliland, Gunness, and Bowles* investigated the hydration of ethylene by placing ethylene, water, and sulfuric acid in an agitated bomb kept at constant temperature by a condensing vapor. After analyzing the liquid phase for ethanol until constant composition indicated equilibrium had been attained, a sample of the vapor phase was withdrawn and analyzed. One run showed the following data:

$$T = 527°\text{K}, \quad P = 264.2 \text{ atm}$$

	Mole fraction	
	Liquid phase	Gas phase
Ethyl alcohol	0.084	0.075
Ethylene	0.0197	0.250
H_2O	0.881	0.675
H_2SO_4	0.015	—

(a) Compute ΔG^0 at 527°K for this reaction and compare it with the value computed from the formula recommended in the article, $\Delta G^0 = 26.9T - 8300$.

(b) The values reported in the literature for absolute entropies, S^0, and heats of formation are as follows:

	S^0 (298°K) cal/g-mol °K	ΔH_f^0 (298°K) kcal/g-mol
$C_2H_5OH(g)$	66.4	−56.6
$C_2H_4(g)$	52.7	+12.5
$H_2O(g)$	45.1	−57.595

Compute ΔG^0 from these data.

10.12. Gaseous nitrogen peroxide consists of a mixture of NO_2 and N_2O_4, and chemical equilibrium between these components is rapidly established. It has been suggested that this gas mixture be employed as a heat transfer medium. To evaluate this proposal, the heat capacity of the *equilibrium mixture* must be determined as a function of temperature.

Calculate and plot the effective heat capacity $(\partial \underline{H}/\partial T)_P$ for the equilibrium mixture between 290°K and 370°K at 1 atm.

Ideal gases may be assumed for this temperature range at 1 atm. $\Delta G^0 = 13{,}693 - 42.21T(°\text{K})$ for the reaction: $N_2O_4 \rightleftharpoons 2\,NO_2$. Unit fugacities are assumed for the standard states and ΔG^0 is in cal/g-mol.

C_p for the *frozen* equilibrium mixture may be approximated as:

C_p(frozen equilibrium) $= 10^{-4}T(°\text{K}) + 0.1735$ cal/g °C

* E. R. Gilliland, R. C. Gunness, and V. O. Bowles, "Free Energy of Ethylene Hydration," *Ind. Eng. Chem.*, **28**, (1936), 370.

10.13. In a laboratory investigation a high-pressure gas reaction $A \rightleftharpoons 2\,B$ is being studied in a flow reactor at 200°C and 100 atm. At the end of the reactor the gases are in chemical equilibrium and their composition is desired.

Unfortunately, to make any analytical measurements on this system it is necessary to bleed off a small side stream through a low-pressure conductivity cell operating at 1 atm. It is found that when the side stream passes through the sampling valve, the temperature drops to 100°C, and the conductivity cell gives compositions of $y_A = 0.5$ and $y_B = 0.5$.

From these experimental data and the physical properties listed below:

(a) Calculate the composition of the gas stream before the sampling valve.

(b) Are the gases in chemical equilibrium after the sampling valve? (Show definite proof for your answer.)

Data and allowable assumptions:

Heat of the reaction, $\Delta H = 3500$ cal/g-mol of A reacting, independent of temperature.

Heat capacity: 7 cal/g-mol °C for B independent of temperature.

The gas mixture is ideal at all pressures, temperatures, and compositions.

Assume no heat transfer in the sampling line or across the sampling valve.

10.14. A rigid, well-insulated gas storage tank of 10-ft³ capacity is filled originally with hydrogen gas at 1 atm and 30°K. Connected to the tank is a large high-pressure manifold containing hydrogen gas at 100 atm and 40°K. The valve is opened and hydrogen is allowed to flow into the tank until the gas inside attains a temperature of 50°K. Assume ideal gases and that the ortho-para equilibrium shift is infinitely rapid (see Problem 10.8 for data on equilibrium composition). Heat capacity data for *pure para*-hydrogen indicate that over the temperature range of interest here, $C_p/R \cong 2.50$. The tank walls and hardware are assumed to have zero heat capacity. Assume that the contents of the tank are well-mixed at all times.

(a) What is the pressure when the temperature of the gas inside the tank reaches 50°K?

(b) What would the temperature be if the manifold and tank pressure were increased without limit?

(c) What is the entropy change of the universe for this process?

10.15. The system N_2O_4—NO_2 remains in chemical equilibrium under essentially all conditions (i.e., if a step change in pressure or temperature is imposed, it takes only about 0.1 to 0.2 microseconds for the system to react and attain equilibrium under the new conditions).

In a portion of a plant design, we are faced with estimating the change in temperature of a high-pressure N_2O_4—NO_2 gas mixture as it is throttled across a valve. The valve is well-insulated and the flow rate is steady. The upstream temperatures and pressures are 80°C and 5 atm. The upstream mole fractions are: $y_{NO_2} = 0.637$ and $y_{N_2O_4} = 0.363$. The downstream pressure is 1 atm.

What is the downstream temperature and composition?

Available Data:

Standard free energies and frozen heat capacities are given in Problem 10.12. The molecular weights of N_2O_4 and NO_2 are 92 and 46, respectively. Although not strictly true, the gas mixture may be assumed to be ideal (i.e., the relation $P\underline{V} = NRT$ is applicable and there is no effect of pressure on the enthalpy of the pure components).

The fraction N_2O_4 dissociated to NO_2 is shown in Figure P10.15.

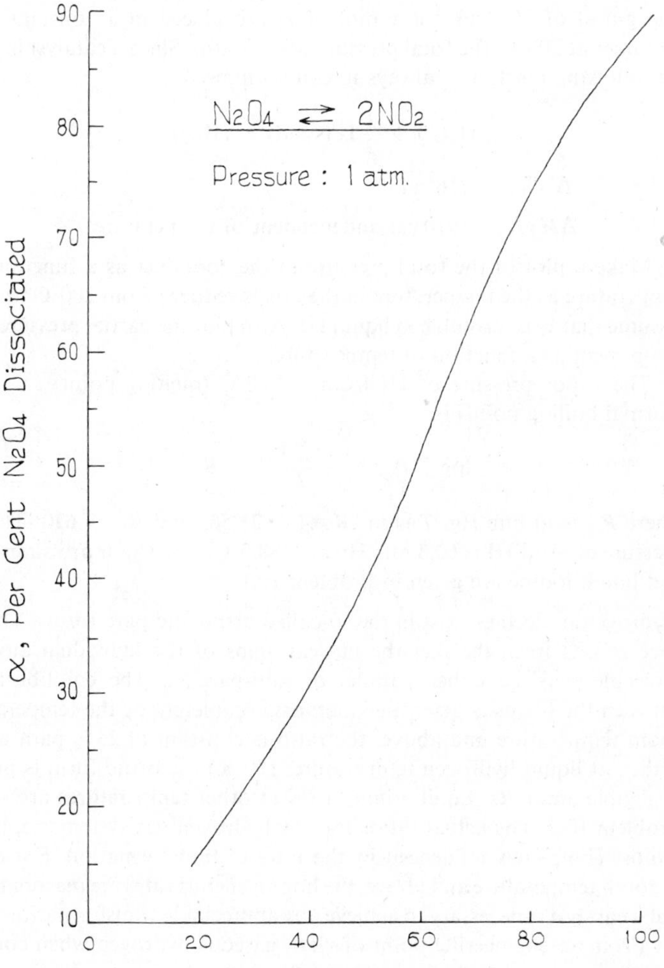

Figure P10.15

10.16. As noted in Problem 10.15, the chemical kinetics are such that gas mixtures of NO_2—N_2O_4 are always in chemical equilibrium.

Suppose that a mixture of these gases is expanded from a pressure and temperature, P_1, T_1 to a lower pressure P_2 in an adiabatic, reversible turbine.

(a) Write a differential equation that expresses how the pressure, temperature, and fraction N_2O_4 dissociated vary during the expansion. (Assume a basis of 1 mole N_2O_4 and let α be the fraction dissociated.)

(b) Show clearly how the adiabatic, reversible work may be calculated.

(c) Repeat (a) and (b) if the turbine were isothermal (at T_1) and reversible. Assume ideal gas mixtures.

10.17. One g-mol of H_2 and one g-mol of I_2 are placed in a constant volume container at 200°C, the total pressure being 1 atm. Since a catalyst is present, the following reaction is always in equilibrium:

$$\frac{1}{2}H_2(g) + \frac{1}{2}I_2\,(\text{solid}) = HI(g)$$

$$\Delta G^0_{298} = 316 \text{ cal}$$

$$\Delta H^0_{298} = 5910 \text{ cal, independent of temperature}$$

Make a plot of the total pressure in the container as a function of the temperature as the temperature in the box is reduced from 200°C to −50°C. Assume that I_2 is insoluble in liquid HI. Also plot the partial pressure of each component as a function of temperature.

The vapor pressure of HI from −50.8°C (melting point) to −35.5°C (normal boiling point) is:

$$\log_{10} P_{vp} = \frac{-52.3\,A}{T} + B$$

where P_{vp} is in mm Hg, T is in °K, $A = 21.58$, and $B = 7.630$. The vapor pressure of solid HI is 60.3 mm Hg at −88.9°C. The vapor pressures of solid and liquid iodine are given in Problem 10.1.

10.18. Hydrogen molecules exist in the so-called ortho and para forms. The difference results from the fact the nuclear spins of the individual atoms in a molecule may be either parallel or anti-parallel. The equilibirum ratio between the forms is determined almost completely by the temperature. At room temperature and above, the ratio is constant at 25% para and 75% ortho; at liquid hydrogen temperatures (20°K) the ortho form is present in negligible amounts. Equilibrium ratios at other temperatures are shown in Problem 10.8. The actual ratios for any hydrogen gas system are, however, almost completely influenced by the rates of transformation. For example, at room temperature and above, the homogeneous rates are reasonably large, but a catalyst is necessary to achieve any appreciable transformation at 20°K.

From an engineering point of view, gaseous hydrogen when cooled and liquefied converts to the lower energy form (para) in an exothermic reaction, and as cooling and liquefaction rates are fast compared to the conversion rate, freshly liquefied hydrogen differs only slightly from the 25%-p, 75%-o composition of the feed gas. Transformation to the para form then occurs at low temperatures, and this requires additional low-temperature refrigeration.

A prospective client is proposing to use stored liquid hydrogen (which is available and fully converted to the para form) to cool a stream of hot helium. The hydrogen gas is to be exhausted around 150°C and fed to another part of the system. Although the size of the heat exchanger is not particularly important, one would like to cool the helium stream as much as possible with a given hydrogen flow rate. The client feels that the residence time of the hydrogen in the system is so short that the exit gas will still exist in a pure para form. It is felt, however, that *if* a good catalyst could be developed, the *endothermic* transformation from p $\longrightarrow$ o could be carried out in the heat exchanger and this additional heat effect used to cool the helium to a lower temperature.

At present he wants us to put some of our money into the concept and develop a catalyst that will yield an equilibrium gas mixture [of H_2(o) and H_2(p)] at all positions in the heat exchanger.

Write a concise memorandum indicating your opinion of the scheme and recommending our course of action.

10.19. Freshly liquefied hydrogen that has not been catalyzed consists of a 3 to 1 ortho-para mixture. On standing, there is a slow shift of the mixture toward the equilibrium concentration, a fact that complicates the problem of storing the liquid for any length of time.

Assume that 1 mole of hydrogen at room temperature and 1 atm is quickly liquefied (in order not to affect the ortho/para ratio) and then allowed to reach an equilibrium state. From the data in Problem 10.8 showing the equilibrium fraction of para-hydrogen as a function of temperature, this should occur when the concentration of para is about 99.8%. During this change the system is chosen to be adiabatic and always vented at 1 atm. How many moles of liquid hydrogen remain after this final equilibrium state is reached?

Data:

Heat of vaporization of hydrogen is assumed independent of ortho-para ratio and equals 195 Btu/lb at 20.6°K, the boiling temperature at 1 atm. There is no appreciable difference between the vapor and liquid phase concentration of the two components (i.e., the relative volatility of ortho- to para-hydrogen is unity over the entire liquid range).

10.20. Some recent experiments have been carried out to determine the equilibrium partial pressure of oxygen over molten potassium oxides. The data were taken in the following way: samples of pure KO_2 were placed in a MgO boat in an evacuated tube. The tube was inserted into an oven at a sufficiently high temperature that the oxide melted. The pressure of the evolved oxygen was measured after equilibrium was attained. From this pressure measurement, the tube volume, and the oven temperature, the moles of oxygen evolved could be ascertained. From this value and also the original sample weight, the atomic O/K ratio of the oxide liquid could be calculated. Next, a known amount of oxygen was bled out of the system, and the system allowed to come to equilibrium again. Again the oxygen pressure was measured, and the liquid O/K ratio calculated. The data indicated that at any given temperature level the oxygen partial pressure depends only on the liquid O/K ratio. The data are shown in the Table of Data at 500, 600, and 650°C for various O/K ratios.

(a) Are these data consistent with the phase rule? Demonstrate. There is essentially nothing known about the structure of the liquid phase. It is black, probably has a high electrical conductivity, and contains some or all of the following species: K^+, O_2^-, K_2O, K_2O_2, KO_2, etc.

(b) Estimate accurately the heat evolved or absorbed if a reaction occurs so that the liquid absorbs oxygen isothermally at 600°C as it changes from an O/K ratio of 1.0 to 1.4. Express your answer on a basis of 1 g-atom of potassium in the liquid.

(c) A "simple" picture of the liquid shows it to be an "ideal" mixture of liquid K_2O, K_2O_2, and KO_2. Demonstrate how you would calculate equilibrium constants for the reactions given below, using only the p_{O_2} and O/K values measured experimentally.

$$2\,KO_2(l) \longrightarrow K_2O_2(l) + O_2(g) \tag{1}$$

$$2\,KO_2(l) \longrightarrow K_2O(l) + \frac{3}{2}O_2(g) \tag{2}$$

(d) How could you test the hypothesis in (c) to see if it were reasonable?

(e) Many thermodynamicists would describe the system as follows:

$$\frac{1}{2}O_2(g) \rightleftharpoons [O(l)]$$

where [O(*l*)] represents the oxygen in the liquid phase.

$$K = \frac{[O(l)]}{p_{O_2}^{1/2}}$$

By defining an activity coefficient such that

$$[O(l)] = (\gamma)(\text{conc. } O_2 \text{ in liquid})$$

with $\gamma \longrightarrow 1.0$ as concentration $\longrightarrow 0$. Show how the activity coefficient and equilibrium constant may be determined from the data at any given temperature and composition. Assume for simplicity that you have at your disposal the partial pressures of oxygen over the entire range of O/K ratios from 0 to 2.

TABLE OF DATA

O/K ratio	Partial pressure of oxygen, atm		
	500°C	600°C	650°C
1.0	0.018	0.070	0.13
1.1	0.10	0.13	0.21
1.2	0.26	0.26	0.31
1.3	0.43	0.40	0.43
1.4	0.61	0.55	0.60

10.21. Fluorine and hydrogen at 25°C are to be fed into a combustion chamber at a weight ratio of 8.00. Assuming that the combustion products are in chemical equilibrium, determine the maximum chamber temperature attainable when the process is conducted at a constant pressure of 29.4 psia.

What would the chamber temperature be if the process were conducted at a pressure of 1,000 psia instead of 29.4 psia?

The gas mixture may be assumed to obey the ideal gas law. The following data are available:

Reactions	*Standard Free Energies (calories); T in °K*
$F_2 = 2\,F$	$\Delta G^0 = 7{,}000$ at $3{,}000°K$
$H_2 = 2\,H$	$\Delta G^0 = 97{,}000 - 3.50T \ln T + 1.17T + 0.00045T^2$
$H + F = HF$	Data not available
$H_2 + F_2 = 2\,HF$	$\Delta G^0 = -129{,}000 + 9.64T$

Species	*Heat capacity: (may be assumed independent of pressure)(cal/g-mol °K)*
F_2	$C_p = 6.50 + 0.0010T$
H_2	$C_p = 6.62 + 0.00081T$
HF	Boiling point: 19.7°C
HF	$P_{vp} = 100$ mm Hg at $-28.2°C$

Atomic weights: F = 19, H = 1.008.

10.22. The thermal conductivity of reacting gas mixtures is often found to be much larger than would be expected from molecular considerations. If a temperature gradient exists in the gas, then, in different temperature regions, the concentration of reactive species may be different; this concentration gradient causes a diffusion flux that adds to the normal thermal conduction heat flux since there is a transport of energy by molecular diffusion.

A convenient system to study this phenomenon utilizes nitrogen dioxide. The rate of the reaction,

$$2\,NO_2 \rightleftharpoons N_2O_4$$

is very rapid in both directions and, for most studies, the mixture may always be assumed to be in chemical equilibrium.

Derive an expression that shows the additional contribution to thermal conductivity because of the diffusion flux for this system. Plot this difference as a function of temperature. Also plot the temperature gradient between two plates one cm apart if the top plate is at 400°K and the bottom plate at 300°K and there is a mixture of NO_2 and N_2O_4 between the plates. The pressure is 1 atm. What is the heat flux between the plates?

Data:

Source: Product Bulletin, *Nitrogen Tetroxide*, Allied Chemical, Nitrogen Division, New York.

BINARY DIFFUSION COEFFICIENTS (AT 1 ATM)

T, °K	D, cm²/sec
300	0.06532
350	0.08843
400	0.1147
500	0.1757
600	0.2467

$$2NO_2 \rightleftharpoons N_2O_4 \qquad \Delta H^0_{298} = -13.87\ k\text{cal/g-mol } N_2O_4$$

Assume that ΔH^0 is independent of temperature.

THERMAL CONDUCTIVITY IF NO REACTION
("FROZEN" CONDUCTIVITY)

T, °K	cal/cm-sec-°K × 10^5
294	3.1
316	3.8
327	4.0
350	4.7
372	5.2
394	5.7

Equilibrium data are shown in Figures P10.15 and P10.22.

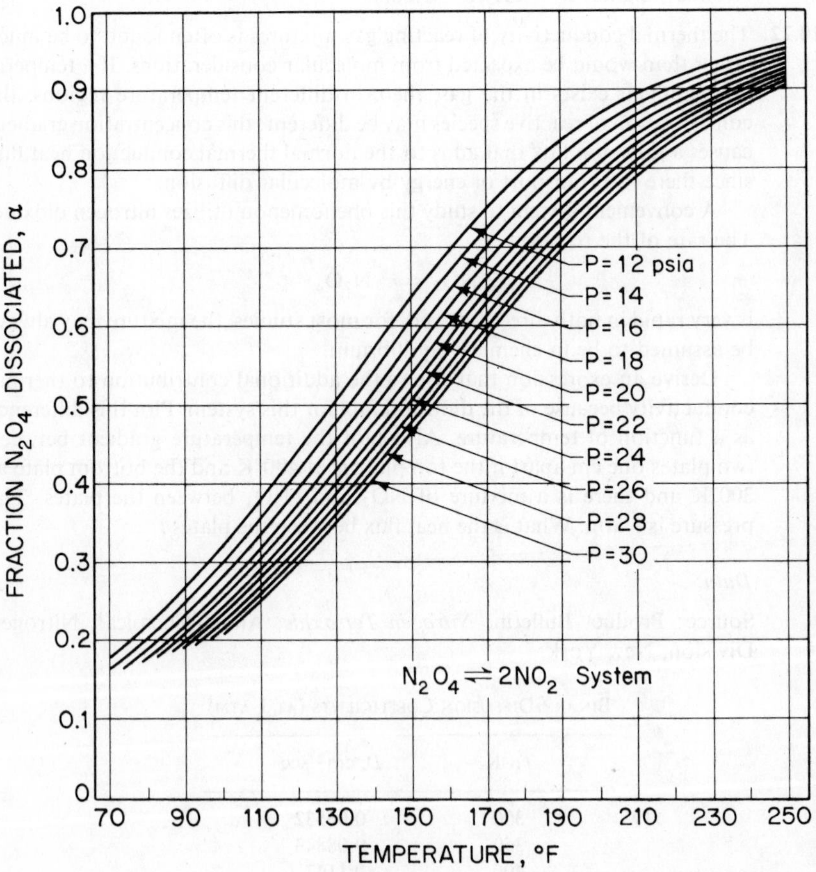

Figure P10.22

10.23. A chromatographic process has been proposed for separating mixtures of hydrogen and deuterium using a bed of palladium-coated particles. The solubilities of the isotopes in palladium are notably different, as shown in Figure P10.23 for the Pd-H_2 (no D) and Pd-D_2 (no H) systems.

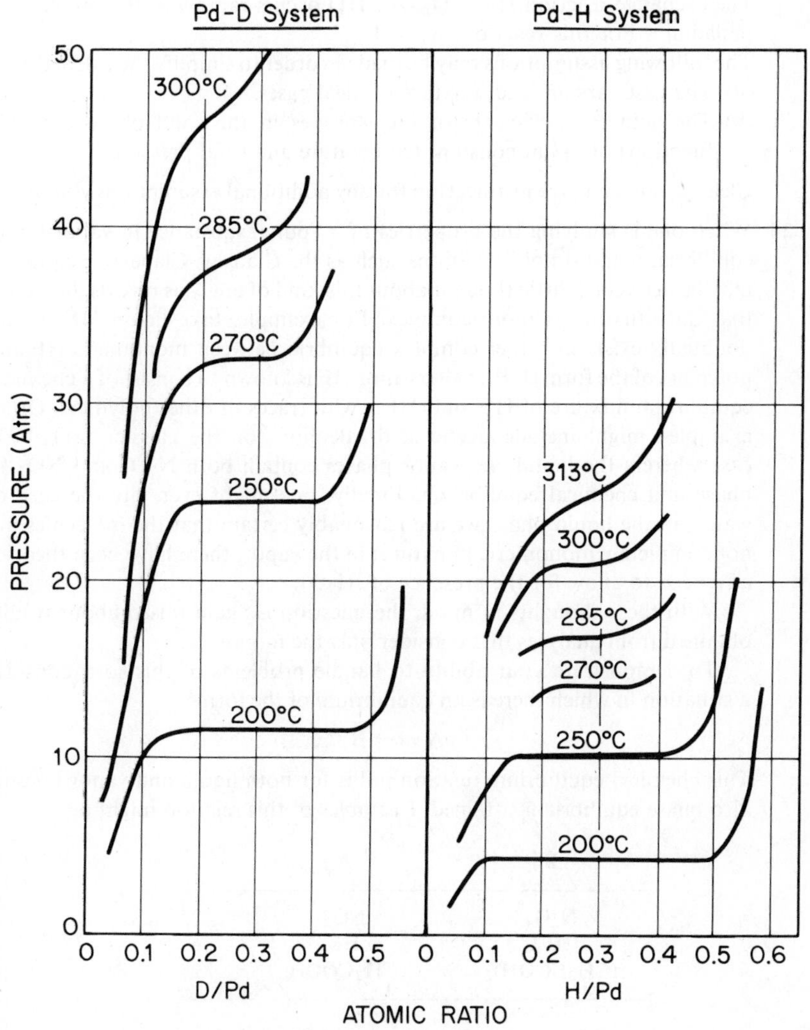

Figure P10.23

Determine the distribution coefficient, $\alpha = (y_{H_2}/x_H)/(y_{D_2}/x_D)$ as a function of Pd atom fraction in the solid phase for palladium in equilibrium with hydrogen *and* deuterium at 250°C and 15 atm total pressure. Clearly outline the procedure and carry through the calculation for $x_{Pd} = 0.80$.

Notes and Additional Data:

Hydrogen isotopes exist in the solid phase as atoms.
Each isotope is known to form several alloys with palladium.
The concentration of *atomic* species in the gas phase is negligible under the conditions of interest.
The exchange reaction $H_2 + D_2 = 2\,HD$ occurs rapidly in the presence of palladium. For this reaction, $K_y = 4$.
The following assumptions may be made in order to simplify the calculations.
(a) The gases are an ideal mixture of ideal gases.
(b) The activity coefficients of the isotopes in the solid phase are only functions of x_{Pd} at constant temperature and total pressure.

Clearly state and give justification for any additional assumptions you make.

10.24. When one is studying the properties of a "pure" substance in vapor-liquid equilibria, many simple equations such as the Clausius-Clapeyron equation may be derived. A little thought about this kind of analysis reveals, however, that the situation is more complex. For example, take liquid HF. It undoubtedly exists in rather complex equilibria between monomeric HF and polymers of the form $(HF)_x$; also vapor HF is known to consist of a chemical equilibrium mixture of HF and $(HF)_6$ with traces of other polymers. Other examples might include acetic acid, alcohols, or the classic N_2O_4-NO_2 case wherein the liquid and vapor phases contain both N_2O_4 and NO_2 in phase and chemical equilibrium. Finally, one might even cite the case of water; in the liquid phase we are reasonably certain that the molecules are not completely monomeric in nature; in the vapor, there have been theories advanced to allow for the presence of $(H_2O)_x$.

With these thoughts in mind, the question is again raised about results obtained from analyses that consider only the monomer.

To demonstrate your ability to handle problems of this sort, consider a situation in which there is an equilibrium of the form:

$$nA \rightleftharpoons B$$

This chemical equilibrium relation holds for both liquid and vapor phases; also phase equilibria is attained. Examples of this relation might be:

B	A	n
N_2O_4	NO_2	2
$(HF)_6$	HF	6
$(CH_3COOH)_2$	CH_3COOH	2

Choose a base of one gram-formula weight of A in each phase to simplify the analysis. Let α^V be the number of moles of monomer reacting in the vapor phase to form α^V/n moles of polymer. Thus, in the vapor phase there are $(1 - \alpha^V)$ moles of monomer and α^V/n moles of polymer. Similarly, let α^L be the comparable parameter in the liquid phase.

(a) Suppose that you could measure experimentally the vapor pressure of this two-phase A-B system as a function of temperature. What would the slope $(dP/dT)_{\text{saturation}}$ represent? Make a rigorous analysis and express your results in the usual thermodynamic nomenclature such as partial molal quantities, α^V, α^L, etc. Also describe in words what your answer means.

(b) If the vapor pressure data were plotted as $\ln P$ against $(1/T)$, the "apparent latent heat" can be calculated as:

$$\Delta H\,(\text{apparent}) = -R\frac{d\ln P}{d(1/T)}$$

How would ΔH (apparent) be related to the enthalpy term found in (a)? In these calculations assume that liquid volumes are negligible compared to vapor volumes and that the vapor phase behaves as an ideal gas mixture.

(c) For the case $6\,HF \rightleftharpoons (HF)_6$, vapor pressure data, plotted as $\ln P$ vs. $(1/T)$ gives an apparent ΔH of 5900 cal/20.1 g HF at 0°C. [R. L. Jarry and Wallace Davis, Jr., "The Vapor Pressure, Association, and Heat of Vaporization of Hydrogen Fluoride," *J. Phys. Chem.*, **57**, (1953), 600.] Also, a calorimetric enthalpy of vaporization gives 1680 cal/20.1 g HF at the same temperature. [Karl Fredenhagen, "Physikalischchemische Messungen am Fluorwasserstoff," *Z. anorg. Chem.*, **210**, (1933), 210.] Estimate the mole fractions of HF and $(HF)_6$ in the vapor phase at 0°C.

Thermodynamics of Surfaces 11

In the absence of gravity or other body forces, free liquids assume a spherical shape. Intermolecular attractive forces operate to pull the molecules together; this attraction results in the formation of spheres. A convenient way to characterize this behavior is to assume that the surface is in tension (i.e., each element of the surface layer experiences tensile forces from neighboring elements); the net result is a surface somewhat similar to that in an elastic balloon.

The analogy to the balloon may be carried further. We know that there exists a pressure difference between the inside and outside of a balloon because of these tensile forces. Similarly, in a bubble or drop of liquid, we will show that the internal pressure also exceeds that outside the drop. To demonstrate this, we will again use the concept of surface tension forces.

In this chapter, we shall first examine surfaces from the standpoint of work interactions (i.e., we shall derive an expression to allow us to determine the work necessary to vary both the *volume and area* of a surface layer). Following this, we shall explore the effects of curvature on small liquid drops and crystal nuclei. In the remainder of the chapter, we shall develop the important thermodynamic expressions for surface layers and also apply the criteria of equilibrium to the nucleation of new phases. The extensive subject matter of surfaces will only be examined from a classical thermodynamic point of view; interesting and very important phenomena such as wetting, surface activity, and adhesion must be left for the reader to enjoy in specialized texts. It is, however, worthwhile to note that although we will treat the surface as a static entity, this is a poor model. As Adam[1] pointed out, some 3×10^{21} molecules of water strike a square centimeter of water surface each second in an equilibrium system of water liquid and vapor near room temperature. The same number flux leaves. With a "parking area" of some 10 A^2, the mean lifetime is about 10^{-6} seconds. In this short time, there is a diffusional exchange between the surface layers and the bulk to a depth of about 100 A. Clearly, this surface is a dynamic entity! Classical thermodynamics is only applicable because we resort to statistical averaging of large numbers of molecules.

11.1 Surface Tension

To illustrate more clearly the mechanical analog of surface forces, consider a plane interface between a vapor phase α and a liquid phase β. The element shown in Figure 11.1 measures x units wide, y units deep, and has a thickness τ such that the top and bottom of the element are located in the homogeneous phases α and β, respectively. Each side of this parallelpiped is subjected to a normal, compressive pressure P that is equal at all points. This equality results from the criteria of phase equilibria derived for *plane* surfaces in

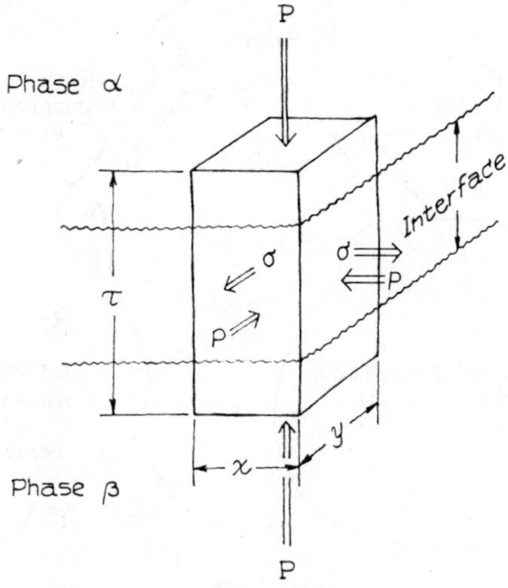

Figure 11.1 Element in plane interface.

[1] N. K. Adam, *The Physics and Chemistry of Surfaces* (New York: Oxford University Press, Inc., 1941), pp. 6, 7.

Chapter 7. Also, in the two directions perpendicular to the thickness, τ, there is the tension force σ *per unit length* (i.e., the tensile force on the front and back faces would be σx; likewise on each side face, σy).

If we choose the parallelpiped as our system, the work done by the system when the volume is increased by $dx\,dy\,d\tau$ is given as:

$$\begin{aligned} đW &= (Pxy)\,d\tau + (P\tau y - \sigma y)\,dx + (P\tau x - \sigma x)\,dy \\ &= P(xy\,d\tau + \tau y\,dx + \tau x\,dy) - \sigma(y\,dx + x\,dy) \qquad (11\text{-}1) \\ &= P\,d\underline{V} - \sigma\,dA \end{aligned}$$

In our later development of the thermodynamics of surface layers, the work term must be that given by Eq. (11-1).

11.2 Equilibrium Considerations

Consider the equilibrium system shown in Figure 11.2. It is composed of two distinct parts: a bulk phase α and a small fragment of a different phase β. Phase α is maintained at constant temperature and pressure by contact with the large isobaric and isothermal reservoirs R_P and R_T. Except for the

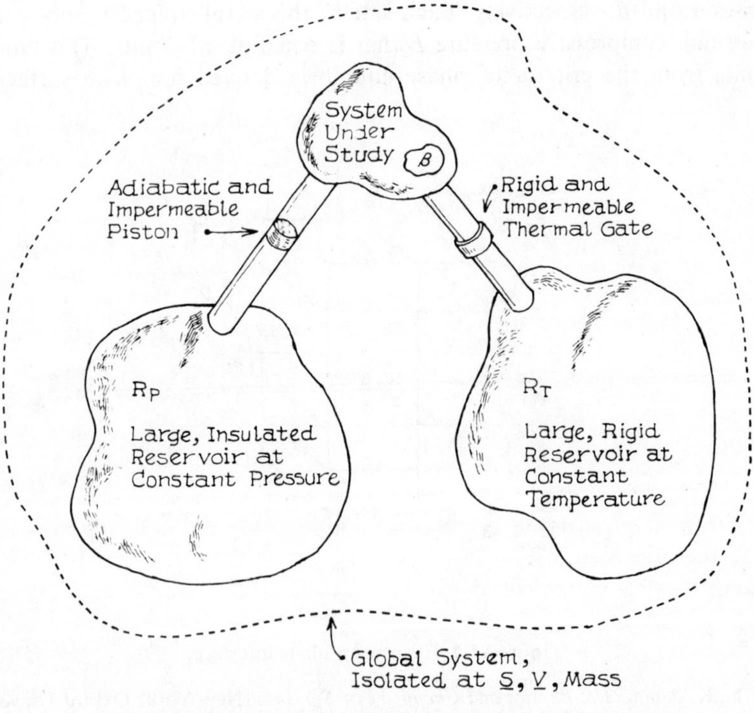

Figure 11.2 System interactions with constant pressure or temperature reservoirs.

presence of the β phase, the situation is identical to that shown in Figure 7.4 and as described as Case (c) in Section 7.2. There, with no β phase, it was shown that the system attained an equilibrium state when the total Gibbs free energy was a minimum.

The present case is, however, somewhat different because the small element of phase β does not necessarily have the same intensive properties as phase α even though both phases are in equilibrium. In addition, we must consider energy terms for the surface phase between α and β.

If the same nomenclature as in Section 7.2 is used, the criterion for stable equilibrium states is:

$$\Delta \underline{U}^{\Sigma} = \Delta \underline{U} + \Delta \underline{U}^{R_P} + \Delta \underline{U}^{R_T} > 0 \tag{7-16}$$

where the superscript Σ represents the global system and terms that are not superscripted refer to the α-β system. As shown before, for variations in volume or entropy,

$$\Delta \underline{U}^{R_P} = -P\,\Delta \underline{V}^{R_P} = P\,\Delta \underline{V} \tag{11-2}$$

$$\Delta \underline{U}^{R_T} = T\,\Delta \underline{S}^{R_T} = -T\,\Delta \underline{S} \tag{11-3}$$

where the global conservation of total volume and entropy has been employed. The P and T values in Eqs. (11-2) and (11-3) are those that characterize phase α but not necessarily phase β or σ. Also,

$$\Delta \underline{U} = \Delta \underline{U}^{\alpha} + \Delta \underline{U}^{\beta} + \Delta \underline{U}^{\sigma} \tag{11-4}$$

where β and σ represent the β and surface phases. If Eqs. (11-2), (11-3), and (11-4) are substituted into Eq. (7-16) with the expansions:

$$\Delta \underline{V} = \Delta \underline{V}^{\alpha} + \Delta \underline{V}^{\beta} + \Delta \underline{V}^{\sigma} \tag{11-5}$$

$$\Delta \underline{S} = \Delta \underline{S}^{\alpha} + \Delta \underline{S}^{\beta} + \Delta \underline{S}^{\sigma} \tag{11-6}$$

then,

$$\begin{aligned}(\Delta \underline{U}^{\alpha} + P\,\Delta \underline{V}^{\alpha} - T\,\Delta \underline{S}^{\alpha}) + (\Delta \underline{U}^{\beta} + P\,\Delta \underline{V}^{\beta} - T\,\Delta \underline{S}^{\beta})\\ + (\Delta \underline{U}^{\sigma} + P\,\Delta \underline{V}^{\sigma} - T\,\Delta \underline{S}^{\sigma}) > 0\end{aligned} \tag{11-7}$$

Let us introduce a term called the availability, $\underline{\mathcal{V}}$, where:

$$\underline{\mathcal{V}} \equiv \underline{U} + P\underline{V} - T\underline{S}; \qquad \begin{aligned} P &= \text{reservoir pressure}\\ T &= \text{reservoir temperature}\end{aligned} \tag{11-8}$$

In this definition, P and T *are constants and equal to the pressure and temperature in the external reservoirs* R_P and R_T. Eq. (11-7) states that availability is the thermodynamic function that attains a minimum value in the α-β-σ system; then, at equilibrium,

$$d\underline{\mathcal{V}} = 0 \tag{11-9}$$

$$\left.\begin{aligned} d^m\underline{\mathcal{V}} &> 0 \quad \text{(stable equilibrium)}\\ &< 0 \quad \text{(unstable equilibrium)}\end{aligned}\right\} \tag{11-10}$$

where $d^m\underline{\mathcal{V}}$ is the lowest order, nonvanishing derivative.

At equilibrium, with Eqs. (11-7), (11-8), and (11-9),

$$d\underline{U}^{\alpha} + d\underline{U}^{\beta} + d\underline{U}^{\sigma} = 0 \tag{11-11}$$

To determine the specific equilibrium criteria for this case, $\underline{U}^{\alpha}$, $\underline{U}^{\beta}$, and $\underline{U}^{\sigma}$ are expanded using the Fundamental Equations in energy representation,

$$d\underline{U}^{\alpha} = T\,d\underline{S}^{\alpha} - P\,d\underline{V}^{\alpha} + \sum_{j=1}^{n} \mu_j^{\alpha}\,dN_j^{\alpha} \tag{11-12}$$

$$d\underline{U}^{\beta} = T^{\beta}\,d\underline{S}^{\beta} - P^{\beta}\,d\underline{V}^{\beta} + \sum_{j=1}^{n} \mu_j^{\beta}\,dN_j^{\beta} \tag{11-13}$$

$$d\underline{U}^{\sigma} = T^{\sigma}\,d\underline{S}^{\sigma} - P^{\sigma}\,d\underline{V}^{\sigma} + \sigma\,dA + \sum_{j=1}^{n} \mu_j^{\sigma}\,dN_j^{\sigma} \tag{11-14}$$

With Eqs. (11-11) through (11-14) and with the understanding that $T = T^{\alpha}$ and $P = P^{\alpha}$ are constants and are determined by the reservoir temperatures and pressures,

$$(T^{\beta} - T)\,d\underline{S}^{\beta} + (T^{\sigma} - T)\,d\underline{S}^{\sigma} + \sum_{j=1}^{n} \mu_j^{\alpha}\,dN_j^{\alpha} + \sum_{j=1}^{n} \mu_j^{\beta}\,dN_j^{\beta}$$
$$+ \sum_{j=1}^{n} \mu_j^{\sigma}\,dN_j^{\sigma} - (P^{\beta} - P)\,d\underline{V}^{\beta} - (P^{\sigma} - P)d\underline{V}^{\sigma} + \sigma\,dA = 0 \tag{11-15}$$

Before examining Eq. (11-15), there is one additional restraint equation that must be used (i.e., there is conservation of mass in the system). For each component,

$$dN_j = dN_j^{\alpha} + dN_j^{\beta} + dN_j^{\sigma} = 0 \tag{11-16}$$

If Eq. (11-16) is multiplied by μ_j^{α} (chosen as a Lagrange multiplier) and added to Eq. (11-15), then each variation is independent and the coefficients must be zero. Therefore,

$$T^{\beta} = T = T^{\alpha} \tag{11-17}$$

$$T^{\sigma} = T = T^{\alpha} \tag{11-18}$$

$$\mu_j^{\beta} = \mu_j^{\alpha} \tag{11-19}$$

$$\mu_j^{\sigma} = \mu_j^{\alpha} \tag{11-20}$$

and

$$(P^{\beta} - P)\,d\underline{V}^{\beta} + (P^{\sigma} - P)\,d\underline{V}^{\sigma} - \sigma\,dA = 0 \tag{11-21}$$

The first four equalities are as we expected: the temperatures and chemical potentials of each component are equal throughout the system. In this regard, our results do not differ from what we found earlier for phase equilibrium criteria. Eq. (11-21) is, however, a different result.

First, let us consider the term $(P^{\sigma} - P)\,d\underline{V}^{\sigma}$. We have some latitude in defining the extent of the surface phase and, depending on our choice, we could have different values of P^{σ} and $d\underline{V}^{\sigma}$. If we were to define our surface phase boundary so that the surface had properties similar to phase α, then $P^{\sigma} \sim P^{\alpha} = P$ and $(P^{\sigma} - P)\,d\underline{V}^{\sigma}$ is essentially zero.

But, if we were to choose the surface phase boundary so that the surface resembled phase β, then this term would simply add to the first term in Eq. (11-21) since $P^\sigma \sim P^\beta$. In either extreme, or in any intermediate case, the contribution of $(P^\sigma - P)\, d\underline{V}^\sigma$ in Eq. (11-21) is probably small and almost all workers in the field have neglected it. If we do likewise, Eq. (11-21) reduces to

$$P^\beta - P = P^\beta - P^\alpha = \sigma\left(\frac{dA}{d\underline{V}^\beta}\right) \tag{11-22}$$

This final relation expresses the pressure difference across a curved interface in an equilibrium system.

Example 11.1

(a) If phase β were spherical in shape, what would be the pressure difference $(P^\beta - P^\alpha)$?

(b) Repeat (a) but assume that the phase boundary between the α and β phases is characterized by two radii of curvature C_1 and C_2.

Solution

(a) Let the radius of the sphere of phase β be r. Then, $A = 4\pi r^2$, $\underline{V}^\beta = (4/3)\pi r^3$ and then, from Eq. (11-22),

$$P^\beta - P^\alpha = \sigma\left(\frac{dA}{d\underline{V}^\beta}\right) = \frac{2\sigma}{r} \tag{11-23}$$

(b) In Figure 11.3 a small element of the surface is shown. The area is xy. If we extend the element by $\delta\tau$ and thereby increase C_1 to $C_1 + \delta\tau$ and C_2 to $C_2 + \delta\tau$, then the increase in area and volume of phase β becomes:

$$\delta A = (y + \delta y)(x + \delta x) - xy = y\,\delta x + x\,\delta y$$
$$\delta \underline{V}^\beta = xy\,\delta\tau$$

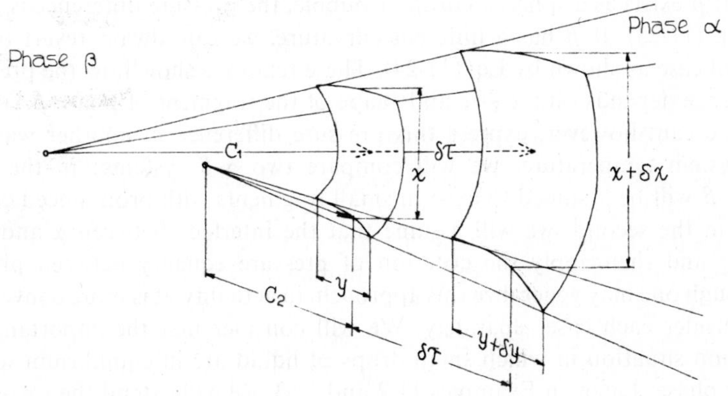

Figure 11.3

But, by similar triangles,

$$\frac{x + \delta x}{C_1 + \delta\tau} = \frac{x}{C_1}$$

$$\frac{y + \delta y}{C_2 + \delta\tau} = \frac{y}{C_2}$$

then,

$$\delta x = \frac{x}{C_1}\,\delta\tau$$

$$\delta y = \frac{y}{C_2}\,\delta\tau$$

Substituting the δx and δy terms into the expression for δA and using Eq. (11-22),

$$P^{\beta} - P^{\alpha} = \sigma\left(\frac{dA}{d\underline{V}^{\beta}}\right) = \frac{\left(\frac{xy\,\delta\tau}{C_1} + \frac{xy\,\delta\tau}{C_2}\right)\sigma}{xy\,\delta\tau}$$

$$P^{\beta} - P^{\alpha} = \sigma\left(\frac{1}{C_1} + \frac{1}{C_2}\right) \tag{11-24}$$

For a sphere, $C_1 = C_2 = r$ and Eq. (11-24) reduces to Eq. (11-23); for a plane surface, $C_1 = C_2 = \infty$ and $P^{\beta} = P^{\alpha}$. This latter case yields the pressure equality we found earlier when we studied phase equilibrium for planar surfaces.

11.3 Pressure Differences Across Curved Interfaces

For an equilibrium system of a bulk phase α containing small fragments of a different phase β, we have shown that the pressure of phase β exceeds that of α. If β exists as a spherical drop or bubble, the pressure difference is given by Eq. (11-23). If β has a different curvature, we can always revert to the general case as shown by Eq. (11-24). These relations show how the pressure difference depends on the size and shape of the fragment of phase β.

One can, however, express this pressure difference in another way. At the system temperature, we will compare two α-β systems: in the first, phase β will be assumed to exist in small fragments with pronounced curvature; in the second, we will assume that the interface between α and β is planar and then apply the criterion of pressure equality between phases. Although one may generalize this approach, in actuality, it is more convenient to consider each case separately. We will consider first the important and common situation in which small drops of liquid are in equilibrium with a vapor phase. Later, in Examples 11.2 and 11.3, we will extend the treatment to finely divided solids in a supersaturated liquid and to vapor bubbles in a superheated liquid.

In Figure 11.4 we show two closed systems for the liquid drop case. In system I, there is a bulk vapor phase at P_e in equilibrium with a liquid drop of radius r and pressure P_D. Both are at the same temperature T. In the nomenclature employed previously, $P_e = P^\alpha = P$ and $P_D = P^\beta$.

In system II, we also have a liquid and gas phase in equilibrium at temperature T. In this case, the pressure is equal in both phases and is, of course, the vapor pressure, P_{vp}. We designate the liquid and vapor phases by superscripts L and V and the systems by subscripts I and II. Since equilibrium is assumed to exist in both systems, from the results of Section 11.3, we can write:

$$\mu_{\text{I}}^L = \mu_{\text{I}}^V; \qquad \mu_{\text{II}}^L = \mu_{\text{II}}^V \tag{11-25}$$

Subtracting,

$$\mu_{\text{I}}^L - \mu_{\text{II}}^L = \mu_{\text{I}}^V - \mu_{\text{II}}^V \tag{11-26}$$

Since the temperatures are the same in I and II, the chemical potential differences depend only on the pressure, thus:

$$\int_{P_{vp}}^{P_D} V^L\,dP = \int_{P_{vp}}^{P_e} V^V\,dP = RT\ln\left(\frac{f_{P_e}}{f_{P_{vp}}}\right) \tag{11-27}$$

Let us make the reasonable assumption that the fugacity ratio in Eq. (11-27) is equal to the pressure ratio (P_e/P_{vp}). This approximation is probably quite good even for nonideal gases since the fugacity coefficients evaluated at P_e and P_{vp} are not greatly different. Also, let us assume that the liquid is incompressible. Then,

$$\begin{aligned} RT\ln\left(\frac{P_e}{P_{vp}}\right) &= V^L(P_D - P_{vp}) \\ &= V^L[(P_D - P_e) + (P_e - P_{vp})] \end{aligned} \tag{11-28}$$

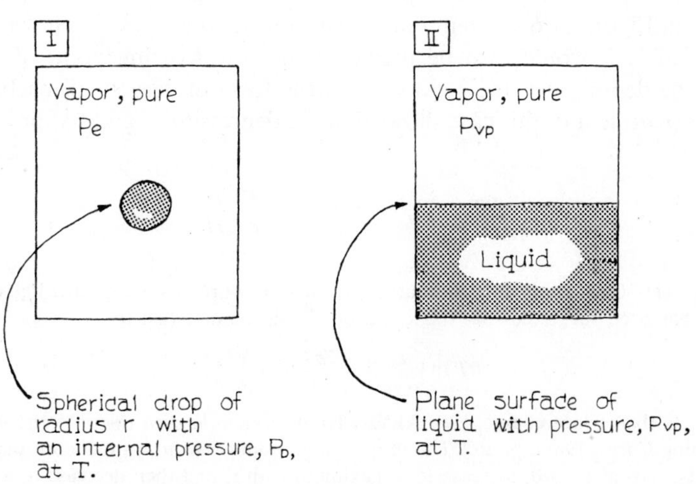

Figure 11.4 Vapor pressure of small drops.

Solving for $(P_D - P_e)$,

$$P_D - P_e = \left(\frac{RT}{V^L}\right) \ln \left(\frac{P_e}{P_{vp}}\right) - (P_e - P_{vp}) \tag{11-29}$$

In almost all cases, the second term on the right of Eq. (11-29) is negligible and can be dropped. Thus, if we call P_e/P_{vp} the saturation ratio, the pressure difference $(P_D - P_e)$ can be directly related to this ratio. For a saturation ratio of unity, $P_D = P_e$ and we simply have system II of Figure 11.4. One can also combine Eq. (11-29) with Eq. (11-23) to obtain:

$$P_e = P_{vp} \exp \left(\frac{2\sigma V^L}{rRT}\right) \tag{11-30}$$

which shows that the pressure of a pure vapor in equilibrium with a small liquid drop is larger than the vapor pressure.[2]

Equation (11-30) allows one to plot the equilibrium vapor pressures of small drops as a function of temperature and drop radii. Such a plot is shown for water between 0 and 100°C in Figure 11.5. For example, at 40°C, the vapor pressure of a plane water surface is 55.3 mm Hg; for spherical drops of radii 10^{-6} and 10^{-7} cm, the calculated vapor pressures are about 60.6 and 147 mm Hg, assuming $\sigma \neq f(r)$. Figure 11.5 may also be interpreted another way. If the partial pressure of water vapor were reported to be, for example, 100 mm Hg, then this vapor would be in equilibrium with a planar water surface at about 52°C. If nucleation occurred on foreign particles of radius 10^{-6} cm, condensation should not occur until a temperature of about 50°C were reached. Likewise, for drops of 10^{-7} cm, the temperature would be only 32°C.

Figure 11.5 is only a small portion of a more general type of plot as is shown in Figure 11.6 wherein liquid, gas, and solid phases are shown. From Figure 11.6, it would also be predicted that the freezing point of a solid would be depressed if the solid were in the form of very small particles. A similar treatment of this case shows that the depression is given approximately by:

$$\Delta T = T_r - T_m = -\frac{2V^S \sigma T_m}{r\, \Delta H_f} \tag{11-31}$$

[2] Eq. (11-30) is applicable to neutral drops. If the drops were electrically charged, then it has been suggested that this equation should be modified to

$$RT \ln \left(\frac{P_e}{P_{vp}}\right) = \frac{2\sigma V^L}{r} - \frac{V^L e^2}{8\pi r^4}$$

where e is the surface charge. [See A. Van Hook, *Crystallization* (New York: Reinhold Publishing Corp., 1961), p. 26.] The second term would lead to a prediction that P_e/P_{vp} would be zero at $r \longrightarrow 0$, increase to a maximum with r, and then decrease to a limit of unity at large r.

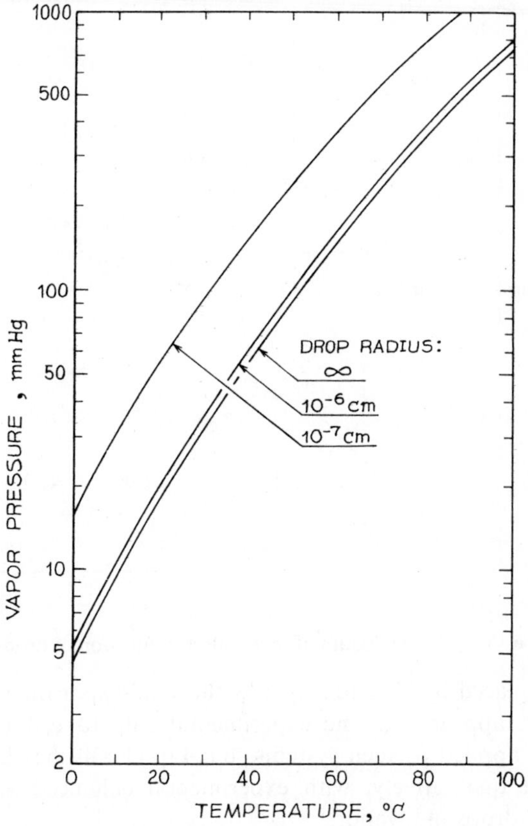

Figure 11.5 Vapor pressure of water as a function of drop size.

where

T_r = freezing point of a drop of radius r at P
T_m = freezing point of a planar surface at P
V^S = molal volume of the solid at T_m, P
σ = solid-liquid interfacial tension
r = particle radius
ΔH_f = heat of fusion

The derivation of Eq. (11-31) is left as a problem at the end of the chapter. Numerical values of ΔT are difficult to obtain since the solid-liquid interfacial tension is not usually known. It might, however, be interesting to test the equation using a system where ΔH_f is exceptionally small (e.g., with a material such as 2,2-dimethylbutane where ΔH_f is only 1.61 cal/g.

It is also interesting to speculate whether or not small, solid particles really do have an enhanced vapor pressure as predicted by Eq. (11-30)

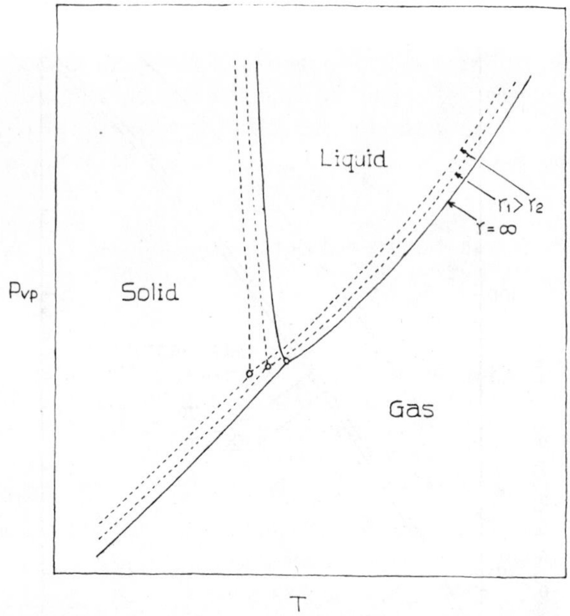

Figure 11.6 Phase diagram of materials as a function of drop size.

where V^L is replaced by V^S and σ is now the solid-vapor interfacial tension. Thus far, there appear to be no experimental data to test the validity of Eq. (11-30) as applied to such systems, but Eq. (11-30) has been shown to agree, at least qualitatively, with experimental enhanced vapor pressure data for liquid drops in a vapor.[3]

Thermodynamics also indicates that the solubility of very small particles should be higher than that for large particles [see Eq. (11-32) in Example 11.2]. This premise, however, was criticized by Harbury[4] who considered the predictive result as unsatisfactory except for particles in the 10-100 A range *and* if the solid-liquid interfacial tension were replaced by an empirical parameter which is numerically much less than typical values measured by more direct methods. Use of a very small σ in Eq. (11-32) leads, of course, to only nominal increases in solubility even for the extremely small particle sizes noted above.

Example 11.2

Derive an expression to relate the solubility of small fine solid particles to the solubility of chemically identical solids with large planar surfaces. Comment on the the assumptions made in the derivation.

[3] V. K. LaMer and Ruth Gruen, "A Direct Test of Kelvin's Equation Connecting Vapor Pressure and Radius of Curvature," *Trans. Faraday Soc.*, **48**, (1952) 410.

[4] Lawrence Harbury, "Solubility and Melting Point as Functions of Particle Size," *J. Phys. Chem.*, **50**, (1946) 190.

Solution

Let II be the equilibrium system consisting of the planar surface and solution. Another equilibrium system, at the same temperature and pressure, but with small particles in the solution is designated as system I. If L represents solution, S, the solid, then

$$\mu_{\text{I}}^{L} = \mu_{\text{I}}^{S}, \qquad \mu_{\text{II}}^{L} = \mu_{\text{II}}^{S}$$

for the component that is present both in the solid and solution. Subtracting these two identities,

$$\mu_{\text{I}}^{L} - \mu_{\text{II}}^{L} = \mu_{\text{I}}^{S} - \mu_{\text{II}}^{S}$$

The term on the right-hand side represents the change in chemical potential of a solid between small solid particles and large bulk crystals. If the particles are modeled as spheres of radius r, then this difference is approximated, as before, as $\int V^S\,dP = 2\sigma V^S/r$.

The term on the left refers to a chemical potential difference (for the component comprising the solid) in two solutions at different concentrations.

$$\mu_{\text{I}}^{L} - \mu_{\text{II}}^{L} = RT\ln\frac{a_{\text{I}}}{a_{\text{II}}} \sim RT\ln\frac{C_{\text{I}}}{C_{\text{II}}}$$

Since C_{I} is the solubility in the presence of small particles, C, and C_{II} is the solubility normally reported for large crystals, C_0, then

$$C = C_0\exp\left(\frac{2\sigma V^S}{rRT}\right) \tag{11-32}$$

Of course, if the solid ionizes during dissolution, then the exponential argument should be divided by i, the increase in the number of particles during dissolution and electrolytic dissociation.

Eq. (11-32) may be criticized from many points of view. Small particles are probably not spherical and a departure from this simple shape can considerably affect C/C_0 values.[5] Also, the correspondence between concentration and activity may not be acceptable for highly soluble compounds. Finally, and most important, it is probably not reasonable to associate σ with a solid-liquid interfacial tension, since, for small aggregates of only a few molecules, this interfacial tension loses any meaning. As was noted above, Harbury suggests that a fictitious interfacial tension, σ', be used where, from the few reliable data available, σ' is much less than σ.

Example 11.3

Following the same approach as utilized to derive Eq. (11-30), determine the pressure inside a small bubble of vapor that is in equilibrium with a mass of superheated liquid. Assume a one-component system.

[5] W. J. Jones, "Uber die Beziehung zwischen geometrischer Form und Dampfdruck, Löslichkeit, und Formenstabilität," *Ann. Physik*, **41**, (1913) 441.

Solution

Eq. (11-26) is still applicable but now, in the integration,

$$\int_{P_{vp}}^{P_e} V^L\,dP = \int_{P_{vp}}^{P_D} V^V\,dP$$

where P_e is the pressure of the superheated liquid and P_D is the pressure inside the bubble of vapor. P_{vp} is the equilibrium vapor pressure at the system temperature. Integrating, assuming ideal gases and incompressible liquids,

$$V^L(P_e - P_{vp}) = RT \ln\left(\frac{P_D}{P_{vp}}\right)$$

or

$$P_D = P_{vp} \exp\left[\frac{V^L(P_e - P_{vp})}{RT}\right] = P_{vp} \exp\left(\frac{2\sigma V^L}{rRT}\right)$$

Thus, the case of a vapor bubble in a superheated liquid leads to the same result as a drop of liquid in a supersaturated vapor, e.g., Eq. (11-30).

11.4 Pure Component Relations

The Fundamental Equations for any two phases α and β, separated by a surface phase σ, have already been given in Eqs. (11-12) through (11-14). Let us assume that the interface between α and β is not highly curved so that $P^\alpha = P^\beta = P^\sigma$ and begin our development by considering a closed system in which there is but a single nonreacting component present. We have also shown that, at equilibrium, $T^\alpha = T^\beta = T^\sigma = T$ and $\mu_j^\alpha = \mu_j^\beta = \mu_j^\sigma = \mu_j$. The total Legendre transform of Eqs. (11-12) through (11-14) yields the Gibbs-Duhem equations [see Eq. (5-69)]. For the α and β phases,

$$-S^\alpha\,dT + V^\alpha\,dP - d\mu = 0 \tag{11-33}$$

$$-S^\beta\,dT + V^\beta\,dP - d\mu = 0 \tag{11-34}$$

For the surface phase, in extensive form,

$$-\underline{S}^\sigma\,dT + \underline{V}^\sigma\,dP - A\,d\sigma - N^\sigma\,d\mu = 0 \tag{11-35}$$

and we divide Eq. (11-35) by A to obtain:

$$-S^\sigma\,dT + \tau\,dP - d\sigma - \Gamma\,d\mu = 0 \tag{11-36}$$

In Eq. (11-36), $S^\sigma = \underline{S}^\sigma/A$ is the entropy per unit area. Likewise, $\tau = \underline{V}^\sigma/A$ is the thickness of the surface and $\Gamma = N^\sigma/A$ is a surface concentration, moles per unit area.

Eliminating dP and $d\mu$ from Eqs. (11-33), (11-34), and (11-36),

$$-\left(\frac{d\sigma}{dT}\right) = (S^\sigma - \Gamma S^\beta) - \left[\frac{(\tau - \Gamma V^\beta)(S^\alpha - S^\beta)}{(V^\alpha - V^\beta)}\right] \tag{11-37}$$

Eq. (11-37) is symmetrical in α and β and terms containing these superscripts may be interchanged. The Euler form of Eq. (11-14) may be written as:

$$\sigma = U^\sigma - TS^\sigma + P\tau - \Gamma\mu \tag{11-38}$$

If Eq. (11-37) is multiplied by T and the result added to Eq. (11-38) and then the Euler form of Eq. (11-13), as multiplied by Γ, is used to eliminate the $\Gamma\mu$ term, there results:

$$\sigma - T\frac{d\sigma}{dT} = (U^\sigma - \Gamma U^\beta) - \left[\frac{(\tau - \Gamma V^\beta)(U^\alpha - U^\beta)}{(V^\alpha - V^\beta)}\right] \tag{11-39}$$

where $U^\sigma \equiv \underline{U}^\sigma/A$

Eqs. (11-37) and (11-39) relate the surface tension and surface tension-temperature gradient to thermodynamic properties such as entropy, energy, and volume. It is not immediately obvious, however, that these relations are invariant with respect to the variations in thickness, τ, of the surface. Suppose for instance, that τ is extended by a slight amount into phase β, then

$$\frac{dS^\sigma}{d\tau} = \frac{S^\beta}{V^\beta}$$

$$\frac{d\Gamma}{d\tau} = \frac{1}{V^\beta}$$

Differentiating say, Eq. (11-37) with respect to a variation in τ into phase β, the right-hand side becomes zero. Thus, the relation is not dependent on this particular variation in τ. Similar results are obtained if τ is increased into phase α.

This invariance leads, of course, to an indefiniteness about terms such as S^σ, U^σ, τ, Γ. As noted above, for a given τ, there is a given S^σ. If τ is increased in phase β, per unit area, the entropy will increase by $S^\beta(d\tau/V^\beta)$. This new S^σ will depend on the value of S^β to a degree proportionate to the increase in τ. We are accustomed to visualizing the properties of the surface layer to be more representative of the less mobile phase or of the discontinuous phase in a two-liquid system. For example, in a liquid-gas system, we often imagine the surface phase to possess nearly liquid-like properties. (Note that this is always possible by choosing the position of the surface layer in a manner to exclude any appreciable gas phase.) By choices of this nature, we may simplify Eqs. (11-37) and (11-39) to more common forms. For example, in a gas-liquid system, if phase β were liquid and phase α gas, then τ/Γ is comparable to $V^\beta = V^L$ which is much less than $(V^\alpha - V^L) = (V^V - V^L)$. Eqs. (11-37) and (11-39) then become:

$$-\frac{d\sigma}{dT} = (S^\sigma - \Gamma S^L) \tag{11-40}$$

$$\sigma - T\left(\frac{d\sigma}{dT}\right) = (U^\sigma - \Gamma U^L) \tag{11-41}$$

Even in this simple form, S^σ, U^σ, and Γ are still indefinite and depend on τ, although neither $(S^\sigma - \Gamma S^L)$ nor $(U^\sigma - \Gamma U^L)$ depends on τ.

Surface tension decreases with temperature and a convenient dimensionless estimation equation for nonpolar materials is given by Brock and Bird.[6]

$$\sigma^* = (1 - T_r)^{11/9} \tag{11-42}$$

where $\sigma^* = \sigma / P_c^{2/3} T^{1/3}$ $(0.113\alpha_c - 0.281)$, where σ is in dyne/cm, P_c in atm, and T_c in °K. α_c is the Riedel factor and is well-represented by Eq. (11-43) where $T_{b_r} = T_b/T_c$ and T_b is the normal boiling point.

$$\alpha_c = 0.9076\left[\frac{1 + T_{b_r} \ln P_c}{(1 - T_{b_r})}\right] \tag{11-43}$$

At the critical point $\sigma = 0$.

Since $d\sigma/dT$ is negative, $(S^\sigma - \Gamma S^L)$ is positive. This quantity represents the difference between the entropy (per unit area) in the surface film and the entropy the same surface would have if it had the properties of the bulk liquid phase. In a similar manner, $\sigma - T(d\sigma/dT)$ may be visulaized as the internal energy difference between the surface film and bulk liquid. Combining these two concepts with the elementary equation $\Delta U = T\,\Delta S - W_{\text{rev}}$, it is easily shown that $-W_{\text{rev}} = \sigma\, dA$ (i.e., the surface tension in a liquid-vapor system is representative of the reversible work to bring material from the bulk liquid to form a unit area of interface). This work term is then indicative of the ease of varying the surface area of a liquid phase in contact with its vapor. Since the minimum, isothermal work is often shown as a Gibbs free energy change, σ is commonly called the Gibbs surface free energy per unit area.

Before leaving this section on pure component surface thermodynamics, it is instructive to review the developments presented thus far. In the initial sections of this chapter, from mechanical considerations only, the concept of surface tension was developed. These same arguments led to some interesting conclusions such as the pressure difference existing across curved interfaces, the enhanced vapor pressure over small convex liquid or solid surfaces, etc. The application of thermodynamics in this development was minimal although it was proved in any equilibrium-multiphase system, that regardless of curvature, the temperature and chemical potentials were equal in all phases.

From these concepts, the surface tension and its temperature gradient were related to fundamental thermodynamic properties of the surface. The utility of this latter development may be questioned. The surface layer emphasized in the foregoing paragraphs is held by some to be completely fictitious and devoid of true physical significance; its depth is indefinite, it is a com-

[6] J. R. Brock and R. B. Bird, "Surface Tension and the Principle of Corresponding States," *AIChE J.*, **1**, (1955), 174.

posite of the properties of the bounding phases, and the absolute values of S^σ, U^σ, etc., are dictated solely by the exact position and thickness of the layer—as defined by the user—not by nature.

11.5 Multicomponent Relations

The thermodynamics of surface layers in multicomponent systems is a straightforward extension from those developed in Section 11.5 for single components although the algebra is considerably more complex. The relations employed are shown below; as before, the surface layer equations are expressed in terms of a unit area of surface (i.e., $S^\sigma = \underline{S}^\sigma/A$; $\tau = \underline{V}^\sigma/A$; $\Gamma_j = N_j^\sigma/A$). The Gibbs-Duhem relation, Eq. (11-36), becomes:

$$-d\sigma = S^\sigma\, dT - \tau\, dP + \sum_{k=1}^{n} \Gamma_k\, d\mu_k \tag{11-44}$$

For both phases α and β one may express the chemical potential of any component j as, for example in phase β,

$$d\mu_j = -\bar{S}_j^\beta\, dT + \bar{V}_j^\beta\, dP + \sum_{k\neq 1} \left(\frac{\partial \mu_j}{\partial x_k^\beta}\right)_{T,\,P,\,x[i,\,k]} dx_k^\beta \tag{11-45}$$

At *constant composition*, in the β and α phases, with Eqs. (11-44) and (11-45),

$$-d\sigma = \left(S^\sigma - \sum_{k=1}^{n} \Gamma_k \bar{S}_k^\beta\right) dT - \left(\tau - \sum_{k=1}^{n} \Gamma_k \bar{V}_k^\beta\right) dP \tag{11-46}$$

and from equations of the form of Eq. (11-45) for any component k, with all $dx = 0$,

$$dP = \frac{\bar{S}_k^\alpha - \bar{S}_k^\beta}{\bar{V}_k^\alpha - \bar{V}_k^\beta}\, dT; \qquad \text{all } x^\beta \text{ constant} \tag{11-47}$$

Eq. (11-47) may be substituted into Eq. (11-46) to eliminate all pressure terms. Applying Eq. (11-46) to a vapor-liquid system where the vapor phase is α and the liquid phase is β, $\tau/\Sigma\Gamma_k$ is comparable to the liquid phase volume $V^L = \sum_{k=1}^{n} x_k^L \bar{V}_k^L$. Assuming ideal solutions where $\bar{V}_k^L = V_k^L$,

$$\tau - \sum_{k=1}^{n} \Gamma_k \bar{V}_k^L \simeq \tau - V^L \sum_{k=1}^{n} \Gamma_k \simeq \sum_{k=1}^{n} \Gamma_k \left(\frac{\tau}{\Sigma\Gamma_i} - V_k^L\right) \sim 0 \tag{11-48}$$

Thus, in this case,

$$-\left(\frac{\partial \sigma}{\partial T}\right)_L = \left(S^\sigma - \sum_{k=1}^{n} \Gamma_k \bar{S}_k^L\right) \quad \text{(constant composition in } L \text{ phase)} \tag{11-49}$$

Eq. (11-48) is a good approximation to eliminate the "dP" term since the small residue $\Sigma\Gamma_i \bar{V}_i^L$ must still be divided [as from Eq. (11-47)] by a large number, $(\bar{V}_k^V - \bar{V}_k^L) \sim \Delta V_{\text{vaporization}}$, to reduce further the magnitude of the multiplier of the "dP" term. The form of Eq. (11-49) is identical to Eq.

(11-40) and the discussion following the latter equation is also applicable to Eq. (11-49). That is, the equation is invariant with respect to the position or thickness of the surface layer as long as each bounding surface of the interfacial layer is in a different phase.

There is, however, a limitation to either Eq. (11-45) or (11-46). That is, the composition of all components in the liquid phase must be held constant. This restriction originates from the elimination of the $d\mu_j$ terms in Eq. (11-44) with Eq. (11-45) as simplified by forcing the terms containing $(\partial\mu_j/\partial x_k^L)\,dx_k^L$ to be zero. There is no thermodynamic inconsistency here since from the phase rule there are n independent variables and if $(n-1)$ compositions in the liquid phase are fixed, then σ can be expressed as a function of temperature only. Similar statements may be made if the derivation of Eq. (11-46) were made by using a relation of the form of Eq. (11-45) but for the α phase. The final resulting equation (11-46) or (11-49) does have this composition restriction and as such, the general applicability is reduced. As in the case of many other multicomponent thermodynamic relations, other composition restrictions may be imposed and different results obtained.

11.6 Variation of Surface Tension with Composition

To illustrate how the surface tension is related to composition, consider a two-phase, two-component system. Suppose we desire to determine $(\partial\sigma/\partial x_1^\beta)_{T,P}$. From Eqs. (11-44) and (11-45),

$$-\left(\frac{\partial\sigma}{\partial x_1^\beta}\right)_{T,P} = \Gamma_1\left(\frac{\partial\mu_1}{\partial x_1^\beta}\right)_{T,P} - \Gamma_2\left(\frac{\partial\mu_2}{\partial x_2^\beta}\right)_{T,P}$$

but, from the Gibbs-Duhem equation,

$$\left(\frac{\partial\mu_2}{\partial x_2^\beta}\right)_{T,P} = \left(\frac{x_1^\beta}{x_2^\beta}\right)\left(\frac{\partial\mu_1}{\partial x_1^\beta}\right)_{T,P}$$

and by definition, $\mu_j = RT\ln\hat{f}_j + \lambda_j(T)$, so

$$-\left(\frac{\partial\sigma}{\partial x_1^\beta}\right)_{T,P} = \left(\Gamma_1 - \frac{x_1^\beta}{x_2^\beta}\Gamma_2\right)RT\left(\frac{\partial\ln\hat{f}_1}{\partial x_1^\beta}\right)_{T,P} \tag{11-50}$$

Expressing Eq. (11-50) in terms of activity coefficients, $\hat{f}_j = f_j\gamma_j x_j$, then

$$-\left(\frac{\partial\sigma}{\partial x_1^\beta}\right)_{T,P} = \left(\Gamma_1 - \frac{x_1^\beta}{x_2^\beta}\Gamma_2\right)RT\left[\frac{1}{x_1^\beta} + \left(\frac{\partial\ln\gamma_1^\beta}{\partial x_1^\beta}\right)\right]_{T,P} \tag{11-51}$$

$$= RT\left(\frac{\Gamma_1}{x_1^\beta} - \frac{\Gamma_2}{x_2^\beta}\right) \qquad \text{(ideal solution)} \tag{11-52}$$

Several points should be noted about the equations derived above for the variation of surface tension with composition. First, as noted, absolute values of Γ_1 and Γ_2 do not exist. The term $[\Gamma_1 - (x_1^\beta/x_2^\beta)\Gamma_2]$ is, however, invariant with respect to variations in the surface layer thickness in the β

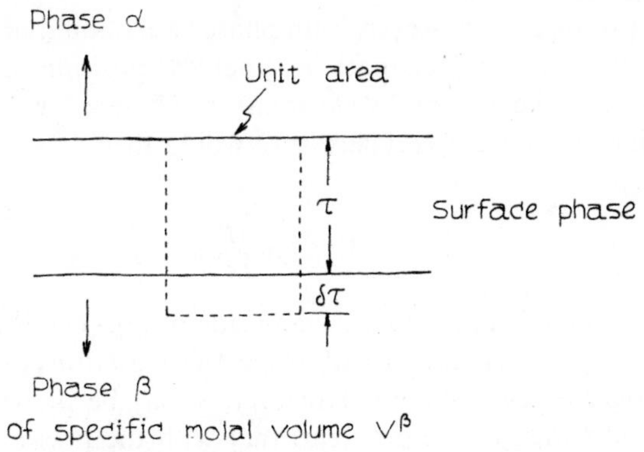

Figure 11.7 Surface phase increase into phase β.

hase. This is illustrated in Figure 11.7. The original moles in the surface ıyer, per unit area, are equal to $\mathbf{\Gamma}_{1_0} + \mathbf{\Gamma}_{2_0}$. With an increase in thickness τ into phase β, the moles in the surface layer become:

$$\mathbf{\Gamma}_{1_f} + \mathbf{\Gamma}_{2_f} = \mathbf{\Gamma}_{1_0} + x_1^\beta \frac{\delta\tau}{V^\beta} + \mathbf{\Gamma}_{2_0} + x_2^\beta \frac{\delta\tau}{V^\beta}$$

'he relationship $[\mathbf{\Gamma}_1 - (x_1^\beta/x_2^\beta)\mathbf{\Gamma}_2]$ is, therefore, invariant for a positive or egative value of $\delta\tau$ into the β phase.

For a variation $\delta\tau$ of the surface layer into phase α the term $[\mathbf{\Gamma}_1 -$ $(x_1^\beta/x_2^\beta)\mathbf{\Gamma}_2]$ is not invariant. Suppose τ is increased by $\delta\tau$ into phase α. Then ıe difference between the final state and the initial state is:

$$\left(\mathbf{\Gamma}_1 - \frac{x_1^\beta}{x_2^\beta}\mathbf{\Gamma}_2\right)_{\text{final}} - \left(\mathbf{\Gamma}_1 - \frac{x_1^\beta}{x_2^\beta}\mathbf{\Gamma}_2\right)_{\text{initial}} = \frac{\delta\tau}{V^\alpha}\,\frac{x_1^\alpha - x_1^\beta x_2^\alpha}{x_2^\beta}$$

'his term is usually considered neglible if phase α is a vapor since V^α is then large number. It may not be negligible if phase α were a liquid phase nmiscible with β. To obtain an expression for $\partial\sigma/\partial x$ that is invariant with ariations of τ into phase α, Eq. (11-51) may be used with all β superscripts eplaced by α; of course, this relation is then not invariant with respect to hase β.

These conclusions are not at all satisfactory since there should be a defiite experimental value of $\partial\sigma/\partial x$ that does not depend on the arbitrary efinition of τ. One of the troubles lies in the fact that we have held both ressure and temperature constant for a two-phase, two-component system nd still allowed a variation in composition in the phases. This procedure iolates the phase rule that allows only two independent variables. We followd the procedure given above to obtain those relations that are commonly ound in other texts and in the literature. Either Eq. (11-50) or Eq. (11-51) is, s we have seen, probably an excellent approximation if α were a gas and β

a liquid at low pressure. If, however, both phases were liquid or if we desired to obtain variations in σ with changes in vapor composition, then the equations would have to be rederived to obtain either $(\partial\sigma/\partial x)_T$ or $(\partial\sigma/\partial x)_P$. The former might be more useful. Pressure would then vary and a term of the form of

$$-(\tau - \Sigma\Gamma_j \bar{V}_j^\beta)\left(\frac{\partial P}{\partial x_1^\beta}\right)_T$$

would have to be included in the right-hand side of Eqs. (11-50) and (11-51). The new term would be invariant with respect to variations of τ in phase β but not for variations into phase α. However, it may be shown (or inferred from the physical nature of the problem) that with the pressure term given above, the $(\partial\sigma/\partial x)_T$ equations are invariant with respect to changes in τ into either phase.

As a final comment in this section, it should be noted that with experimental data showing how the surface tension of a liquid varies with composition and how the fugacity (or partial pressure) of a component similarly varies, Eq. (11-50) allows one to make the following rearrangements:

$$\Psi \equiv -\frac{\left(\frac{\partial\sigma}{\partial x_1}\right)_T}{RT\left(\frac{\partial \ln p_1}{\partial x_1}\right)_T} = \Gamma_1 - \frac{x_1}{x_2}\Gamma_2 \tag{11-53}$$

The term x in Eq. (11-53) refers to the mole fraction in the bulk liquid. Furthermore, if we recall that Γ represents the moles per unit area in the surface layer, then for this unit area, there should be some molecular "parking area" for each component such that:

$$\Gamma_1\bar{A}_1 + \Gamma_2\bar{A}_2 = 1 \tag{11-54}$$

where the parking areas for 1 and 2, $\bar{A}_1$ and $\bar{A}_2$, have been more aptly termed *partial areas* by Guggenheim.[7] They may often be approximated as the molecular cross section. Solving Eqs. (11-53) and (11-54) for Γ_1 and Γ_2, one may obtain an estimate of the "surface" concentrations z, i.e.,

$$z_1 \equiv \frac{\Gamma_1}{\Gamma_1 + \Gamma_2} = \frac{g + \bar{A}_2\Psi}{(1+g) + \Psi(\bar{A}_2 - \bar{A}_1)} \tag{11-55}$$

$$z_2 \equiv \frac{\Gamma_2}{\Gamma_1 + \Gamma_2} = \frac{1 - \bar{A}_1\Psi}{(1+g) + \Psi(\bar{A}_2 - \bar{A}_1)} \tag{11-56}$$

$$g \equiv \frac{x_1}{x_2}$$

Surface concentrations might be more useful parameters when one is dealing with the surface layer although Eq. (11-54) has certainly not been proved

[7] E. A. Guggenheim, *Thermodynamics*, 4th ed. (Amsterdam: North-Holland Publishing Company, 1959), pp. 194–201, 261–74, 296–99.

and values of $\bar{A}_1$ or $\bar{A}_2$ cannot be determined in an *a priori* manner. Eqs. (11-55) and (11-56) are illustrated in the problems at the end of the chapter.

11.7 Nucleation

Nucleation refers to the birth process of a new phase. It obviously has many important ramifications in science and engineering.

As is well-known, if a homogeneous, particle-free, pure vapor is cooled until the pressure is equal to the vapor pressure, thermodynamics would indicate that a liquid (or solid) phase could form. The actual facts show that this does not occur. In such a case, the temperature must be decreased (and often by a considerable amount) below the normal dew-point temperature before the second phase appears. Such undercooling is, however, often not found if the gas contains many "dust-like" particles or if the gas were "seeded" with a small quantity of the liquid (or solid) phase when the dew point was reached. Similar statements apply to the precipitation of solids from liquids or the freezing of liquids.

The reason for these phenomena lies in the fact, previously developed in Section 11.4, that small "nuclei" of the new phase have higher vapor pressures (or more accurately, higher chemical potentials) than a bulk, planar phase. These nuclei will only be stable when present in a supersaturated mother phase.

The equilibrium and stability criteria that are applicable to the formation of a new phase have already been given in Eqs. (11-9) and (11-10), that is, at equilibrium, the availability function, $\mathfrak{V}$, is either a minimum for stable equilibrium or a maximum for unstable equilibrium.

Let us assume that we have initially a β-free phase, α, which is always at T and P as dictated by isothermal and isobaric reservoirs (see Figure 11.1). We shall allow the formation of a β phase but at the same time insist that phase α remain at T and P. The moles in the α-β system are constant. We wish to determine the change in availability during this process. From the definition of $\mathfrak{V}$ in Eq. (11-8),

$$\Delta\mathfrak{V} = \mathfrak{V} - \mathfrak{V}_i \tag{11-57}$$

$$\begin{aligned}\Delta\mathfrak{V} = {} & (\underline{U}^\alpha + P\underline{V}^\alpha - T\underline{S}^\alpha) + (\underline{U}^\beta + P\underline{V}^\beta - T\underline{S}^\beta) \\ & + (\underline{U}^\sigma + P\underline{V}^\sigma - T\underline{S}^\sigma) - (\underline{U}_i^\alpha + P\underline{V}_i^\alpha - T\underline{S}_i^\alpha)\end{aligned} \tag{11-58}$$

Now,

$$N_i^\alpha = N^\alpha + N^\beta + N^\sigma \tag{11-59}$$

and,

$$\left.\begin{aligned} U^\alpha &= TS^\alpha - PV^\alpha + \mu^\alpha \\ U^\beta &= TS^\beta - P^\beta V^\beta + \mu^\beta \\ U^\sigma &= T\frac{\underline{S}^\sigma}{N^\sigma} - P^\sigma \frac{\underline{V}^\sigma}{N_\sigma} + \sigma\frac{A}{N^\sigma} + \mu^\sigma \end{aligned}\right\} \tag{11-60}$$

so that substitution of Eqs. (11-59) and (11-60) into Eq. (11-58) yields

$$\Delta\underline{\mho} = N^\beta[(\mu^\beta - \mu^\alpha) + (P - P^\beta)V^\beta] + \sigma A + N^\sigma[(\mu^\sigma - \mu^\alpha) + (P - P^\sigma)V^\sigma] \tag{11-61}$$

Eq. (11-61) is, of course, very similar to Eq. (11-15); the principal difference is that in the former we assumed that the temperatures were everywhere equal in the final system. A finite, rather than infinitesimal change was also proposed.

If we proceed one step further and assume that the final state is an equilibrium state, then $\Delta\underline{\mho}$ equals the change in availability when a homogeneous phase α changes to an α-β system with small fragments of the β phase in equilibrium with the residual α phase. Also, at this terminal state, as we have shown in Eqs. (11-19) and (11-20), there is equality of chemical potentials throughout the system so that Eq. (11-61) simplifies to:

$$\Delta\underline{\mho} = N^\beta(P - P^\beta)V^\beta + \sigma A \tag{11-62}$$

where we have also neglected the very small contribution due to $N^\sigma(P - P^\sigma)V^\sigma$. If phase β is a sphere, then from Eq. (11-23) with $A = 4\pi r^2$ and $N^\beta = (4/3)\pi r^3/V^\beta$,

$$\left.\begin{aligned} \Delta\underline{\mho} &= \frac{4}{3}\pi\sigma r^2 \\ &= \frac{\frac{16}{3}\pi\sigma^3}{(P^\beta - P)^2} \end{aligned}\right\} \tag{11-63}$$

Expressions for $(P^\beta - P)$ that were derived in Section 11.4 may be used to determine $\Delta\underline{\mho}$ in terms of P^α and other system variables.

Example 11.4

If a drop of liquid water were in equilibrium with water vapor at 100°C and 1.8 atm, estimate the size of the water drop and the pressure inside the drop. At 100°C, σ(water) = 58.9 dyne/cm and $V^L = 1.04$ cm³/g. What is the change in availability in the formation of this drop?

Solution

From Eq. (11-29), with $R = 82.06$ atm-cm³/g-mol °K, $T = 373$°K, $V^L = (1.04)(18)$ cm³/g-mol, $P_e = 1.8$ atm, $P_{vp} = 1$ atm, then

$$P_D - P_e = \left[\frac{(82.06)(373)}{(1.04)(18)}\right] \ln(1.8) - (1.8 - 1) = 961 - 0.8 = 960 \text{ atm}$$

(Note that the term $(P_e - P_{vp})$ was assumed small relative to the ln term in deriving Eq. (11-30); this example illustrates that this simplification is indeed reasonable.)

$$P_D \sim 961 \text{ atm}$$

Then from Eq. (11-23),

$$r = \frac{2\sigma}{P_D - P_e}$$

Now, $P_D - P_e = 961$ atm $= 97.3$ MN/m^2 and $\sigma = 58.9$ dyne/cm $=$ (58.9×10^{-3})N/m, thus

$$r = \frac{(2)(58.9)(10^{-3})}{(97.3)(10^6)} = 12.1 \times 10^{-10}\text{ m} = 12\text{ A}$$

From Eq. (11-63),

$$\Delta \underline{\mathcal{U}} = \frac{\left(\frac{16}{3}\right)\pi(0.0589)^3}{(97.3)^2(10^6)^2} = 3.6 \times 10^{-19}\text{ J}$$

Example 11.5

Suppose that a bubble of steam were in equilibrium with liquid water at 100°C. Let there be essentially a vacuum on the liquid phase. Using the data given in Example 11.4, estimate the pressure inside the bubble, the bubble size, and the change in availability in forming the bubble.

Solution

From the relation given in Example 11.3, with P_B the pressure inside the bubble, $P_{vp} = 1$ atm, and $P^L \sim 0$, then,

$$P_B = 1 \exp\left[\frac{(1.04)(18)(0 - 1)}{(82.06)(373)}\right] = \exp[-(6.1)(10^{-4})] \sim 1\text{ atm}$$

$$= 0.101\text{ MN/m}^2$$

$$r = \frac{2\sigma}{(P_B - P^L)^2} = \frac{(2)(58.9)(10^{-3})}{(0.101)(10^6)}\text{m} = \sim 10^4\text{ A}$$

$$\Delta \underline{\mathcal{U}} = \frac{\left(\frac{16}{3}\right)\pi(0.0589)^3}{(0.101)^2(10^6)^2} = 3.3 \times 10^{-13}\text{ J}$$

Returning now to the question of stability, we note from Eq. (11-62),

$$\Delta \underline{\mathcal{U}} = f(N^\beta, r) \tag{11-64}$$

since $\underline{V}^\beta$ and A depend on r. If we expand the availability in a Taylor series around the state of equilibrium,

$$\underline{\mathcal{U}} - \underline{\mathcal{U}}_{\text{eq}} = \mathcal{U}_r\,\delta r + \mathcal{U}_N\,\delta N^\beta + \mathcal{U}_{rr}\,\delta r^2 + 2\mathcal{U}_{rN}\,\delta r\,\delta N^\beta + \mathcal{U}_{NN}\,\delta N^{\beta^2} + \cdots \tag{11-65}$$

where we have adopted a short-hand notation for partial derivatives (e.g., $\mathcal{U}_r = (\partial\underline{\mathcal{U}}/\partial r)_{N^\beta}$, $\mathcal{U}_{NN} = (\partial^2\underline{\mathcal{U}}/\partial N^{\beta^2})_r$, etc.). The first-order terms are zero at equilibrium, i.e., from Eq. (11-61), neglecting the surface phase since it contributes negligibly, and noting that μ^α, P, and σ are constants,

$$\mathcal{V}_r = N^\beta\left(\frac{\partial \mu^\beta}{\partial r}\right)_{N^\beta} - \frac{4}{3}\pi r^3\left(\frac{\partial P^\beta}{\partial r}\right)_{N^\beta} + (P - P^\beta)(4\pi r^2) + 8\pi\sigma r \tag{11-66}$$

The first two terms cancel since:

$$\left(\frac{\partial \mu^\beta}{\partial r}\right)_{N^\beta} = V^\beta\left(\frac{\partial P^\beta}{\partial r}\right)_{N^\beta} = \frac{\frac{4}{3}\pi r^3}{N^\beta}\left(\frac{\partial P^\beta}{\partial r}\right)_{N^\beta} \tag{11-67}$$

Thus,

$$\mathcal{V}_r = (P - P^\beta)(4\pi r^2) + 8\pi\sigma r \tag{11-68}$$

and, as expected, *at equilibrium*, using Eq. (11-23),

$$\mathcal{V}_r = 0$$

Next, we will show that $\mathcal{V}_N$ is also zero at equilibrium. Beginning again with Eq. (11-61),

$$\mathcal{V}_N = (\mu^\beta - \mu^\alpha) + N^\beta\left(\frac{\partial \mu^\beta}{\partial N^\beta}\right)_r - \frac{4}{3}\pi r^3\left(\frac{\partial P^\beta}{\partial N^\beta}\right)_r \tag{11-69}$$

but

$$\left(\frac{\partial \mu^\beta}{\partial N^\beta}\right)_r = V^\beta\left(\frac{\partial P^\beta}{\partial N^\beta}\right)_r = \frac{\frac{4}{3}\pi r^3}{N^\beta}\left(\frac{\partial P^\beta}{\partial N^\beta}\right)_r \tag{11-70}$$

Therefore,

$$\mathcal{V}_N = \mu^\beta - \mu^\alpha = 0 \tag{11-71}$$

by Eq. (11-19). Next, considering the second-order terms, they may be rearranged by forming a sum-of-squares:

$$\underline{\mathcal{V}} - \underline{\mathcal{V}}_{\text{eq}} = \mathcal{V}_{NN}\left[\delta N^\beta + \left(\frac{\mathcal{V}_{rN}}{\mathcal{V}_{NN}}\right)\delta r\right]^2 + \frac{(\mathcal{V}_{NN}\mathcal{V}_{rr} - \mathcal{V}_{rN}^2)(\delta r)^2}{\mathcal{V}_{NN}} \tag{11-72}$$

We need to develop expressions for $\mathcal{V}_{NN}$, $\mathcal{V}_{rN}$, and $\mathcal{V}_{rr}$.

$\mathcal{V}_{NN}$: Beginning with Eq. (11-71) and differentiating with respect to N^β, with μ^α a constant,

$$\mathcal{V}_{NN} = \left(\frac{\partial \mu^\beta}{\partial N^\beta}\right)_r = V^\beta\left(\frac{\partial P^\beta}{\partial N^\beta}\right)_r \tag{11-73}$$

To obtain a more convenient expression for $\mathcal{V}_{NN}$, since phase β is at constant T,

$$P^\beta = f(N^\beta, \underline{V}^\beta) \tag{11-74}$$

Applying Euler's theorem,

$$0 = \underline{V}^\beta\left(\frac{\partial P^\beta}{\partial \underline{V}^\beta}\right)_{N^\beta} + N^\beta\left(\frac{\partial P^\beta}{\partial N^\beta}\right)_{V^\beta} \tag{11-75}$$

Let

$$\kappa_T = -\left(\frac{\partial \ln \underline{V}}{\partial P}\right)_{T,N} \tag{11-76}$$

Then, with Eqs. (11-75), (11-76), and $\underline{V}^\beta = (4/3)\pi r^3$,

$$\mathcal{V}_{NN} = \frac{\frac{4}{3}\pi r^3}{\kappa_T N^{\beta^2}} \tag{11-77}$$

$\mathcal{V}_{rr}$: Differentiating Eq. (11-68) with respect to r, and using Eq. (11-23),

$$\mathcal{V}_{rr} = -8\pi\sigma - \frac{16\pi\sigma^2}{(P^\beta - P)^2}\left(\frac{\partial P^\beta}{\partial r}\right)_{N^\beta} \tag{11-78}$$

But,

$$\left(\frac{\partial P^\beta}{\partial r}\right)_{N^\beta} = \left(\frac{\partial P^\beta}{\partial \underline{V}^\beta}\right)_{N^\beta}\left(\frac{\partial \underline{V}^\beta}{\partial r}\right)_{N^\beta} = -\frac{3}{\kappa_T r} \tag{11-79}$$

So,

$$\mathcal{V}_{rr} = \frac{24\pi\sigma}{(P^\beta - P)\kappa_T} - 8\pi\sigma = \frac{12\pi r}{\kappa_T} - 8\pi\sigma \tag{11-80}$$

$\mathcal{V}_{rN}$: Beginning with Eq. (11-68) and differentiating with respect to N^β,

$$\mathcal{V}_{rN} = -4\pi r^2\left(\frac{\partial P^\beta}{\partial N^\beta}\right)_r = -\frac{4\pi r^2}{\kappa_T N^\beta} \tag{11-81}$$

where Eqs. (11-75) and (11-76) have also been used.

Clearly, $\mathcal{V}_{NN}$ is positive since $\kappa_T > 0$. Thus, the stability of our system is determined solely by the coefficient $(\mathcal{V}_{NN}\mathcal{V}_{rr} - \mathcal{V}_{rN}^2)$ in Eq. (11-72).

$$\mathcal{V}_{NN}\mathcal{V}_{rr} - \mathcal{V}_{rN}^2 = \frac{\frac{4}{3}\pi r^3}{\kappa_T N^{\beta^2}}\left[\frac{12\pi r}{\kappa_T} - 8\pi\sigma\right] - \frac{16\pi^2 r^4}{\kappa_T^2 N^{\beta^2}} = \frac{(-8\pi\sigma)\left(\frac{4}{3}\right)(\pi r^3)}{\kappa_T N^{\beta^2}} \tag{11-82}$$

Also,

$$\frac{\mathcal{V}_{NN}\mathcal{V}_{rr} - \mathcal{V}_{rN}^2}{\mathcal{V}_{NN}} = -8\pi\sigma < 0 \tag{11-83}$$

The interpretation of this result is that a phase β embryo, in equilibrium with a mother phase α, is in a state of *unstable* equilibrium. The fact that $\mathcal{V}_{NN}$ and $\mathcal{V}_{rr}$ are both positive is not significant; as illustrated in Chapter 7, it is the higher order determinant that specifies the stability of a system.

In this examination, the formation of a new phase β from a mother phase α actually proceeds by increasing the availability of a system—a progression that is counter to our intuition and to usual trends in thermodynamics. Normally, systems seek a state of lowest availability. In a molecular sense, by fluctuations in density, small fragments of a new phase are formed from a mother phase α which is in a metastable condition. All these new fragments are unstable and disappear until a fragment of a critical size is formed. Then, and only then, can the fragment grow to form a bulk new phase β.

The change in availability to form an equilibrium-sized sphere of phase β [Eq. (11-63)] is often employed in nucleation theory wherein it is assumed that the rate of nucleation is proportional to $\exp(-\Delta\mathcal{V}/RT)$. An interesting study of the superheating of liquids which demonstrates the utility of the availability concept is summarized in the work of Apfel.[8]

PROBLEMS

11.1. Derive Eq. (11-30) for the system shown in Figure 11.4, except let the total pressure be fixed at P_T in both systems by the addition of an inert gas, insoluble in the liquid. In this case, what would happen if the boundary wall were removed?

11.2. In a distillation tray there is considerable spray of very fine liquid droplets. Since small droplets have higher vapor pressures than plane liquid-vapor interfaces, we would like to know whether or not the small spray mists have significantly different relative volatilities than plane interfaces. Take, for example, ethanol-water at 25°C. Using the data given below, estimate the relative volatility of alcohol-to-water for 0.1 micron drops and compare this to the volatility for plane surfaces.

(a) Liquid composition, 50 mole percent water.

(b) Planar vapor-liquid equilibrium at 25°C.

x(mole % water)	p(mm Hg) water	p(mm Hg) ethanol
0	23.75	0.0
10	21.7	17.8
20	20.4	26.8
30	19.4	31.2
40	18.35	34.2
50	17.3	36.9
60	15.8	40.1
70	13.3	43.9
80	10.0	48.3
90	5.5	53.3
100	0.0	59.0

(c) Surface tension of 50 mole percent ethanol-water solution at 25°C = 25.4 erg/cm^2.

(d) Densities of liquid at 25°C.

[8] R. E. Apfel, "Vapor Cavity Formation in Liquids," Tech. Memo. 62 (Cambridge, Mass.: Acoustics Research Laboratory, Harvard University, Feb. 1970); see also *J. Chem. Phys.*, **54**, 62 (1971).

Wt. % water	Density, g/cm^3
0	0.785
10	0.814
20	0.839
30	0.863
40	0.887
50	0.910
60	0.929
70	0.951
80	0.966
90	0.980
100	0.997

(e) Assume that the surface tension is independent of drop size.

(f) Assume that the partial molal volumes of both components in the liquid are independent of pressure.

11.3. In the expression for surface work, Eq. (11-1), the pressure-volume term is given as $P\,d\underline{V}$. If one were dealing with a curved surface, the internal pressure differs from the external pressure. Which pressure should one use in Eq. (11-1)—or does it make any difference? Why?

11.4. Derive Eq. (11-31) and state any assumptions made.

11.5. Assuming that curves such as shown in Figure 11.6 do exist, what is the general equation showing the approximate slope of a plot of ln P (equilibrium) versus $1/T$ for small droplets of a liquid? For water at 20°C where $\sigma = 72.75$ dyne/cm, $\Delta H_v = 10{,}000$ cal/g-mol, $V^L = 18$ cm^3/g-mol; for $r = 10^{-7}$ cm, what is the slope in °K? How does the heat of vaporization vary with drop size?

11.6. A Ph.D. thesis at Syracuse University carried out by Dr. Fernandez has studied the growth rates of water-ice thin platelets in a subcooled water-salt solution. It was postulated that the platelets grew by extension of scalloped edges of the crystal perpendicular to the basal plane. His physical model studied two-dimensional growth of a single scallop as shown. (See Figure P11.6.)

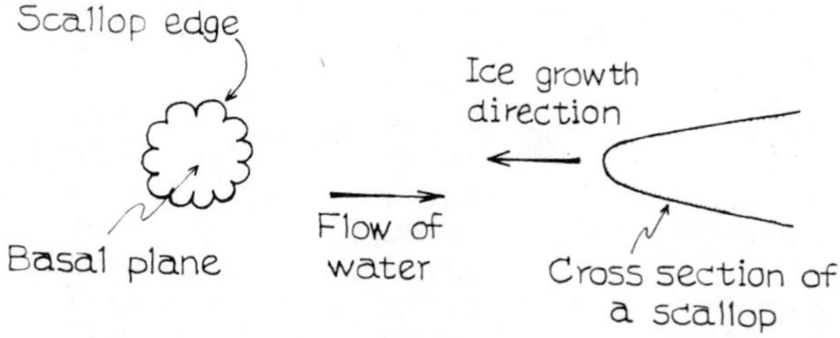

Figure P11.6

The geometric shape of a scallop is believed to be nearly parabolic, but the curvature near the tip is almost constant; thus, it is postulated that Eq. (11-31) may be used to estimate the tip surface temperature. In this case, Dr. Fernandez postulated that the temperature at the ice tip T_e is given by:

$$T_e = T_m - \frac{2\sigma T_m V^S}{\Delta H_f r}$$

where σ is the interfacial tension between solution and pure water-ice, T_m is the freezing point of pure water, 273.2°K, V^S is the molal volume of ice, ΔH_f is the heat of fusion of ice, and r is the tip curvature.

(a) Derive this relationship for a salt solution containing x weight percent salt. What approximations are necessary?

Assuming that the equation given above is applicable, let the growth process be controlled by the rate of heat transfer away from the tip, i.e.,

$$q = h(T_e - T_\infty) \quad \text{and} \quad q = \frac{R\,\Delta H_f}{V^S}$$

where q is the heat flux, h the heat transfer coefficient, T_∞ the bulk temperature in the salt bath, and R the linear growth rate of the tip.

The heat transfer coefficient for flow normal to a parabolic cylinder may be approximated as:

$$\frac{hr}{k} = A\left(\frac{Vr}{\nu}\right)^{1/2}$$

where r is the tip radius, k and ν the conductivity and kinematic viscosity of liquid water, V the impinging liquid velocity, and A is a constant.

(b) From these relations and the concept of a steady-state value of R during growth, determine how R depends on the experimental variable $\Delta T \equiv T_m - T_\infty$, e.g., does R double if ΔT is doubled?

In these calculations, assume no heat conduction through the ice.

11.7. The surface tension of very dilute aqueous solutions of butanol has been measured at 25°C and reported by Harkins and Wampler (*J. Am. Chem. Soc.*, **53**, (1931), 850). These authors also report the activities of dilute solutions as a function of molality as follows:

Molality	Activity*	Surface tension erg/cm^2
0.00329	0.00328	72.80
0.00658	0.00654	72.26
0.01320	0.01304	70.82
0.0264	0.02581	68.00
0.0536	0.05184	63.14
0.1050	0.09892	56.31
0.2110	0.19277	48.08
0.4330	0.37961	38.87
0.8540	0.71189	29.87

*Standard state, $a = m$ as $m \to 0$.

Estimate the mole fraction butanol in the interface at a bulk molality of 0.1050.

If you wish, you may assume that, on the average, one molecule of butanol occupies about 27 A^2 and one molecule of water, 7 A^2.

11.8. A spherical bubble is enclosed by a thin ethanol film; the bubble is 1 mm in diameter and the film is 1×10^{-4} mm thick. The bubble is in equilibrium with surrounding air which is at 1 atm and 25°C.

(a) Evaluate the surface free energy of the ethanol film, $(\partial \underline{U}^\sigma/\partial A)_{P,T}$, erg/cm^2.

(b) A small tube is inserted into the bubble and air (at 25°C) blown in so that it is expanded rapidly to a diameter of 3 mm. Neglect any heat or mass transfer from the air to the liquid film and assume that the temperature of the film is uniform throughout. What is the temperature in this film after the expansion?

Data:

For ethanol, the surface tension is believed to be insensitive to pressure variations but is related to temperature as

$$\sigma = 21.0\left(1 - \frac{T - 298}{250}\right) \qquad T \text{ in °K, } \sigma \text{ in dyne/cm}$$

The surface heat capacity is a constant.

$$C_A = T\left(\frac{\partial S}{\partial T}\right)_A = 0.65 \text{ cal/g °C}$$

The density of ethanol is assumed independent of temperature or pressure and equals 0.79 g/cm^3.

ΔH_v, ethanol = 400 Btu/lb at 25°C.

11.9. Consider the hypothetical system shown in Figure P11.9. A "microscopic cylinder" with piston is immersed in a constant temperature bath at $T_0 = 20$°C. In the space enclosed by the piston is a small quantity of helium gas (10^{-16} g-mol), 5 mole percent of which is contained within a bubble of radius r formed from a very thin film of liquid. The liquid has a surface tension of 40 dyne/cm at 20°C.

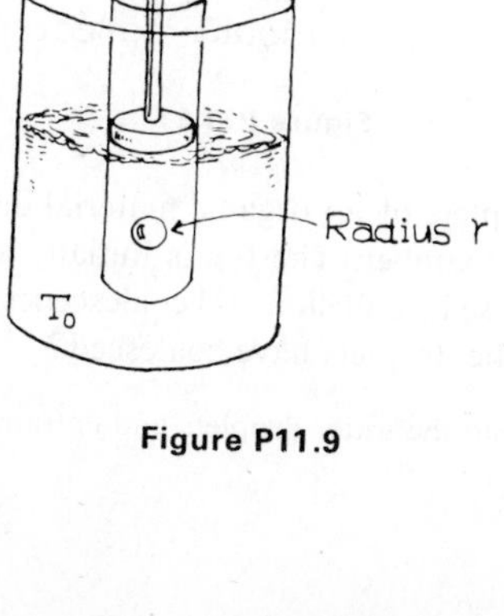

Figure P11.9

(a) Calculate equilibrium values of the bubble radius, internal and external pressures, and volume fraction gas inside the bubble for two cases:

Case A. The piston is held so that the total volume of the helium gas is constant at $V_0 = 10^{-12}$ cm^3.

Case B. The piston is released, permitting the external pressure to go to one atmosphere.

(b) Show in both cases that the equilibrium attained is stable with respect to isothermal variations in the bubble radius.

Assume that the surface tension is essentially a function of temperature only and the liquid has a negligible vapor pressure at 20°C. Helium is a perfect gas.

11.10. If two-dimensional nucleation occurs on a substrate and the nucleus is in the form of a circular pillbox, derive a relationship to estimate the critical radius of the pillbox to allow further growth to occur. Assume that the height of the pillbox remains constant.

11.11. It has been proposed to separate a mixture of liquid water and air by impinging it upon an inclined screen that is composed of a hydrophobic material.

The water droplets are larger than the screen openings and, if the flow is below a critical value, the liquid water is said not to penetrate the screen. Suppose that the screen were made of Teflon coated wire so that the wire thickness was about 0.0014 in. and the openings were square with sides 0.0017 in. (0.043 mm). This would correspond to a standard 325 mesh.

If the air and water are at atmospheric pressure and room temperature, what would be your best approximation as to the velocity of the stream* above which breakthrough might result?

Assume that the contact angle between a Teflon, water, and air interface is 105°. The angle ϕ in Figure P11.11 is 45°. Also, the surface tension of water is about 72 dyne/cm.

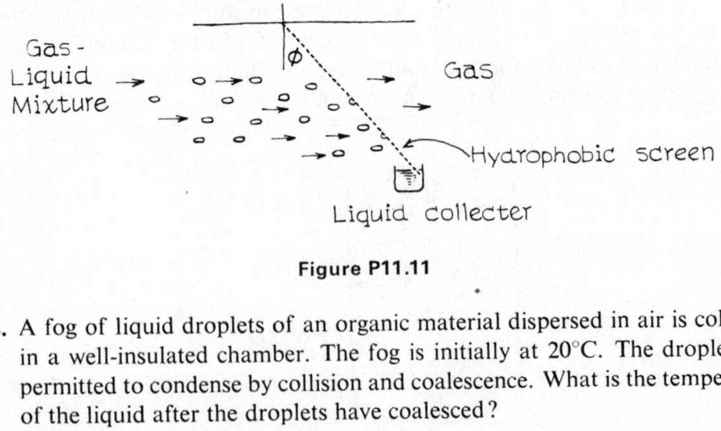

Figure P11.11

11.12. A fog of liquid droplets of an organic material dispersed in air is collected in a well-insulated chamber. The fog is initially at 20°C. The droplets are permitted to condense by collision and coalescence. What is the temperature of the liquid after the droplets have coalesced?

*Assuming no slip between the water droplets and entraining air.

Additional data:

The organic liquid can be assumed nonvolatile.
The pressure change can be neglected.

$$\left(\frac{\partial \sigma}{\partial T}\right)_A = -3 \times 10^{-9} \text{ cal/cm}^2\text{-}^\circ\text{C}$$

The specific heat at constant area (and constant liquid specific volume),

$$C_A \equiv T\left(\frac{\partial S}{\partial T}\right)_A = 1 \text{ cal/g }^\circ\text{C}$$

$\sigma = 6 \times 10^{-7}$ cal/cm^2 at 20°C (σ is independent of surface area). The surface area of the droplets is initially 10^7 cm^2/g.

Systems Under Stress or in Electric, Magnetic, or Potential Fields

12

To this point we have dealt almost entirely with simple systems in which work done on or by the system could be associated with a change in system boundaries (i.e., $P\,d\underline{V}$ or $\sigma\,dA$). With these limitations we expressed $\underline{U}$ as a function of ($\underline{S}$, $\underline{V}$, A, $N_1, \ldots, N_n$) in a Fundamental Equation, and from this starting point developed useful Legendre transforms, various equilibrium relationships, etc.

There exist other types of work that may also be done on or by a system and it is occasionally necessary to take these into account. For example, if our system were in an electric field, then, by changing the field or by moving the system from one point in the field to another, there is a *work interaction* between the system and the environment. In this chapter we will establish Fundamental Equations that include these new variables. To minimize the number of terms in the derived equations, we shall consider each new work form separately, although it should be obvious that the general form would contain all.

We first develop the general equations for electrostatic and electromagnetic work and then in succession treat individually electrostatic and electromagnetic systems, systems under (one-dimensional) stress, and conclude with a treatment of potential and kinetic energies. In all cases, we begin with the Fundamental Equation for intrinsic energy expressed in differential form and include the new work term of interest:

$$d\underline{U} = T\,d\underline{S} - P\,d\underline{V} - [đW_{\text{rev}}] + \sum_j \mu_j\,dN_j \tag{12-1}$$

12.1 Electrostatic and Electromagnetic Work

In the theory of electrodynamics, Maxwell's equations occupy the same pre-eminent position reserved for Newton's laws in classical mechanics (i.e., a vast amount of empirical evidence accumulated over the past century has led scientists to believe that all macroscopic electromagnetic phenomena are governed by these equations). They are adopted here as the basis of an expression for electromagnetic work. Following the treatment of Stratton,[1] the final result is a relation for the work *done on the system by external sources to establish the field:*

$$W = \int_{\underline{V}_s}\left[\int_0^{\mathfrak{D}} \mathcal{E}\cdot d\mathfrak{D} + \int_0^{\mathfrak{B}} \mathcal{H}\cdot d\mathfrak{B}\right] d\underline{V}_s \tag{12-2}$$

where[2]:

$\mathcal{E}$ = electric field strength, V/m

$\mathfrak{D}$ = electric flux density, A s/m^2

$\mathcal{H}$ = magnetic field strength, A/m

$\mathfrak{B}$ = magnetic flux density, V s/m^2

The integral is taken over the entire volume subject to the field, $\underline{V}_s$, and includes the free space as well as any volume occupied by the system. $\underline{V}_s$ is a constant, independent of the temperature and pressure. Also, for the work to be given by Eq. (12-2), the electric and magnetic fields must vanish at the boundaries of $\underline{V}_s$.

The electric vectors $\mathcal{E}$ and $\mathfrak{D}$ are related by the permittivity, ξ:

$$\mathfrak{D} = \xi\mathcal{E} \tag{12-3}$$

where in free space $\xi = \xi_0 = (8.854)(10^{-12})$F/m.[3] For real substances ξ is a function of the material, temperature, pressure, and electric field strength.

In a similar manner $\mathfrak{B}$ is related to $\mathcal{H}$ by the permeability, μ:

$$\mathfrak{B} = \mu\mathcal{H} \tag{12-4}$$

[1] J. A. Stratton, *Electromagnetic Theory* (New York: McGraw-Hill Book Company, 1941).

[2] Rational mks units are used. Note that in these definitions, A s ≡ coulomb and V s/m^2 ≡ tesla = weber/m^2.

[3] F = farad = A s/V

where in free space $\mu = \mu_0 = (4\pi)(10^{-7})$H/m.[4] For real substances, the permeability, μ, is a function of the material, temperature, pressure, and magnetic field strength.

The constants ξ_0 and μ_0 are related to the velocity of electromagnetic waves in free space by Eq. (12-5):

$$c = (\xi_0\mu_0)^{-1/2} = (2.9979)(10^8)\text{m/s} \tag{12-5}$$

In general, ξ and μ are symmetric tensors of rank two[5] but as used here, we will only consider the cases in which $\mathfrak{D}$ and $\mathcal{E}$ are parallel as are $\mathcal{B}$ and $\mathcal{H}$. Thus, ξ and μ can be visualized as scalar multipliers. This simplification limits the treatment of electric systems to simple geometries (e.g., to parallel-plate condensers or long uniformly wound solenoids), but the principal concepts and results are not greatly affected.

Several other parameters are commonly used to describe electric systems. For example, the dielectric constant, ϵ, and magnetic susceptibility, χ, are defined as:

$$\epsilon \equiv \frac{\xi}{\xi_0} \tag{12-6}$$

$$\chi \equiv \frac{\mu}{\mu_0} - 1 \tag{12-7}$$

It is also convenient to define two other quantities. We shall refer to $\mathcal{P}$ as the *electric polarization* and $\mathfrak{M}$ as the *magnetization* and define them as:

$$\mathcal{P} \equiv \mathfrak{D} - \xi_0\mathcal{E} = \xi_0(\epsilon - 1)\mathcal{E} \tag{12-8}$$

$$\mathfrak{M} \equiv \frac{\mathcal{B}}{\mu_0} - \mathcal{H} = \chi\mathcal{H} \tag{12-9}$$

It is obvious that, as defined, $\mathcal{P}$ and $\mathfrak{M}$ vanish in free space.

Since we are limiting our treatment to isotropic, one-dimensional systems, the vectorial notation may be dropped. Also, since we are interested primarily in systems in which the electric and magnetic fields do not vary over the volume ($\underline{V}_s$) or if an average field can be ascertained, Eq. (12-2) can be integrated over $\underline{V}_s$. Finally, let us replace $\mathfrak{D}$ and $\mathcal{B}$ by using Eqs. (12-8) and (12-9). The resulting work is:

$$W = \underline{V}_s\left(\int_0^{\mathcal{P}} \mathcal{E}\, d\mathcal{P} + \xi_0 \int_0^{\mathcal{E}} \mathcal{E}\, d\mathcal{E} + \mu_0 \int_0^{\mathfrak{M}} \mathcal{H}\, d\mathfrak{M} + \mu_0 \int_0^{\mathcal{H}} \mathcal{H}\, d\mathcal{H}\right) \tag{12-10}$$

The first and third terms represent the reversible work required to polarize or magnetize a material system in $\underline{V}_s$. The second and fourth terms represent the reversible work required to establish an electric and magnetic field in $\underline{V}_s$.

[4] H = henry = V s/A.

[5] For example:

$$\mathfrak{D}_x = \xi_{xx}\mathcal{E}_x + \xi_{xy}\mathcal{E}_y + \xi_{xz}\mathcal{E}_z, \text{ etc.}$$

These latter terms are independent of the material in the field. This convenient separation of work terms was one of the principal reasons for defining $\mathcal{P}$ and $\mathfrak{M}$.

12.2 Electrostatic Systems

To write the Fundamental Equation we use Eq. (12-1) with $(đW_{\text{rev}})$ as the *negative* of the first two terms in Eq. (12-10). The sign change is necessary since we are interested in the work done *by* the system. Thus:

$$d\underline{U} = T\,d\underline{S} - P\,d\underline{V} + \underline{V}_s\left(\mathcal{E}\,d\mathcal{P} + \frac{\xi_0\,d\mathcal{E}^2}{2}\right) + \sum_j \mu_j\,dN_j \qquad (12\text{-}11)$$

As noted previously, the term $(\underline{V}_s\xi_0\mathcal{E}^2/2)$ does not depend on the material within the system. Consequently, it is usually incorporated as a constant in the definition of the internal energy. Following this procedure:

$$\underline{U}' = \underline{U} - \frac{\underline{V}_s\xi_0\mathcal{E}^2}{2} \qquad (12\text{-}12)$$

Then:

$$d\underline{U}' = T\,d\underline{S} - P\,d\underline{V} + \underline{V}_s\mathcal{E}\,d\mathcal{P} + \sum_j \mu_j\,dN_j \qquad (12\text{-}13)$$

Again we emphasize that $\mathcal{P}$ is the component of the electric polarization vector parallel to the field $\mathcal{E}$.

Eq. (12-13) is the Fundamental Equation that we were seeking. It is immediately obvious that a number of Legendre transforms may be written in order to obtain independent variable sets other than $\underline{U}'(\underline{S}, \underline{V}, \mathcal{P}, N_1, \ldots, N_n)$. Also, since the Fundamental Equation is homogeneous, then the Euler form is:

$$\underline{U}' = T\underline{S} - P\underline{V} + \underline{V}_s\mathcal{E}\mathcal{P} + \sum_j \mu_j N_j \qquad (12\text{-}14)$$

and the Gibbs-Duhem relation becomes:

$$\underline{S}\,dT - \underline{V}\,dP + \underline{V}_s\mathcal{P}\,d\mathcal{E} + \sum_j N_j\,d\mu_j = 0 \qquad (12\text{-}15)$$

These equations are identical to those derived earlier with the exception of the new electrostatic terms. The chemical potential now, however, becomes:

$$\mu_k = \bar{U}'_k - T\bar{S}_k + P\bar{V}_k - \underline{V}_s\mathcal{E}\left(\frac{\partial\mathcal{P}}{\partial N_k}\right)_{T,P,N_j[k]} \qquad (12\text{-}16)$$

$$= \bar{U}'_k - T\bar{S}_k + P\bar{V}_k - \underline{V}_s\mathcal{E}\bar{\mathcal{P}}_k \qquad (12\text{-}17)$$

To maintain our previous contention that μ_k is a partial molal free energy, we see that we should define the Gibbs free energy as the transform:

$$\underline{G}(T, P, \mathcal{E}, N_j) = \underline{U}' - T\underline{S} + P\underline{V} - \underline{V}_s\mathcal{E}\mathcal{P} \qquad (12\text{-}18)$$

It is important to recognize that this free energy function is no different from the Gibbs function used previously in this book. The designation of the variable $\mathcal{E}$ was not noted earlier since we implicitly assumed that our systems were either not dielectrics ($\epsilon = 1.0$), the field was zero, or there was no interaction between an imposed field and our system. This same reasoning holds for all other transforms and, in addition, to the equilibrium and stability criteria developed earlier.

For dielectric materials in an electric field, the concept of the pressure is somewhat different from the same material in the absence of a field. Pressures within a body are modified by the presence of the field and to apply the Fundamental Equation, one should consider that the pressure term refers to that value acting on the boundaries of the system. Solids, however, often present a complex problem in the theory of elasticity. A detailed discussion of this case is given by Landau and Lifshitz.[6]

Several examples are presented below to illustrate the application of the Fundamental Equation including an electrostatic term.

Example 12.1

For a dielectric material present in an electrostatic field, the variation in the system volume with changes in field strength at constant temperature, pressure, and mole numbers is called *electrostriction*. Estimate the fractional change in volume of hydrogen gas at 293°K and 20.2 MN/m² when the electric field on the system is increased from 0 to 10^6 V/m. Experimental data show that for hydrogen under these conditions a plot of ln P vs. the dielectric constant yields essentially a straight line with a slope of 24.8, i.e., at 293°K and for fields between 0 and 10^6 V/m,

$$\left(\frac{\partial \ln P}{\partial \epsilon}\right)_{T,\mathcal{E},N} = 24.8$$

Solution

We first wish to obtain the partial derivative $(\partial \underline{V}/\partial \mathcal{E})_{T,P,N}$. The Legendre transform of Eq. (12-14) is:

$$\psi = \psi(T, P, \mathcal{E}, N) = \underline{U}' - T\underline{S} + P\underline{V} - \underline{V}_s \mathcal{E} \mathcal{P}$$

$$d\psi = -\underline{S}\, dT + \underline{V}\, dP - \underline{V}_s \mathcal{P}\, d\mathcal{E} + \sum_j \mu_j\, dN_j$$

Assuming that the field is constant and taking $\mathcal{P}$ as the component of $\mathcal{P}$

[6] L. P. Landau and E. M. Lifshitz, "Electrodynamics of Continuous Media," Vol. 8, *Course of Theoretical Physics*, translated by G. B. Sykes and G. S. Bell, (New York: Pergamon Press, 1960).

parallel to $\mathcal{E}$, Eq. (12-8) leads to:

$$d\underline{\psi} = -\underline{S}\,dT + \underline{V}\,dP - \underline{V}_s\xi_0(\epsilon - 1)\mathcal{E}\,d\mathcal{E} + \sum_j \mu_j\,dN_j$$

and, therefore,

$$\left(\frac{\partial \underline{V}}{\partial \mathcal{E}}\right)_{T,P,N} = -\frac{\partial}{\partial P}[\underline{V}_s\xi_0(\epsilon - 1)\mathcal{E}]_{T,\mathcal{E},N} = -\underline{V}_s\xi_0\mathcal{E}\left(\frac{\partial \epsilon}{\partial P}\right)_{T,\mathcal{E},N}$$

Separating variables and integrating,

$$\frac{\underline{V}_{\mathcal{E}} - \underline{V}_{\mathcal{E}=0}}{\underline{V}_s} = -\frac{\xi_0}{P(24.8)(2)}\mathcal{E}^2$$

ξ_0 is a constant equal to $(8.85)(10^{-12})$ A s/V m. P is 20.2 MN/m², and $\mathcal{E} = 10^6$ V/m.

$$\frac{\underline{V}_{\mathcal{E}} - \underline{V}_{\mathcal{E}=0}}{\underline{V}_s} = \frac{-(8.85)(10^{-12})(10^6)^2}{(20.2)(10^6)(2)(24.8)} \sim -(9)(10^{-9})$$

In this case, the contraction in volume is extremely small. Normally, for macroscopic systems, electrostriction produces negligible changes in volume, although for microscopic systems in which local fields can be very large, there can be significant effects. For example, applied to ions in solution, electrostriction is often invoked to account qualitatively for the decrease in total volume when ionic solutions are mixed.

Example 12.2

Indicate how the entropy of a dielectric changes as an electric field is applied at constant temperature, pressure, and mole numbers.

Solution

If we use the same Legendre transform developed in Example 12.1,

$$\left(\frac{\partial \underline{S}}{\partial \mathcal{E}}\right)_{T,P,N} = \frac{\partial}{\partial T}[\underline{V}_s\xi_0(\epsilon - 1)\mathcal{E}]_{P,\mathcal{E},N} = \xi_0\mathcal{E}\underline{V}_s\left(\frac{\partial \epsilon}{\partial T}\right)_{P,\mathcal{E},N}$$

For most materials, ξ and, therefore, ϵ decreases with an increase in temperature; thus the entropy decreases as the material is polarized. Perhaps this result could have been anticipated from a consideration of the fact that, in an electric field, the molecules would tend to align their dipoles with the field and the overall randomness of the system would decrease. At constant $\underline{V}$ and N, during polarization there also would be a heat interaction of $T\,d\underline{S}$ and, since $d\underline{S} < 0$, heat would be evolved.

Example 12.3

Estimate the work required to establish a field when a dielectric is present within a parallel-plate capacitor and the field is increased from 0 to $\mathcal{E}$.

Solution

This work is given by the first two terms in Eq. (12-10). This work is easily shown to be:

$$W = \frac{\underline{V}_s \xi_0 \epsilon \mathcal{E}^2}{2}$$

To estimate W, choose a basis of 1 kg-mol. For a field of 10^6 V/m, selecting as examples, air, methyl alcohol, and fused silica:

Material	Conditions	ϵ	Volume, $\underline{V}_s$ m^3/kg-mol	Work J/kg-mol	Approx. breakdown V/m
Air	273°K, 1 atm	1.0006	22.4	100	$(3)(10^6)$
Methyl alcohol (liquid)	293°K	31.2	.0402	5.5	—
Fused silica	293°K	4	.027	.48	10^7

Compared to the value of RT at 300°K [$\sim(2.5)(10^6)$ J/kg-mol], these work terms are indeed insignificant; more energy could be stored at higher field strengths, but the value of 10^6 V/m used in the example is close to the breakdown strength of most materials.

Example 12.4

A flat plate capacitor is placed inside a system containing a pure dielectric material. As the capacitor is charged, show how the chemical potential and concentration vary with field strength.

Solution

Let us first find the derivative:

$$\left(\frac{\partial \mu_i}{\partial \mathcal{E}}\right)_{T,\underline{V},N}$$

For this derivative we desire the Legendre transform,

$$\psi = \psi(T, \underline{V}, \mathcal{E}, N)$$

$$\psi = U' - T\underline{S} - \underline{V}_s \mathcal{E} \mathcal{P}$$

$$d\psi = -\underline{S}\,dT - P\,d\underline{V} - \underline{V}_s \mathcal{P}\,d\mathcal{E} + \sum_j \mu_j\,dN_j$$

and

$$\left(\frac{\partial \mu_i}{\partial \mathcal{E}}\right)_{T,\underline{V},N} = -\underline{V}_s\left(\frac{\partial \mathcal{P}}{\partial N_i}\right)_{T,\underline{V},\mathcal{E},N_j[i]} = -\underline{V}_s \xi_0 \mathcal{E}\left(\frac{\partial \epsilon}{\partial N_i}\right)_{T,\underline{V},\mathcal{E},N_j[i]}$$

where Eq. (12-8) has been used. Define the variation of ϵ with N_i as ϵ' and assume that it is independent of field strength; then, integrating:

$$\mu_i(\mathcal{E}) - \mu_i(\mathcal{E} = 0) = -\frac{\underline{V}_s \xi_0 \mathcal{E}^2 \epsilon'}{2}$$

This equation indicates the variation of μ_i with $\mathcal{E}$. As suggested by Guggenheim,[7] the value of μ_i ($\mathcal{E} = 0$) can be related to a standard-state value at the same temperature by:

$$\mu_i = \mu_i^0 + RT \ln \frac{\hat{f}_i}{\hat{f}_i^0} \qquad (\mathcal{E} = 0)$$

where the superscript represents some arbitrarily chosen standard state at T. Also, for simplicity, let us assume that the material is an ideal gas:

$$\hat{f}_i = y_i P = P = \frac{N_i RT}{\underline{V}}$$

Substituting:

$$\mu_i(\mathcal{E}) = \mu_i^0 + RT \ln \frac{N_i RT}{\hat{f}_i^0 \underline{V}} - \frac{\underline{V}_s \xi_0 \mathcal{E}^2 \epsilon'}{2}$$

The dielectric constant can be expressed by the Debye equation as:

$$\frac{\epsilon - 1}{\epsilon + 2} = \frac{NL}{3\underline{V}}\left(\alpha + \frac{\mu^{+2}}{3\xi_0 kT}\right)$$

ϵ is the dielectric constant, α the molecular polarizability, μ^+ the dipole moment, and L is Avogadro's number. For ideal gases $\epsilon \sim 1$; thus (for a pure gas):

$$\epsilon' = \left(\frac{\partial \epsilon}{\partial N}\right)_{T,\underline{V}} = \frac{L}{\underline{V}}\left(\alpha + \frac{\mu^{+2}}{3\xi_0 kT}\right)$$

Then, assuming that $\underline{V}_s \sim \underline{V}$,

$$\mu_i(\mathcal{E}) = \mu_i^0 + RT \ln \frac{CRT}{\hat{f}_i^0} - \frac{\xi_0 L \mathcal{E}^2 \left[\alpha + \frac{\mu^{+2}}{3\xi_0 kT}\right]}{2}$$

where C is the concentration.

Assume now that this gas in the field $\mathcal{E}$ is in equilibrium with more pure gas outside the field, but at temperature T. For this external gas at $\mathcal{E} = 0$, we could again write a similar expression for μ_i:

$$\mu_i(\text{outside}) = \mu_i^0 + RT \ln \frac{C^{\text{outside}} RT}{\hat{f}_i^0}$$

At equilibrium, since $\mu_i(\text{outside}) = \mu_i(\text{inside})$,

$$RT \ln \frac{C^{\text{inside}}}{C^{\text{outside}}} = \frac{\xi_0 L \mathcal{E}^2 \left[\alpha + \frac{\mu^{+2}}{3\xi_0 kT}\right]}{2}$$

[7] E. A. Guggenheim, *Thermodynamics* (Amsterdam: North-Holland Publishing Co., 1959), Chap. 11.

Suppose that our system consisted of HCL. $\alpha \sim 2.6 \times 10^{-30}$ m³/molecule and $\mu^+ \sim 1$ Debye $= [1/(3)(10^{29})]$ C m/molecule. k is $(1.38)(10^{-23})$ J/molecule °K and let $T = 300$°K. For the other terms $\xi_0 = (8.854)(10^{-12})$A s/V m, $L = (6.032)(10^{23})$ molecules/mol, and assume that $E = 10^7$ V/m. Then

$$RT \ln \frac{C^{\text{inside}}}{C^{\text{outside}}} = (8.854)(10^{-12})(6.023)(10^{23})(10^{14})$$

$$\times \{2.6 \times 10^{-30} + [(3)^2(10^{58})(8.854)(10^{-12})(1.38)(10^{-23})(300)(3)]^{-1}\}0.5$$

$$= (52.3)(10^{25})(2.6 + 101)(10^{-30})(0.5) = 2.8 \times 10^{-2} \text{ J/mol}$$

$$\frac{C^{\text{inside}}}{C^{\text{outside}}} = \exp\left[\frac{(2.8)(10^{-2})}{(8.314)(300)}\right] = \exp(1.1 \times 10^{-5})$$

The enhancement is obviously very small. Only at much higher field strengths would one be able to show an appreciable difference between the concentrations inside and outside the field.

It is perhaps clear from the examples shown above that in most cases electric fields affect the thermodynamic properties of a system very little. To obtain significant effects, extremely large field strengths are required, but before they may be attained, breakdown normally occurs. Nevertheless, in the immediate vicinity of ions in a solution, very high field strengths do exist and significantly affect the microscopic properties of matter.

12.3 Electromagnetic Systems

We can develop the thermodynamics of systems in electromagnetic fields in much the same way as we did for systems in electrostatic fields. (See Section 12.2.) In this case, however, we must deal with the magnetic flux density, $\mathcal{B}$, the magnetic field strength, $\mathcal{H}$, and the magnetization (or magnetic moment per unit volume or magnetic polarization vector), $\mathfrak{M}$. They are related as shown in Eq. (12-9). $\mathfrak{M}$ is zero unless there is material in the field.

As before, the work done *by* the system is given by the negative of the last two terms in Eq. (12-10), and the Fundamental Equation then becomes:

$$d\underline{U} = T\,d\underline{S} - P\,d\underline{V} + \underline{V}_s[(\mu_0\mathcal{H}\,d\mathfrak{M} + \mu_0\mathcal{H}\,d\mathcal{H}] + \sum_j \mu_j\,dN_j \tag{12-19}$$

The second term in the electromagnetic work is normally separated and combined with $\underline{U}$ in a manner analogous to Eq. (12-12) to define or modify an internal energy $\underline{U}'$.

$$\underline{U}' = \underline{U} - \frac{\underline{V}_s\mu_0\mathcal{H}^2}{2} \tag{12-20}$$

so that:

$$d\underline{U}' = T\,d\underline{S} - P\,d\underline{V} + \underline{V}_s\mu_0\mathcal{H}\,d\mathfrak{M} + \sum_j \mu_j\,dN_j \tag{12-21}$$

This Fundamental Equation is analogous to Eq. (12-13) for electrostatic systems[8] and as noted there the Euler form, the Gibbs-Duhem, and various Legendre transforms are readily obtained. To illustrate the use of such forms, several examples are given below.

Example 12.5

Holding the temperature, pressure, and mole numbers constant, how is the entropy of a material affected by changes in the magnetic field strength?

Solution

We shall want to find a Legendre transform to obtain $\psi = \psi(T, P, \mathcal{M}, N)$. Thus,

$$\psi = \underline{U}' - T\underline{S} + P\underline{V} - \mu_0 \underline{V}_s \mathcal{H}\mathcal{M}$$

$$d\psi = -\underline{S}\,dT + \underline{V}\,dP - \mu_0 \underline{V}_s \mathcal{M}\,d\mathcal{H} + \sum_j \mu_j\,dN_j$$

and,

$$\left(\frac{\partial \underline{S}}{\partial \mathcal{H}}\right)_{T,P,N} = \mu_0 \underline{V}_s \left(\frac{\partial \mathcal{M}}{\partial T}\right)_{P,\mathcal{H},N} = \mu_0 \underline{V}_s \mathcal{M}\delta_I$$

where δ_I is equal to $(\partial \ln \mathcal{M}/\partial T)$ and is called the thermal magnetization coefficient. Values of δ_I are not often known. For water, Camp and Johnson[9] report that $\delta_I \sim (5)(10^{-4})°K^{-1}$. For water, $\mu \sim (1.3)(10^{-6})$ V s/m A and $\chi = (-9.06)(10^{-6})$. Then, for a value of $\mathcal{B} = 1000$ gauss[10] $= 0.1$ Wb/m² $= 0.1$ V s/m², from Eq. (12-4),

$$\mathcal{H} = \frac{\mathcal{B}}{\mu} = \frac{0.1}{(1.3)(10^{-6})} = (7.7)(10^4)\text{ A/m}$$

$$\mathcal{M} = \chi\mathcal{H} = -(9.06)(10^{-6})(7.7)(10^4) \sim -0.7\text{ A/m.}$$

Let $\underline{V} = 1$ m³ so that

$$\mathcal{M}\underline{V}_s = -0.7\text{ A m}^2$$

Then

$$\left(\frac{\partial \underline{S}}{\partial \mathcal{H}}\right)_{T,P,N} = (4\pi)(10^{-7})(-0.7)(5)(10^{-4}) \sim -3 \times 10^{-10}\text{ J/(A/m)°K}$$

[8] See, however, F. W. Camp and E. F. Johnson, "The Effect of Strong Magnetic Fields on Chemical Engineering Systems," MATT-67, Plasma Physics Laboratory, Princeton University, Princeton, N. J. It is argued in this report that there should be an additional term in Eq. (12-21) to account for the expansion of the system at constant $\mathcal{M}$. This term is: $\mu_0 \mathcal{H}\mathcal{M}\,d\underline{V}_s$ and should be included with the $P\,d\underline{V}$ term to give $-(P - \mu_0\mathcal{H}\mathcal{M})\,d\underline{V}$. If this is true, then to Eq. (12-13) one should correct the $-P\,d\underline{V}$ term to $-(P - \mathcal{E}\mathcal{P})\,d\underline{V}$ to allow the expansion of a system in an electrostatic field at constant P. In our treatment we have considered $\underline{V}_s$ to be a constant and equal to the total volume affected by electric or magnetic fields whereas $\underline{V}$ is taken as the volume of the material system under consideration.

[9] *op. cit.*

[10] In S. I. units, this would be abbreviated as 0.1 tesla.

For one m³ of water, with an increase in $\mathcal{B}$ from 0 to 1000 gauss, the entropy decreases, but very slightly.

Example 12.6

For a pure component, at constant pressure, show how the boiling point varies with the magnetic intensity, $\mathcal{H}$.

Solution

For phase equilibrium, the chemical potential is equal in the vapor and liquid. From Eq. (12-21), a total Legendre transform to $\psi(T, P, \mathcal{H}, \mu)$, for a pure component, yields the Gibbs-Duhem equation:

$$0 = -\underline{S}\,dT + \underline{V}\,dP - \mu_0 \underline{V}_s \mathfrak{M}\,d\mathcal{H} - N\,d\mu$$

If we divide by the number of moles, N, and solve for $d\mu$:

$$d\mu = -S\,dT + V\,dP - \mu_0 V_s \mathfrak{M}\,d\mathcal{H}$$

Holding the pressure constant and using:

$$\mu^V = \mu^L$$

then,

$$\left(\frac{\partial T}{\partial \mathcal{H}}\right)_{[L-V],P} = -\mu_0 \mathfrak{M} \frac{(V^V - V^L)}{(S^V - S^L)}$$

Since entropy and volume are weak function of $\mathcal{H}$, with $\mathfrak{M} = \chi \mathcal{H}$,

$$\Delta T = T_{\mathcal{H}} - T_{\mathcal{H}=0} = -\mu_0 \chi \frac{(V^V - V^L)\mathcal{H}^2}{2(S^V - S^L)}$$

To illustrate the use of this relation, consider water at 100°C[11] and determine ΔT for the case in which $\mathcal{H}$ increases from zero to a field of $(8)(10^5)$ A/m. This corresponds to 10,000 gauss. Also:

$$\mu_0 = (4\pi)(10^{-7}) \text{ V s/A m}$$

$$\chi = (-9.06)(10^{-6})$$

$$V^V - V^L = 30.1 \text{ m}^3\text{/kg-mol}$$

$$S^V - S^L = (1.09)(10^5) \text{ J/kg-mol°K}$$

$$\Delta T = -\frac{(4\pi)(10^{-7})(-9.06)(10^{-6})(30.1)(8)^2(10^5)^2}{(1.09)(10^5)(2)} = 10^{-3}\text{°K}$$

The increase is inconsequential.[12]

---

[11] F. W. Camp and E. F. Johnson, *op. cit.*

[12] A similar relation can be derived for calculating the change in boiling point of a material in an electrostatic field. [R. K. Lyon, *Nature*, **192**, (1961), 1285.] Though the calculated values for ΔT for methyl, ethyl, and propyl alcohols were about $(1.3)(10^{-5})$°K for $\mathcal{E} = (5)(10^5)$ V/m, experimental data indicated the boiling temperature *decreased* 2.0, 0.7, and 0.4°K in the three cases. These results are in complete disagreement with theory.

It is clear from the examples shown, that electromagnetic fields affect thermodynamic properties only slightly. The phenomenon of superconductivity at low temperatures is, however, one important exception. Another is the magnetocaloric effect that is utilized for cooling below 1°K.[13] Referring to Example 12.5, we showed:

$$\left(\frac{\partial \underline{S}}{\partial \mathcal{H}}\right)_{T,P,N} = \mu_0 \underline{V}_s \left(\frac{\partial \mathcal{M}}{\partial T}\right)_{P,\mathcal{H},N} \tag{12-22}$$

From Eq. (12-9), $\mathcal{M} = \chi \mathcal{H}$ and at constant N Eq. (12-22) becomes:

$$\left(\frac{\partial S}{\partial \mathcal{H}^2}\right)_{T,P,N} = \mu_0 \frac{V_s}{2}\left(\frac{\partial \chi}{\partial T}\right)_{P,\mathcal{H},N}$$

For many paramagnetic materials at low temperatures the relation between χ and T is given by Curie's Law:

$$\chi = \frac{C}{T}$$

Therefore,

$$\left(\frac{\partial S}{\partial \mathcal{H}^2}\right)_{T,P,N} = -\frac{\mu_0 C V_s}{2T^2} \tag{12-23}$$

At low temperatures, the entropy will decrease considerably when the material is isothermally magnetized. The next step is to isolate thermally the sample and remove it from the field. A temperature drop occurs on demagnetization as the interested reader can verify by determining the sign of $(\partial T/\partial \mathcal{H})_{\underline{S},P,N}$.

12.4 Thermodynamics of Systems Under Stress

It is rare for chemical engineers to become involved with thermodynamic analyses of systems that are in tension or compression. Significant problems arising in this area usually fall heir to applied mechanicists who are interested in the behavior of materials under static or dynamic loads. The materials involved are solid instead of fluid since, for the latter, the consequences of the imposition of a force can be treated by pressure-volume terms introduced earlier in the book. In fact, solids are almost always considered to be elastic (i.e., although they may be stressed and deformed, the total volume is invariant). In the thermodynamics of such systems, the system volume is then assumed constant and no $P\underline{V}$ terms are used.

[13] W. F. Giauque, "Paramagnetism and the Third Law of Thermodynamics. Interpretation of the Low-Temperature Magnetic Susceptibility of Gadolinium Sulfate," *J. Am. Chem. Soc.*, **49**, (1927), 1870. P. Debye, "Einige Bemerkungen zur Magnetisierung bei tiefer Temperatur," *Ann. Physik*, **81**, (1926), 1154. W. F. Giauque and D. P. MacDougall, "Attainment of Temperatures Below 1° Absolute by Demagnetization of $Gd_2(SO_4)_3 \cdot 8H_2O$," *Phys. Rev.*, **43**, (1933), 768.

In the discussion of stressed systems, we will refer to the *strain* resulting from an applied *stress*. These terms are readily visualized in a qualitative way since it is common experience to apply a force to, for example, rubber and expect a change in the dimensions of the specimen. A careful definition of terms is, however, necessary before any quantitative relations can be meaningful. Such definitions will be our first task although the ultimate objective is to formulate a Fundmental Equation for stressed systems.

When we use the term stress, we obviously imply a force/area. Stress is, therefore, a vectorial quantity with both a magnitude and direction. Similarly, for strain we describe the movement of an element of the system in a particular direction. Consider Figure 12.1. A bar of an elastic material is fixed to a rigid plate and subjected to a tensile force, F_x. The sides are unconstrained. Because of F_x, the bar is elongated by Δx and is in tension. The stress in this case, F_x/A, produces a strain, Δx. Also, as shown, there are induced strains Δy and Δz as well as induced stresses in the y and z directions. Thus, the situation is complicated even in this relatively simple example. For any body under stress, we must consider the strain in the three coordinate directions. Also, one should allow for *shear* (i.e. , a stress so directed that there is a rotation or a twisting of the system).

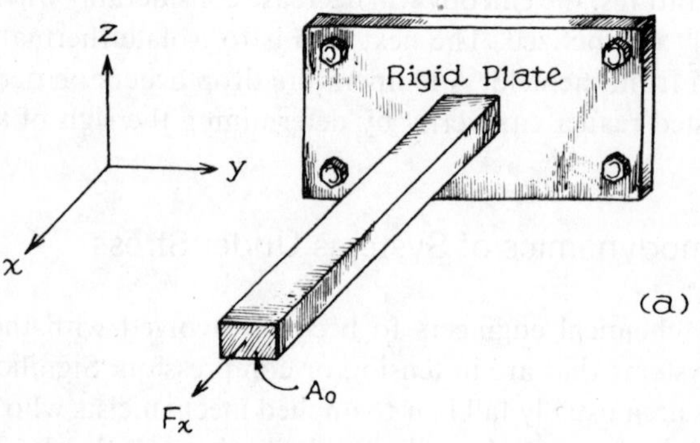

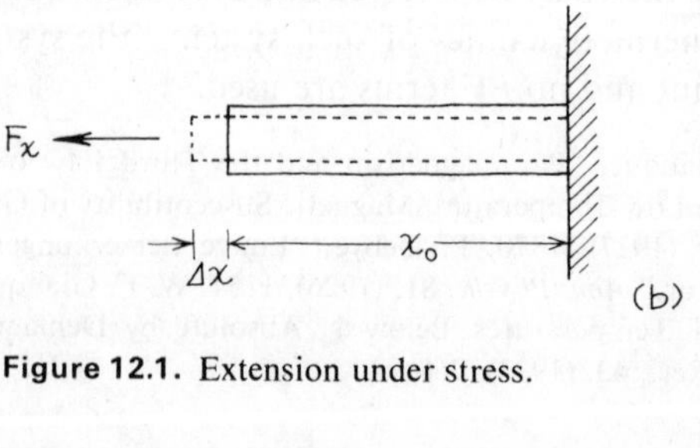

Figure 12.1. Extension under stress.

It is not within the scope of this book to develop in detail three-dimensional elastic theory as applications rarely occur in chemical engineering. The interested reader is referred to the excellent presentation of Callen.[14] We will limit our treatment to a one-dimensional case in which there is but one applied stress or force. Thus we return to Figure 12.1. Here the work done on the system is $F_x \, \Delta x$. The fact that there is a finite Δy or Δz is immaterial since we are only interested in force-distance interactions operating over the system boundaries and F_x is the only force to fit this criterion.[15]

As used in a thermodynamics study, the designation of work as simply $F_x \, dx$ is inconvenient since the terms do not blend neatly with other extensive properties in a Fundamental Equation. Let us, therefore, multiply and divide by the total system volume $\underline{V}$,

$$\text{đ}W = \frac{\underline{V}}{\underline{V}} F_x \, dx = \frac{\underline{V} F_x \, dx}{Ax} \tag{12-24}$$

$$= \underline{V} \mathscr{s} \, d\Omega$$

The stress is $F_x/A = \mathscr{s}$ and $d\Omega$, dx/x, is the fractional extension (or contraction) in the same direction as $\mathscr{s}$. This work is then an extensive property. Since this is the work done by the environment on the system, the negative of the work in Eq. (12-24) is the desired quantity to be used in Eq. (12-1). The Fundamental Equation in this case (with $\underline{V}$ = constant) is then:

$$d\underline{U} = T \, d\underline{S} + \mathscr{s} \underline{V} \, d\Omega + \sum_j \mu_j \, dN_j \tag{12-25}$$

Once we have reached this point, the Euler relation, the Gibbs-Duhem equation, and many Legendre transforms can be written immediately. The Euler form is, for example,

$$\underline{U} = T\underline{S} + \mathscr{s} \underline{V} \Omega + \sum_j \mu_j N_j \tag{12-26}$$

Various coefficients may also be defined with the $\mathscr{s}$-Ω notation. When a system is acted upon by a single force (as in Figure 12.1), the stress-strain ratio is called *Young's modulus, Y:*

$$Y \equiv \left(\frac{\partial \mathscr{s}}{\partial \Omega} \right)_T \tag{12-27}$$

[14] H. B. Callen, *Thermodynamics* (New York: John Wiley and Sons, Inc., 1960), Chap. 13.

[15] In the general case with both linear movement and rotation we should have to consider that the work done on the system consisted of six terms. Three terms would involve the linear movement in the three rectangular coordinates due to stresses in these directions. There would, however, be three other terms to account for motion in, for example, the y and z direction from an x-directed stress. There are not nine terms because of the conservation of angular momentum (i.e., the x-movement due to a y-stress is equal to the y-movement due to an x-stress). See Callen, *op. cit.*

If, however, the sides are constrained from expanding or contracting in the other directions, then the derivative,

$$\left(\frac{\partial \mathscr{s}}{\partial \Omega}\right)_{T,\,\text{restrained}} \equiv \textit{isothermal elastic stiffness coefficient} \qquad (12\text{-}28)$$

Many other solid properties can be introduced, including heat capacities at constant length (or stress), etc. We illustrate one application of thermodynamics to such systems in the example below and provide several other problems at the end of the chapter.

Example 12.7

An adiabatic, elastic deformation of a body is normally accompanied by a change in temperature. This phenomenon is known as the *thermoelastic* effect. Estimate the magnitude of this effect for a specimen of Armco iron which is stressed from 0 to 15,000 psi at 500°C. Under these conditions, the coefficient of linear thermal expansion is:

$$\alpha_{\mathscr{s}} = \left(\frac{1}{L_0}\right)\left(\frac{\partial L_0}{\partial T}\right)_{\mathscr{s}} = (16.8)(10^{-6})°\text{K}^{-1}$$

and the heat capacity at constant stress is:

$$C_{\mathscr{s}} = T\left(\frac{\partial S}{\partial T}\right)_{\mathscr{s}} \sim 9.1 \text{ cal/g-mol°K}$$

Solution

We desire the partial derivative $(\partial T/\partial \mathscr{s})_{\underline{S},N}$. Thus, we employ a Legendre transform, $\psi = \psi(\underline{S}, \mathscr{s}, N)$. From Eq. (12-25),

$$d\psi = T\, d\underline{S} - \Omega \underline{V}\, d\mathscr{s} + \sum_j \mu_j\, dN_j$$

$$\left(\frac{\partial T}{\partial \mathscr{s}}\right)_{\underline{S},N} = -\underline{V}\left(\frac{\partial \Omega}{\partial \underline{S}}\right)_{\mathscr{s},N}$$

$$= \frac{\underline{V}\left(\dfrac{\partial \Omega}{\partial T}\right)_{\mathscr{s},N}}{\left(\dfrac{\partial \underline{S}}{\partial T}\right)_{\mathscr{s},N}}$$

From the definition of Ω,

$$\alpha_{\mathscr{s}} = \left(\frac{\partial \Omega}{\partial T}\right)_{\mathscr{s},N}$$

Thus,

$$\int_{T_1}^{T_2} \frac{dT}{T} = -\int_{\mathscr{s}_1}^{\mathscr{s}_2} \frac{\alpha_{\mathscr{s}}}{\dfrac{C_{\mathscr{s}} N}{\underline{V}}}\, d\mathscr{s}$$

The term $(C_{\mathscr{s}} N/\underline{V})$ is simply the heat capacity at constant stress, on a unit volume basis. Assuming both it and α_s are not strong functions of s (see below), then

$$\ln \frac{T_2}{T_1} = -\frac{\alpha_s(\Delta s)}{\frac{C_s N}{\underline{V}}}$$

For small differences between T_2 and T_1,

$$\ln \frac{T_2}{T_1} \sim \frac{T_2 - T_1}{T} = \frac{\Delta T}{T}$$

and

$$\Delta T = -\frac{\alpha_s T(\Delta s)}{\frac{C_s N}{\underline{V}}}$$

α_s has been given as $(16.8)(10^{-6})°K^{-1}$. Δs is 15,000 psi $= (1.5)(10^4)(6894) = 103$ MN/m². T is 773°K, and

$$\frac{C_s N}{\underline{V}} = \frac{(9.1)(4.19)}{(55.8)}(7.6)(10^6) = 5.18 \text{ MJ/m}^3$$

where the atomic weight was chosen as 55.8 and the density as 7.6 g/cm³. Thus,

$$\Delta T = -\frac{(16.8)(10^{-6})(103)(773)}{(5.18)} = -.26°K$$

The expected temperature drop would be about 0.26°K for this adiabatic stress. This value is in excellent agreement with the value of −0.25°K reported by Rocca and Bever.[16] These same authors discuss the variation of $\alpha_s/(C_s N/\underline{V})$ with s and show that for stresses of the order of 10,000 to 15,000 psi, this group varies only 2 to 3%.

It is also interesting to note that a temperature *drop must occur upon stretching* if $\alpha > 0$. For some substances in which there is a contraction upon heating, the *thermoelastic effect* will be positive (i. e., the material will heat when adiabatically stretched).

12.5 Systems in Body-Force Fields or Under Acceleration Forces

We have now treated several systems in which there were work terms other than of the $P\,d\underline{V}$ type. Once we expressed this work (in an extensive manner), we simply included it in the general Fundamental Equation and from there proceeded to derive the desired Legendre transforms and partial derivatives.

[16] Rocca, Robert and M. B. Bever, "The Thermoelastic Effect," *Trans. AIME*, **88**, Feb. 1950, *Journal of Metals*, p. 327.

Implicit in this treatment was the supposition that the internal energy $\underline{U}$ was a function of variables other than $\underline{S}$, $\underline{V}$, $N_1, \ldots, N_n$ (e.g., for systems in electrostatic fields $\underline{U} = \underline{U}(\underline{S}, \underline{V}, \mathcal{P}, N_1, \ldots, N_n)$ where $\mathcal{P}$ is the component of the electric polarization parallel to the electric field $\mathcal{E}$.

It was then a simple task to develop any desired partial derivative from the appropriate Fundamental Equation. We also indicated in Section 12.3 that it was logical to expand the definitions of $\underline{H}$, $\underline{A}$, $\underline{G}$, etc., to include new system variables. Some authors, however, retain the earlier definitions for these well-known properties and define new properties if electrostatic, electromagnetic, etc., terms are to be included. Either approach is quite satisfactory, although the duality is often confusing. It is preferable to avoid definitions if possible and only consider the necessary independent variables. A Legendre transform is then employed to obtain the desired new Fundamental Equation.

Except for learning new notations introduced in Sections 12.2 through 12.4, there are no new principles introduced. When, however, one wishes to study systems in potential (body-force) fields or as acted upon to produce acceleration (or deceleration), then a somewhat different approach is taken. As we will show below, the work can readily be expressed in terms appropriate for these systems, but it is not immediately obvious how to use these terms to develop the Fundamental Equation. We must carefully consider the question whether the internal energy of a system depends on the position of the system in a potential field or on the velocity. Although we could, if we desired, extend the definition of $\underline{U}$ to include these variables, the results would run counter to our intuitive desire to preserve the character of $\underline{U}$ to be independent of position and velocity as much as possible. True, if we have a nonhomogeneous electromagnetic field, $\underline{U}$ is already a function of position, but we shall overlook such inconsistencies. Although this is a text on classical thermodynamics, and little or no note has been taken of the constituent molecules in a system, we must admit that the desire to elminate potential and acceleration fields from the definition of $\underline{U}$ relates to the fact that we would like to feel that those variables that affect $\underline{U}$ also modify some molecular properties. Perhaps the molecular velocity or spacing, or rotational frequency, etc., might be affected. These properties cannot, however, be related neatly to the overall system velocity or, for example, to the position in a potential field.

The net result of these admittedly intuitive feelings is to arrive at the decision to express the total energy of a system as a sum of the familiar internal energy and other energies that may be associated with potential or acceleration fields, i.e.,

$$\text{system energy} = (\text{internal} + \text{potential} + \text{kinetic}) \text{ energy}$$

$$\underline{E} = \underline{U} + \underline{PE} + \underline{KE} \tag{12-29}$$

The $\underline{PE}$ term is found by considering the work effects necessary to move a system in a force field. If a system is moved a distance $d\mathbf{L}$ against or with a colinear force $\mathbf{F}$, then:

$$d(\underline{PE}) = -\text{đ}W = -\mathbf{F} \cdot d\mathbf{L} = -M\mathbf{a} \cdot d\mathbf{L} = M\, d\phi \tag{12-30}$$

where the potential ϕ is defined as:

$$d\phi \equiv -\mathbf{a} \cdot d\mathbf{L} \tag{12-31}$$

The potential energy of the system is then:

$$\underline{PE} = M\phi \tag{12-32}$$

since the mass M was considered constant.

To illustrate, consider a system in a gravitational field with an acceleration $\mathbf{g}$. Measure $\mathbf{L}$ in the *up* direction and $\mathbf{g}$ in the *down* direction. Then, since $\mathbf{L}$ and $\mathbf{g}$ are colinear vectors but with opposite directions,

$$d\phi = g\, dL \tag{12-33}$$

$$\underline{PE} = M \int d\phi = M\phi = M \int g\, dL \tag{12-34}$$

No limits have been placed on the integration since one may arbitrarily define the potential energy to be zero at some reference value of L and then relate all other values to this base state.

As another example, consider a solid-body rotation. Let $\mathbf{L} = \mathbf{r} =$ the radius measured *outward* from the center of rotation. Here the centrifugal acceleration is $\boldsymbol{\omega}^2\mathbf{r}$ and is also directed radially *outward*. Then,

$$d\phi = -\boldsymbol{\omega}^2\mathbf{r} \cdot d\mathbf{r} \tag{12-35}$$

$$\underline{PE} = M \int d\phi = M\phi = -M \int \omega^2 r\, dr \tag{12-36}$$

Example 12.8

What is the potential of a synchronous satellite at a distance r from the center of the earth and at an altitude where $g = g_r$?

Solution

From the definitions above,

$$d\phi = (g_r - \omega^2 r)\, dr.$$

For synchronous operation, $d\phi = 0$ or,

$$g_r = \omega^2 r$$

The kinetic energy of a system is, of course,

$$\underline{KE} = \frac{1}{2} M v^2 \tag{12-37}$$

and thus:

$$\underline{E} = \underline{U} + \underline{PE} + \underline{KE} = \underline{U} + M\phi + \frac{1}{2} M v^2 \tag{12-38}$$

To express the differential form for $\underline{E}$, and still employ the customary expansion for $\underline{U}$, we see immediately that there must be terms on the right-hand-side involving M, ϕ, and v. It we use Eq. (12-1) with no electrical, etc., work terms, for simplicity, with the molecular weight of j being m_j,

$$\begin{aligned} d\underline{E} &= d\underline{U} + d\,(\underline{PE}) + d\,(\underline{KE}) \\ &= T\,d\underline{S} - P\,d\underline{V} + M\,d\phi + M\,d\frac{v^2}{2} \\ &\quad + \sum_j \left(\mu_j + m_j\phi + \frac{m_j v^2}{2}\right) dN_j \end{aligned} \tag{12-39}$$

Let us now examine these terms more carefully. The $M\,d\phi$ and $M\,d(v^2/2)$ terms indicate the negative value of the work the system *does on the environment* in undergoing a change in ϕ or v. For example, let the process be the fall of the system from a higher to a lower elevation in a gravitational field and, in so falling, a weight is raised in the environment. Now $d\phi = g\,dL < 0$ (since L is decreasing) and thus the work done by the system on the environment is positive as indicated by the rise of the weight.

Within the summation we also find terms involving ϕ and v. The interpretation is simple. As the mole numbers change because of mass addition or loss, there is introduced or removed potential and kinetic energy from the system. Of course, since Eq. (12-39) represents a quasi-static process, all such interchange must be such that the specific potential and kinetic energies of the mass entering and leaving the system must be equal to that present in the system. This is rather obvious since only one ϕ or v is noted. The analogy is the Fundamental Equation $\underline{U} = \underline{U}(\underline{S}, \underline{V}, N)$ compared to the open-system First Law, Eq. (3-62). We could write the comparable open-system relation for this case by inspection:

$$\begin{aligned} d\underline{E} &= d\left(\underline{U} + \phi M + \frac{Mv^2}{2}\right) \\ &= đQ - đW + \left[\sum_{j=1}^{n} \left(\bar{h}_j + \phi m_j + \frac{m_j v^2}{2}\right) dn_j\right]_e \\ &\quad - \left[\sum_{j=1}^{n} \left(\bar{h}_j + \phi m_j + \frac{m_j v^2}{2}\right) dn_j\right]_\ell \end{aligned} \tag{12-40}$$

Then it is obvious that without the potential and kinetic energy terms, Eq. (12-40) reduces to the general open-system First Law expression, Eq. (3-62). To show the connection between Eqs. (12-39) and (12-40) in another way, assume that the interchange of mass, heat, and work with the enviroment is reversible, i.e.,

$$\bar{h}_{je} = \bar{h}_{j\ell} = \bar{H}_j \text{ (system)}$$

$$\phi_e = \phi_\ell = \phi \text{ (system)}$$

$$v_e = v_\ell = v \text{ (system)}$$

then:

$$d\underline{S} = \frac{đQ}{T} + \left[\sum_{j=1}^{n} \bar{S}_j \, dn_j\right]_e - \left[\sum_{j=1}^{n} \bar{S}_j \, dn_j\right]_\ell \tag{12-41}$$

and,

$$đW = P\,d\underline{V} - M\,d\phi - Mv\,dv \tag{12-42}$$

$$d\underline{E} = T\,d\underline{S} - P\,d\underline{V} + M\,d\phi + Mv\,dv + \left[\sum_{j=1}^{n}\left(\mu_j + m_j\phi + \frac{m_j v^2}{2}\right) dn_j\right]_e$$
$$- \left[\sum_{j=1}^{n}\left(\mu_j + m_j\phi + \frac{m_j v^2}{2}\right) dn_j\right]_\ell \tag{12-43}$$

or, since $(dn_j)_e - (dn_j)_\ell = dN_j$, by substitution, Eq. (12-39) is obtained.

Applying Euler's theorm to Eq. (12-39) and noting that ϕ and v are invariant with changes in mass,

$$\left.\begin{aligned} \underline{E} &= T\underline{S} - P\underline{V} + \sum_{j=1}^{n}\left(\mu_j + m_j\phi + \frac{m_j v^2}{2}\right)N_j \\ &= \underline{U} + \sum_{j=1}^{n}\left(m_j\phi + \frac{m_j v^2}{2}\right)N_j \end{aligned}\right\} \tag{12-44}$$

Since $\underline{E} = \underline{E}(\underline{S}, \underline{V}, \phi, v, N_j)$, it is now possible to employ Legendre transforms as developed in Chapter 5 to obtain other fundamental representations and relationships between parital derivatives.

Example 12.9

Obtain the transforms (a) $\psi(T, P, \phi, v, N_1, \ldots, N_n)$ and (b) $\psi(T, P, M, v, N_1, \ldots, N_n)$ and show how the pressure varies with potential at constant T and velocity in a closed system.

Solution

(a) $\psi(T, P, \phi, v, N_1, \ldots, N_n)$

$$\psi = \underline{E} - T\underline{S} + P\underline{V}$$

$$d\psi = -\underline{S}\,dT + \underline{V}\,dP + M\,d\phi + Mv\,dv + \sum_{j=1}^{n}\left(\mu_j + m_j\phi + \frac{m_j v^2}{2}\right) dN_j$$

(b) $\psi(T, P, M, v, N_1, \ldots, N_n)$

$$\psi = \underline{E} - T\underline{S} + P\underline{V} - M\phi$$

$$d\psi = -\underline{S}\,dT + \underline{V}\,dP - \phi\,dM + Mv\,dv + \sum_{j=1}^{n}\left(\mu_j + m_j\phi + \frac{m_j v^2}{2}\right) dN_j$$

From this last relation,

$$\left(\frac{\partial P}{\partial \phi}\right)_{T,v,N} = -\left(\frac{\partial M}{\partial \underline{V}}\right)_{T,P,v,N} = -\rho$$

where ρ is the mass density. This relation is, of course, the usual equation to determine a hydrostatic pressure.

The final point of interest in this section involves the criteria of equilibrium in body force fields. We shall only develop such criteria for a system in a potential energy field because no useful criteria are found for systems with kinetic energy (i.e., it is readily shown that systems with the lowest velocities are most stable).

For a system in a potential energy field, one criterion often given is that:

$$dP = -\rho\, d\phi \tag{12-45}$$

but this was already derived in Example 12.9. Furthermore, it may easily be shown that, at equilibrium, there can be no gradients of temperature for a system in a potential field. The more interesting criteria result from an examination of the variation of chemical potential with ϕ.

Consider Figure 12.2. We show a system at equilibrium in a potential field. That is, there is some difference in ϕ between points A and B. (One might consider the mixture to be in a vertical pipe with A and B representing two heights.) The temperature of the system is uniform although the pressure varies with ϕ [see Eq. (12-45)]. At A and B we have semipermeable membranes that allow component j to pass freely into side tube C. At A the chemical potential of j is equal on both sides of the membrane; a similar statement

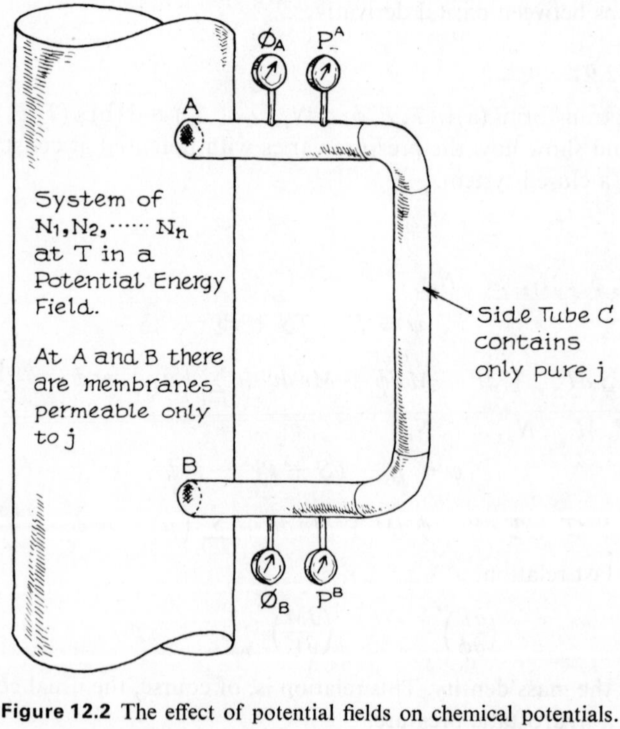

Figure 12.2 The effect of potential fields on chemical potentials.

may be made at B. Thus,

$$\mu_{j,\text{mix}}^A = \mu_{j,\text{pure}}^A$$

$$\mu_{j,\text{mix}}^B = \mu_{j,\text{pure}}^B$$

subtracting

$$\mu_{j,\text{mix}}^A - \mu_{j,\text{mix}}^B = \mu_{j,\text{pure}}^A - \mu_{j,\text{pure}}^B \tag{12-46}$$

But in the side tube, where only pure j is present,

$$\mu_{j,\text{pure}}^A - \mu_{j,\text{pure}}^B = \int_{P^B}^{P^A} V_j\, dP = \int_{\phi^B}^{\phi^A} (-V_j \rho_j)\, d\phi = -m_j(\phi^A - \phi^B) \tag{12-47}$$

Combining Eqs. (12-46) and (12-47) and noting that A and B could have been chosen at random,

$$\mu_{j,\text{mix}}^A + m_j\phi^A = \mu_{j,\text{mix}}^B + m_j\phi^B = \text{constant} \tag{12-48}$$

Eq. (12-48) is the desired result. It shows that the chemical potential is not a constant in a potential energy field but rather the sum of the chemical potential *and* the product of the molecular weight times the potential is constant.

Example 12.10

A deep, small well hole has been capped and left undisturbed for years. It is believed that the temperature of the gas does not vary significantly with depth and that convection currents are negligible. The pressure at the well top is 2 atm, the temperature 500°R, and the gas composition is 70 mole percent helium and 30 mole percent methane. The hole is one mile deep. What is the helium mole fraction at the bottom? Assume ideal gases.

Solution

With Z measured down from the surface and $\phi = 0$ on the surface, $\phi = -gZ$. Then from Eq. (12-47)

$$\mu_{\text{He},b} - \mu_{\text{He},t} = -m_{\text{He}}(\phi_b - \phi_t) = m_{\text{He}}gZ_b$$

where the subscripts b and t represent bottom and top conditions. $m_{\text{He}} =$ 4 kg/kg-mol, $g = 9.81$ m/s², and $Z_b = 5280$ ft $= 1609$ m. Also, for ideal gases, with $T = 278$°K, $R = 8.314$ kJ/kg-mol°K,

$$\mu_{\text{He},b} - \mu_{\text{He},t} = RT \ln \frac{\hat{f}_{\text{He},b}}{\hat{f}_{\text{He},t}} = RT \ln \frac{p_{\text{He},b}}{p_{\text{He},t}}$$

Thus,

$$p_{\text{He},b} = p_{\text{He},t} \exp\left[\frac{(4)(9.81)(1609)}{(8314)(278)}\right] = (2)(0.7)(1.028) = 1.44 \text{ atm}$$

In a similar manner $p_{\text{CH}_4,b} = 0.67$ atm, so the fraction helium at the bottom is $1.44/(1.44 + 0.67) = 0.68$.

PROBLEMS

12.1. Show that for an equilibrium system that exists both within and external to an electric field there is no gradient in chemical potential. Assume that the temperature is maintained constant throughout.

12.2. For a mixture of reacting gases, show how the chemical equilibrium constant varies with the electric field strength.

12.3. Present a similar analysis as requested in Problem 12.2 but for the magnetic field strength.

12.4. What is the work required to establish a magnetic field of 10,000 gauss in a long solenoid that contains one kg-mol of air at standard conditions? Assume that $\mu_{\text{air}} = \mu_0$.

12.5. A student is carrying out an experiment in which he desires to hang a balance pan a definite distance above a table top. (See Figure P12.5.) To accomplish this, he decides to use a rubber band that he has conveniently found in his pocket. He readily obtains the desired separation distance by adding mass to the pan in order to stretch the rubber band the correct amount. The student has, however, just come in from the outside and when used, the rubber band is at the temperature outside the laboratory. He works rapidly and when the pan and weights are hung on the rubber band, it quickly stretches to a length of 12 in.

Continuing the experiment, as the rubber band approaches room temperature, the student notices with irritation that the pan is rising. To counteract this, he gradually adds weights to the pan to maintain the same stretched length (12 in.). A total of 14 g has to be added until the rubber band is at room temperature and no more movement occurs.

Is it summer or winter outside?

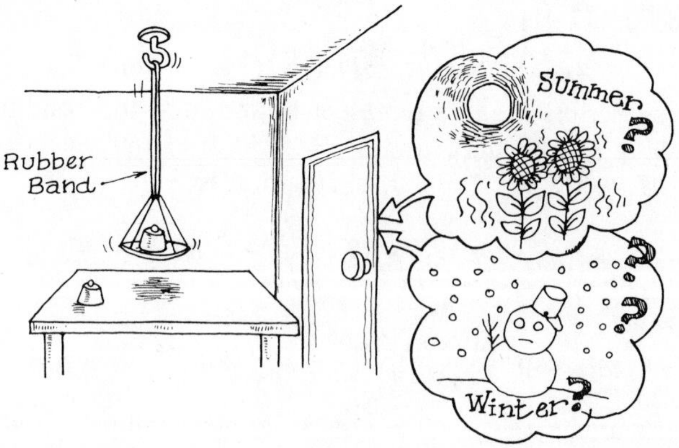

Figure P12.5

Assume that the original extension of the rubber band is adiabatic and reversible. Initially, the rubber band was 3 in. long and had an (unstretched) cross section measuring $\frac{1}{16} \times \frac{1}{16}$ in. It does not crystallize when stretched and data for this particular rubber band show that $(\partial \mathcal{F}/\partial T)_L = 4410$ N/m² °K, and $T(\partial S/\partial T)_L \sim 1.9$ kJ/kg °K. The rubber band has a specific gravity of 0.95 and the room temperature is 298°K.

12.6. Crackpot Inventions, Inc. is currently designing an elastic-rod Carnot engine. (See Figure P12.6.) In this device, work will be produced by heat transfer from a high-temperature (T_A) to a low-temperature (T_B) bath by the extension and relaxation of an elastic rod. This device will operate in a four-step cycle:

a. Isothermally stretching the elastic rod from L_1 to L_2 while transferring Q_B to the low-temperature bath.
b. Adiabatically and reversibly stretching the rod from L_2, T_B to L_3, T_A.
c. Isothermally relaxing the rod from L_3 to L_4, while absorbing Q_A from the high-temperature bath.
d. Adiabatically and reversibly relaxing the rod to its initial length.

As it now stands, the device is fitted for a 10-in. rod that is capable of stretching to 12 in. Crackpot wants to know what temperature they should use in their high-temperature bath so that their device can produce at least 25 ft-lb of work per cycle while using the room environment as the cold bath.

Their elastic rod has the following characteristics:

Size: 10 by 0.1 by 0.1 in. (unstretched)

$(\partial \mathcal{F}/\partial T)_L = 566$ lb/in.² °R

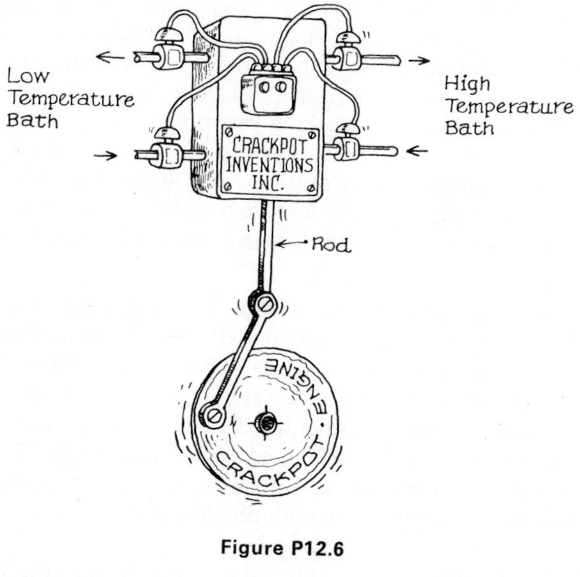

Figure P12.6

$C_L \equiv T(\partial S/\partial T)_L = 591$ ft-lb/lb °R

Specific gravity: 1.3

Room temperature is 75°F. What must be the temperature of the high-temperature bath? What is the efficiency of the engine?

12.7. A manned spacecraft is to be the payload of a large, multistage rocket. At liftoff, the rocket will accelerate vertically with a constant acceleration. Only later will a trajectory be programmed.

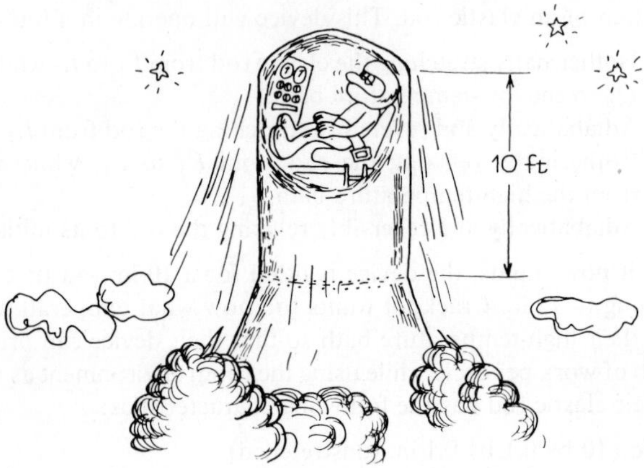

Figure P12.7

Prior to launch, the manned capsule is filled with 20% O_2 and 80% He at 60°F and 1 atm and is sealed until the high-acceleration launch phase is completed. During this same period, any heat transfer to or within the manned capsule may be neglected (i.e., it will remain isothermal), and assume that there is no composition variation due to respiration.

It is necessary to insure that during the launch acceleration period the oxygen partial pressure never drops below 130 mm Hg. Assuming that the capsule is 10 ft high and of constant cross section, what is the total pressure at the leading edge (top) of the capsule at the maximum tolerable acceleration?

12.8. We have just received a proposal to support a project designed to desalinate water for submerged submarines. Attached to the skin of the submarine would be a membrane that is semipermeable to water. When the submarine is submerged to the correct depth, water would pass into the submarine.

We assume that the sea water density and temperature remain constant with depth and are 63.93 lb/ft³ and 60°F, respectively. The submarine interior is at 1 atm and has a humidity equivalent to saturation at 60°F (0.2563 psia). The partial pressure over the ocean at the surface is 0.2424 psia.

At what depth should the submarine cruise to make the system work?

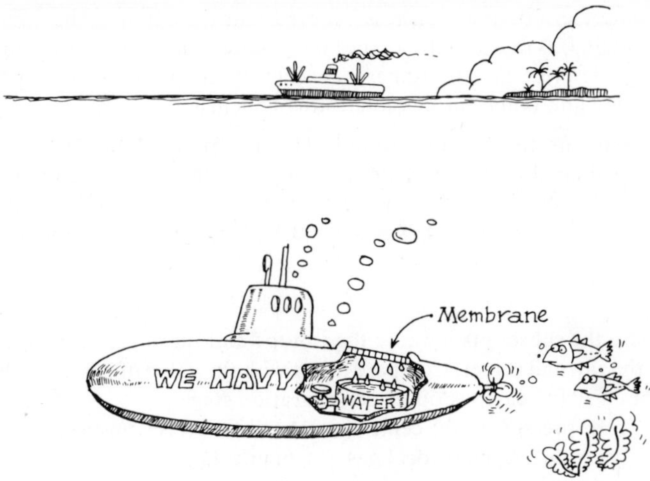

Figure P12.8

12.9. An equimolar mixture of He^3 and He^4 is pumped into a small diameter, 100 ft long, well-insulated, vertical tube. The initial pressure is 1 atm and the temperature 5°K. The gas mixture is ideal and no convection currents are present.

Derive a relation to express the equilibrium composition as a function of height and determine the fraction of He^4 and the total pressure at the bottom of the tube. How would your results change if the tube length were allowed to increase without limit?

12.10. Preliminary reports indicate that the atmosphere of Saturn consists primarily of methane and hydrogen. (See Figure P12.10.) At the outer boundary,

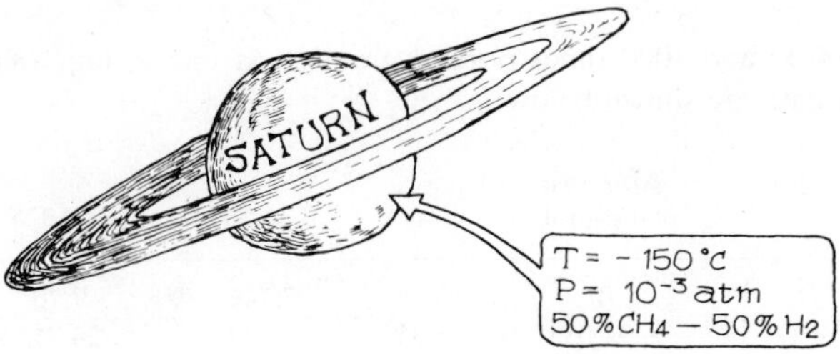

Figure P12.10

skimpy data show:

$T \sim -150°C$

$P \sim 10^{-3}$ atm

50% CH_4 and 50% H_2

From other observations, it is believed that the depth of the atmosphere is about 16,000 miles. With a constant gravitational acceleration equal to 1.1 g_{earth}, describe the atmospheric concentration and pressure variations with altitude and point out any interesting conclusions.

12.11. To separate the uranium isotope $U^{235}F_6$ from $U^{238}F_6$ by centrifugation one is limited by the mechanical strength of the material on the periphery. Recent reports, however, have indicated that some new alloys will withstand tangential velocities of 400 m/s at 400°K. Estimate a separation factor α_{8-5}:

$$\alpha_{8-5} = \frac{y_8^r y_5^0}{y_5^r y_8^0}$$

where the subscripts indicate the isotope and the superscripts r and 0 refer to the rim and center. Compare this α with the value of 1.0043 as determined from a single gaseous diffusion separation stage.

If the pressure at the center were 1 atm, what is your best estimate of the rim pressure? Assume ideal gases. Comment.

12.12. A centrifuge 10 in. in radius rotating at 15,000 rpm contains a mixture of benzene and toluene at 150°F. After equilibrium is reached, the mixture at the center is sampled and found to be a vapor with 50 mm Hg partial pressures of both benzene and toluene.

Plot the concentration and total pressure profile as a function of radius. Assume that the gas is ideal and that any liquid phase present forms an ideal solution. At 150°F, the vapor pressures of the benzene and toluene are 468 and 175 mm Hg, respectively.

12.13. In some of the new high-thrust rockets extreme acceleration fields are expected. Combustion experts are already analyzing problems expected to be encountered in such an environment. You are asked to aid in the program to estimate the equilibrium constant for the reaction:

$$CO + \frac{1}{2}O_2 = CO_2$$

at 1000°K and 1000 times normal gravity. At one g, $\log_{10} K_{1000} = 10.3$. Other data are shown below:

	$\Delta H_{f,298}$ cal/g-mol	A	$B \times 10^3$	$C \times 10^7$
O_2	0	6.26	2.746	−7.70
CO_2	−94,052	6.85	8.533	−24.75
CO	−26,416	6.25	2.091	−4.59

C_p(cal/g-mol °K) $= A + BT/2 + CT^2/3$, where T is in °C.

12.14. Tritium separation from deuterium and hydrogen is not easily accomplished. Yet there are now suggestions that it can be more readily carried out in ultracentrifuges. Estimate what increase in tritium concentration you might expect at the rim draw-off cock if the feed is at atmospheric pressure and is comprised of 5% T_2, 50% D_2, and 45% H_2. The unit is to operate isothermally at 43°K and with a rim velocity of 400 m/s. Do you visualize any practical problems?

12.15. A graduate student adept in the application of thermodynamics was overheard disputing the reasoning of a pragmatic meteorologist. They were discussing the pressure and composition profile in the earth's atmosphere. Both agreed that as a basis, they would choose the earth's surface where:

$N_2/O_2 = 4.0$

Air is saturated with water (= 6.3 mm Hg at 40°F)

$T = 40°F$

$P = 760$ mm Hg

The student would prefer to model the atmosphere as quiescent and isothermal in an equilibrium state. Using this model, estimate the N_2/O_2 ratio, the partial pressure of water, and the total pressure at an altitude of one mile. (Neglect CO_2.)

The meteorlogist claims that this model is sheer nonsense. His own model involves visualizing a unit mass of air at sea level expanding polytropically as it is moved up above the surface of the earth. By polytropic expansion, he means that the air expands by a relation $PV^n =$ constant. He has selected n to be 1.2. No compositional changes are allowed. Again calculate the pressure, the water partial pressure, and the temperature at an altitude of one mile. Compare and contrast both models.

12.16. As a final problem we present a variation on a scheme that has surfaced several times in the last few years. See, for example, *Scientific American*, Dec. 1971, p. 100 and April, 1972, p. 110 as well as *Science*, June 2, 1972, p. 1011 and Dec. 15, 1972, p. 1199.*

We suggest below a simple scheme to reclaim fresh water from the ocean. A long pipe is lowered into the ocean and on the bottom of this pipe we install a membrane permeable to water. (See Figure P12.16.) As we all know, the density of sea water exceeds that of fresh water; therefore, at any depth, the pressure in the ocean would exceed that which would be exerted by a comparable column of fresh water. If we make the pipe of sufficient length, the pressure difference at the bottom would exceed the osmotic pressure and water from the sea would flow into the pipe.

Comment on the feasibility of this process. Assume that the ocean is an equilibrium system.

How would your evaluation change if we took a more realistic ocean and assume that currents mix the ocean to such a degree that there is neither concentration nor temperature variation with depth?

*Regarding the article first mentioned, we enjoyed and profited by a series of letters with Professor Octave Levenspiel at Oregon State University and Professor Noel de Nevers at the University of Utah.

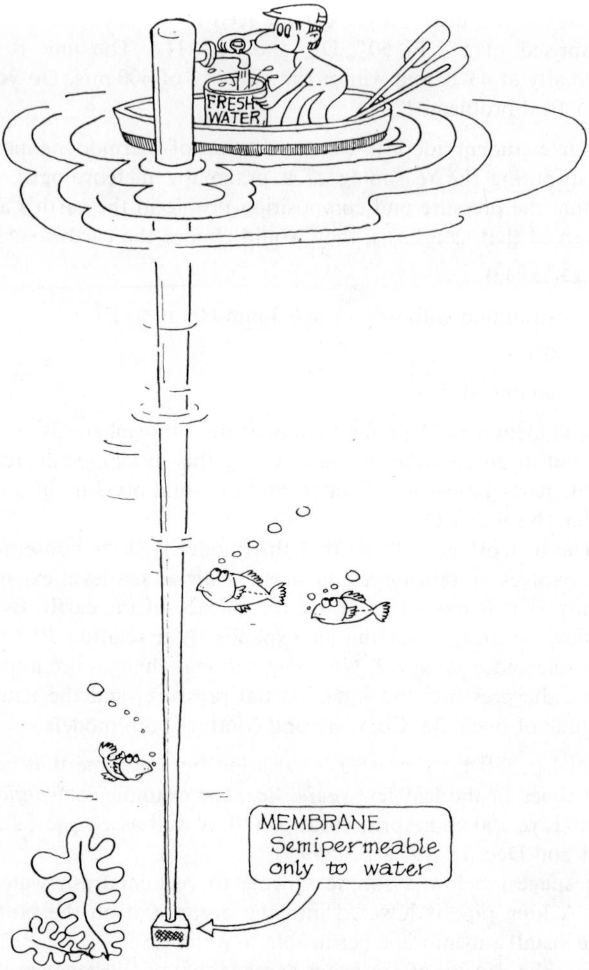

Figure P12.16

Some data are given below to aid in your evaluation:

Assume that the ocean temperature = 10°C.

P_{vp} for water at 10°C = 9.21 mm Hg.

The partial pressure of water over the ocean, at the surface, is 9.05 mm Hg (this value assumes a 3.5% NaCl solution).

Density of sea water = 1.0255 g/cm³.

Sea water and fresh water are incompressible.

The partial molal volume of H_2O in sea water is equal to the pure molal volume of H_2O.

Summary of the Postulates A

I. For closed simple systems with given internal restraints, there exist stable equilibrium states that can be characterized completely by two independently variable properties in addition to the masses of the particular chemical species initially charged.

II. In processes for which there is no net effect on the environment, all systems (simple and composite) with given internal restraints will change in such a way that they approach one and only one stable equilibrium state for each simple subsystem. In the limiting condition, the entire system is said to be at equilibrium.

III. For any states, (1) and (2), in which a closed system is at equilibrium, the change of state represented by (1) $\longrightarrow$ (2) and/or the reverse change (2) $\longrightarrow$ (1) can occur by at least one adiabatic process and the adiabatic work interaction between this system and its surroundings is determined uniquely by specifying the end states (1) and (2).

IV. If the sets of systems A,B and A,C each have no heat interaction when connected across nonadiabatic walls, then there will be no heat interaction if systems B and C are also so connected.

Mathematical Relations of Functions of State

B

Let B be any property, primitive or derived, of a system and let x, y, and z be independently variable properties of a single-component system. (The results can readily be generalized to $n + 2$ independent variables for a n-component system.) Since B is a function of state, a function f exists such that

$$B = f(x, y, z) \tag{B-1}$$

The function $f(x, y, z)$ is usually specified to within an arbitrary constant because derived properties are usually defined in terms of measured *differences* between two states. If the function f is known, differences in the value of B between two stable equilibrium states can be calculated as

$$\Delta B = B_2 - B_1 = f(x_2, y_2, z_2) - f(x_1, y_1, z_1) \tag{B-2}$$

In many cases, we may not know $f(x, y, z)$ explicitly, but we may have the differential form of Eq. (B-1):

$$dB = \left(\frac{\partial f}{\partial x}\right)_{y,z} dx + \left(\frac{\partial f}{\partial y}\right)_{x,z} dy + \left(\frac{\partial f}{\partial z}\right)_{x,y} dz \tag{B-3}$$

If all three partial derivatives are known, $f(x, y, z)$ can be evaluated to within an arbitrary constant by the method of indefinite integrals. Integrating first with respect to x,

$$f(x, y, z) = \int \left(\frac{\partial f}{\partial x}\right)_{y,z} dx + g(y, z) \tag{B-4}$$

where y and z are held constant in the integration and g is a function of y and z only. If we differentiate Eq. (B-4) with respect to y at constant x and z, and if we equate to the known function $(\partial f/\partial y)_{x,z}$

$$\left(\frac{\partial g}{\partial y}\right)_z = \left(\frac{\partial f}{\partial y}\right)_{x,z} - \frac{\partial}{\partial y}\left[\int \left(\frac{\partial f}{\partial x}\right)_{y,z} dx\right]_{x,z} \tag{B-5}$$

Integrating with respect to y while holding x and z constant, we obtain

$$g(y, z) = \int \left(\frac{\partial f}{\partial y}\right)_{x,z} dy - \int \frac{\partial}{\partial y}\left[\int \left(\frac{\partial f}{\partial x}\right)_{y,z} dx\right]_{x,z} dy + g'(z) \tag{B-6}$$

where g' is a function of z only. Substituting Eq. (B-6) into Eq. (B-4), we obtain

$$\begin{aligned} f(x, y, z) = &\int \left(\frac{\partial f}{\partial x}\right)_{y,z} dx + \int \left(\frac{\partial f}{\partial y}\right)_{x,z} dy \\ &- \int \frac{\partial}{\partial y}\left[\int \left(\frac{\partial f}{\partial x}\right)_{y,z} dx\right]_{x,z} dy + g'(z) \end{aligned} \tag{B-7}$$

The function $g'(z)$ can be evaluated to within an arbitrary constant by repeating the procedure.

For functions of more than two variables, the method of indefinite integrals is somewhat laborious, and it is only worth the effort if an analytical solution of $f(x, y, z)$ is desired.

In general, we will be interested in evaluating numerical differences in B, and a somewhat simpler solution can usually be obtained by integrating Eq. (B-3) over a specific path. Since B is a state function, the value of ΔB will be independent of the path chosen for integration; hence, any convenient path will suffice. One such path is to proceed from x_1, y_1, z_1 to x_2, y_1, z_1 to x_2, y_2, z_1 and then to the final state x_2, y_2, z_2. In this case, it can be readily shown that

$$\Delta B = \int_{x_1}^{x_2} \left(\frac{\partial f}{\partial x}\right)_{y_1,z_1} dx + \int_{y_1}^{y_2} \left(\frac{\partial f}{\partial y}\right)_{x_2,z_1} dy + \int_{z_1}^{z_2} \left(\frac{\partial f}{\partial z}\right)_{x_2,y_2} dz \tag{B-8}$$

Finally, if we are faced with the problem of evaluating the difference in a function ϕ given a differential equation of the type

$$d\phi = M\,dx + N\,dy + Q\,dz \tag{B-9}$$

We can use any of the methods described above if it can be shown that ϕ is a state function of the variables, x, y, and z. It can be shown that the necessary and sufficient requirement for this condition is that each of the following equations are satisfied:

$$\left(\frac{\partial M}{\partial y}\right)_{x,z} = \left(\frac{\partial N}{\partial x}\right)_{y,z}$$
$$\left(\frac{\partial M}{\partial z}\right)_{x,y} = \left(\frac{\partial Q}{\partial x}\right)_{y,z} \qquad \text{(B-10)}$$
$$\left(\frac{\partial N}{\partial z}\right)_{x,y} = \left(\frac{\partial Q}{\partial y}\right)_{x,z}$$

Alternatively, if any of Eq. set (B-10) is not satisfied, $d\phi$ is not an exact differential (e.g., either ϕ is not a state function or ϕ is a state function of variables other than x, y, and z). In this case, $\int d\phi$ is called a line integral because the value of $\Delta\phi$ will depend on the specific path used for integration.

Some important thermodynamic variables are not state functions or properties (i.e., work and heat interactions). Differentials of these functions will be denoted by a bar drawn through the d of the differential sign as $đ\phi$. Such functions are sometimes referred to as Pfaffians.

Derivation of Euler's Theorem

C

Consider a function $f(a, b, x, y)$ which is homogeneous to the degree h in x and y. By definition, if the variables x and y are each multiplied by a factor k, the value of $f(a, b, kx, ky)$ will be increased by a factor of k^h. Thus, for any value of k, we have

$$f(a, b, X, Y) = k^h f(a, b, x, y) \tag{C-1}$$

where

$$X = kx \quad \text{and} \quad Y = ky$$

Equating the total differentials of Eq. (C-1) and treating k as a variable since Eq. (C-1) is valid for all k, we obtain

$$\begin{aligned}
&\frac{\partial}{\partial a}[f(a, b, X, Y)]_{b,X,Y}\,da + \frac{\partial}{\partial b}[f(a, b, X, Y)]_{a,X,Y}\,db \\
&\quad + \frac{\partial}{\partial X}[f(a, b, X, Y)]_{a,b,Y}\,dX + \frac{\partial}{\partial Y}[f(a, b, X, Y)]_{a,b,X}\,dY \\
&= (k^h)\frac{\partial}{\partial a}[f(a, b, x, y)]_{b,x,y}\,da + (k^h)\frac{\partial}{\partial b}[f(a, b, x, y)]_{a,x,y}\,db \\
&\quad + (k^h)\frac{\partial}{\partial x}[f(a, b, x, y)]_{a,b,y}\,dx + (k^h)\frac{\partial}{\partial y}[f(a, b, x, y)]_{a,b,x}\,dy \\
&\quad + (hk^{h-1})[f(a, b, x, y)]\,dk
\end{aligned} \tag{C-2}$$

but

$$dX = k\,dx + x\,dk \quad \text{and} \quad dY = k\,dy + y\,dk \tag{C-3}$$

Substituting Eq. (C-3) into Eq. (C-2), and collecting terms, we obtain

$$\begin{aligned}
&\left\{\frac{\partial}{\partial a}[f(a, b, X, Y)] - (k^h)\frac{\partial}{\partial a}[f(a, b, x, y)]\right\} da \\
&+ \left\{\frac{\partial}{\partial_c}[f(a, b, X, Y)] - (k^h)\frac{\partial}{\partial_b}[f(a, b, x, y)]\right\} db \\
&+ \left\{(k)\frac{\partial}{\partial X}[f(a, b, X, Y)] - (k^h)\frac{\partial}{\partial x}[f(a, b, x, y)]\right\} dx \\
&+ \left\{(k)\frac{\partial}{\partial Y}[f(a, b, X, Y)] - (k^h)\frac{\partial}{\partial y}[f(a, b, x, y)]\right\} dy \\
&+ \left\{(x)\frac{\partial}{\partial X}[f(a, b, X, Y)] + (y)\frac{\partial}{\partial Y}[f(a, b, X, Y)] \right. \\
&\quad \left. - hk^{h-1}f(a, b, x, y)\right\} dk = 0
\end{aligned} \tag{C-4}$$

Since a, b, x, y, and k are independent, Eq. (C-4) is valid only if the coefficients of da, db, dx, dy, and dk are each zero. Thus,

$$\frac{\partial}{\partial a}[f(a, b, X, Y)] = (k^h)\frac{\partial}{\partial a}[f(a, b, x, y)] \tag{C-5}$$

$$\frac{\partial}{\partial b}[f(a, b, X, Y)] = (k^h)\frac{\partial}{\partial b}[f(a, b, x, y)] \tag{C-6}$$

$$\frac{\partial}{\partial X}[f(a, b, X, Y)] = (k^{h-1})\frac{\partial}{\partial x}[f(a, b, x, y)] \tag{C-7}$$

$$\frac{\partial}{\partial Y}[f(a, b, X, Y)] = (k^{h-1})\frac{\partial}{\partial y}[f(a, b, x, y)] \tag{C-8}$$

$$(x)\frac{\partial}{\partial X}[f(a, b, X, Y)] + (y)\frac{\partial}{\partial Y}[f(a, b, X, Y)] = (hk^{h-1})[f(a, b, x, y)] \tag{C-9}$$

Substituting Eqs. (C-7) and (C-8) into (C-9), we obtain

$$(x)\frac{\partial}{\partial x}[f(a, b, x, y)] + (y)\frac{\partial}{\partial y}[f(a, b, x, y)] = h[f(a, b, x, y)] \tag{C-10}$$

Eq. (C-10) is the general form of Euler's theorem. Note that it contains terms only in those variables for which f is homogeneous to degree h.

Applications of Euler's Theorem

The thermodynamic functions of interest to us are special cases of homogeneous functions. In particular, all of our functions are either homogeneous to the first degree in mass (extensive) or homogeneous to the zeroth degree in mass (intensive). Thus, the arbitrary multiplier, k, will always be n (or $1/n$ as the case may be), and h will be either one or zero.

Energy

$$\underline{U} = f(\underline{S}, \underline{V}, N)$$

Since $\underline{U}$ is first order in mass, and since $\underline{S}$, $\underline{V}$, N are all proportional to mass, we have

$$\underline{U}(n\underline{S}, n\underline{V}, nN) = n\underline{U}(\underline{S}, \underline{V}, N)$$

Therefore,

$$x = \underline{S}$$

$$y = \underline{V}$$

$$z = N$$

so that Eq. (C-10) leads to,

$$\underline{U} = \left(\frac{\partial \underline{U}}{\partial \underline{S}}\right)_{\underline{V},N} \underline{S} + \left(\frac{\partial \underline{U}}{\partial \underline{V}}\right)_{\underline{S},N} \underline{V} + \left(\frac{\partial \underline{U}}{\partial N}\right)_{\underline{S},\underline{V}} N$$

or

$$\underline{U} = T\underline{S} - P\underline{V} + \mu N$$

Enthalpy

$$\underline{H} = f(\underline{S}, P, N)$$

If we multiply the mass by n, we will increase $\underline{S}$ and N by a factor of n, but P will remain unchanged. That is,

$$\underline{H}(n\underline{S}, P, nN) = n\underline{H}(\underline{S}, P, N)$$

Thus,

$$a = P$$

$$x = \underline{S}$$

$$y = N$$

so that from Eq. (C-10),

$$\underline{H} = \left(\frac{\partial \underline{H}}{\partial \underline{S}}\right)_{P,N} \underline{S} + \left(\frac{\partial \underline{H}}{\partial N}\right)_{\underline{S},P} N$$

or

$$\underline{H} = T\underline{S} + \mu N$$

Estimation of the Thermodynamic Properties of Pure Materials

D

Heat Capacities in the Ideal Gas State

The calculation of C_v^* from theory was outlined in Section 6.5. It was pointed out in that section that engineers have utilized the results of theoretical calculations for many compounds to allow them to develop more generalized, but approximate, estimation methods. One such technique is shown below.

$$C_p^* = a + bT + cT^2 + dT^3 \tag{D-1}$$

Let T be in °K and C_p^* in cal/g-mol°K. The coefficients a, b, c, and d are determined by summing the group contributions applicable to the molecule in question. Table D.1 lists values suggested by Rihani and Doraiswamy.[1] Errors for most organic compounds are usually less than 4% for the temperature range between 300°K and 1000°K. For inorganic materials, other estimation methods are recommended.[2]

[1] R. N. Rihani and L. K. Doraiswamy, "Estimation of Heat Capacity of Organic Compounds from Group Contributions," *Ind. Eng. Chem. Fund.*, **4**, (1965), 17.

[2] R. C. Reid and T. K. Sherwood, *Properties of Gases and Liquids*, 2nd ed. (New York: McGraw-Hill Book Company, 1966), Chap. 5.

TABLE D.1
RIHANI AND DORAISWAMY GROUP CONTRIBUTIONS TO C_p^*

Aliphatic Hydrocarbon Groups

Group	*a*	*b* × 10²	*c* × 10⁴	*d* × 10⁶
$-CH_3$	0.6087	2.1433	−0.0852	0.001135
$-\overset{\vert}{C}H_2$	0.3945	2.1363	−0.1197	0.002596
$=CH_2$	0.5266	1.8357	−0.0954	0.001950
$-\overset{\vert}{\underset{\vert}{C}}-H$	−3.5232	3.4158	−0.2816	0.008015
$-\overset{\vert}{\underset{\vert}{C}}-$	−5.8307	4.4541	−0.4208	0.012630
$H\!>\!C{=}CH_2$	0.2773	3.4580	−0.1918	0.004130
$>\!C{=}CH_2$	−0.4173	3.8857	−0.2783	0.007364
$H\!>\!C{=}C\!<\!H$ (H, H above)	−3.1210	3.8060	−0.2359	0.005504
$H\!>\!C{=}C\!<\!H$ (H above, H below)	0.9377	2.9904	−0.1749	0.003918
$>\!C{=}C\!<\!H$	−1.4714	3.3842	−0.2371	0.006063
$>\!C{=}C\!<$	0.4736	3.5183	−0.3150	0.009205
$H\!>\!C{=}C{=}CH_2$	2.2400	4.2896	−0.2566	0.005908
$>\!C{=}C{=}CH_2$	2.6308	4.1658	−0.2845	0.007277
$H\!>\!C{=}C{=}C\!<\!H$	−3.1249	6.6843	−0.5766	0.017430
$\equiv CH$	2.8443	1.0172	−0.0690	0.001866

Aromatic Hydrocarbon Groups

Group	*a*	*b* × 10²	*c* × 10⁴	*d* × 10⁶
$HC\!<$	−1.4572	1.9147	−0.1233	0.002985
$-C\!<$	−1.3883	1.5159	−0.1069	0.002659
$\leftrightarrow C\!<$	0.1219	1.2170	−0.0855	0.002122

TABLE D.1 (Continued)

RIHANI AND DORAISWAMY GROUP CONTRIBUTIONS TO C_p^*

Contributions due to Ring Formation

Ring	*a*	$b \times 10^2$	$c \times 10^4$	$d \times 10^6$
3-membered ring	−3.5320	−0.0300	0.0747	−0.005514
4-membered ring	−8.6550	1.0780	0.0425	−0.000250
5-membered ring:				
Pentane	−12.2850	1.8609	−0.1037	0.002145
Pentene	−6.8813	0.7818	−0.0345	0.000591
6-membered ring:				
Hexane	−13.3923	2.1392	−0.0429	−0.001865
Hexene	−8.0238	2.2239	−0.1915	0.005473

Oxygen Containing Groups

Group	*a*	$b \times 10^2$	$c \times 10^4$	$d \times 10^6$
—OH	6.5128	−0.1347	0.0414	−0.001623
—O—	2.8461	−0.0100	0.0454	−0.002728
$-\overset{H}{\overset{\vert}{C}}=O$	3.5184	0.9437	0.0614	−0.006978
>C=O	1.0016	2.0763	−0.1636	0.004494
$-\overset{O}{\overset{\Vert}{C}}-O-H$	1.4055	3.4632	−0.2557	0.006886
—C(=O)O—	2.7350	1.0751	0.0667	−0.009230
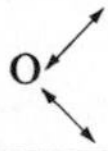	−3.7344	1.3727	−0.1265	0.003789

Sulfur Containing Groups

Group	*a*	$b \times 10^2$	$c \times 10^4$	$d \times 10^6$
—SH	2.5597	1.3347	−0.1189	0.003820
—S—	4.2256	0.1127	−0.0026	−0.000072
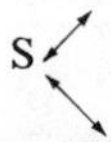	4.0824	−0.0301	0.0731	−0.006081
$-SO_3H$	6.9218	2.4735	0.1776	−0.022445

TABLE D.1 (Continued)

RIHANI AND DORAISWAMY GROUP CONTRIBUTIONS TO C_p^*

Halogen Containing Groups

Group	a	$b \times 10^2$	$c \times 10^4$	$d \times 10^6$
—F	1.4382	0.3452	−0.0106	−0.000034
—Cl	3.0660	0.2122	−0.0128	0.000276
—Br	2.7605	0.4731	−0.0455	0.001420
—I	3.2651	0.4901	−0.0539	0.001782

Nitrogen Containing Groups

Group	a	$b \times 10^2$	$c \times 10^4$	$d \times 10^6$
—C≡N	4.5104	0.5461	0.0269	−0.003790
—N≡C	5.0860	0.3492	0.0259	−0.002436
$—NH_2$	4.1783	0.7378	0.0679	−0.007310
>NH	−1.2530	2.1932	−0.1604	0.004237
>N—	−3.4677	2.9433	−0.2673	0.007828
N↗↖	2.4458	0.3436	0.0171	−0.002719
$—NO_2$	1.0898	2.6401	−0.1871	0.004750

P-V-T Properties of Pure Gases

Eq. (6-91) with Figures 6.1 and 6.2 constitute a convenient and accurate method to estimate the *P-V-T* behavior of pure component gases. The acentric factor, critical pressure, and critical temperature are required. A convenient tabulation of these constants is found in Appendix A of Reid and Sherwood, *op. cit.*

P-V-T Properties of Pure Liquids

A convenient generalized correlation relating the reduced density, temperature, and pressure is shown in Eq. (D-2)

$$\frac{V_c}{V} = \rho_r = f(Z_c, T_r, P_r) \tag{D-2}$$

The functionality is shown in Table D.2.[3] The critical volume must be

[3] A. L. Lydersen, R. A. Greenkorn, and O. A. Hougen, *Generalized Properties of Pure Fluids*, Coll. Eng., University of Wisconsin, Eng. Exp. Sta. Rpt. 4, Madison, Wisconsin, Oct., 1955.

TABLE D.2
REDUCED DENSITY OF LIQUIDS*

T_r	Saturated liquid				$P_r = 1.0$				$P_r = 2.0$		
	W $Z_c = 0.23$	I $Z_c = 0.25$	II $Z_c = 0.27$	III $Z_c = 0.29$	W $Z_c = 0.23$	I $Z_c = 0.25$	II $Z_c = 0.27$	III $Z_c = 0.29$	I $Z_c = 0.25$	II $Z_c = 0.27$	III Z_c = 0.2[illegible]
0.30	.....	3.487	3.287	3.081	.....	3.490	3.290	3.084	3.494	3.294	3.088
0.32	.....	3.450	3.253	3.049	.....	3.454	3.256	3.052	3.458	3.260	3.056
0.34	.....	3.419	3.223	3.021	.....	3.423	3.227	3.025	3.427	3.231	3.029
0.36	.....	3.383	3.189	2.989	.....	3.387	3.193	2.993	3.392	3.198	2.998
0.38	.....	3.348	3.156	2.959	.....	3.354	3.162	2.964	3.358	3.170	2.970
0.40	.....	3.306	3.118	2.922	.....	3.313	3.123	2.928	3.322	3.132	2.936
0.42	3.140	3.271	3.084	2.891	3.181	3.278	3.090	2.897	3.287	3.099	2.905
0.44	3.138	3.234	3.049	2.858	3.174	3.239	3.054	2.863	3.251	3.065	2.873
0.46	3.130	3.195	3.012	2.824	3.164	3.203	3.020	2.831	3.215	3.031	2.841
0.48	3.118	3.156	2.975	2.789	3.149	3.165	2.984	2.797	3.177	2.995	2.808
0.50	3.101	3.115	2.937	2.753	3.132	3.126	2.947	2.763	3.136	2.957	2.772
0.52	3.082	3.076	2.900	2.719	3.115	3.088	2.911	2.729	3.099	2.922	2.739
0.54	3.060	3.036	2.862	2.683	3.099	3.050	2.875	2.696	3.063	2.888	2.707
0.56	3.032	2.996	2.825	2.648	3.071	3.012	2.840	2.662	3.028	2.855	2.676
0.58	3.005	2.956	2.787	2.613	3.040	2.974	2.800	2.630	2.990	2.823	2.646
0.60	2.973	2.913	2.746	2.574	3.007	2.932	2.764	2.591	2.952	2.783	2.609
0.61	2.957	2.893	2.727	2.556	2.989	2.913	2.746	2.574	2.936	2.768	2.595
0.62	2.940	2.868	2.704	2.535	2.965	2.888	2.723	2.553	2.916	2.749	2.577
0.63	2.923	2.849	2.686	2.518	2.954	2.868	2.704	2.535	2.897	2.731	2.560
0.64	2.904	2.825	2.663	2.496	2.938	2.845	2.682	2.514	2.877	2.712	2.542
0.65	2.889	2.800	2.640	2.475	2.919	2.824	2.660	2.494	2.852	2.689	2.521
0.66	2.868	2.781	2.622	2.458	2.900	2.800	2.640	2.475	2.836	2.671	2.507
0.67	2.848	2.757	2.599	2.436	2.882	2.784	2.625	2.461	2.816	2.655	2.489
0.68	2.827	2.733	2.577	2.416	2.864	2.761	2.603	2.440	2.797	2.637	2.472
0.69	2.810	2.709	2.554	2.394	2.846	2.737	2.580	2.419	2.777	2.618	2.451
0.70	2.785	2.686	2.532	2.374	2.828	2.718	2.562	2.402	2.757	2.599	2.436
0.71	2.768	2.661	2.509	2.352	2.805	2.693	2.539	2.380	2.733	2.577	2.416
0.72	2.741	2.637	2.486	2.330	2.782	2.673	2.520	2.362	2.711	2.555	2.395
0.73	2.717	2.614	2.460	2.310	2.759	2.650	2.498	2.342	2.687	2.533	2.376
0.74	2.693	2.586	2.438	2.285	2.736	2.621	2.471	2.316	2.662	2.512	2.351
0.75	2.667	2.557	2.411	2.260	2.714	2.598	2.449	2.296	2.640	2.490	2.333
0.76	2.643	2.534	2.389	2.240	2.690	2.573	2.426	2.271	2.620	2.473	2.317
0.77	2.617	2.505	2.363	2.215	2.668	2.546	2.400	2.250	2.594	2.445	2.292
0.78	2.593	2.478	2.336	2.190	2.644	2.522	2.378	2.229	2.571	2.423	2.271
0.79	2.566	2.450	2.310	2.168	2.621	2.494	2.351	2.204	2.546	2.400	2.250
0.80	2.535	2.420	2.284	2.145	2.597	2.470	2.329	2.183	2.524	2.377	2.230
0.81	2.502	2.390	2.257	2.121	2.577	2.446	2.306	2.160	2.500	2.354	2.206
0.82	2.478	2.359	2.231	2.096	2.553	2.418	2.280	2.137	2.472	2.330	2.183
0.83	2.442	2.327	2.201	2.070	2.526	2.387	2.250	2.109	2.447	2.306	2.161
0.84	2.407	2.295	2.171	2.044	2.498	2.359	2.224	2.085	2.420	2.281	2.137
0.85	2.370	2.263	2.141	2.014	2.468	2.327	2.194	2.057	2.394	2.256	2.114
0.86	2.340	2.227	2.107	1.984	2.436	2.290	2.161	2.038	2.358	2.231	2.098
0.87	2.297	2.191	2.077	1.957	2.402	2.253	2.131	2.002	2.330	2.204	2.070
0.88	2.256	2.155	2.043	1.925	2.364	2.217	2.098	1.972	2.302	2.177	2.049
0.89	2.216	2.116	2.006	1.891	2.324	2.179	2.063	1.941	2.274	2.150	2.022
0.90	2.191	2.076	1.969	1.859	2.285	2.140	2.027	1.911	2.243	2.122	1.998
0.91	2.131	2.032	1.932	1.824	2.232	2.094	1.990	1.877	2.211	2.092	1.970
0.92	2.077	1.989	1.890	1.789	2.174	2.051	1.948	1.843	2.180	2.061	1.943
0.93	2.020	1.910	1.846	1.747	2.113	2.000	1.904	1.802	2.145	2.033	1.913
0.94	1.965	1.888	1.797	1.707	2.057	1.948	1.855	1.762	2.104	2.001	1.887
0.95	1.898	1.829	1.745	1.657	1.994	1.889	1.803	1.713	2.063	1.965	1.856
0.96	1.784	1.765	1.685	1.605	1.920	1.821	1.743	1.661	2.028	1.931	1.825
0.97	1.729	1.689	1.617	1.545	1.850	1.740	1.667	1.594	1.988	1.892	1.790
0.98	1.628	1.508	1.535	1.469	1.748	1.644	1.580	1.513	1.916	1.852	1.755
0.99	1.475	1.470	1.420	1.368	1.624	1.450	1.450	1.397	1.902	1.810	1.719
1.00	1.000	1.000	1.000	1.000	1.000	1.000	1.000	1.000	1.854	1.764	1.676

*A. L. Lydersen, R. A. Greenkorn, and O. A. Hougen, Generalized Thermodynamic Properties of Pure Fluids, *Coll. Eng., Univ. Wisconsin, Eng. Expt. Sta. Rept.* 4, Madison, Wis., October, 1955.

TABLE D.2 (Continued)

REDUCED DENSITY OF LIQUIDS

T_r	$P_r = 4.0$			$P_r = 6.0$			$P_r = 10$			$P_r = 15$		
	I $Z_c = 0.25$	II $Z_c = 0.27$	III $Z_c = 0.29$	I $Z_c = 0.25$	II $Z_c = 0.27$	III $Z_c = 0.29$	I $Z_c = 0.25$	II $Z_c = 0.27$	III $Z_c = 0.29$	I $Z_c = 0.25$	II $Z_c = 0.27$	III $Z_c = 0.29$
0.30	3.500	3.300	3.094	3.506	3.305	3.098	3.512	3.320	3.112	3.527	3.325	3.116
0.32	3.465	3.267	3.063	3.471	3.272	3.067	3.484	3.285	3.079	3.495	3.295	3.088
0.34	3.437	3.240	3.037	3.442	3.245	3.041	3.453	3.255	3.051	3.463	3.265	3.060
0.36	3.402	3.207	3.006	3.407	3.212	3.011	3.421	3.225	3.028	3.431	3.235	3.032
0.38	3.373	3.180	2.981	3.378	3.185	2.986	3.389	3.195	2.995	3.401	3.206	3.005
0.40	3.334	3.143	2.946	3.339	3.148	2.951	3.357	3.165	2.967	3.370	3.177	2.978
0.42	3.301	3.112	2.917	3.306	3.117	2.922	3.325	3.135	2.938	3.340	3.147	2.950
0.44	3.267	3.080	2.887	3.273	3.086	2.894	3.292	3.104	2.909	3.307	3.118	2.923
0.46	3.232	3.047	2.856	3.239	3.054	2.863	3.262	3.075	2.882	3.278	3.090	2.896
0.48	3.195	3.012	2.824	3.208	3.024	2.835	3.230	3.045	2.854	3.242	3.068	2.876
0.50	3.156	2.975	2.789	3.171	2.990	2.803	3.198	3.015	2.826	3.214	3.030	2.840
0.52	3.120	2.941	2.757	3.140	2.960	2.775	3.166	2.985	2.798	3.182	3.000	2.812
0.54	3.088	2.911	2.729	3.104	2.926	2.743	3.134	2.955	2.770	3.153	2.973	2.787
0.56	3.056	2.881	2.701	3.072	2.896	2.715	3.103	2.925	2.742	3.120	2.949	2.764
0.58	3.020	2.847	2.669	3.040	2.870	2.691	3.072	2.896	2.714	3.093	2.916	2.733
0.60	2.984	2.813	2.637	3.008	2.836	2.659	3.044	2.870	2.690	3.063	2.888	2.707
0.61	2.964	2.794	2.619	2.996	2.825	2.649	3.028	2.855	2.676	3.050	2.875	2.695
0.62	2.945	2.776	2.602	2.980	2.809	2.634	3.013	2.841	2.663	3.036	2.862	2.683
0.63	2.929	2.761	2.588	2.964	2.794	2.620	2.998	2.828	2.651	3.022	2.849	2.670
0.64	2.913	2.746	2.574	2.948	2.779	2.606	2.985	2.814	2.638	3.008	2.836	2.660
0.65	2.893	2.727	2.556	2.932	2.764	2.591	2.970	2.800	2.624	2.995	2.824	2.647
0.66	2.877	2.712	2.542	2.916	2.749	2.577	2.951	2.782	2.602	2.982	2.811	2.635
0.67	2.856	2.693	2.524	2.900	2.734	2.563	2.940	2.772	2.598	2.968	2.798	2.623
0.68	2.836	2.676	2.507	2.881	2.716	2.546	2.918	2.751	2.579	2.954	2.785	2.610
0.69	2.820	2.660	2.494	2.865	2.701	2.532	2.910	2.743	2.571	2.940	2.776	2.602
0.70	2.802	2.642	2.477	2.849	2.686	2.518	2.890	2.730	2.560	2.928	2.760	2.589
0.71	2.784	2.625	2.461	2.833	2.671	2.500	2.872	2.716	2.547	2.915	2.748	2.577
0.72	2.765	2.607	2.444	2.816	2.655	2.489	2.860	2.701	2.533	2.901	2.735	2.565
0.73	2.747	2.590	2.428	2.799	2.639	2.474	2.851	2.688	2.521	2.894	2.720	2.559
0.74	2.729	2.573	2.412	2.781	2.622	2.458	2.837	2.675	2.509	2.874	2.710	2.542
0.75	2.709	2.554	2.394	2.761	2.603	2.441	2.812	2.661	2.495	2.861	2.697	2.530
0.76	2.689	2.535	2.376	2.745	2.588	2.426	2.800	2.646	2.480	2.818	2.685	2.518
0.77	2.672	2.518	2.360	2.729	2.573	2.412	2.781	2.631	2.468	2.833	2.671	2.505
0.78	2.652	2.500	2.344	2.709	2.554	2.394	2.770	2.617	2.454	2.820	2.659	2.494
0.79	2.631	2.480	2.325	2.693	2.539	2.380	2.750	2.602	2.440	2.807	2.646	2.482
0.80	2.609	2.460	2.306	2.673	2.520	2.362	2.735	2.587	2.432	2.790	2.634	2.476
0.81	2.588	2.440	2.287	2.656	2.504	2.347	2.719	2.572	2.418	2.778	2.621	2.464
0.82	2.567	2.420	2.269	2.638	2.487	2.331	2.704	2.558	2.405	2.762	2.609	2.452
0.83	2.546	2.400	2.250	2.619	2.470	2.315	2.687	2.542	2.389	2.750	2.595	2.439
0.84	2.524	2.380	2.232	2.660	2.449	2.295	2.671	2.527	2.375	2.740	2.583	2.428
0.85	2.503	2.360	2.214	2.580	2.433	2.281	2.655	2.512	2.361	2.720	2.571	2.417
0.86	2.482	2.340	2.195	2.562	2.416	2.265	2.639	2.496	2.346	2.708	2.559	2.405
0.87	2.461	2.320	2.176	2.543	2.398	2.248	2.621	2.479	2.330	2.698	2.545	2.392
0.88	2.438	2.299	2.156	2.524	2.380	2.232	2.606	2.465	2.317	2.682	2.532	2.380
0.89	2.415	2.277	2.136	2.505	2.362	2.216	2.590	2.450	2.303	2.670	2.519	2.368
0.90	2.390	2.257	2.119	2.486	2.344	2.200	2.572	2.434	2.282	2.658	2.506	2.349
0.91	2.365	2.235	2.100	2.466	2.325	2.182	2.560	2.418	2.267	2.644	2.493	2.340
0.92	2.342	2.214	2.080	2.440	2.307	2.165	2.540	2.402	2.252	2.632	2.481	2.330
0.93	2.316	2.191	2.060	2.420	2.288	2.117	2.532	2.387	2.238	2.620	2.470	2.316
0.94	2.292	2.168	2.039	2.400	2.268	2.129	2.514	2.370	2.222	2.606	2.457	2.303
0.95	2.267	2.145	2.018	2.378	2.249	2.113	2.498	2.355	2.208	2.593	2.445	2.292
0.96	2.240	2.120	1.995	2.356	2.229	2.096	2.480	2.338	2.192	2.581	2.433	2.281
0.97	2.211	2.095	1.973	2.334	2.208	2.077	2.463	2.322	2.177	2.567	2.420	2.269
0.98	2.184	2.072	1.950	2.313	2.188	2.059	2.446	2.306	2.162	2.555	2.409	2.258
0.99	2.155	2.043	1.925	2.289	2.165	2.038	2.429	2.290	2.147	2.541	2.396	2.246
1.00	2.127	2.016	1.900	2.266	2.143	2.018	2.412	2.274	2.132	2.532	2.383	2.234

Table D.2 (Continued)

Reduced Density of Liquids

T_r	$P_r = 20$			$P_r = 25$			$P_r = 30$		
	I $Z_c = 0.25$	II $Z_c = 0.27$	III $Z_c = 0.29$	I $Z_c = 0.25$	II $Z_c = 0.27$	III $Z_c = 0.29$	I $Z_c = 0.25$	II $Z_c = 0.27$	III $Z_c = 0.29$
0.30	3.535	3.333	3.124	3.540	3.337	3.128	3.546	3.343	3.133
0.32	3.506	3.305	3.098	3.511	3.310	3.102	3.517	3.316	3.108
0.34	3.474	3.275	3.070	3.481	3.282	3.076	3.489	3.289	3.083
0.36	3.442	3.245	3.042	3.453	3.255	3.051	3.459	3.261	3.057
0.38	3.410	3.215	3.013	3.421	3.225	3.023	3.430	3.233	3.030
0.40	3.378	3.185	2.985	3.390	3.196	2.996	3.400	3.205	3.004
0.42	3.350	3.158	2.960	3.360	3.168	2.969	3.370	3.177	2.978
0.44	3.319	3.129	2.933	3.306	3.140	2.943	3.341	3.150	2.952
0.46	3.288	3.100	2.906	3.302	3.113	2.918	3.310	3.121	2.925
0.48	3.257	3.071	2.878	3.274	3.085	2.892	3.282	3.094	2.900
0.50	3.226	3.041	2.850	3.245	3.059	2.867	3.252	3.066	2.874
0.52	3.197	3.014	2.825	3.214	3.030	2.840	3.225	3.040	2.849
0.54	3.167	2.986	2.799	3.186	3.004	2.816	3.198	3.015	2.826
0.56	3.139	2.959	2.773	3.157	2.976	2.789	3.170	2.989	2.802
0.58	3.109	2.931	2.750	3.130	2.951	2.766	3.145	2.965	2.779
0.60	3.081	2.905	2.723	3.103	2.925	2.742	3.120	2.941	2.757
0.61	3.070	2.894	2.713	3.090	2.913	2.730	3.108	2.930	2.746
0.62	3.056	2.881	2.700	3.076	2.900	2.718	3.096	2.919	2.736
0.63	3.044	2.870	2.690	3.064	2.889	2.708	3.082	2.906	2.724
0.64	3.031	2.858	2.679	3.052	2.877	2.697	3.065	2.890	2.709
0.65	3.018	2.845	2.667	3.040	2.866	2.686	3.060	2.885	2.704
0.66	3.005	2.833	2.655	3.028	2.855	2.676	3.050	2.875	2.695
0.67	2.991	2.820	2.643	3.016	2.843	2.665	3.038	2.864	2.684
0.68	2.980	2.809	2.633	3.004	2.832	2.654	3.027	2.854	2.675
0.69	2.967	2.797	2.622	2.912	2.820	2.643	3.016	2.843	2.665
0.70	2.955	2.786	2.613	2.981	2.810	2.635	3.005	2.833	2.657
0.71	2.943	2.775	2.600	2.969	2.799	2.625	2.994	2.823	2.648
0.72	2.931	2.763	2.591	2.956	2.787	2.614	2.985	2.814	2.639
0.73	2.917	2.750	2.579	2.945	2.776	2.604	2.974	2.804	2.630
0.74	2.906	2.740	2.570	2.933	2.765	2.593	2.964	2.794	2.620
0.75	2.896	2.730	2.560	2.921	2.754	2.583	2.953	2.784	2.610
0.76	2.886	2.721	2.552	2.911	2.744	2.574	2.942	2.774	2.602
0.77	2.870	2.706	2.538	2.899	2.733	2.563	2.932	2.764	2.592
0.78	2.859	2.695	2.528	2.887	2.722	2.553	2.920	2.753	2.582
0.79	2.847	2.684	2.517	2.877	2.712	2.544	2.910	2.743	2.578
0.80	2.830	2.673	2.513	2.864	2.702	2.540	2.900	2.734	2.570
0.81	2.820	2.661	2.501	2.852	2.693	2.531	2.890	2.725	2.561
0.82	2.810	2.650	2.490	2.844	2.683	2.522	2.880	2.715	2.552
0.83	2.798	2.639	2.481	2.832	2.673	2.513	2.870	2.706	2.544
0.84	2.784	2.628	2.470	2.820	2.664	2.504	2.860	2.698	2.536
0.85	2.772	2.616	2.459	2.810	2.655	2.496	2.852	2.689	2.528
0.86	2.762	2.605	2.449	2.800	2.645	2.486	2.840	2.680	2.519
0.87	2.752	2.595	2.439	2.792	2.635	2.477	2.829	2.672	2.512
0.88	2.740	2.585	2.430	2.780	2.626	2.468	2.821	2.669	2.509
0.89	2.730	2.574	2.420	2.770	2.616	2.459	2.816	2.655	2.496
0.90	2.719	2.563	2.403	2.760	2.608	2.445	2.808	2.647	2.482
0.91	2.707	2.552	2.392	2.756	2.598	2.435	2.798	2.638	2.473
0.92	2.695	2.541	2.382	2.745	2.588	2.426	2.790	2.630	2.466
0.93	2.685	2.531	2.373	2.736	2.579	2.418	2.781	2.622	2.458
0.94	2.674	2.521	2.363	2.726	2.570	2.409	2.773	2.614	2.451
0.95	2.663	2.511	2.354	2.717	2.561	2.400	2.763	2.605	2.442
0.96	2.652	2.500	2.344	2.706	2.551	2.391	2.753	2.596	2.434
0.97	2.640	2.489	2.333	2.696	2.542	2.383	2.745	2.588	2.426
0.98	2.628	2.480	2.323	2.686	2.532	2.373	2.736	2.580	2.419
0.99	2.617	2.467	2.313	2.676	2.523	2.365	2.727	2.571	2.410
1.00	2.606	2.457	2.303	2.667	2.514	2.357	2.720	2.563	2.403

known. Alternatively, if the liquid density is available at one T_r and P_r, Eq. (D-2) may be written

$$\rho_2 = \rho_1\left[\frac{f(Z_c, T_{r_2}, P_{r_2})}{f(Z_c, T_{r_1}, P_{r_1})}\right] \tag{D-3}$$

Analytical correlations are also available.[4,5,6]

Vapor Pressures and Heats of Vaporization

The simplest possible vapor pressure correlation results from integration of the Clapeyron equation (see Chap. 9) between P_1, T_1 and P_2, T_2.

$$\ln\frac{P_2}{P_1} = \frac{\Delta H_v}{\Delta Z_v}\left(\frac{1}{T_1} - \frac{1}{T_2}\right) \tag{D-4}$$

The term $(\Delta H_v/\Delta Z_v)$ has been treated as a constant. $\Delta Z_v \equiv Z$ (saturated vapor) $-$ Z(saturated liquid) may be determined from the P-V-T property correlations of the individual phases.

It is more common, however, to integrate the Clapeyron equation as an indefinite integral, again assuming that $(\Delta H_v/\Delta Z_v)$ = constant,

$$\ln P_{vp} = A + \frac{B}{T} \tag{D-5}$$

Applying Eq. (D-5) at both the normal boiling point and at the critical point to obtain A and B, we have

$$\ln P_{vp_r} = \ln\frac{P_{vp}}{P_c} = h\left(1 - \frac{1}{T_r}\right) \tag{D-6}$$

$$h = \ln P_c\left[\frac{T_{b_r}}{(1 - T_{b_r})}\right] \tag{D-7}$$

Eqs. (D-6) and (D-7) allow a rapid, and usually reasonably accurate, estimation of the vapor pressure of a pure material at any temperature betwen T_b and T_c given P_c, T_c and T_b. More accurate, although more complex, correlations are, however, available.[7,8]

[4] L. C. Yen and S. S. Woods, "A Generalized Equation for Computer Calculation of Liquid Densities," *AIChE J.*, **12**, (1966), 95.

[5] E. W. Lyckman, C. A. Eckert and J. M. Prausnitz, "Generalized Liquid Volumes and Solubility Parameters for Regular Solution Application," *Chem. Eng. Sci.*, **20**, (1965), 703.

[6] P. L. Chueh and J. M. Praunsnitz, "Vapor-Liquid Equilibria at High Pressures: Calculation of Partial Molal Volumes in Nonpolar Liquid Mixtures," *AIChE J.*, **13**, (1967), 1099; ibid., "A Generalized Correlation for the Compressibilities of Normal Liquids," *AIChE J.*, **15**, (1969), 471.

[7] R. C. Reid and T. K. Sherwood, *op. cit.*, Chap. 4.

[8] R. E. Thek and L. I. Stiel, "A New Reduced Vapor Pressure Equation," *AIChE J.*, **12**, (1966), 599; ibid., **13**, (1967), 626.

For latent heats of vaporization, the empirical equations shown below are as accurate as any available.

$$\Delta H_v = GRT_c[1 + T_{b_r}^2 + k(1 + 2T_{b_r})]\frac{1 - 0.97}{P_c T_{b_r}}$$

$$G = 0.5691 + 0.4525h$$

$$k = \frac{h}{G} - (1 + T_{b_r}) \qquad \text{(D-8)}$$

h is defined in Eq. (D-7).

In these equations, P_c is in atmospheres and ΔH_v has the same units as RT_c.

Free Energies and Enthalpies of Formation

Although empirical estimation methods are available,[9] it is recommended that the excellent compilation of Stull, Westrum, and Sinke[10] be used if at all possible.

Effect of Pressure on Enthalpy and Entropy

Sections 6.6 and 6.7 show how isothermal enthalpy and entropy changes may be calculated. To carry out the integrations, P-V-T data must be available or an applicable equation of state can be used instead. We choose to illustrate one type of correlation where Eq. (6-91) is employed in Eqs. (6-96) and (6-100). The results are expressed as:

$$\frac{(H^* - H)}{RT_c} = \left[\frac{(H^* - H)}{RT_c}\right]^{(0)} + \omega\left[\frac{(H^* - H)}{RT_c}\right]^{(1)} \qquad \text{(D-9)}$$

$$\left[\frac{(S^* - S)}{R}\right] - \ln P = \left[\frac{(S^* - S)}{R}\right]^{(0)} + \omega\left[\frac{(S^* - S)}{R}\right]^{(1)} \qquad \text{(D-10)}$$

The square-bracket terms on the right-hand side have been plotted as a function of T_r and P_r by Edmister[11] from Pitzer's tables and are shown in Figures D.1 through D.4. In these figures H^* is the enthalpy of the gas at low pressure (actually an ideal gas). S^* is the hypothetical entropy of the gas as an *ideal* gas at one atmosphere. Eq. (D-10) is obtained from Eq. (6-100) by letting the reference pressure P_s be 1 atm.

[9] R. C. Reid and T. K. Sherwood, *op. cit.*, Chap. 5.

[10] D. R. Stull, E. F. Westrum, and G. C. Sinke, *The Chemical Thermodynamics of Organic Compounds* (New York: John Wiley & Sons, Inc., 1969).

[11] W. C. Edmister, "Isothermal Pressure Corrections to the Enthalpy and Entropy," *Hydro. Proc.*, **46**, 4 (1967), 165.

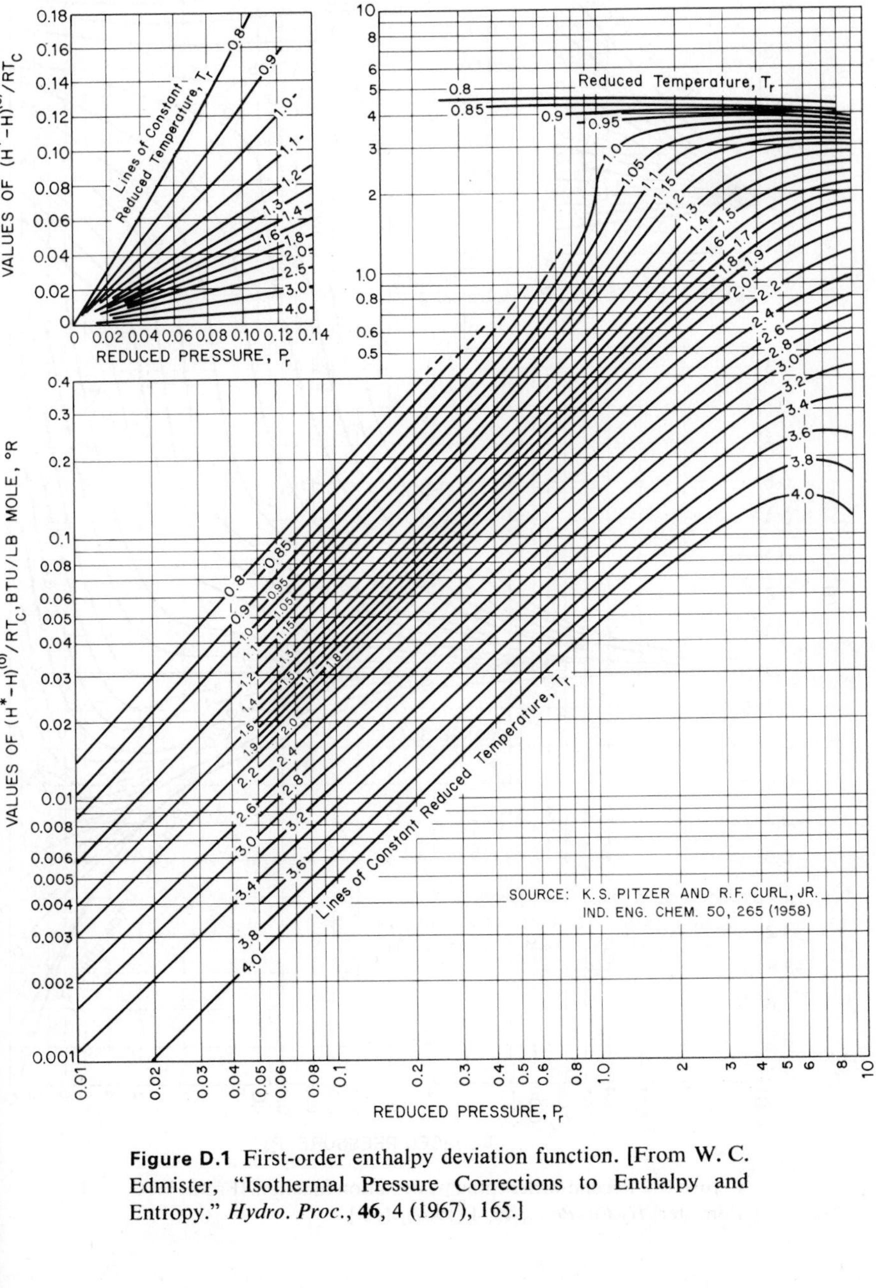

Figure D.1 First-order enthalpy deviation function. [From W. C. Edmister, "Isothermal Pressure Corrections to Enthalpy and Entropy." *Hydro. Proc.*, **46**, 4 (1967), 165.]

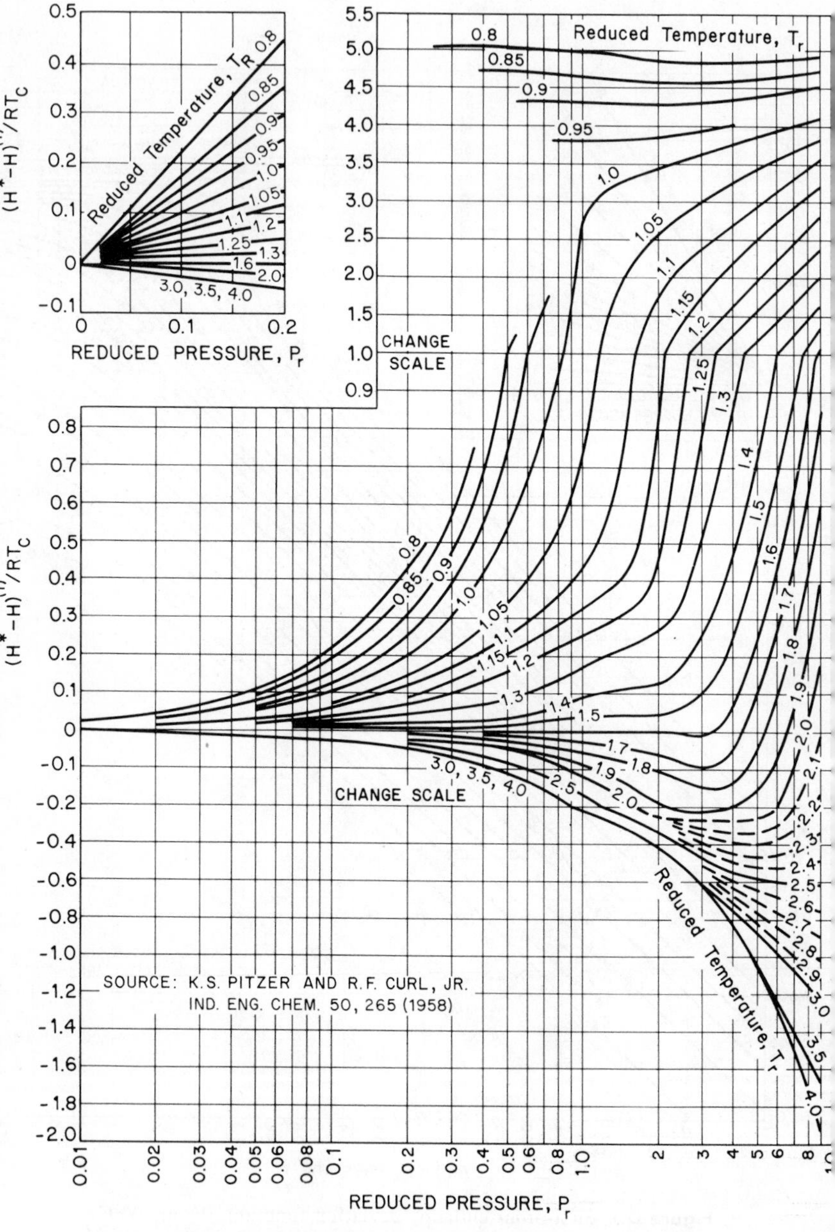

Figure D.2 Second-order enthalpy deviation function. [From: W. C. Edmister, *Hydro. Proc.*, **46**, 4 (1967), 165.]

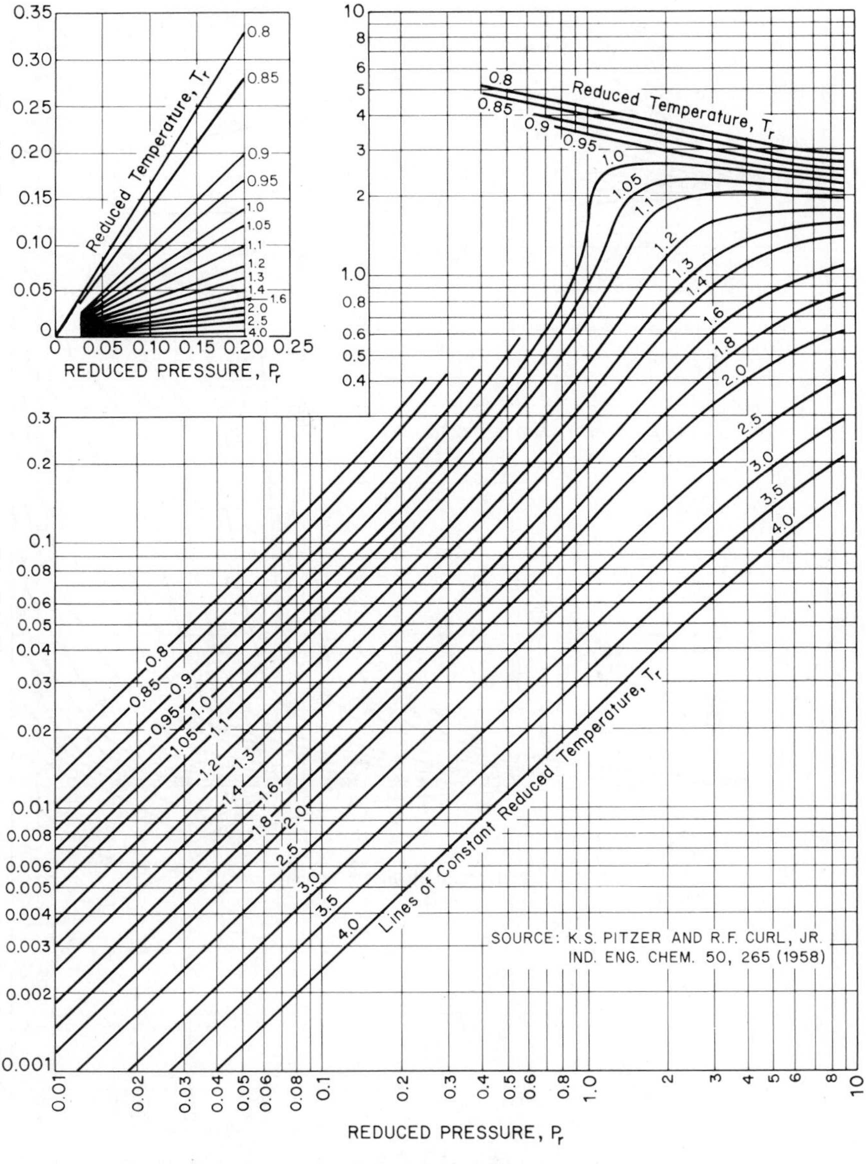

Figure D.3 First-order entropy deviation function. [From W C. Edmister, *Hydro. Proc.*, **46**, 4 (1967), 165.]

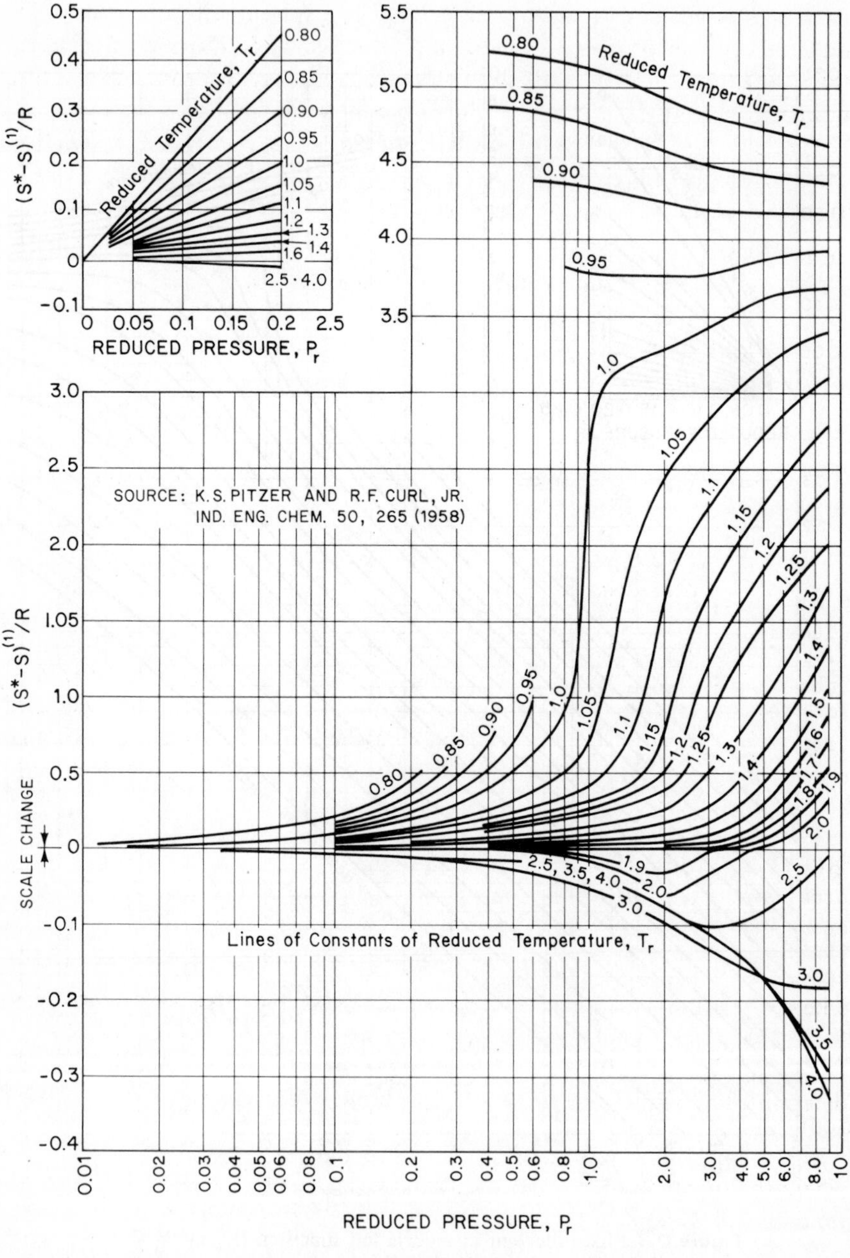

Figure D.4 Second-order entropy deviation function. [From W. C. Edmister, *Hydro. Proc.*, **46**, 4 (1967), 165.]

Evaluation of enthalpy and entropy changes

It is often necessary to determine the change in the enthalpy or entropy of a material between two specified states. If thermodynamic data are not available, this evaluation can be carried out with the use of the generalized correlations presented here. By the nature of these correlations, the "zero-pressure" or "hypothetical ideal-gas" state must be employed as a reference state. For example, if the change in enthalpy or entropy were desired between P_1, T_1, and P_2, T_2, the most convenient calculational procedure would be as follows:

(a) Determine ΔH and ΔS at constant T_1 from P_1 to the ideal gas or zero-pressure state with Eqs. (D-9) and (D-10).

(b) Determine ΔH and ΔS in this ideal gas or zero-pressure state from T_1 to T_2 by

$$\Delta H = \int_{T_1}^{T_2} C_p^* \, dT \tag{D-11}$$

$$\Delta S = \int_{T_1}^{T_2} \left(\frac{C_p^*}{T}\right) dT \tag{D-12}$$

where C_p^* is the ideal gas heat capacity.

(c) Determine ΔH and ΔS at constant T_2 from the ideal gas or zero-pressure state to P_2 with Eqs. (D-9) and (D-10).

(d) Add the three separate ΔH and ΔS values.

This procedure emphasizes that enthalpy and entropy are state functions and their difference between definite initial and end states is path-independent. Of course, if in the isothermal steps there should be a phase change, the enthalpy and entropy changes for the phase change must also be included.

Heat Capacities for Nonideal Gases

From the identity,

$$C_p = C_p^* + \Delta C_p = C_p^* + (C_p - C_p^*) \tag{D-13}$$

where C_p^* is the heat capacity of an ideal gas at T. C_p is the nonideal gas heat capacity at T and P and $\Delta C_p = C_p - C_p^*$ is found from Eq. (6-102).

Again employing the Pitzer expansion, Eq. (6-91), we see that ΔC_p can be determined as a function of T_r and P_r. The relation used is:

$$\Delta C_p = \Delta C_p^{(0)} + \omega \Delta C_p^{(1)} \tag{D-14}$$

where the functions $\Delta C_p^{(0)}$ and $\Delta C_p^{(1)}$ are shown in Figures D.5 and D.6 as functions of T_r and P_r.[12]

[12] W. C. Edmister, "Effect of Pressure on Heat Capacity and Joule-Thomson Coefficient," *Hydro. Proc.*, **46**, 5 (1967), 187.

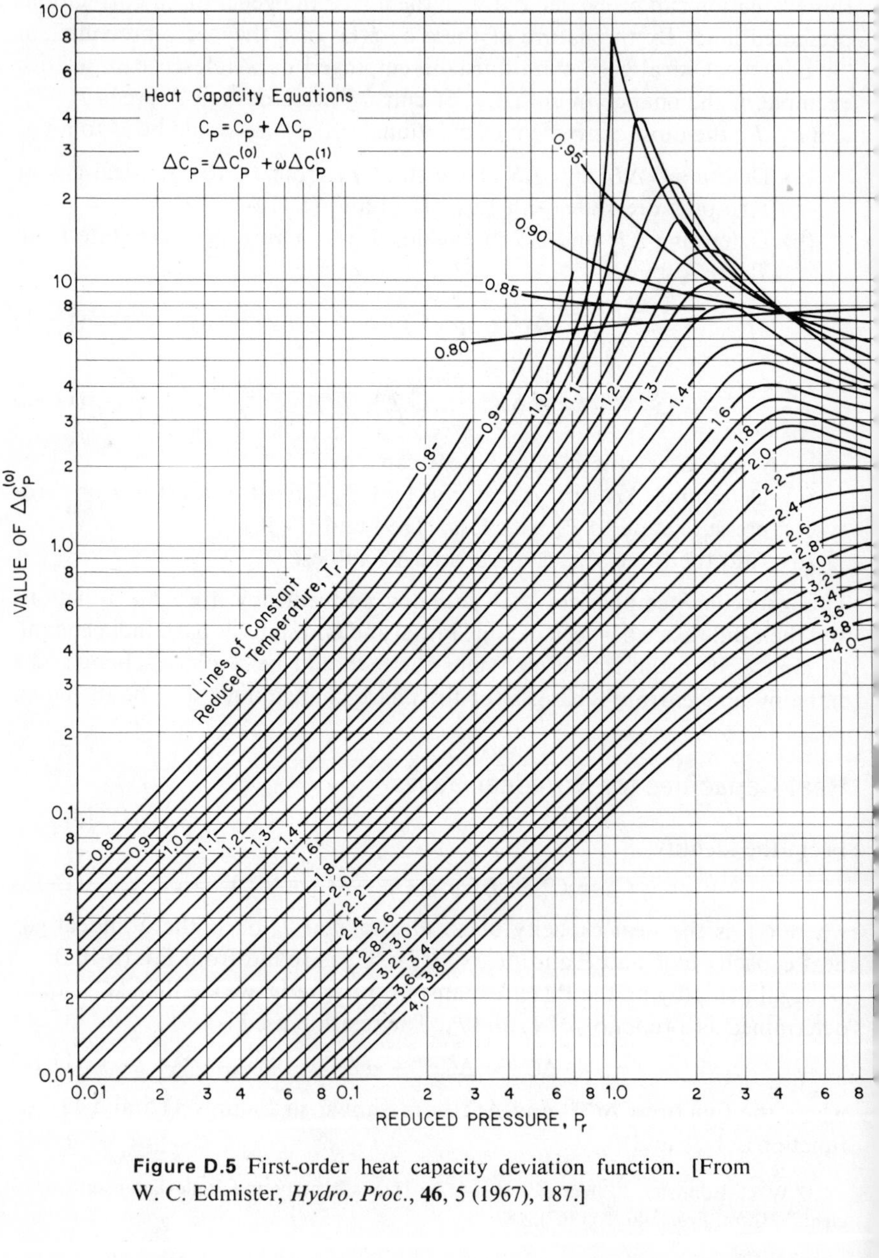

Figure D.5 First-order heat capacity deviation function. [From W. C. Edmister, *Hydro. Proc.*, **46**, 5 (1967), 187.]

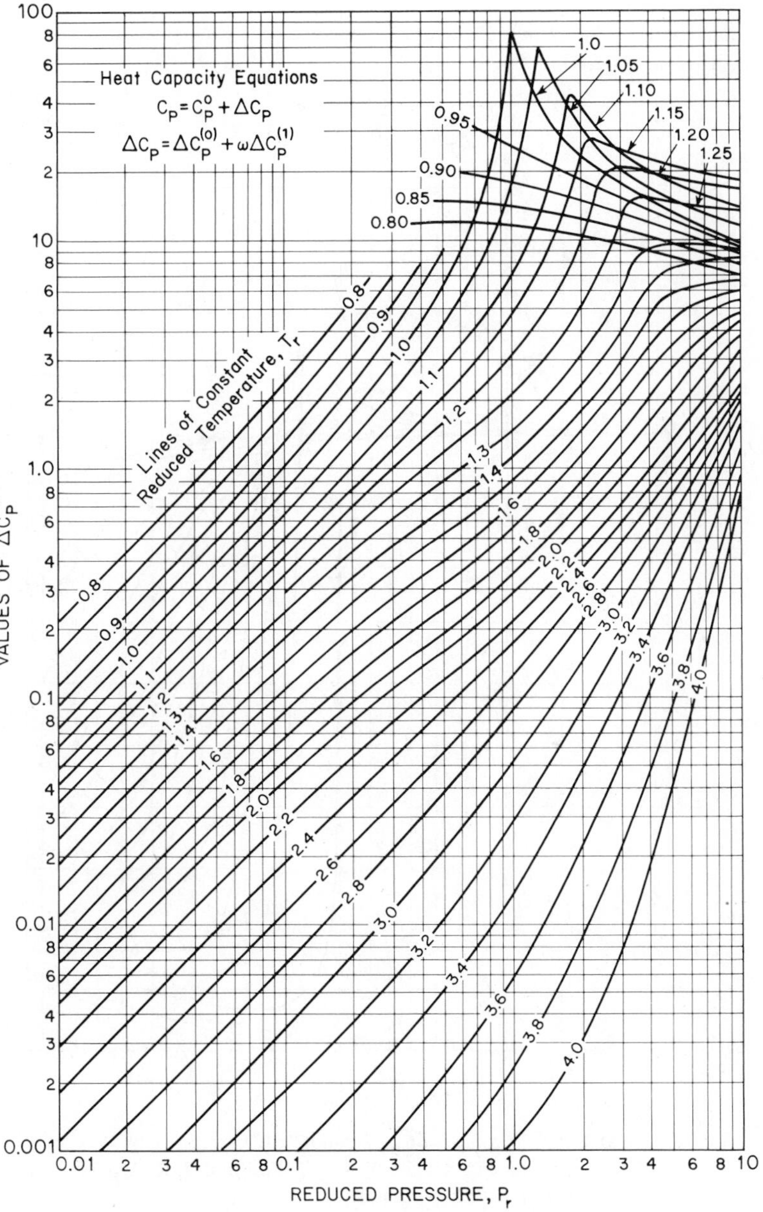

Figure D.6 Second-order heat capacity deviation function. [From W. C. Edmister, *Hydro. Proc.*, **46**, 5 (1967), 187.

Derivative Properties

The derivative properties, Z_p and Z_T, discussed in Section 6.9, may be expanded in the form:[13]

$$Z_p = Z_p^{(0)} + \omega Z_p^{(1)} \tag{D-15}$$

$$Z_T = Z_T^{(0)} + \omega Z_T^{(1)} \tag{D-16}$$

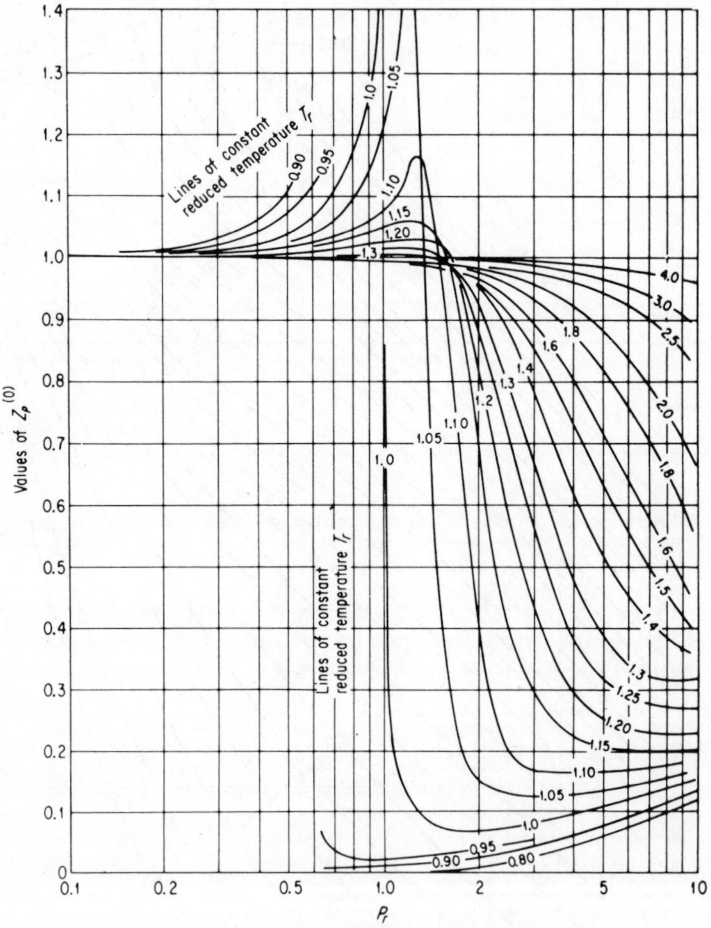

Figure D.7 $Z_p^{(0)}$ as a function of P_r and T_r. [Chart drawn by W. C. Edmister from data in article by R. C. Reid and J. R. Valbert, *Ind. Eng. Chem. Fund.*, **1**, (1962), 292.]

[13] R. C. Reid and J. R. Valbert, "Derivative Compressibility Factors," *Ind. Eng. Chem. Fund.*, **1**, (1962), 292.

where the $Z_p^{(0)}$, $Z_p^{(1)}$, $Z_T^{(0)}$, $Z_T^{(1)}$ functions are plotted as a function of T_r and P_r in Figures D.7 through D.10.

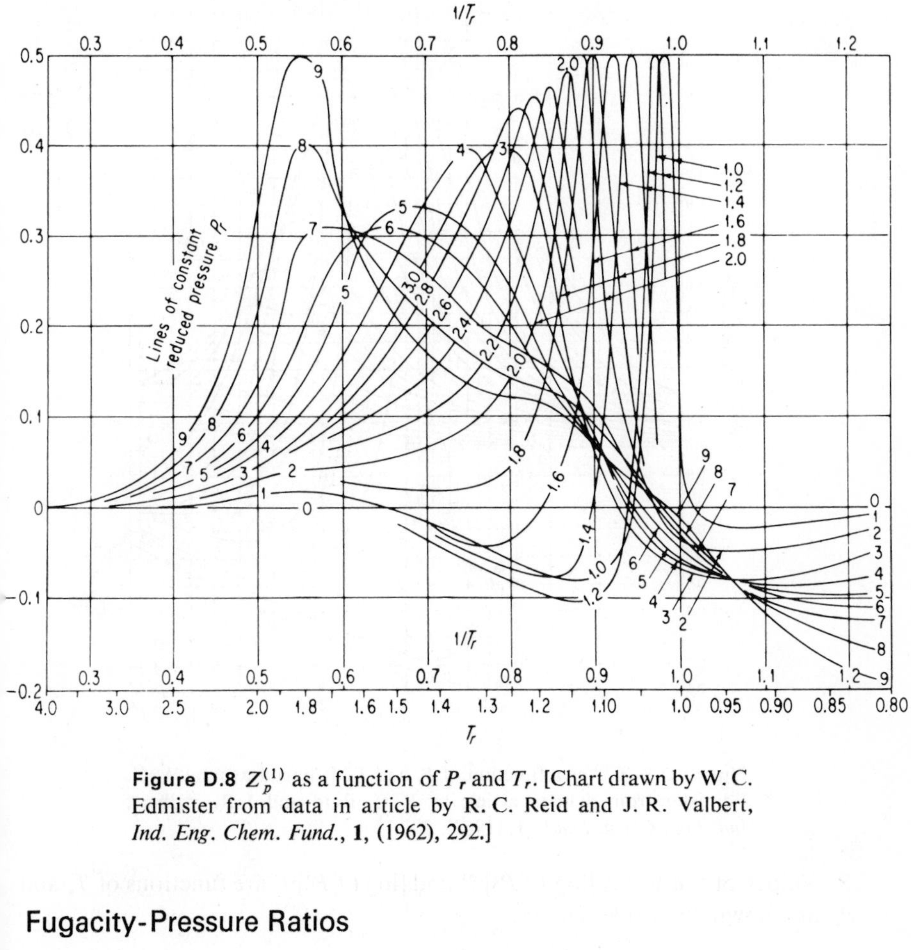

Figure D.8 $Z_p^{(1)}$ as a function of P_r and T_r. [Chart drawn by W. C. Edmister from data in article by R. C. Reid and J. R. Valbert, *Ind. Eng. Chem. Fund.*, **1**, (1962), 292.]

Fugacity-Pressure Ratios

For a pure material, the fugacity-pressure ratio is given as:

$$\ln \frac{f_j}{P} = \int_0^P \frac{(Z-1)}{P}\, dP \tag{D-17}$$

as described in Section 6.2. This ratio may be generalized by using Eq. (6-91) to express Z. If this is done, the final expression may be written, using the common logarithm instead of the natural base,

$$\log \frac{f_j}{P} = \left(\log \frac{f}{P}\right)^{(0)} + \omega\left(\log \frac{f}{P}\right)^{(1)} \tag{D-18}$$

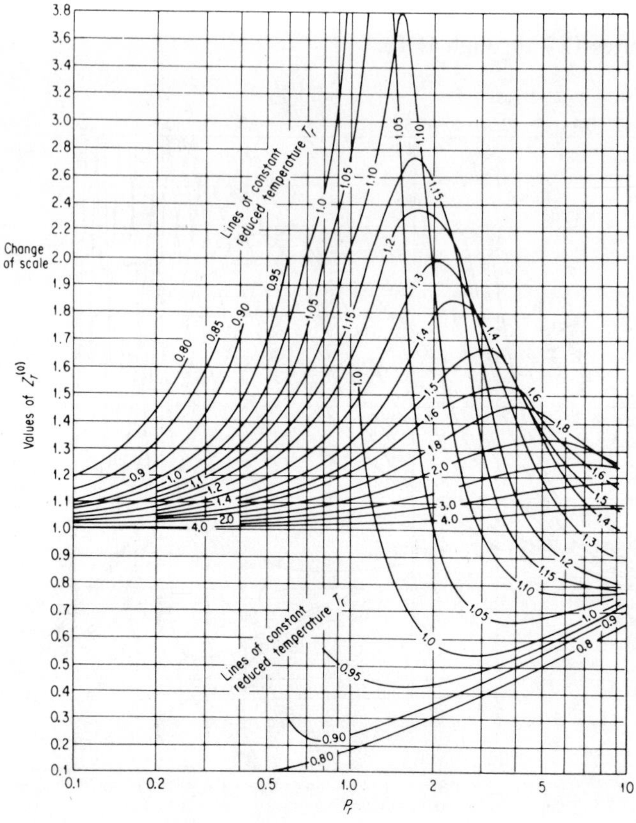

Figure D.9 $Z_T^{(0)}$ as a function of P_r and T_r. [Chart drawn by W. C. Edmister from data in article by R. C. Reid and J. R. Valbert, *Ind. Eng. Chem. Fund.*, **1**, (1962), 292.]

Values of the terms $[\log (f/P)]^{(0)}$ and $[\log (f/P)]^{(1)}$ are functions of T_r and P_r as shown in Table D.3.[14]

[14] K. S. Pitzer, D. Z. Lippmann, R. F. Curl, C. M. Huggins, and D. E. Petersen, "The Volumetric and Thermodynamic Properties of Fluids. II. Compressibility Factor, Vapor Pressure, and Entropy of Vaporization," *J. Am. Chem. Soc.*, **77**, (1955), 3433.

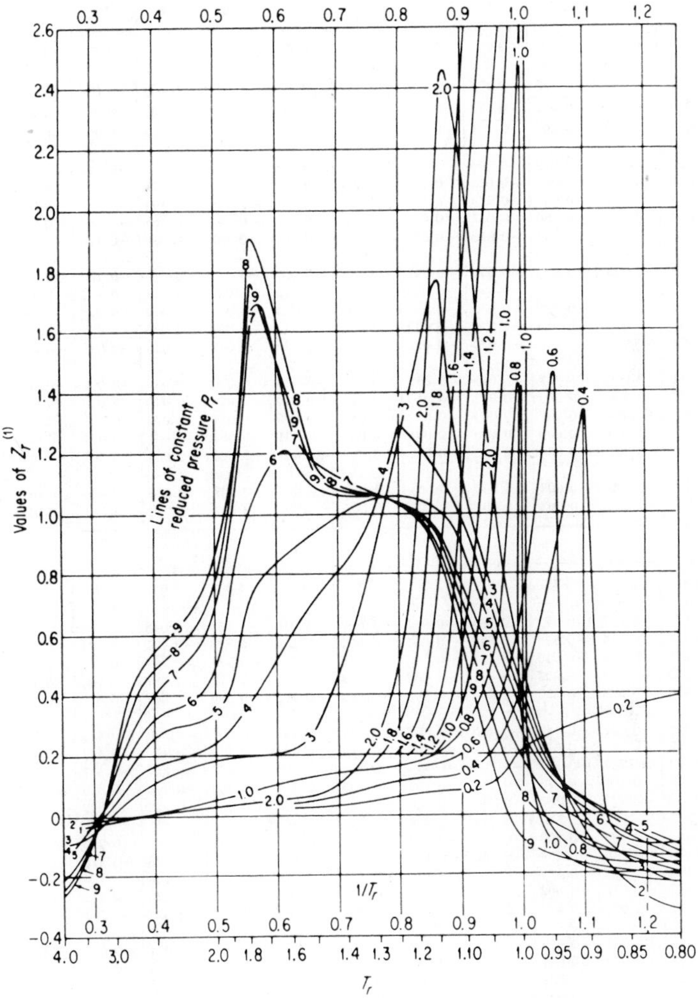

Figure D.10 $Z_T^{(1)}$ as a function of T_r and P_r [Chart drawn by W. C. Edmister from data in article by R. C. Reid and J. R. Valbert, *Ind. Eng. Chem. Fund.*, **1**, (1962), 292.]

TABLE D.3

VALUES OF [LOG $(f/P)]^{(0)}$

T_r	P_r 0.2	0.4	0.6	0.8	1.0	1.2	1.4	1.6	1.8	2.0	2.2	2.4	2
0.80	−0.060	−0.262	−0.425	−0.535	−0.618	−0.683	−0.736	−0.780	−0.817	−0.849	−0.877	−0.901	−0
0.85	−0.046	−0.120	−0.281	−0.392	−0.474	−0.539	−0.592	−0.636	−0.673	−0.705	−0.733	−0.757	−0
0.90	-0.042_5	-0.087_5	−0.163	−0.273	−0.356	−0.421	−0.474	−0.517	−0.554	−0.587	−0.614	−0.639	−0
0.95	−0.033	−0.070	−0.112	−0.173	−0.255	−0.319	−0.372	−0.415	−0.452	−0.483	−0.511	−0.535	−0
1.00	−0.028	−0.059	−0.094	−0.131	−0.175	−0.237	−0.287	−0.330	−0.367	−0.398	−0.425	−0.449	−0
1.05	−0.024	−0.051	−0.079	−0.109	−0.142	−0.178	−0.218	−0.257	−0.292	−0.322	−0.349	−0.372	−0
1.10	−0.021	−0.044	−0.067	−0.093	−0.120	−0.147	−0.177	−0.207	−0.237	−0.264	−0.289	−0.311	−0
1.15	−0.018	−0.037	−0.058	−0.079	−0.101	−0.123	−0.146	−0.170	−0.194	−0.217	−0.238	−0.258	−0
1.20	−0.016	−0.032	−0.050	−0.067	−0.086	−0.104	−0.124	−0.143	−0.163	−0.182	−0.200	−0.217	−0
1.25	−0.014	−0.029	−0.044	−0.059	−0.075	−0.091	−0.107	−0.123	−0.139	−0.155	−0.171	−0.186	−0
1.3	−0.012	−0.025	−0.038	−0.051	−0.065	−0.078	−0.092	−0.106	−0.119	−0.133	−0.146	−0.159	−0.
1.4	−0.010	−0.021	−0.031	−0.041	−0.052	−0.062	−0.072	−0.082	−0.092	−0.102	−0.111	−0.120	−0.
1.5	−0.008	−0.016	−0.024	−0.032	−0.040	−0.047	−0.055	−0.063	−0.070	−0.078	−0.085	−0.092	−0
1.6	−0.007	−0.013	−0.019	−0.026	−0.032	−0.038	−0.044	−0.050	−0.056	−0.062	−0.067	−0.072	−0
1.7	−0.005	−0.010	−0.015	−0.020	−0.025	−0.030	−0.034	−0.039	−0.043	−0.047	−0.051	−0.056	−0
1.8	−0.004	−0.008	−0.012	−0.015	−0.019	−0.022	−0.026	−0.030	−0.033	−0.036	−0.039	−0.042	−0.
1.9	−0.003	−0.006	−0.009	−0.012	−0.015	−0.018	−0.020	−0.023	−0.025	−0.028	−0.030	−0.033	−0
2.0	−0.002	−0.004	−0.007	−0.009	−0.011	−0.013	−0.015	−0.017	−0.019	−0.021	−0.023	−0.025	−0
2.5	0.000	0.000	0.000	0.000	−0.001	−0.001	−0.001	−0.001	−0.001	−0.001	−0.001	−0.001	−0.
3.0	0.000	+0.001	+0.001	+0.002	+0.002	+0.003	+0.003	+0.004	+0.004	+0.005	+0.005	+0.006	+0.
3.5	+0.001	0.002	0.003	0.003	0.004	0.005	0.006	0.007	0.008	0.009	0.010	0.011	0.
4.0	0.001	0.002	0.003	0.005	0.006	0.007	0.008	0.009	0.010	0.011	0.012	0.013	0.

T_r	P_r 2.8	3.0	3.2	3.4	3.6	3.8	4.0	4.5	5.0	6.0	7.0	8.0	9
0.80	−0.941	−0.957	−0.972	−0.985	−0.997	−1.007	−1.016	−1.035	−1.048	−1.064	−1.067	−1.063	−1
0.85	−0.797	−0.814	−0.829	−0.842	−0.854	−0.864	−0.874	−0.893	−0.907	−0.924	−0.929	−0.926	−0
0.90	−0.679	−0.696	−0.710	−0.724	−0.736	−0.746	−0.756	−0.775	−0.789	−0.807	−0.814	−0.813	−0
0.95	−0.575	−0.592	−0.607	−0.621	−0.632	−0.643	−0.652	−0.672	−0.687	−0.706	−0.713	−0.713	−0
1.00	−0.489	−0.505	−0.520	−0.534	−0.545	−0.556	−0.566	−0.586	−0.601	−0.620	−0.629	−0.630	−0
1.05	−0.411	−0.428	−0.442	−0.455	−0.467	−0.478	−0.488	−0.508	−0.523	−0.543	−0.552	−0.553	−0
1.10	−0.348	−0.364	−0.378	−0.391	−0.403	−0.413	−0.422	−0.442	−0.457	−0.477	−0.487	−0.489	−0.
1.15	−0.293	−0.307	−0.321	−0.333	−0.344	−0.354	−0.363	−0.383	−0.397	−0.417	−0.427	−0.429	−0.
1.20	−0.247	−0.261	−0.273	−0.285	−0.295	−0.305	−0.314	−0.332	−0.346	−0.366	−0.375	−0.378	−0.
1.25	−0.212	−0.224	−0.236	−0.246	−0.256	−0.264	−0.273	−0.290	−0.304	−0.322	−0.331	−0.334	−0.
1.3	−0.182	−0.193	−0.203	−0.212	−0.221	−0.229	−0.237	−0.253	−0.266	−0.283	−0.292	−0.295	−0.
1.4	−0.138	−0.146	−0.154	−0.162	−0.169	−0.175	−0.181	−0.194	−0.205	−0.220	−0.228	−0.231	−0.
1.5	−0.104	−0.112	−0.117	−0.124	−0.129	−0.134	−0.139	−0.149	−0.158	−0.170	−0.176	−0.178	−0.
1.6	−0.082	−0.087	−0.092	−0.096	−0.100	−0.104	−0.108	−0.116	−0.123	−0.132	−0.137	−0.138	−0.
1.7	−0.063	−0.067	−0.071	−0.074	−0.077	−0.080	−0.083	−0.089	−0.094	−0.101	−0.105	−0.105	−0.
1.8	−0.048	−0.051	−0.053	−0.056	−0.058	−0.060	−0.063	−0.067	−0.071	−0.076	−0.078	0.077	−0.
1.9	−0.037	−0.039	−0.041	−0.043	−0.045	−0.046	−0.048	−0.051	−0.054	−0.057	−0.057	−0.055	−0
2.0	−0.028	−0.029	−0.031	−0.032	−0.033	−0.034	−0.035	−0.037	−0.039	−0.040	−0.039	−0.038	−0
2.5	−0.001	−0.001	−0.001	−0.001	0.000	0.000	0.000	+0.001	+0.003	+0.006	+0.011	+0.016	+0.
3.0	+0.007	+0.008	+0.009	+0.010	+0.011	+0.012	+0.012	0.015	0.017	0.023	0.028	0.035	0.
3.5	0.013	0.014	0.015	0.016	0.017	0.018	0.020	0.022	0.025	0.031	0.038	0.044	0.
4.0	0.015	0.016	0.017	0.019	0.020	0.021	0.022	0.025	0.028	0.034	0.040	0.047	0.

From: Pitzer, K.S., D.Z. Lippmann, R.F. Curl, C.M. Huggins, and D.E. Peterson, J.Am. Chem. Soc. 77, 3433 (1955).

TABLE D.3 (Continued)

VALUES OF [LOG (f/P)][1]

r	P_r 0.2	0.4	0.6	0.8	1.0	1.2	1.4	1.6	1.8	2.0	2.2	2.4	2.6	2.8	3.0	4.0	5.0	6.0	7.0	8.0	9.0
80	−0.04	−0.47	−0.48	−0.48	−0.48	−0.49	−0.50	−0.50	−0.51	−0.51	−0.52	−0.52	−0.53	−0.53	−0.54	−0.56	−0.59	−0.61	−0.63	−0.65	−0.67
85	−0.03	−0.31	−0.31	−0.32	−0.33	−0.33	−0.34	−0.35	−0.35	−0.36	−0.37	−0.37	−0.38	−0.38	−0.39	−0.41	−0.44	−0.46	−0.48	−0.50	−0.51
90	−0.02	−0.04	−0.18	−0.20	−0.20	−0.21	−0.21	−0.22	−0.23	−0.23	−0.24	−0.24	−0.25	−0.26	−0.26	−0.29	−0.31	−0.33	−0.35	−0.36	−0.38
95	−0.01	−0.02	−0.03	−0.09	−0.10	−0.11	−0.12	−0.12	−0.13	−0.13	−0.14	−0.15	−0.15	−0.16	−0.16	−0.18	−0.20	−0.22	−0.24	−0.26	−0.27
00	−0.01	−0.01	−0.01	−0.02	−0.03	−0.03	−0.04	−0.05	−0.05	−0.06	−0.06	−0.07	−0.07	−0.08	−0.08	−0.10	−0.12	−0.13	−0.15	−0.17	−0.18
05	0.00	0.00	0.00	0.00	+0.01	+0.01	+0.01	+0.01	0.00	0.00	0.00	0.00	−0.01	−0.01	−0.01	−0.03	−0.05	−0.06	−0.07	−0.09	−0.11
10	0.00	0.00	0.00	+0.01	0.01	0.02	0.02	0.03	+0.03	+0.03	+0.03	+0.03	+0.03	+0.03	+0.03	+0.02	0.00	−0.01	−0.02	−0.03	−0.05
15	0.00	0.00	0.00	0.01	0.02	0.02	0.03	0.04	0.04	0.05	0.05	0.05	0.06	0.06	0.06	0.05	+0.05	+0.04	+0.02	+0.01	0.00
20	0.00	+0.01	+0.01	0.01	0.02	0.03	0.04	0.05	0.05	0.06	0.07	0.07	0.08	0.08	0.08	0.09	0.09	0.08	0.07	0.07	+0.06
25	0.00	0.01	0.01	0.02	0.03	0.03	0.04	0.05	0.06	0.07	0.07	0.08	0.09	0.09	0.10	0.11	0.11	0.11	0.10	0.10	0.09
.3	+0.01	0.01	0.02	0.02	0.03	0.04	0.04	0.05	0.06	0.07	0.08	0.08	0.09	0.10	0.10	0.12	0.13	0.13	0.13	0.12	0.12
.4	0.01	0.01	0.02	0.03	0.04	0.04	0.05	0.06	0.07	0.08	0.08	0.09	0.10	0.11	0.11	0.13	0.15	0.15	0.16	0.16	0.16
.5	0.01	0.02	0.02	0.03	0.04	0.05	0.05	0.06	0.06	0.07	0.08	0.08	0.09	0.10	0.11	0.13	0.15	0.16	0.17	0.17	0.18
.6	0.01	0.02	0.02	0.03	0.04	0.05	0.05	0.06	0.06	0.07	0.08	0.08	0.09	0.10	0.11	0.14	0.16	0.18	0.19	0.20	0.21
.7	0.01	0.02	0.02	0.03	0.04	0.05	0.05	0.06	0.06	0.07	0.08	0.08	0.09	0.10	0.11	0.14	0.16	0.18	0.20	0.21	0.23
.8	0.01	0.02	0.02	0.03	0.04	0.05	0.05	0.06	0.06	0.07	0.08	0.08	0.09	0.10	0.11	0.14	0.16	0.19	0.21	0.23	0.24
.9	0.01	0.02	0.02	0.03	0.04	0.05	0.05	0.06	0.06	0.07	0.08	0.08	0.09	0.10	0.11	0.14	0.16	0.19	0.21	0.23	0.25
0	0.01	0.01	0.02	0.03	0.04	0.05	0.05	0.06	0.07	0.07	0.08	0.08	0.09	0.09	0.10	0.13	0.16	0.19	0.21	0.23	0.26
5	0.01	0.01	0.02	0.02	0.03	0.04	0.04	0.05	0.06	0.06	0.07	0.07	0.08	0.08	0.09	0.12	0.14	0.17	0.19	0.22	0.24
0	0.00	0.01	0.01	0.02	0.02	0.03	0.04	0.04	0.05	0.05	0.05	0.06	0.06	0.07	0.07	0.10	0.12	0.15	0.17	0.20	0.22
5	0.00	0.01	0.01	0.02	0.02	0.02	0.03	0.03	0.04	0.04	0.04	0.05	0.05	0.06	0.06	0.08	0.10	0.13	0.15	0.17	0.19
0	0.00	0.01	0.01	0.02	0.02	0.02	0.02	0.03	0.03	0.03	0.04	0.04	0.04	0.05	0.05	0.07	0.09	0.10	0.12	0.14	0.15

Estimation of the Thermodynamic Properties of Mixtures

E

The thermodynamics of mixtures are treated in Chapter 8. To use the relations that were developed, property values must be available. While good experimental data are always most desirable, it is a fact that for most systems, such data are not available and recourse must be made to estimation techniques.

In this brief appendix we suggest a few techniques to estimate both mixture quantities (i.e., partial molal properties) and properties of a mixture. For a more complete discussion, we refer to current articles in the literature and to books.[1]

Gas Mixture Equations of State

The general relation,

$$\phi(P, V, T, y_1, \ldots, y_{n-1}) = 0 \tag{E-1}$$

[1] R. C. Reid and T. K. Sherwood, *Properties of Gases and Liquids*, 2nd ed. (New York: McGraw-Hill Book Company, 1966), Chap. 7.

constitutes a mixture equation of state. Many have been proposed but few have been thoroughly tested. Most were developed as pure-component relations and the constants then related to composition.

Ideal gas mixture

At low pressures and under conditions far removed from the saturation curve, it is often satisfactory to assume the mixture to behave as an ideal gas, i.e.,

$$P\underline{V} = NRT \tag{E-2}$$

where N is the total number of moles.

Dalton's law

If one assumes that partial pressures are additive when the components are mixed at constant volume and temperature, it can be shown that this is equivalent to the mixture rule:

$$Z_m = \sum_{i=1}^{n} y_i Z_i \tag{E-3}$$

where Z_i refers to the pure component compressibility factors evaluated at the system temperature and at the *partial pressure* of the component in the mixture. Eq. (E-3) is only slightly less restrictive than assuming ideal gases.

Amagat's law

If the gas mixture forms an ideal solution (i.e., if the volumes are additive when mixing pure components at constant temperature and total pressure), then Z_m is still calculated as in Eq. (E-3) except that Z_i is now evaluated at the *total* pressure of the mixture. Amagat's law usually has a wider range of applicability than Dalton's law.

Pseudocritical concept

In many instances, the pure-component correlations, such as the Pitzer expansion for the compressibility factor [Eq. (6-91)], may be used for mixtures. In such cases some pseudocritical rules must, however, be devised to define T_{c_m}, P_{c_m}, and ω_m. It is ordinarily not satisfactory to employ true mixture critical constants even if they were available. The pseudocritical concept is based on the presumption that there exists a single pure fluid that has the same *P-V-T* properties as the mixture—and the critical properties of this pseudofluid can be related to the composition of the mixture and to the critical constants of the constituents making up the mixture. The simplest set of pseudocritical rules yet devised was suggested by Kay:[2]

$$T_{c_m} = \sum_{i=1}^{n} y_i T_{c_i} \tag{E-4}$$

$$P_{c_m} = \sum_{i=1}^{n} y_i P_{c_i} \tag{E-5}$$

$$\omega_m = \sum_{i=1}^{n} y_i \omega_i \tag{E-6}$$

Many other rules have been suggested. All are more complex than Kay's rules, but they do allow estimation of pseudocritical properties which, when used to define T_{r_m} and P_{r_m}, permit corresponding states correlations developed for pure components to be employed for mixtures of known composition. We show only one of these more complex rules below; others are discussed elsewhere.[3]

Joffe-Stewart, Burkhardt and Voo pseudocritical rules[4,5]

$$T_{c_m} = \frac{K^2}{J} \tag{E-7}$$

$$P_{c_m} = \frac{K^2}{J^2} \tag{E-8}$$

$$K = \sum_{i=1}^{n} y_i \left(\frac{T_{c_i}}{P_{c_i}^{1/2}} \right) \tag{E-9}$$

$$J = \frac{1}{8} \sum_{i=1}^{n} \sum_{j=1}^{n} y_i y_j \left[\left(\frac{T_{c_i}}{P_{c_i}} \right)^{1/3} + \left(\frac{T_{c_j}}{P_{c_j}} \right)^{1/3} \right]^3 \tag{E-10}$$

In essentially all pseudocritical rules, if ω_m or Z_{c_m} should appear, it is defined as a mole fraction average [e.g., as in Eq. (E-6)].

In summary, then, if the pseudocritical concept were to be employed to estimate any mixture property, one would treat the mixture as a hypothetical pure component, determine the mixture critical properties with rules such as Eqs. (E-4) through (E-6) or Eqs. (E-7) and (E-8), and then proceed to use any pure component property correlation such as those given in Chapter 6 or Appendix D. This procedure usually yields results of engineering accuracy for such quantities as Z_m, $(H - H^*)_m$, $(C_p - C_p^*)_m$, etc., but it is often not convenient to calculate partial molal quantities such as $\bar{V}_j$, $\ln(\hat{f}_j/y_j P)$, etc., since

[2] W. B. Kay, "Density of Hydrocarbon Gases and Vapors," *Ind. Eng. Chem.*, **28**, (1936), 1014.

[3] R. C. Reid and T. K. Sherwood, *op. cit.*, pp. 313–19.

[4] J. Joffe, "Compressibilities of Gas Mixtures," *Ind. Eng. Chem.*, **39**, (1947), 837.

[5] W. E. Stewart, S. F. Burkhardt, and D. Voo, Paper presented at the Kansas City National AIChE Meeting, May 18, 1959.

in these cases differentiation of total mixture properties with respect to mole numbers is necessary. [For example, $\bar{V}_j \equiv (\partial \underline{V}_m/\partial N_j)_{T,P,N_k[j]}$.] This difficulty, coupled with the desirability of employing an analytical form of Eq. (E-1) in machine computation has led to many papers proposing analytical equations.

Analytical equations of state

It is, perhaps, not in the spirit of a text on thermodynamic principles to become too deeply involved in a discussion of analytical equations of state to estimate gas-phase *P-V-T* properties. Yet, this text has as a principal objective the application of thermodynamics; and, to apply many of the mixture relations given in Chapter 8 (and succeeding chapters), one must have access either to data or to an analytical equation of state (or use a hand-calculational method such as the pseudocritical concept). We, therefore, do not feel that it is out of place to treat briefly mixture equations of state. Also, any of the relations given below could be used for pure components by letting all the mole fractions be zero except one which is set equal to unity.

Redlich-Kwong

Almost all property estimation systems in use today employ as the "simplest" analytical equation of state one suggested by Redlich and Kwong.[6] In its original form it may be written as

$$Z^3 - Z^2 + (A^* - B^{*2} - B^*)Z - A^*B^* = 0 \tag{E-11}$$

where Z is the compressibility factor and,

$$A^* = \frac{\Omega_a P_r}{T_r^{2.5}} \tag{E-12}$$

$$B^* = \frac{\Omega_b P_r}{T_r} \tag{E-13}$$

$$\Omega_a = [9(2^{1/3} - 1)]^{-1} = 0.427480\ldots \tag{E-14}$$

$$\Omega_b = (27\Omega_a)^{-1} = 0.086640\ldots \tag{E-15}$$

The numerical values of Ω_a and Ω_b were found by applying $(\partial P/\partial V)_T = (\partial^2 P/\partial V^2)_T = 0$ at the critical point.

For mixtures, A^* and B^* are determined from:

$$A_m^* = \left(\sum_{i=1}^{n} y_i A_i^{*1/2}\right)^2 \tag{E-16}$$

$$B_m^* = \sum_{i=1}^{n} y_i B_i^* \tag{E-17}$$

[6] O. Redlich and J. N. S. Kwong, "On the Thermodynamics of Solutions V." *Chem. Rev.*, **44**, (1949), 233.

where A_i^* and B_i^* are then found from Eqs. (E-12) through (E-15) using pure component criticals.

Eq. (E-11) is used to obtain the compressibility factor. It may also be employed to determine other thermodynamic properties. The two of most use are the enthalpy deviation from an ideal gas and the fugacity of a component. From Eq. (6-96):

$$\frac{(H_m - H_m^*)}{RT} = Z_m - 1 - \frac{3}{2}\frac{A_m^*}{B_m^*}\ln\left(1 + \frac{B_m^*}{Z_m}\right) \tag{E-18}$$

while from Eq. (8-150):

$$\begin{aligned}\ln \phi_j = \ln\left(\frac{\hat{f}_j}{Py_j}\right) &= (Z - 1)\left(\frac{B_j^*}{B_m^*}\right) - \ln(Z_m - B_m^*)\\ &\quad - \frac{A_m^*}{B_m^*}\left[2\left(\frac{A_j^*}{A_m^*}\right)^{1/2} - \frac{B_j^*}{B_m^*}\right]\ln\left[1 + \left(\frac{B_m^*}{Z_m}\right)\right]\end{aligned} \tag{E-19}$$

This original Redlich-Kwong form, although relatively successful in treating pure components and mixtures, has been modified by a number of authors[7-17]. Most of these modifications increase the accuracy of the Redlich-Kwong equation but at the price of a significant increase in complexity.

[7] J. M. Estes and P. C. Tully, "A Modified Redlich-Kwong Equation for Helium from 30° to 1473°K," *AIChE J.*, **13**, (1967), 192.

[8] O. Redlich and A. K. Dunlop, "Thermodynamics of Solutions: VII. An Improved Equation of State," *Chem. Eng. Prog. Sym. Series*, **59**, (1963), 95.

[9] F. J. Ackerman, M. S. Thesis, Chemical Engineering, University of California, Berkeley, 1963; UCRL-10650.

[10] O. Redlich, F. J. Ackerman, R. D. Gunn, M. Jacobson, and S. Lau, "Thermodynamics of Solutions," *Ind. Eng. Chem. Fund.*, **4**, (1965), 369.

[11] R. D. Gray, Jr., N. H. Rent, and D. Zudkevitch, "A Modified Redlich-Kwong Equation of State," *AIChE J.*, **16**, (1970), 991.

[12] G. M. Wilson, "Vapor-Liquid Equilibria, Correlation by Means of a Modified Redlich-Kwong Equation of State," *Adv. Cryo. Eng.*, **9**, (1964), 168; *ibid.*, **11**, (1966), 392.

[13] P. L. Chueh and J. M. Prausnitz, "Vapor-Liquid Equilibria at High Pressures: Vapor Phase Fugacity Coefficients in Nonpolar and Quantum Gas Mixtures," *Ind. Eng. Chem. Fund.*, **6**, (1967), 492.

[14] P. L. Chueh, and J. M. Prausnitz, "Vapor-Liquid Equilibria at High Pressures: Calculation of Partial Molal Volumes in Nonpolar Liquid Mixtures," *AIChE J.*, **13**, (1967), 1099; Ibid., "Vapor-Liquid Equilibria at High Pressures: Calculation of Critical Temperatures, Volumes, and Pressures of Nonpolar Mixtures," *AIChE J.*, **13**, (1967), 1107.

[15] O. Redlich and V. B. T. Ngo, "An Improved Equation of State," *Ind. Eng. Chem. Fund.*, **9**, (1970), 287.

[16] G. M. Wilson, "A Modified Redlich-Kwong Equation of State, Application to General Physical Data Calculation," Paper 15c presented at the Cleveland National Meeting of the AIChE, May 4–7, 1970.

[17] W. F. Vogl and K. R. Hall, "Generalized Temperature Dependence of the Redlich-Kwong Constants," *AIChE J.*, **16**, (1970), 1103.

Other analytical equations

For those who require accurate equations of state and will employ machine computation, it is recommended that the Barner-Adler[18,19] form be used for systems containing any nonhydrocarbons and the Lee-Erbar-Edmister or Soave equation[20,21] for systems containing only hydrocarbons (with perhaps some CO_2, H_2S, N_2 or similar materials). In the original references, techniques are given to indicate how they may be applied to mixtures to calculate volumes, enthalpy deviations, and fugacity coefficients. Other important equations of state not noted here (e.g., the Benedict-Webb-Rubin, Virial, etc.) are discussed elsewhere.[22]

Liquid Mixture Equations of State

There are no generally applicable liquid equations of state for mixtures. To obtain *P-V-T* relations for liquid mixtures, pure component liquid corresponding-states estimation techniques are employed (see Appendix D) and the critical constants obtained from pseudocritical rules such as those given in the previous section. Partial molal volumes are usually estimated by special methods[23] and not by differentiation of a generalized liquid equation of state.

Heat Capacities of Mixtures

Heat capacities of mixtures are normally required to determine the change of enthalpy (or entropy) with temperature. Since there are no reliable techniques to obtain mixture heat capacities under conditions when the mixture is a nonideal gas or a liquid, temperature variations are accounted for in the

[18] H. E. Barner and S. B. Adler, "Three-Parameter Formulation of the Joffe Equation of State," *Ind. Eng. Chem. Fund.*, **9**, (1970), 521.

[19] H. E. Barner and C. W. Quinlan, "Interaction Parameters for Kay's Pseudocritical Temperature," *Ind. Eng. Chem. Proc. Des. Dev.*, **8**, (1969), 407.

[20] B.-I. Lee, J. H. Erbar, and W. C. Edmister, "Prediction of Thermodynamic Properties for Low Temperature Hydrocarbon Process Calculations," *AIChE J.*, **19**, (1973), 349.

[21] G. Soave, "Equilibrium Constants from a Modified Redlich-Kwong Equation of State," *Chem. Eng. Sci.*, **27**, (1972), 1197.

[22] R. C. Reid and T. K. Sherwood, *op. cit.*, Chap. 7.

[23] Chueh, P. L. and J. M. Prausnitz "Vapor-Liquid Equilibria at High Pressures: Calculation of Partial Molar Volumes in Nonpolar Liquid Mixtures," *AIChE J.*, **13**, (1967), 1099.

ideal gas state and pressure effects by a separate estimation of the isothermal enthalpy deviation from an applicable equation of state [e.g., Eq. (6-96)].

For an ideal gas mixture, the mixture heat capacity is simply:

$$C_{p_m}^* = \sum_{i=1}^{n} y_i C_{p_i}^* \tag{E-20}$$

where $C_{p_i}^*$, the heat capacity of component i in the ideal gas state, is estimated by methods described in Appendix D.

Liquid-phase Activity Coefficients

F

As of this writing, liquid-phase activity coefficients is an active area of research and, consequently, many of the current developments may be outdated very soon. In this appendix we shall not attempt to provide more than an abbreviated summary of the more successful correlations now in use.[1] It is anticipated that for the foreseeable future the reader should refer to current literature in periodicals and journals to keep pace with new developments.

Many functional forms have been suggested to express the activity coefficient of a condensed phase mixture as a function of composition. Many of the early forms were entirely empirical, while some of the more recently developed forms are based in part on semitheoretical analysis of molecular models of liquid mixtures.

Developments have progressed from specific algebraic equations to more complex generalizations. Thus, the two-suffix Margules equation,

$$\ln \gamma_1 = Ax_2^2; \qquad \ln \gamma_2 = Ax_1^2 \tag{F-1}$$

[1] A number of specialized texts provide a comprehensive treatment and critical survey of current developments. See, e.g., references (1) and (2) of the citations tabulated at the end of this appendix.

and the three-suffix Margules equation,

$$\ln \gamma_1 = Ax_2^3 + Bx_2^2; \qquad \ln \gamma_2 = Ax_1^3 + Bx_1^2 \tag{F-2}$$

have been shown to be special cases of the Redlich-Kister expansion [see Section 8.7, Eq.(8-207)]. Similarly, the van Laar, Margules, and Scatchard-Hamer equations can be shown to be special cases of the Wohl expansion (3).

Instead of discussing these correlations further, we will emphasize more recent forms developed by Wilson, Renon and Prausnitz, and others. It has been well-documented that the latter group are generally more accurate in correlating γ-x data than those derived from a Wohl expansion and, more important, they can be used for multicomponent systems without the necessity of introducing ternary or higher order constants. The simplest of this group is the Wilson form.

Wilson's equations (4)

For a system containing n components,

$$\frac{\Delta G^{EX}}{RT} = -\sum_{i=1}^{n} x_i \ln\left(\sum_{j=1}^{n} x_j \Lambda_{ij}\right) \tag{F-3}$$

From Eq. (8-193),

$$\ln \gamma_k = -\ln\left(\sum_{j=1}^{n} x_j \Lambda_{kj}\right) + 1 - \sum_{i=1}^{n} \frac{x_i \Lambda_{ik}}{\sum_{j=1}^{n} x_j \Lambda_{ij}} \tag{F-4}$$

For a binary system of 1 and 2, Eq. (F-4) reduces to:

$$\ln \gamma_1 = -\ln(x_1 + \Lambda_{12}x_2) + x_2\left(\frac{\Lambda_{12}}{x_1 + \Lambda_{12}x_2} - \frac{\Lambda_{21}}{\Lambda_{21}x_1 + x_2}\right) \tag{F-5}$$

$$\ln \gamma_2 = -\ln(x_2 + \Lambda_{21}x_1) - x_1\left(\frac{\Lambda_{12}}{x_1 + \Lambda_{12}x_2} - \frac{\Lambda_{21}}{\Lambda_{21}x_1 + x_2}\right) \tag{F-6}$$

For each binary there are two parameters, Λ_{ij} and Λ_{ji}. Although these two parameters can be treated as empirical constants, they can be given some physical significance by following the semitheoretical development of the Wilson equation from a molecular model of a liquid mixture. Thus, these constants are defined as:

$$\Lambda_{ij} = \frac{V_j}{V_i} \exp\left(-\frac{\lambda_{ij} - \lambda_{ii}}{RT}\right) \tag{F-7}$$

$$\Lambda_{ji} = \frac{V_i}{V_j} \exp\left(-\frac{\lambda_{ij} - \lambda_{jj}}{RT}\right) \tag{F-8}$$

V_i and V_j are molal liquid volumes of i and j at T. The binary parameters now become the difference quantities $(\lambda_{ij} - \lambda_{ii})$ and $(\lambda_{ij} - \lambda_{jj})$. The qualitative physical significance given to these λ terms is that λ_{ij} represents an energy of interaction between molecules i and j whereas λ_{ii} and λ_{jj} are related, in a similar manner, to the interaction between pure i and j molecules. For a multicomponent system, these binary parameters must be known for each possible binary.

The Wilson equation has been shown to correlate well the activity coefficients of many types of miscible systems including highly nonideal binaries with polar and nonpolar components (5, 6, 7). It can be shown, however, that $(\partial^2 \Delta G/\partial x^2)_{T,P}$ can never be less than zero irrespective of the values of Λ_{12} or Λ_{21} so that Wilson's correlation can never represent well a case in which a homogeneous liquid phase becomes unstable and splits into two phases.[2] Also, it cannot predict $\ln \gamma$ vs. x curves where there is a maximum or minimum. Except for these limitations, it is one of the most widely used correlations and values of the Wilson parameters have been determined for many binaries [e.g. , see (8)]. Computation methods to obtain the best set of parameters are also available (9).

NRTL equation

Another activity coefficient correlation that is currently as widely used as the Wilson equation is called the NRTL equation, i.e. , the Non-Random Two Liquid equation. Developed by Renon and Prausnitz in 1968 (10), it may be written as:

$$\frac{\Delta G^{EX}}{RT} = \sum_{i=1}^{n} x_i \frac{\sum_{j=1}^{n} \tau_{ji} G_{ji} x_j}{\sum_{r=1}^{n} G_{ri} x_r} \tag{F-9}$$

and from Eq. (8-193),

$$RT \ln \gamma_i = \frac{\sum_{j=1}^{n} \tau_{ji} x_j}{\sum_{k=1}^{n} G_{ki} x_k} + \sum_{j=1}^{n} \frac{x_j G_{ij}}{\sum_{k=1}^{n} G_{kj} x_k} \left(\tau_{ij} - \frac{\sum_{r=1}^{n} x_r \tau_{rj} G_{rj}}{\sum_{k=1}^{n} G_{kj} x_k} \right) \tag{F-10}$$

For a binary system of 1 and 2,

$$\ln \gamma_1 = x_2^2 \left[\frac{\tau_{21} G_{21}^2}{(x_1 + x_2 G_{21})^2} + \frac{\tau_{12} G_{12}}{(x_2 + x_1 G_{12})^2} \right] \tag{F-11}$$

$$\ln \gamma_2 = x_1^2 \left[\frac{\tau_{12} G_{12}^2}{(x_2 + x_1 G_{12})^2} + \frac{\tau_{21} G_{21}}{(x_1 + x_2 G_{21})^2} \right] \tag{F-12}$$

In these relations,

$$G_{ji} = \exp(-\alpha \tau_{ji}) \tag{F-13}$$

$$\tau_{ji} = \frac{g_{ji} - g_{ii}}{RT} \qquad (g_{ji} = g_{ij}) \tag{F-14}$$

[2] $\Delta G = \Delta G^{ID} + \Delta G^{EX}$

The adjustable parameters in this case are the difference quantities $(g_{ji} - g_{ii})$ and α. There are, therefore, three parameters per binary. This equation can predict the γ-x relation for immiscible systems (if $\alpha < 0.426$), and it is normally even more accurate than the Wilson form although one might have expected this since three parameters are required. Although τ_{ij}, τ_{ji}, and α can be treated as empirical constants, in the molecular model used to develop the NRTL equation, the terms g_{ji}, g_{ii}, and g_{jj} represent the Gibbs free energy of interaction between molecule pairs.

The Wilson and NRTL correlations are the best available at the present time to relate activity coefficients with composition. For a binary mixture, the former has two adjustable parameters and the latter three. In actuality, the NRTL form is usually used as a two-parameter equation since the predictive results are not particularly sensitive to α and one can usually estimate a satisfactory value from the system type. Usually $\alpha \sim 0.3$ except for some cases in which there is a strongly self-associating polar material (such as an alcohol) with a nonpolar substance. In this case, $\alpha \sim 0.47$, but Wilson's equation is more accurate for this special case.

Several investigators have proposed that both relations can, in effect, be reduced to one unknown or empirically determined parameter if the terms λ_{ii} or g_{ii} are related to the internal energy vaporization (6, 11, 12). This correspondence is carried out in different ways. Tassios (11), who first suggested the idea, simply uses

$$\lambda_{jj} = -\Delta U_{v_j} = -(\Delta H_{v_j} - RT) \tag{F-15}$$

but Schreiber and Eckert have introduced the coordination number z (the number of nearest neighbors in a lattice cell),

$$\lambda_{jj} = -\frac{2}{z}(\Delta H_{v_j} - RT) \tag{F-16}$$

They chose, however, $z = 10$ in their comparison. Finally, Bruin and Prausnitz, in testing the NRTL form, employ

$$g_{jj} = -\beta(U_{\text{SAT LIQ}} - U^*)_j \tag{F-17}$$

with β as a constant. Other forms could, of course, be suggested. The net result is that the Wilson and NRTL equations are effectively simplified to a single unknown parameter (e.g., λ_{ij} for the Wilson or g_{ij} for the NRTL). Except for very nonideal solutions, these one-parameter forms often yield a good fit between calculated and experimental γ-x relations.

Literature Cited

1. PRAUSNITZ, J. M. *Molecular Thermodynamics of Fluid Phase Equilibria.* (Englewood Cliffs, N.J.: Prentice-Hall, Inc., 1969).

2. NULL, H. R., *Phase Equilibria in Process Design.* (New York: John Wiley & Sons, 1970),

3. WOHL, K., "Thermodynamic Evaluation of Binary and Ternary Liquid Systems," *Trans. AIChE*, **42**, (1946), 215.

4. WILSON, G. M., "Vapor-Liquid Equilibrium, XI. A New Expression for the Excess Free Energy of Mixing," *J. Am. Chem. Soc.*, **86**, (1964), 127.

5. ORYE, R. V. and J. M. PRAUSNITZ "Multicomponent Equilibria with the Wilson Equation, *Ind. Eng. Chem.*, **57** (5), (1965), 19.

6. SCHREIBER, L. B. and C. A. ECKERT, "Use of Infinite Dilution Activity Coefficients with Wilson's Equation," *Ind. Eng. Chem. Proc. Des. Develop.*, **10**, (1971), 572.

7. MESNAGE, JOSETTE, and A. A. MARSAN, "Vapor-Liquid Equilibrium at Atmospheric Pressure," *J. Chem. Eng. Data*, **16**, (1971), 434.

8. HOLMES, M. J. and M. VAN WINKLE, "Prediction of Ternary Vapor-Liquid Equilibria from Binary Data," *Ind. Eng. Chem.*, **62** (1), (1970), 21.

9. NAGAHAMA, KUNIO, ISAO SUZUKI, and MITSUHO HIRATI, "Estimation of Wilson Parameters," *J. Chem. Eng.* (Japan), **4**, (1971), 1.

10. RENON, HENRI, and J. M. PRAUSNITZ, "Local Compositions in Thermodynamic Excess Functions for Liquid Mixtures," *AIChE J.*, **14**, (1968), 135.

11. TASSIOS, DIMITRIOS, "A Single-Parameter Equation for Isothermal Vapor-Liquid Correlations," *AIChE J.*, **17**, (1971), 1367.

12. BRUIN, SOLKE and J. M. PRAUSNITZ, "One-Parameter Equation for Excess Energy of Strongly Nonideal Liquid Mixtures," *Ind. Eng. Chem. Proc. Des. Develop.*, **10**, (1971), 562.

Index

A

B

C

D

E

Q

R

S

T

U

V

W

Y

Z